LIE ALGEBRAS

LIE ALGEBRAS

by

NATHAN JACOBSON
Henry Ford II Professor of Mathematics
Yale University, New Haven, Connecticut

Dover Publications, Inc.
New York

Published in Canada by General Publishing Company, Ltd., 30 Lesmill Road, Don Mills, Toronto, Ontario.
Published in the United Kingdom by Constable and Company, Ltd., 10 Orange Street, London WC2H 7EG.

This Dover edition, first published in 1979, is an unabridged and corrected republication of the work originally published in 1962 by Interscience Publishers, a division of John Wiley & Sons, Inc.

International Standard Book Number: 0-486-63832-4
Library of Congress Catalog Card Number: 79-52006

Manufactured in the United States of America
Dover Publications, Inc.
180 Varick Street
New York, N.Y. 10014

PREFACE

The present book is based on lectures which the author has given at Yale during the past ten years, especially those given during the academic year 1959–1960. It is primarily a textbook to be studied by students on their own or to be used for a course on Lie algebras. Besides the usual general knowledge of algebraic concepts, a good acquaintance with linear algebra (linear transformations, bilinear forms, tensor products) is presupposed. Moreover, this is about all the equipment needed for an understanding of the first nine chapters. For the tenth chapter, we require also a knowledge of the notions of Galois theory and some of the results of the Wedderburn structure theory of associative algebras.

The subject of Lie algebras has much to recommend it as a subject for study immediately following courses on general abstract algebra and linear algebra, both because of the beauty of its results and its structure, and because of its many contacts with other branches of mathematics (group theory, differential geometry, differential equations, topology). In this exposition we have tried to avoid making the treatment too abstract and have consistently followed the point of view of treating the theory as a branch of linear algebra. The general abstract notions occur in two groups: the first, adequate for the structure theory, in Chapter I; and the second, adequate for representation theory, in Chapter V. Chapters I through IV give the structure theory, which culminates in the classification of the so-called "split simple Lie algebras." The basic results on representation theory are given in Chapters VI through VIII. In Chapter IX the automorphisms of semi-simple Lie algebras over an algebraically closed field of characteristic zero are determined. These results are applied in Chapter X to the problem of sorting out the simple Lie algebras over an arbitrary field.

No attempt has been made to indicate the historical development of the subject or to give credit for individual contributions to it. In this respect we have confined ourselves to brief indications here and there of the names of those responsible for the main ideas. It is well to record here the author's own indebtedness to one of the great creators of the theory, Professor Hermann Weyl, whose lectures at the Institute for Advanced Study in 1933–

1934 were truly inspiring and led to the author's research in this field. It should be noted also that in these lectures Professor Weyl, although primarily concerned with the Lie theory of continuous groups, set the subject of Lie algebras on its own independent course by introducing for the first time the term "Lie algebra" as a substitute for "infinitesimal group," which had been used exclusively until then.

A fairly extensive bibliography is included; however, this is by no means complete. The primary aim in compiling the bibliography has been to indicate the avenues for further study of the topics of the book and those which are immediately related to it.

I am very much indebted to my colleague George Seligman for carefully reading the various versions of the manuscript and offering many suggestions for improving the exposition. Drs. Paul Cohn and Ancel Mewborn have also made valuable comments, and all three have assisted with the proofreading. I take this opportunity to offer all three my sincere thanks.

May 28, 1961
New Haven, Connecticut

Nathan Jacobson

CONTENTS

CHAPTER I

Basic Concepts

CHAPTER II

Solvable and Nilpotent Lie Algebras

CHAPTER III

Cartan's Criterion and Its Consequences

Chapter VIII

Characters of the Irreducible Modules

Chapter IX

Automorphisms

Chapter X

Simple Lie Algebras over an Arbitrary Field

LIE ALGEBRAS

CHAPTER I

Basic Concepts

The theory of Lie algebras is an outgrowth of the Lie theory of continuous groups. The main result of the latter is the reduction of "local" problems concerning Lie groups to corresponding problems on Lie algebras, thus to problems in linear algebra. One associates with every Lie group a Lie algebra over the reals or complexes and one establishes a correspondence between the analytic subgroups of the Lie group and the subalgebras of its Lie algebra, in which invariant subgroups correspond to ideals, abelian subgroups to abelian subalgebras, etc. Isomorphism of the Lie algebras is equivalent to local isomorphism of the corresponding Lie groups. We shall not discuss these matters in detail since excellent modern accounts of the Lie theory are available. The reader may consult one of the following books: Chevalley's *Theory of Lie Groups*, Cohn's *Lie Groups*, Pontrjagin's *Topological Groups*.

More recently, two other types of group theory have been aided by the introduction of appropriate Lie algebras in their study. The first of these is the theory of free groups which can be studied by means of free Lie algebras using a method which was originated by Magnus. Although the connection here is not so close as in the Lie theory, significant results on free groups and other types of discrete groups have been obtained using Lie algebras. Particularly noteworthy are the results on the so-called restricted Burnside problem: Is there a bound for the orders of the finite groups with a fixed number r of generators and satisfying the relation $x^m = 1$, m a fixed positive integer? It is worth mentioning that Lie algebras of prime characteristic play an important role in these applications to discrete group theory. Again we shall not enter into the details but refer the interested reader to two articles which give a good account of this method in group theory. These are: Lazard [2] and Higman [1].

The type of correspondence between subgroups of a Lie group and subalgebras of its Lie algebra which obtains in the Lie theory has a counterpart in Chevalley's theory of linear algebraic groups. Roughly speaking, a linear algebraic group is a subgroup of the group of non-singular $n \times n$ matrices which is specified by a set of polynomial equations in the entries of the matrices. An example is the orthogonal group which is defined by the set of equations $\sum_i \alpha_{ij}^2 = 1$, $\sum_i \alpha_{ij}\alpha_{ik} = 0$, $j \neq k$, $j, k = 1, \cdots, n$, on the entries α_{ij} of the matrix (α_{ij}). With each linear algebraic group Chevalley has defined a corresponding Lie algebra (see Chevalley [2]) which gives useful information on the group and is decisive in the theory of linear algebraic groups of characteristic zero.

In view of all this group theoretic background it is not surprising that the basic concepts in the theory of Lie algebras have a group-theoretic flavor. This should be kept in mind throughout the study of Lie algebras and particularly in this chapter, which gives the foundations that are adequate for the main structure theory to be developed in Chapters II to IV. Questions on foundations are taken up again in Chapter V. These concern some concepts that are necessary for the representation theory, which will be treated in Chapters VI and VII.

1. Definition and construction of Lie and associative algebras

We recall the definition of a non-associative algebra (=not necessarily associative algebra) $\mathfrak{A}$ over a field Φ. This is just a vector space $\mathfrak{A}$ over Φ in which a bilinear composition is defined. Thus for every pair (x, y), x, y in $\mathfrak{A}$, we can associate a product $xy \in \mathfrak{A}$ and this satisfies the bilinearity conditions

$$(1) \qquad (x_1 + x_2)y = x_1 y + x_2 y\,, \qquad x(y_1 + y_2) = xy_1 + xy_2\,.$$

$$(2) \qquad \alpha(xy) = (\alpha x)y = x(\alpha y)\,, \qquad \alpha \in \Phi\,.$$

A similar definition can be given for a non-associative algebra over a commutative ring Φ having an identity element (unit) 1. This is a left Φ-module with a product $xy \in \mathfrak{A}$ satisfying (1) and (2). We shall be interested mainly in the case of algebras over fields and, in fact, in such algebras which are finite-dimensional as vector spaces. For such an algebra we have a basis $(e_1, e_2, \cdots, e_n)$ and we can write $e_i e_j = \sum_{k=1}^n \gamma_{ijk} e_k$ where the γ's are in Φ. The n^3 γ_{ijk} are called the *constants of multiplication* of the algebra (relative to the chosen basis). They give the values of every product $e_i e_j$, $i, j =$

$1, 2, \cdots, n$. Moreover, these products determine every product in $\mathfrak{A}$. Thus let x and y be any two elements of $\mathfrak{A}$ and write $x = \sum \xi_i e_i$, $y = \sum \eta_j e_j$, $\xi_i, \eta_j \in \Phi$. Then, by (1) and (2),

$$\begin{aligned} xy &= (\sum_i \xi_i e_i)(\sum_j \eta_j e_j) = \sum_{i,j} (\xi_i e_i)(\eta_j e_j) \\ &= \sum_{i,j} \xi_i (e_i(\eta_j e_j)) = \sum_{i,j} \xi_i \eta_j (e_i e_j) \,, \end{aligned}$$

and this is determined by the $e_i e_j$.

This reasoning indicates a universal construction for finite-dimensional non-associative algebras. We begin with any vector space $\mathfrak{A}$ and a basis (e_i) in $\mathfrak{A}$. For every pair (i, j) we define in any way we please $e_i e_j$ as an element of $\mathfrak{A}$. Then if $x = \sum_1^n \xi_i e_i$, $y = \sum_1^n \eta_j e_j$ we define

$$(3) \qquad xy = \sum_{i,\, j=1}^{n} \xi_i \eta_j (e_i e_j) \,.$$

One checks immediately that this is bilinear in the sense that (1) and (2) are valid. The choice of $e_i e_j$ is equivalent to the choice of the elements γ_{ijk} in Φ such that $e_i e_j = \sum \gamma_{ijk} e_k$.

The notion of a non-associative algebra is too general to lead to interesting structural results. In order to obtain such results one must impose some further conditions on the multiplication. The most important ones—and the ones which will concern us here, are the associative laws and the Lie conditions.

DEFINITION 1. A non-associative algebra $\mathfrak{A}$ is said to be *associative* if its multiplication satisfies the associative law

$$(4) \qquad (xy)z = x(yz) \,.$$

A non-associative algebra $\mathfrak{A}$ is said to be a *Lie algebra* if its multiplication satisfies the Lie conditions

$$(5) \qquad x^2 = 0 \,, \qquad (xy)z + (yz)x + (zx)y = 0 \,.$$

The second of these is called the *Jacobi identity*.

Since these types of non-associative algebras are defined by identities, it is clear that subalgebras and homomorphic images are of the same type, i.e., associative or Lie. If $\mathfrak{A}$ is a Lie algebra and $x, y \in \mathfrak{A}$, then $0 = (x + y)^2 = x^2 + xy + yx + y^2 = xy + yx$ so that

$$(6) \qquad xy = -yx$$

holds in any Lie algebra. Conversely, if this condition holds then $2x^2 = 0$, so that, if the characteristic is not two, then $x^2 = 0$. Hence for algebras of characteristic $\neq 2$ the condition (6) can be used for the first of (5) in the definition of a Lie algebra.

PROPOSITION 1. *A non-associative algebra* $\mathfrak{A}$ *with basis* $(e_1, e_2, \cdots, e_n)$ *over* Φ *is associative if and only if* $(e_ie_j)e_k = e_i(e_je_k)$ *for* $i, j, k = 1, 2, \cdots, n$. *If* $e_ie_j = \sum_r \gamma_{ijr}e_r$ *these conditions are equivalent to*

$$(7) \qquad \sum_r \gamma_{ijr}\gamma_{rks} = \sum_r \gamma_{irs}\gamma_{jkr}, \qquad i, j, k, s = 1, 2, \cdots, n .$$

The algebra $\mathfrak{A}$ *is Lie if and only if* $e_i^2 = 0$, $e_ie_j = -e_je_i$,

$$(e_ie_j)e_k + (e_je_k)e_i + (e_ke_i)e_j = 0$$

for $i, j, k = 1, 2, \cdots, n$. *These conditions are equivalent to*

$$(8) \qquad \begin{gathered} \gamma_{iik} = 0, \qquad \gamma_{ijk} = -\gamma_{jik}, \\ \sum_r (\gamma_{ijr}\gamma_{rks} + \gamma_{jkr}\gamma_{ris} + \gamma_{kir}\gamma_{rjs}) = 0 . \end{gathered}$$

Proof: If $\mathfrak{A}$ is associative, then $(e_ie_j)e_k = e_i(e_je_k)$. Conversely, assume these conditions hold for the e_i. If $x = \sum \xi_ie_i$, $y = \sum \eta_je_j$, $z = \sum \zeta_ke_k$, then $(xy)z = \sum \xi_i\eta_j\zeta_k(e_ie_j)e_k$ and $x(yz) = \sum \xi_i\eta_j\zeta_ke_i(e_je_k)$. Hence $(xy)z = x(yz)$ and $\mathfrak{A}$ is associative. If $e_ie_j = \sum \gamma_{ijr}e_r$, then $(e_ie_j)e_k = \sum_{r,s}\gamma_{ijr}\gamma_{rks}e_s$ and $e_i(e_je_k) = \sum_{r,s}\gamma_{irs}\gamma_{jkr}e_s$. Hence the linear independence of the e_i implies that the conditions $(e_ie_j)e_k = e_i(e_je_k)$ are equivalent to (7). The proof in the Lie case is similar to the foregoing and will be omitted.

In actual practice the general procedure we have indicated is not often used in constructing examples of associative and of Lie algebras except for algebras of low dimensionalities. We shall employ this in determining the Lie algebras of one, two, and three dimensions in §4. There are a couple of simplifying remarks that can be made in the Lie case. First, we note that if $e_i^2 = 0$ and $e_ie_j = -e_je_i$ in an algebra, then the validity of $(e_ie_j)e_k + (e_je_k)e_i + (e_ke_i)e_j = 0$ for a particular triple i, j, k implies $(e_je_i)e_k + (e_ie_k)e_j + (e_ke_j)e_i = 0$. Since cyclic permutations of i, j, k are clearly allowed it follows that the Jacobi identity for $(e_{i'}e_{j'})e_{k'}$ is valid for i', j', k', a permutation of i, j, k. Next let $i = j$. Then $e_i^2e_k + (e_ie_k)e_i + (e_ke_i)e_i = 0 + (e_ie_k)e_i - (e_ie_k)e_i = 0$. Hence $e_i^2 = 0$, $e_ie_j = -e_je_i$ or, what is the same thing, $x^2 = 0$ in $\mathfrak{A}$ implies that the Jacobi identities are satisfied for e_i, e_i, e_j. In particular, the Jacobi identities are consequences of $x^2 = 0$ if $\dim \mathfrak{A} \leqq 2$

and if dim $\mathfrak{A} = 3$, then the only identity we have to check is $(e_1e_2)e_3 + (e_2e_3)e_1 + (e_3e_1)e_2 = 0$.

2. *Algebras of linear transformations. Derivations*

Actually, it is unnecessary to sit down and construct examples of associative and Lie algebras by the method of bases and multiplication tables since these algebras occur "in nature." The prime examples of associative algebras are obtained as follows. Let $\mathfrak{M}$ be a vector space over a field Φ and let $\mathfrak{E}$ denote the set of linear transformations of $\mathfrak{M}$ into itself. We recall that if $A, B \in \mathfrak{E}$ and $\alpha \in \Phi$, then $A + B$, αA and AB are defined by $x(A + B) = xA + xB$, $x(\alpha A) = \alpha(xA)$, $x(AB) = (xA)B$ for x in $\mathfrak{M}$. Then it is well known that $\mathfrak{E}$ is a vector space relative to $+$ and the scalar multiplication and that multiplication is associative and satisfies (1) and (2). Hence $\mathfrak{E}$ is an associative algebra. It is well known also that if $\mathfrak{M}$ is m-dimensional, $m < \infty$, then $\mathfrak{E}$ is m^2-dimensional over Φ. If $(e_1, e_2, \cdots, e_m)$ is a basis for $\mathfrak{M}$ over Φ, then the linear transformations E_{ij} such that $e_iE_{ij} = e_j$, $e_rE_{ij} = 0$ if $r \neq i$, $i, j = 1, \cdots, m$, form a basis for $\mathfrak{E}$ over Φ. If $A \in \mathfrak{E}$, then we can write $e_iA = \sum_j \alpha_{ij}e_j$, $i = 1, \cdots, m$, and $(\alpha) = (\alpha_{ij})$ is the matrix of A relative to the basis (e_i). The correspondence $A \rightarrow (\alpha)$ is an isomorphism of $\mathfrak{E}$ onto the algebra Φ_m of $m \times m$ matrices with entries α_{ij} in Φ.

The algebra $\mathfrak{E}$ is called *the* (*associative*) *algebra of linear transformations* in $\mathfrak{M}$ over Φ. Any subalgebra $\mathfrak{A}$ of $\mathfrak{E}$, that is, a subspace of $\mathfrak{E}$ which is closed under multiplication, is called *an algebra of linear transformations*.

If $\mathfrak{A}$ is an arbitrary non-associative algebra and $a \in \mathfrak{A}$, then the mapping a_R which sends any x into xa is a linear transformation. It is well known and easy to check that $(a + b)_R = a_R + b_R$, $(\alpha a)_R = \alpha a_R$ and if $\mathfrak{A}$ is associative, $(ab)_R = a_Rb_R$. Hence if $\mathfrak{A}$ is an associative algebra, the mapping $a \rightarrow a_R$ is a homomorphism of $\mathfrak{A}$ into the algebra $\mathfrak{E}$ of linear transformations in the vector space $\mathfrak{A}$. If $\mathfrak{A}$ has an identity (or unit) 1, then $a \rightarrow a_R$ is an isomorphism of $\mathfrak{A}$ into $\mathfrak{E}$. Hence $\mathfrak{A}$ is isomorphic to an algebra of linear transformations. If $\mathfrak{A}$ does not have an identity, we can adjoin one in a simple way to get an algebra, $\mathfrak{A}^*$ with an identity such that dim $\mathfrak{A}^* =$ dim $\mathfrak{A} + 1$ (cf. Jacobson [2], vol. I, p. 84). Since $\mathfrak{A}^*$ is isomorphic to an algebra of linear transformations, the same is true for $\mathfrak{A}$. If $\mathfrak{A}$ is finite-dimensional, the argument shows that

$\mathfrak{A}$ is isomorphic to an algebra of linear transformations in a finite-dimensional vector space.

Lie algebras arise from associative algebras in a very simple way. Let $\mathfrak{A}$ be an associative algebra. If $x, y \in \mathfrak{A}$, then we define the *Lie product* or (additive) *commutator* of x and y as

$$[xy] = xy - yx \,. \tag{9}$$

One checks immediately that

$$[x_1 + x_2, y] = [x_1 y] + [x_2 y] \,,$$
$$[x, y_1 + y_2] = [x y_1] + [x y_2] \,,$$
$$\alpha[xy] = [\alpha x, y] = [x, \alpha y] \,.$$

Moreover,

$$[xx] = x^2 - x^2 = 0 \,,$$
$$[[xy]z] + [[yz]x] + [[zx]y]$$
$$= (xy - yx)z - z(xy - yx) + (yz - zy)x$$
$$- x(yz - zy) + (zx - xz)y - y(zx - xz) = 0 \,.$$

Thus the product $[xy]$ satisfies all the conditions on the product in a Lie algebra. The Lie algebra obtained in this way is called *the Lie algebra* of the associative algebra $\mathfrak{A}$. We shall denote this Lie algebra as $\mathfrak{A}_L$. In particular, we have the Lie algebra $\mathfrak{E}_L$ obtained from $\mathfrak{E}$. Any subalgebra $\mathfrak{L}$ of $\mathfrak{E}_L$ is called *a Lie algebra of linear transformations*. We shall see later that every Lie algebra is isomorphic to a subalgebra of a Lie algebra $\mathfrak{A}_L$, $\mathfrak{A}$ associative. In view of the result just proved on associative algebras this is equivalent to showing that every Lie algebra is isomorphic to a Lie algebra of linear transformations.

We shall consider now some important instances of subalgebras of Lie algebras $\mathfrak{E}_L$, $\mathfrak{E}$ the associative algebra of linear transformations in a vector space $\mathfrak{M}$ over a field Φ.

Orthogonal Lie algebra. Let $\mathfrak{M}$ be equipped with a non-degenerate symmetric bilinear form (x, y) and assume $\mathfrak{M}$ finite-dimensional. Then any linear transformation A in $\mathfrak{M}$ has an adjoint A^* relative to (x, y); that is, A^* is linear and satisfies: $(xA, y) = (x, yA^*)$. The mapping $A \to A^*$ is an anti-automorphism in the algebra $\mathfrak{E}$: $(A + B)^* = A^* + B^*$, $(\alpha A)^* = \alpha A^*$, $(AB)^* = B^* A^*$. Let $\mathfrak{S}$ denote the set of $A \in \mathfrak{E}$ which are skew in the sense that $A^* = -A$. Then $\mathfrak{S}$ is a subspace of $\mathfrak{E}$ and if $A^* = -A$, $B^* = -B$, then $[AB]^* =$

$(AB - BA)^* = B^*A^* - A^*B^* = BA - AB = [BA] = -[AB]$. Hence $[AB] \in \mathfrak{S}$ and $\mathfrak{S}$ is a subalgebra of $\mathfrak{E}_L$.

If Φ is the field of real numbers, then the Lie algebra $\mathfrak{S}$ is the Lie algebra of the orthogonal group of $\mathfrak{M}$ relative to (x, y). This is the group of linear transformations O in $\mathfrak{M}$ which are orthogonal in the sense that $(xO, yO) = (x, y)$, x, y in $\mathfrak{M}$. For this reason we shall call $\mathfrak{S}$ the *orthogonal Lie algebra* relative to (x, y).

Symplectic Lie algebra. Here we suppose (x, y) is a non-degenerate alternate form: $(x, x) = 0$ and again $\dim \mathfrak{M} < \infty$. We recall that these conditions imply that $\dim \mathfrak{M} = 2l$ is even. Again let A^* be the adjoint of $A (\in \mathfrak{E})$ relative to (x, y). Then the set $\mathfrak{S}$ of skew $(A^* = -A)$ linear transformations is a subalgebra of $\mathfrak{E}_L$. This is related to the symplectic group and so we shall call it the *symplectic Lie algebra* $\mathfrak{S}$ of the alternate form (x, y).

Triangular linear transformations. Let $0 \subset \mathfrak{M}_1 \subset \mathfrak{M}_2 \subset \cdots \subset \mathfrak{M}_m = \mathfrak{M}$ be a chain of subspaces of $\mathfrak{M}$ such that $\dim \mathfrak{M}_i = i$ and let $\mathfrak{T}$ be the set of linear transformations T such that $\mathfrak{M}_i T \subseteq \mathfrak{M}_i$. It is clear that $\mathfrak{T}$ is a subalgebra of the associative algebra $\mathfrak{E}$: hence $\mathfrak{T}_L$ is a subalgebra of $\mathfrak{E}_L$. We can choose a basis $(x_1, x_2, \cdots, x_m)$ for $\mathfrak{M}$ so that $(x_1, x_2, \cdots, x_i)$ is a basis for $\mathfrak{M}_i$. Then if $T \in \mathfrak{T}$, $\mathfrak{M}_i T \subseteq \mathfrak{M}_i$ implies that the matrix of T relative to $(x_1, x_2, \cdots, x_m)$ is of the form

$$(\tau) = \begin{bmatrix} \tau_{11} & 0 & \cdot & \cdot & \cdot & 0 \\ \tau_{21} & \tau_{22} & 0 & \cdot & \cdot & \cdot \\ & \cdot & \cdot & & & \cdot \\ & \cdot & & \cdot & & \cdot \\ & \cdot & & & \cdot & \cdot \\ \tau_{m1} & \tau_{m2} & \cdot & \cdot & \cdot & \tau_{mm} \end{bmatrix}. \tag{10}$$

Such a matrix is called *triangular* and correspondingly we shall call any $T \in \mathfrak{T}$ a triangular linear transformation.

Derivation algebras. Let $\mathfrak{A}$ be an arbitrary non-associative algebra. A *derivation* D in $\mathfrak{A}$ is a linear mapping of $\mathfrak{A}$ into $\mathfrak{A}$ satisfying

$$(xy)D = (xD)y + x(yD) . \tag{11}$$

Let $\mathfrak{D}(\mathfrak{A})$ denote the set of derivations in $\mathfrak{A}$. If $D_1, D_2 \in \mathfrak{D}(\mathfrak{A})$, then

$$\begin{aligned}(xy)(D_1 + D_2) = (xy)D_1 + (xy)D_2 &= (xD_1)y \\ &+ x(yD_1) + (xD_2)y + x(yD_2) \\ &= (x(D_1 + D_2))y + x(y(D_1 + D_2)) .\end{aligned}$$

Hence $D_1 + D_2 \in \mathfrak{D}(\mathfrak{A})$. Similarly, one checks that $\alpha D_1 \in \mathfrak{D}(\mathfrak{A})$ if $\alpha \in \Phi$. We have

$$\begin{aligned}(xy)D_1D_2 &= ((xD_1)y + x(yD_1))D_2\\ &= (xD_1D_2)y + (xD_1)(yD_2) + (xD_2)(yD_1) + x(yD_1D_2)\,.\end{aligned}$$

Interchange of 1, 2 and subtraction gives

$$(xy)[D_1D_2] = (x[D_1D_2])y + x(y[D_1D_2])\,.$$

Hence $[D_1D_2] \in \mathfrak{D}(\mathfrak{A})$ and so $\mathfrak{D}(\mathfrak{A})$ is a subalgebra of $\mathfrak{E}_L$, where $\mathfrak{E}$ is the algebra of linear transformations in the vector space $\mathfrak{A}$. We shall call this the *Lie algebra of derivations* or *derivation algebra of* $\mathfrak{A}$.

The Lie algebra $\mathfrak{D}(\mathfrak{A})$ is the Lie algebra of the group of automorphisms of $\mathfrak{A}$ if $\mathfrak{A}$ is a finite-dimensional algebra over the field of real numbers. We shall not prove any of our assertions on the relation between Lie groups and Lie algebras but refer the reader to the literature on Lie groups for this. However, in the present instance we shall indicate the link between the group of automorphisms and the Lie algebra of derivations.

Let D be a derivation. Then induction on n gives the Leibniz rule:

$$(xy)D^n = \sum_{i=0}^{n}\binom{n}{i}(xD^i)(yD^{n-i})\,. \tag{12}$$

If the characteristic of Φ is 0 we can divide by $n!$ and obtain

$$(xy)\frac{D^n}{n!} = \sum_{i=0}^{n}\left(\frac{1}{i!}xD^i\right)\left(\frac{1}{(n-i)!}yD^{n-i}\right). \tag{12$'$}$$

If $\mathfrak{A}$ is finite-dimensional over the field of reals, then it is easy to prove (cf. Jacobson [2], vol. II, p. 197) that the series

$$1 + D + \frac{D^2}{2!} + \frac{D^3}{3!} + \cdots \tag{13}$$

converges for every linear mapping D in $\mathfrak{A}$, and the linear mapping $\exp D$ defined by (13) is $1:1$. Also it is easy to see, using (12′), that if D is a derivation, then $G = \exp D$ satisfies $(xy)G = (xG)(yG)$. Hence G is an automorphism of $\mathfrak{A}$.

A connection between automorphisms and derivations can be established in a purely algebraic setting which has important applications. Here we suppose the base field of $\mathfrak{A}$ is arbitrary of characteristic 0. Let D be a nilpotent derivation, say, $D^N = 0$. Consider the mapping

$$G \equiv \exp D = 1 + D + \frac{D^2}{2!} + \cdots + \frac{D^{N-1}}{(N-1)!}\,. \tag{14}$$

We write this as $G = 1 + Z$, $Z = D + (D^2/2!) + \cdots + (D^{N-1}/(N-1)!)$ and note that $Z^N = 0$. Hence $G = 1 + Z$ has the inverse $1 - Z + Z^2 + \cdots \pm Z^{n-1}$ and so G is 1 : 1 of $\mathfrak{A}$ onto $\mathfrak{A}$. We have

$$\begin{aligned}(xG)(yG) &= \left(\sum_{i=0}^{N-1} \frac{xD^i}{i!}\right)\left(\sum_{j=0}^{N-1} \frac{yD^j}{j!}\right)\\ &= \sum_{n=0}^{2N-2}\left(\sum_{i=0}^{n}\left(\frac{xD^i}{i!}\right)\left(\frac{yD^{n-i}}{(n-i)!}\right)\right)\\ &= \sum_{n=0}^{2N-2}(xy)\frac{D^n}{n!} \qquad \text{(by 12')}\\ &= \sum_{n=0}^{N-1}(xy)\frac{D^n}{n!}\\ &= (xy)G\,.\end{aligned}$$

Hence G is an automorphism of $\mathfrak{A}$.

3. *Inner derivations of associative and Lie algebras*

If a is any element of a non-associative algebra $\mathfrak{A}$, then a determines two mappings a_L: $x \rightarrow ax$ and a_R: $x \rightarrow xa$ of $\mathfrak{A}$ into itself. These are called the *left multiplication* and *right multiplication* determined by a. The defining conditions (1) and (2) for an algebra show that a_L and a_R are linear mappings and the mappings $a \rightarrow a_L$, $a \rightarrow a_R$ are linear of $\mathfrak{A}$ into the space $\mathfrak{E}$ of linear transformations in $\mathfrak{A}$. Now let $\mathfrak{A}$ be associative and set $D_a = a_R - a_L$. Hence D_a is the linear mapping $x \rightarrow xa - ax$. We have

$$xya - axy = (xa - ax)y + x(ya - ay)\,; \tag{15}$$

hence D_a is a derivation in the associative algebra $\mathfrak{A}$. We shall call this the *inner derivation* determined by a.

Next let $\mathfrak{L}$ be a Lie algebra. Because of the way Lie algebras arise from associative ones it is customary to denote the product in $\mathfrak{L}$ by $[xy]$ and we shall do this from now on. Also, it is usual to denote the right multiplication a_R $(= -a_L$ since $[xa] = -[ax])$ by ad a and to call this the *adjoint* mapping determined by a. We have

$$[[xy]a] + [[ya]x] + [[ax]y] = 0 ,$$
$$[[xy]a] = - [[ya]x] - [[ax]y] = [x[ya]] + [[xa]y] ;$$

hence ad a: $x \to [xa]$ is a derivation. We call this also the *inner derivation* determined by $a \in \mathfrak{L}$.

A subset $\mathfrak{B}$ of a non-associative algebra $\mathfrak{A}$ is called an *ideal* if (1) $\mathfrak{B}$ is a subspace of the vector space $\mathfrak{A}$, (2) ab, $ba \in \mathfrak{B}$ for any a in $\mathfrak{A}$, b in $\mathfrak{B}$. Consider the set of elements of the form $\sum a_i b_i$, a_i, b_i in $\mathfrak{A}$. We denote this set as $\mathfrak{A}^2$ and we can check that this is an ideal in $\mathfrak{A}$. If $\mathfrak{A} = \mathfrak{L}$ is a Lie algebra, then it is customary to write $\mathfrak{L}'$ for $\mathfrak{L}^2$ and to call this the *derived algebra* (or ideal) of $\mathfrak{L}$. If $\mathfrak{L}$ is a Lie algebra, then the skew symmetry of the multiplication implies that a subspace $\mathfrak{B}$ of $\mathfrak{L}$ is an ideal if and only if $[ab]$ (or $[ba]$) is in $\mathfrak{B}$ for every $a \in \mathfrak{L}$, $b \in \mathfrak{B}$. It follows that the subset $\mathfrak{C}$ of elements c such that $[ac] = 0$ for all $a \in \mathfrak{L}$ is an ideal. This is called the *center* of $\mathfrak{L}$. $\mathfrak{L}$ is called *abelian* if $\mathfrak{L} = \mathfrak{C}$, which is equivalent to $\mathfrak{L}' = 0$.

Proposition 2. *If $\mathfrak{A}$ is associative or Lie, then the inner derivations form an ideal $\mathfrak{J}(\mathfrak{A})$ in the derivation algebra $\mathfrak{D}(\mathfrak{A})$.*

Proof: In any non-associative algebra we have $(a + b)_L = a_L + b_L$, $(\alpha a)_L = \alpha a_L$, $(a + b)_R = a_R + b_R$, $(\alpha a)_R = \alpha a_R$. Hence if $D_a = a_R - a_L$ then $D_{a+b} = D_a + D_b$, $D_{\alpha a} = \alpha D_a$ and the inner derivations of an associative or of a Lie algebra form a subspace of $\mathfrak{D}(\mathfrak{A})$. Let D be a derivation in $\mathfrak{A}$. Then $(ax)D = (aD)x + a(xD)$, or $(ax)D - a(xD) = (aD)x$. In operator form this reads $(xa_L)D - (xD)a_L = x(aD)_L$, or $[a_L D] = a_L D - Da_L = (aD)_L$. Similarly, $[a_R D] = (aD)_R$ and consequently also $[D_a D] = D_{aD}$. These formulas show that if $\mathfrak{A}$ is associative or Lie and I is an inner derivation and D any derivation, then $[ID]$ is an inner derivation. Hence $\mathfrak{J}(\mathfrak{A})$ is an ideal in $\mathfrak{D}(\mathfrak{A})$.

Example. Let $\mathfrak{L}$ be the algebra with basis (e, f) such that $[ef] = e = - [fe]$ and all other products of base elements are 0. Then $[aa] = 0$ in $\mathfrak{L}$ and since dim $\mathfrak{L} = 2$, $\mathfrak{L}$ is a Lie algebra. The derived algebra $\mathfrak{L}' = \Phi e$. If D is a derivation in any algebra $\mathfrak{A}$, then $\mathfrak{A}^2 D \subseteq \mathfrak{A}^2$. Hence if D is a derivation in $\mathfrak{L}$ then $eD = \delta e$. Also ad(δf) has the property $e(\text{ad}\, \delta f) = [e, \delta f] = \delta e$. Hence if $E = D - \text{ad}\, \delta f$, then E is a derivation and $eE = 0$. Then $e = [ef]$ gives $0 = [e, fE]$. It follows that $fE = \gamma e$. Now ad$(- \gamma e)$ satisfies $e\,\text{ad}(- \gamma e) = 0$, $f\,\text{ad}(- \gamma e) = [f, - \gamma e] = \gamma [ef] = \gamma e$. Hence $E = \text{ad}(- \gamma e)$ is inner and $D = E + \text{ad}\, \delta f$ is inner. Thus every derivation of $\mathfrak{L} = \Phi e + \Phi f$ is inner.

In group theory one defines a group to be complete if all of its automorphisms are inner and its center is the identity. If H is complete and invariant in G then H is a direct factor of G. By analogy we shall call a Lie algebra *complete* if its derivations are all inner and its center is 0.

PROPOSITION 3. *If* $\mathfrak{K}$ *is complete and an ideal in* $\mathfrak{L}$, *then* $\mathfrak{L} = \mathfrak{K} \oplus \mathfrak{B}$ *where* $\mathfrak{B}$ *is an ideal.*

Proof: We note first that if $\mathfrak{K}$ is an ideal in $\mathfrak{L}$, then the centralizer $\mathfrak{B}$ of $\mathfrak{K}$, that is, the set of elements b such that $[kb] = 0$ for all $k \in \mathfrak{K}$ is an ideal. $\mathfrak{B}$ is evidently a subspace and if $b \in \mathfrak{B}$ and $a \in \mathfrak{L}$, then $[k[ba]] = -[a[kb]] - [b[ak]] = 0 - [b, k']$, $k' = [ak] \in \mathfrak{K}$; hence $[k[ba]] = 0$ for all $k \in \mathfrak{K}$ and $[ba] \in \mathfrak{B}$. Hence $\mathfrak{B}$ is an ideal. Now let $\mathfrak{K}$ be complete. If $c \in \mathfrak{K} \cap \mathfrak{B}$, then c is in the center of $\mathfrak{K}$ and so $c = 0$. Hence $\mathfrak{K} \cap \mathfrak{B} = 0$. Next let $a \in \mathfrak{L}$. Since $\mathfrak{K}$ is an ideal in $\mathfrak{L}$, ad a maps $\mathfrak{K}$ into itself and hence it induces a derivation D in $\mathfrak{K}$. This is inner and so we have a $k \in \mathfrak{K}$ such that $xD = [xa] = [xk]$ for every $x \in \mathfrak{K}$. Then $b = a - k \in \mathfrak{B}$ and $a = b + k$, $b \in \mathfrak{B}$, $k \in \mathfrak{K}$. Thus $\mathfrak{L} = \mathfrak{K} + \mathfrak{B} = \mathfrak{K} \oplus \mathfrak{B}$ as required.

Example. The algebra $\Phi e + \Phi f$ of the last example is complete.

4. *Determination of the Lie algebras of low dimensionalities*

We shall now determine all the Lie algebras $\mathfrak{L}$ such that dim $\mathfrak{L} \leqq 3$. If $(e_1, e_2, \cdots, e_n)$ is a basis for a Lie algebra $\mathfrak{L}$, then $[e_i e_i] = 0$ and $[e_i e_j] = -[e_j e_i]$. Hence in giving the multiplication table for the basis, it suffices to give the products $[e_i e_j]$ for $i < j$. We shall use these abbreviated multiplication tables in our discussion.

I. dim $\mathfrak{L} = 1$. Then $\mathfrak{L} = \Phi e$, $[ee] = 0$.

II. dim $\mathfrak{L} = 2$.

(a) $\mathfrak{L}' = 0$, $\mathfrak{L}$ is abelian.

(b) $\mathfrak{L}' \neq 0$. Since $\mathfrak{L} = \Phi e + \Phi f$, $\mathfrak{L}' = \Phi[ef]$ is one-dimensional. We may choose e so that $\mathfrak{L}' = \Phi e$. Then $[ef] = \alpha e \neq 0$ and replacement of f by $\alpha^{-1} f$ permits us to take $[ef] = e$. Then $\mathfrak{L}$ is the algebra of the example of § 3. This can now be characterized as *the* non-abelian two-dimensional Lie algebra.

III. dim $\mathfrak{L} = 3$.

(a) $\mathfrak{L}' = 0$, $\mathfrak{L}$ abelian.

(b) dim $\mathfrak{L}' = 1$, $\mathfrak{L}' \subseteqq \mathfrak{C}$ the center. If $\mathfrak{L}' = \Phi e$ we write $\mathfrak{L} = \Phi e + \Phi f + \Phi g$. Then $\mathfrak{L}' = \Phi[fg]$. Hence we may suppose $[fg] = e$. Thus

$\mathfrak{L}$ has basis (e, f, g), with multiplication table

$$[fg] = e\,, \qquad [ef] = 0\,, \qquad [eg] = 0\,. \tag{16}$$

We have only one Lie algebra satisfying our conditions. (If we have (16), then the Jacobi condition is satisfied.)

(c) $\dim \mathfrak{L}' = 1$, $\mathfrak{L}' \not\subseteq \mathfrak{C}$ the center. If $\mathfrak{L}' = \Phi e$, then there is an f such that $[ef] \neq 0$. Then $[ef] = \beta e \neq 0$ and we may suppose $[ef] = e$. Hence $\Phi e + \Phi f$ is the non-abelian two-dimensional algebra $\mathfrak{K}$. Since $\mathfrak{K} \supseteq \mathfrak{L}'$, $\mathfrak{K}$ is an ideal and since $\mathfrak{K}$ is complete, $\mathfrak{L} = \mathfrak{K} \oplus \mathfrak{B}$, $\mathfrak{B} = \Phi g$. Hence $\mathfrak{L}$ has basis (e, f, g) with multiplication table

$$[ef] = e\,, \qquad [eg] = 0\,, \qquad [fg] = 0\,. \tag{17}$$

(d) $\dim \mathfrak{L}' = 2$. $\mathfrak{L}'$ cannot be the non-abelian two-dimensional Lie algebra $\mathfrak{K}$. For then $\mathfrak{L} = \mathfrak{K} \oplus \mathfrak{B}$ and $\mathfrak{L}' = \mathfrak{K}' = \mathfrak{K}$. But $\mathfrak{K}' \subset \mathfrak{K}$. Hence we have $\mathfrak{L}'$ abelian. Let $\mathfrak{L}' = \Phi e + \Phi f$ and $\mathfrak{L} = \Phi e + \Phi f + \Phi g$. Then $\mathfrak{L}' = \Phi[eg] + \Phi[fg]$ and so $\operatorname{ad} g$ induces a $1:1$ linear mapping in $\mathfrak{L}'$. Hence we have basis (e, f, g) with

$$[ef] = 0\,, \qquad [eg] = \alpha e + \beta f\,, \qquad [fg] = \gamma e + \delta f \tag{18}$$

where $A = \begin{pmatrix} \alpha & \beta \\ \gamma & \delta \end{pmatrix}$ is a non-singular matrix. Conversely, in any space $\mathfrak{L}$ with basis (e, f, g) we can define a product $[ab]$ so that $[aa] = 0$ and (18) holds. Then $[[ef]g] + [[fg]e] + [[ge]f] = 0$ and hence $\mathfrak{L}$ is a Lie algebra. What changes can be made in the multiplication table (18)? Our choice of basis amounts to this: We have chosen a basis (e, f) for $\mathfrak{L}'$ and supplemented this with a g to get a basis for $\mathfrak{L}$. A change of basis in $\mathfrak{L}'$ will change A to a similar matrix $M^{-1}AM$. The type of change allowable for g is to replace it by $\rho g + x$, $\rho \neq 0$ in Φ, x in $\mathfrak{L}'$. Then $[e, \rho g + x] = \rho[eg]$, $[f, \rho g + x] = \rho[fg]$ so this changes A to ρA. Hence the different matrices A which can be used in (18) are the non-zero multiples of the matrices similar to A. This means that we have a $1:1$ correspondence between the algebras $\mathfrak{L}$ satisfying $\dim \mathfrak{L} = 3$, $\dim \mathfrak{L}' = 2$ and the conjugacy classes in the two dimensional collineation group.

If the field is algebraically closed we can choose A in one of the following forms:

$$\begin{pmatrix} 1 & 0 \\ 0 & \alpha \end{pmatrix} \quad \alpha \neq 0\,, \qquad \begin{pmatrix} 1 & \beta \\ 0 & 1 \end{pmatrix} \quad \beta \neq 0\,.$$

These give the multiplication tables

$$[ef] = 0\,, \qquad [eg] = e\,, \qquad [fg] = \alpha f$$
$$[ef] = 0\,, \qquad [eg] = e + \beta f\,, \qquad [fg] = f\,.$$

Different choices of α give different algebras unless $\alpha\alpha' = 1$. Hence we get an infinite number of non-isomorphic algebras.

(e) dim $\mathfrak{L}' = 3$. Let (e_1, e_2, e_3) be a basis and set $[e_2e_3] = f_1$, $[e_3e_1] = f_2$, $[e_1e_2] = f_3$. Then (f_1, f_2, f_3) is a basis. Write $f_i = \sum_{j=1}^{3}\alpha_{ij}e_j$, $A = (\alpha_{ij})$ non-singular. The only Jacobi condition which has to be imposed is that $[f_1e_1] + [f_2e_2] + [f_3e_3] = 0$. This gives

$$\begin{aligned} 0 &= \alpha_{12}[e_2e_1] + \alpha_{13}[e_3e_1] + \alpha_{21}[e_1e_2] + \alpha_{23}[e_3e_2] + \alpha_{31}[e_1e_3] + \alpha_{32}[e_2e_3] \\ &= -\alpha_{12}f_3 + \alpha_{13}f_2 + \alpha_{21}f_3 - \alpha_{23}f_1 - \alpha_{31}f_2 + \alpha_{32}f_1\,. \end{aligned}$$

Hence $\alpha_{ij} = \alpha_{ji}$ and so A is a symmetric matrix. Let $(\bar{e}_1, \bar{e}_2, \bar{e}_3)$ be a second basis where $\bar{e}_i = \sum\mu_{ij}e_j$, $M = (\mu_{ij})$ non-singular. Set $\bar{f}_1 = [\bar{e}_2\bar{e}_3]$, $\bar{f}_2 = [\bar{e}_3\bar{e}_1]$, $\bar{f}_3 = [\bar{e}_1\bar{e}_2]$. We have for (i, j, k) any cyclic permutation of $(1, 2, 3)$

$$\begin{aligned} \bar{f}_i = [\bar{e}_j\bar{e}_k] &= [\textstyle\sum\mu_{jr}e_r, \sum\mu_{ks}e_s] = \displaystyle\sum_{r,s}\mu_{jr}\mu_{ks}[e_re_s] \\ &= (\mu_{j2}\mu_{k3} - \mu_{j3}\mu_{k2})f_1 + (\mu_{j3}\mu_{k1} - \mu_{j1}\mu_{k3})f_2 + (\mu_{j1}\mu_{k2} - \mu_{j2}\mu_{k1})f_3 \\ &= \textstyle\sum\nu_{ir}f_r\,. \end{aligned}$$

The matrix $N = (\nu_{ij}) = \operatorname{adj} M' = (M')^{-1}\det M'$. The matrix relating the f's to the e's is A and that relating the e's to the $\bar{e}$'s is M^{-1}. Hence if $\bar{A}$ is the matrix $(\bar{\alpha}_{ij})$ such that $\bar{f}_i = \sum\bar{\alpha}_{ij}\bar{e}_j$, then

$$\bar{A} = (\det M')(M')^{-1}AM^{-1}\,. \tag{19}$$

Two matrices A, B are called *multiplicatively cogredient* if $B = \rho N'AN$ where N is non-singular and $\rho \neq 0$ in Φ. In this case we may write $B = \rho\sigma^2(\sigma^{-1}N)'A(\sigma^{-1}N)$, $\sigma = \rho$ det N and if the matrices are of three rows and columns, then we take $M = \sigma N^{-1}$ and $B = \mu(M^{-1})'AM^{-1}$, $\mu = \rho\sigma^2 = \det M$. Thus we have the relation (19). Thus the conditions on A and $\bar{A}$ are that these symmetric matrices are multiplicatively cogredient. Hence with each $\mathfrak{L}$ satisfying our conditions we can associate a unique class of non-singular multiplicatively cogredient symmetric matrices. We have as many algebras as there are classes of such matrices. For the remainder of this section we assume the characteristic is not two. Then each cogredience class contains a diagonal matrix of the form diag $\{\alpha, \beta, 1\}$, $\alpha\beta \neq 0$. This implies that the basis can be chosen so that

$$[e_1e_2] = e_3\,, \qquad [e_2e_3] = \alpha e_1\,, \qquad [e_3e_1] = \beta e_2\,. \tag{20}$$

If the base field is the field of reals, then we have two different algebras obtained by taking $\alpha = \beta = 1$ and $\alpha = -1$, $\beta = 1$. If the field is algebraically closed we can take $\alpha = \beta = 1$.

We shall now single out a particular algebra in the family of algebras satisfying $\dim \mathfrak{L} = 3 = \dim \mathfrak{L}'$. We impose the condition that $\mathfrak{L}$ contains an element h such that $\mathrm{ad}\, h$ has a characteristic root $\alpha \neq 0$ belonging to Φ. Then we have a vector $e \neq 0$ such that $[eh] = e\,\mathrm{ad}\, h = \alpha e \neq 0$ and since $[hh] = 0$, e and h are linearly independent and are part of a basis $(e_1, e_2, e_3) \equiv (e, h, f)$. If (f_1, f_2, f_3) are defined as before, the symmetric matrix (α_{ij}) is now

$$\begin{pmatrix} \alpha_{11} & \alpha_{12} & \alpha \\ \alpha_{12} & \alpha_{22} & 0 \\ \alpha & 0 & 0 \end{pmatrix}. \tag{21}$$

Then we have $[eh] = \alpha e$, $[hh] = 0$, $[fh] = -\alpha f - \alpha_{11} e - \alpha_{12} h$, which implies that the characteristic roots of $\mathrm{ad}\, h$ are 0, α and $-\alpha$. We may replace f by a characteristic vector belonging to the root $-\alpha$ of $\mathrm{ad}\, h$. This is linearly independent of (e, h) and may be used for f. Hence we may suppose that $[eh] = \alpha e$, $[fh] = -\alpha f$. If we replace h by $2\alpha^{-1} h$ we obtain $[eh] = 2e$, $[fh] = -2f$. The form of (21) now gives $[ef] = \beta h \neq 0$. If we replace f by $\beta^{-1} f$, then we obtain the basis (e, f, h) such that

$$[eh] = 2e\,, \qquad [fh] = -2f\,, \qquad [ef] = h\,. \tag{22}$$

Thus we see that there is a unique algebra satisfying our condition. We shall see soon that any $\mathfrak{L}$ such that $\dim \mathfrak{L} = \dim \mathfrak{L}' = 3$ is simple, that is, $\mathfrak{L}$ has no ideals other than itself and 0 and $\mathfrak{L}' \neq 0$. The particular algebra we have singled out by the condition that it contains h with $\mathrm{ad}\, h$ having a non-zero characteristic root in Φ is called *the split three-dimensional simple Lie algebra*. It will play an important role in the sequel.

5. Representations and modules

If $\mathfrak{A}$ is an associative algebra over a field Φ, then a *representation* of $\mathfrak{A}$ is a homomorphism of $\mathfrak{A}$ into an algebra $\mathfrak{E}$ of linear transformations of a vector space $\mathfrak{M}$ over Φ. If $a \rightarrow A$, $b \rightarrow B$ in the representation, then, by definition, $a + b \rightarrow A + B$, $\alpha a \rightarrow \alpha A$, $\alpha \in \Phi$, and $ab \rightarrow AB$. A *right* $\mathfrak{A}$*-module* for the associative algebra $\mathfrak{A}$ is a vector space $\mathfrak{M}$ over Φ together with a binary product of $\mathfrak{M} \times \mathfrak{A}$

into $\mathfrak{M}$ mapping (x, a), $x \in \mathfrak{M}$, $a \in \mathfrak{A}$, into an element $xa \in \mathfrak{M}$ such that

1. $(x_1 + x_2)a = x_1a + x_2a$, $\qquad x(a_1 + a_2) = xa_1 + xa_2$,
2. $\alpha(xa) = (\alpha x)a = x(\alpha a)$, $\qquad \alpha \in \Phi$,
3. $x(ab) = (xa)b$.

If $a \rightarrow A$ is a representation of $\mathfrak{A}$, then the representation space $\mathfrak{M}$ can be made into a right $\mathfrak{A}$-module by defining $xa \equiv xA$. Thus we will have

$$(x_1 + x_2)a = (x_1 + x_2)A = x_1A + x_2A = x_1a + x_2a$$

$$x(a_1 + a_2) = x(A_1 + A_2) = xA_1 + xA_2 = xa_1 + xa_2$$

$$\alpha(xa) = \alpha(xA) = (\alpha x)A = (\alpha x)a$$

$$\alpha(xA) = x(\alpha A) = x(\alpha a)$$

$$x(ab) = x(AB) = (xA)B = (xa)b \, .$$

Conversely, if $\mathfrak{M}$ is any right $\mathfrak{A}$-module, then for any $a \in \mathfrak{A}$ we let A denote the mapping $x \rightarrow xa$. Then the first part of 1 and the first part of 2 show that A is a linear transformation in $\mathfrak{M}$ over Φ. The rest of the conditions in 1, 2, and 3 imply that $a \rightarrow A$ is a representation.

In the theory of representations and in other parts of the theory of associative algebras, algebras with an identity play a preponderant role. In fact, for most considerations it is convenient to confine attention to these algebras and to consider only those homomorphisms which map the identity into the identity of the image algebra. In the sequel we shall find it useful at times to deal with associative algebras which need not have identities. We shall therefore adopt the following conventions on terminology: "*Algebra*" *without any modifier will mean* "*associative algebra with an identity*." For these "subalgebra" will mean subalgebra in the usual sense containing the identity, and "homomorphism" will mean homomorphism in the usual sense mapping 1 into 1. In particular, this will be understood for representations. The corresponding notion of a module is defined by 1 through 3 above, together with the condition

4. $x1 = x$, $\qquad x \in \mathfrak{M}$.

If we wish to allow the possibility that $\mathfrak{A}$ does not have an identity then we shall speak of the "associative algebra" $\mathfrak{A}$ and if we wish to drop 4, then we shall speak of a "module for associative algebra" rather than a module for "the algebra $\mathfrak{A}$."

The algebra $\mathfrak{A}$ can be considered as a right $\mathfrak{A}$-module by taking xa to be the product as defined in $\mathfrak{A}$. Then 1, 2, and 3 hold as a consequence of the axioms for an algebra and 4. holds since 1 is the identity. The representation $a \rightarrow A$ where A is the linear transformation $x \rightarrow xa$ is called the *regular representation.* We have seen (§ 2) that the regular representation is *faithful,* that is, an isomorphism.

Now let $\mathfrak{L}$ be a Lie algebra. Then we define a *representation* of $\mathfrak{L}$ to be a homomorphism $l \rightarrow L$ of $\mathfrak{L}$ into a Lie algebra $\mathfrak{E}_L$, $\mathfrak{E}$ the algebra of linear transformations of a vector space $\mathfrak{M}$ over Φ. The conditions here are that if $l_1 \rightarrow L_1$, $l_2 \rightarrow L_2$, then

$$\begin{aligned} &l_1 + l_2 \rightarrow L_1 + L_2, \ \alpha l_1 \rightarrow \alpha L_1 \\ &[l_1 l_2] \rightarrow [L_1 L_2] \equiv L_1 L_2 - L_2 L_1 \,. \end{aligned} \tag{23}$$

We now define xl for $x \in \mathfrak{M}$, $l \in \mathfrak{L}$ by $xl = xL$. Then (23) and the linearity of L gives the following conditions:

1. $(x_1 + x_2)l = x_1 l + x_2 l \,, \qquad x(l_1 + l_2) = xl_1 + xl_2 \,,$
2. $\alpha(xl) = (\alpha x)l = x(\alpha l) \,,$
3. $x[l_1 l_2] = (xl_1)l_2 - (xl_2)l_1 \,.$

We shall now use these conditions to define the concept of an $\mathfrak{L}$-module, $\mathfrak{L}$ a Lie algebra. This is a vector space $\mathfrak{M}$ over Φ with a mapping of $\mathfrak{M} \times \mathfrak{L}$ to $\mathfrak{M}$ such that the result xl satisfies 1, 2, and 3 above.

As in the associative case, the concepts of module and representation are equivalent. Thus we have indicated that if $l \rightarrow L$ is a representation, then the representation space $\mathfrak{M}$ can be considered as a module. On the other hand, if $\mathfrak{M}$ is any module, then for any $l \in \mathfrak{L}$ we let L denote the mapping $x \rightarrow xl$. Then L is linear in $\mathfrak{M}$ over Φ and $l \rightarrow L$ is a representation of $\mathfrak{L}$ by linear transformations in $\mathfrak{M}$ over Φ.

We note next that $\mathfrak{L}$ itself can be considered as an $\mathfrak{L}$-module by taking xl to be the product $[xl]$ in $\mathfrak{L}$. Then 1 and 2 are consequences of the axioms for an algebra and 3. follows from the Jacobi identity and skew symmetry. We have denoted the representing transformation of l: $x \rightarrow [xl]$ by $\operatorname{ad} l$. The representation $l \rightarrow \operatorname{ad} l$ determined by this module is called the *adjoint representation* of

$\mathfrak{L}$. We recall that the mappings ad l are derivations in $\mathfrak{L}$.

If $\mathfrak{M}$ is a module for the Lie algebra $\mathfrak{L}$ then we can consider $\mathfrak{M}$ as an abelian Lie algebra with product $[xy] = 0$. Then the mappings $x \to xl$ are linear transformations in $\mathfrak{M}$ and because of the triviality of the multiplication in $\mathfrak{M}$, these are derivations. More generally, we suppose now that $\mathfrak{M}$ is an $\mathfrak{L}$-module which is at the same time a Lie algebra, and we assume that the module mappings $x \to xl$ are derivations in $\mathfrak{M}$. Thus, in addition to the axioms for a Lie algebra in $\mathfrak{L}$ and in $\mathfrak{M}$ and 1, 2, and 3 above, we have also

4. $[x_1x_2]l = [x_1l, x_2] + [x_1, x_2l]$.

Now let $\mathfrak{K}$ be the direct sum of the two vector spaces $\mathfrak{L}$ and $\mathfrak{M}$. We introduce in $\mathfrak{K}$ a multiplication $[uv]$ by means of the formula

$$[x_1 + l_1, x_2 + l_2] = [x_1x_2] + x_1l_2 - x_2l_1 + [l_1l_2] \; . \tag{24}$$

It is clear that this composition is bilinear, so it makes the vector space $\mathfrak{K}$ into a non-associative algebra. Moreover,

$$[x + l, x + l] = 0 + xl - xl + 0 = 0 \; .$$

$$\begin{aligned}
[[x_1 + l_1, x_2 + l_2]x_3 + l_3] &= [[x_1x_2]x_3] + [[x_1l_2]x_3] \\
&\quad + [[l_1x_2]x_3] + [[l_1l_2]x_3] + [[x_1x_2]l_3] \\
&\quad + [[x_1l_2]l_3] + [[l_1x_2]l_3] + [[l_1l_2]l_3] \\
&= [[x_1x_2]x_3] + [x_1l_2, x_3] - [x_2l_1, x_3] - x_3[l_1l_2] \\
&\quad + [x_1x_2]l_3 + (x_1l_2)l_3 - (x_2l_1)l_3 + [[l_1l_2]l_3] \\
&= [[x_1x_2]x_3] + [x_1l_2, x_3] - [x_2l_1, x_3] - (x_3l_1)l_2 \\
&\quad + (x_3l_2)l_1 + [x_1l_3, x_2] + [x_1, x_2l_3] \\
&\quad + (x_1l_2)l_3 - (x_2l_1)l_3 + [[l_1l_2]l_3] \; .
\end{aligned}$$

If we permute 1, 2, and 3 cyclically and add, then the terms involving three x's or three l's add up to 0 by the Jacobi identity in $\mathfrak{L}$ and in $\mathfrak{M}$. The terms involving two x's and one l are

$$\begin{aligned}
&[x_1l_2, x_3] - [x_2l_1, x_3] + [x_1l_3, x_2] + [x_1, x_2l_3] \\
&\quad + [x_2l_3, x_1] - [x_3l_2, x_1] + [x_2l_1, x_3] + [x_2, x_3l_1] \\
&\quad + [x_3l_1, x_2] - [x_1l_3, x_2] + [x_3l_2, x_1] + [x_3, x_1l_2] = 0 \; .
\end{aligned}$$

The terms involving two l's and one x are

$$\begin{aligned}
&- (x_3l_1)l_2 + (x_3l_2)l_1 + (x_1l_2)l_3 - (x_2l_1)l_3 \\
&\quad - (x_1l_2)l_3 + (x_1l_3)l_2 + (x_2l_3)l_1 - (x_3l_2)l_1 \\
&\quad - (x_2l_3)l_1 + (x_2l_1)l_3 + (x_3l_1)l_2 - (x_1l_3)l_2 = 0 \; .
\end{aligned}$$

This shows that $\mathfrak{K} = \mathfrak{L} \oplus \mathfrak{M}$ is a Lie algebra. It is immediate from (24) and the fact that $[xl] \in \mathfrak{M}$ for x in $\mathfrak{M}$, l in $\mathfrak{L}$, that $\mathfrak{L}$ is a subalgebra of $\mathfrak{K}$ and $\mathfrak{M}$ is an ideal in $\mathfrak{K}$. We shall call $\mathfrak{K}$ the *split extension* of $\mathfrak{L}$ by $\mathfrak{M}$.

An important special case of this is obtained by taking $\mathfrak{M} = \mathfrak{L}$ to be any Lie algebra and $\mathfrak{L} = \mathfrak{D}$, the derivation algebra. Since $\mathfrak{D}$ is, by definition, a Lie algebra of linear transformations, the identity mapping is a representation of $\mathfrak{D}$ in $\mathfrak{L}$. $\mathfrak{L}$ becomes a $\mathfrak{D}$-module by defining the module product $lD = lD$, $l \in \mathfrak{L}$, $D \in \mathfrak{D}$. The split extension of $\mathfrak{D}$ by $\mathfrak{L}$, $\mathfrak{H} = \mathfrak{D} \oplus \mathfrak{L}$ is called the *holomorph* of $\mathfrak{L}$. This is the analogue of the holomorph of a group which is an extension of the group of automorphisms of a group by the group.

We can make the same construction $\mathfrak{K} = \mathfrak{D}_1 \oplus \mathfrak{L}$ where $\mathfrak{L}$ is any Lie algebra and $\mathfrak{D}_1$ is a subalgebra of the derivation algebra. In particular, it is often useful to do this for $\mathfrak{D}_1 = \Phi D$, the subalgebra of multiples of one derivation D of $\mathfrak{L}$.

Another important special case of a split extension is obtained by taking $\mathfrak{L}$ and $\mathfrak{M}$ to be arbitrary Lie algebras and considering $\mathfrak{M}$ as a trivial module for $\mathfrak{L}$ by defining $ml = 0$, $m \in \mathfrak{M}$, $l \in \mathfrak{L}$. The Lie algebra $\mathfrak{K} = \mathfrak{L} \oplus \mathfrak{M}$ is the *direct sum* of $\mathfrak{L}$ and $\mathfrak{M}$. More generally, if $\mathfrak{L}_1, \mathfrak{L}_2, \cdots, \mathfrak{L}_r$ are Lie algebras then the direct sum $\mathfrak{L} = \mathfrak{L}_1 \oplus \mathfrak{L}_2 \oplus \cdots \oplus \mathfrak{L}_r$ is the vector space direct sum of the $\mathfrak{L}_i$ with the Lie product $[\sum_1^r l_i, \sum_1^r m_i] = \sum_1^r [l_i m_i]$. As in group theory, if $\mathfrak{L}$ is a Lie algebra and $\mathfrak{L}$ contains ideals $\mathfrak{L}_i$ such that $\mathfrak{L} = \mathfrak{L}_1 \oplus \mathfrak{L}_2 \oplus \cdots \oplus \mathfrak{L}_r$, as vector space, then $[l_i l_j] \in \mathfrak{L}_i \cap \mathfrak{L}_j = 0$ if $l_i \in \mathfrak{L}_i$, $l_j \in \mathfrak{L}_j$ and $i \neq j$. Then $\mathfrak{L}$ is isomorphic to the direct sum of the Lie algebras $\mathfrak{L}_i$ and we shall say that $\mathfrak{L}$ is a *direct sum of the ideals* $\mathfrak{L}_i$ of $\mathfrak{L}$.

The kernel $\mathfrak{K}$ of a homomorphism η of a Lie algebra $\mathfrak{L}$ into a Lie algebra $\mathfrak{M}$ is an ideal in $\mathfrak{L}$ and the image $\mathfrak{L}\eta$ is a subalgebra of $\mathfrak{M}$. The fundamental theorem of homomorphisms states that $\mathfrak{L}\eta \cong \mathfrak{L}/\mathfrak{K}$ under the correspondence $l + \mathfrak{K} \rightarrow l\eta$. We recall that $\mathfrak{L}/\mathfrak{K}$ is the vector space $\mathfrak{L}/\mathfrak{K}$ considered as an algebra relative to the multiplication $[l_1 + \mathfrak{K}, l_2 + \mathfrak{K}] = [l_1 l_2] + \mathfrak{K}$. This is a Lie algebra. The kernel of the adjoint representation is the set of elements c such that $[xc] = 0$ for all x. This is just the center $\mathfrak{C}$ of $\mathfrak{L}$. The image $\mathfrak{I} = \mathrm{ad}\,\mathfrak{L}$ is the algebra of inner derivations and we have $\mathfrak{L}/\mathfrak{C} \cong \mathfrak{I}$. If $\mathfrak{C} = 0$, $\mathrm{ad}\,\mathfrak{L}$ is a Lie algebra of linear transformations isomorphic to $\mathfrak{L}$. Thus in this case we obtain in an easy way a faithful representation of $\mathfrak{L}$. We shall see later that every $\mathfrak{L}$ has a faithful representation and that every finite-dimensional $\mathfrak{L}$ has a faithful finite-dimensional representation, that is, a faithful representation acting in a finite-dimensional space.

Examples. We shall now determine the matrices of the adjoint representations of two of our examples.

(a) $\mathfrak{L}$ the Lie algebra with basis (e, f), $[ef] = e$. We have $e \operatorname{ad} e = 0$, $f \operatorname{ad} e = -e$; $e \operatorname{ad} f = e$, $f \operatorname{ad} f = 0$. Hence relative to the basis (e, f) the matrices are

$$e \rightarrow \begin{pmatrix} 0 & 0 \\ -1 & 0 \end{pmatrix}, \qquad f \rightarrow \begin{pmatrix} 1 & 0 \\ 0 & 0 \end{pmatrix}.$$

(b) $\mathfrak{L}$ the Lie algebra with basis (e_1, e_2, e_3) such that $[e_1e_2] = e_3$, $[e_2e_3] = e_1$, $[e_3e_1] = e_2$. Here $e_1 \operatorname{ad} e_1 = 0$, $e_2 \operatorname{ad} e_1 = -e_3$, $e_3 \operatorname{ad} e_1 = e_2$; $e_1 \operatorname{ad} e_2 = e_3$, $e_2 \operatorname{ad} e_2 = 0$, $e_3 \operatorname{ad} e_2 = -e_1$; $e_1 \operatorname{ad} e_3 = -e_2$, $e_2 \operatorname{ad} e_3 = e_1$, $e_3 \operatorname{ad} e_3 = 0$. Hence the representation by matrices is

$$e_1 \rightarrow \begin{pmatrix} 0 & 0 & 0 \\ 0 & 0 & -1 \\ 0 & 1 & 0 \end{pmatrix}, \qquad e_2 \rightarrow \begin{pmatrix} 0 & 0 & 1 \\ 0 & 0 & 0 \\ -1 & 0 & 0 \end{pmatrix},$$

$$e_3 \rightarrow \begin{pmatrix} 0 & -1 & 0 \\ 1 & 0 & 0 \\ 0 & 0 & 0 \end{pmatrix}.$$

Note that the matrices are skew symmetric and form a basis for the space of skew symmetric matrices. Hence we see that $\mathfrak{L}$ is isomorphic to the Lie algebra of skew symmetric matrices in the matrix algebra Φ_3.

6. *Some basic module operations*

The notion of a submodule $\mathfrak{N}$ of a module $\mathfrak{M}$ for a Lie or associative algebra is clear: $\mathfrak{N}$ is a subspace of $\mathfrak{M}$ closed under the composition by elements of the algebra. If $\mathfrak{N}$ is a submodule, then we obtain the factor module $\mathfrak{M}/\mathfrak{N}$ which is the coset space $\mathfrak{M}/\mathfrak{N}$ with the module compositions $(x + \mathfrak{N})a = xa + \mathfrak{N}$, a in the algebra. If $\mathfrak{M}_1$ and $\mathfrak{M}_2$ are two modules for an associative or a Lie algebra, then the space $\mathfrak{M}_1 \oplus \mathfrak{M}_2$ is a module relative to the composition $(x_1 + x_2)a = x_1a + x_2a$, $x_i \in \mathfrak{M}_i$. This module is called the *direct sum* $\mathfrak{M}_1 \oplus \mathfrak{M}_2$ of the two given modules. A similar construction can be given for any number of modules.

The module concepts we have just indicated are applicable to both associative and Lie algebras. We shall now consider some notions which apply only in the Lie case. These are analogous to well-known concepts of the representation theory of groups. The principal ones we shall consider are the tensor product of modules and the contragredient module.

We suppose first that $l \to L_1$ and $l \to L_2$ are two representations of a Lie algebra $\mathfrak{L}$ by linear transformations acting in the same vector space $\mathfrak{M}$ over Φ. We assume, moreover, that if L_1 is any representing transformation from the first representation and M_2 is any representing transformation from the second representation, then $[L_1M_2] = L_1M_2 - M_2L_1 = 0$. We shall now define a new mapping of $\mathfrak{L}$ into the algebra $\mathfrak{E}_L$ of linear transformations in $\mathfrak{M}$ by

$$l \to L_1 + L_2 \,. \tag{25}$$

Since this is the sum of the linear mappings $l \to L_1$ and $l \to L_2$ it is linear. Now let $m \to M_1$, $m \to M_2$, so that in the new mapping we have $m \to M_1 + M_2$. Then

$$\begin{aligned}[L_1 + L_2, M_1 + M_2] &= [L_1M_1] + [L_1M_2] + [L_2M_1] + [L_2M_2] \\ &= [L_1M_1] + [L_2M_2]\end{aligned} \tag{26}$$

and since $[lm] \to [L_1M_1] + [L_2M_2]$, the new mapping is a representation. (*Note*: Nothing like this holds in the associative case.)

We suppose next that $\mathfrak{M}_1$ and $\mathfrak{M}_2$ are any two modules for a Lie algebra $\mathfrak{L}$. Let $\mathfrak{M} = \mathfrak{M}_1 \otimes \mathfrak{M}_2$ ($\equiv \mathfrak{M}_1 \otimes_\Phi \mathfrak{M}_2$) the tensor (or Kronecker or direct) product of $\mathfrak{M}_1$ and $\mathfrak{M}_2$. We recall that if A_i is a linear transformation in $\mathfrak{M}_i$, then the mapping $\sum x_j \otimes y_j \to \sum x_jA_1 \otimes y_jA_2$, $x_j \in \mathfrak{M}_1$, $y_j \in \mathfrak{M}_2$, is a linear transformation $A_1 \otimes A_2$ in $\mathfrak{M}_1 \otimes \mathfrak{M}_2$. We have the rules

$$\begin{aligned}&(A_1 + B_1) \otimes A_2 = A_1 \otimes A_2 + B_1 \otimes A_2 \\ &A_1 \otimes (A_2 + B_2) = A_1 \otimes A_2 + A_1 \otimes B_2 \\ &\alpha(A_1 \otimes B_1) = \alpha A_1 \otimes B_1 = A_1 \otimes \alpha B_1 \,, \qquad \alpha \in \Phi \\ &(A_1 \otimes A_2)(B_1 \otimes B_2) = A_1B_1 \otimes A_2B_2 \,.\end{aligned} \tag{27}$$

It is clear from these relations that the mapping $A_1 \to A_1 \otimes 1_2$, 1_2, the identity in $\mathfrak{M}_2$, is a homomorphism of the algebra $\mathfrak{E}(\mathfrak{M}_1)$ of linear transformations in $\mathfrak{M}_1$ into the algebra $\mathfrak{E}(\mathfrak{M}_1 \otimes \mathfrak{M}_2)$. Similarly, $A_2 \to 1_1 \otimes A_2$ is a homomorphism of $\mathfrak{E}(\mathfrak{M}_2)$ into $\mathfrak{E}(\mathfrak{M}_1 \otimes \mathfrak{M}_2)$. Now we have the representation R_i determined by the $\mathfrak{M}_i$. The linear transformation l^{R_i} associated with $l \in \mathfrak{L}$ is $x_i \to x_il$. The resultant of the Lie homomorphism $l \rightarrow l^{R_i}$ with the associative (hence Lie) homomorphism of $\mathfrak{E}(\mathfrak{M}_i)$ into $\mathfrak{E}(\mathfrak{M}_1 \otimes \mathfrak{M}_2)$ is a representation of $\mathfrak{L}$ acting in $\mathfrak{M}_1 \otimes \mathfrak{M}_2$. The two representations of $\mathfrak{L}$ obtained in this way are

$$l \rightarrow l^{R_1} \otimes 1_2 \quad \text{and} \quad l \rightarrow 1_1 \otimes l^{R_2} . \tag{28}$$

If l, $m \in \mathfrak{L}$, then

$$(l^{R_1} \otimes 1_2)(1_1 \otimes m^{R_2}) = l^{R_1} \otimes m^{R_2} = (1_1 \otimes m^{R_2})(l^{R_1} \otimes 1_2) . \tag{29}$$

Hence the commutativity condition of the last paragraph holds. It now follows that

$$l \rightarrow l^{R_1} \otimes 1_2 + 1_1 \otimes l^{R_2} \tag{30}$$

is a representation of $\mathfrak{L}$ acting in $\mathfrak{M}_1 \otimes \mathfrak{M}_2$. In this way $\mathfrak{M}_1 \otimes \mathfrak{M}_2$ is an $\mathfrak{L}$-module with the module composition defined by

$$(\sum x_j \otimes y_j)l = \sum x_jl \otimes y_j + \sum x_j \otimes y_jl . \tag{31}$$

The module $\mathfrak{M}_1 \otimes \mathfrak{M}_2$ obtained in this way is called the *tensor product* of the two modules $\mathfrak{M}_i$. The same terminology is applied to the representation, which we denote as $R_1 \otimes R_2$.

We consider next a Lie module $\mathfrak{M}$ and the dual space $\mathfrak{M}^*$ of linear functions on $\mathfrak{M}$ (to the base field). We shall denote the value of a linear function y^* at the vector $x \in \mathfrak{M}$ by $\langle x, y^* \rangle$. Then $\langle x, y^* \rangle \in \Phi$ and this product is bilinear:

$$\begin{aligned} &\langle x_1 + x_2, y^* \rangle = \langle x_1, y^* \rangle + \langle x_2, y^* \rangle \\ &\langle x, y_1^* + y_2^* \rangle = \langle x, y_1^* \rangle + \langle x, y_2^* \rangle \\ &\langle \alpha x, y^* \rangle = \alpha \langle x, y^* \rangle = \langle x, \alpha y^* \rangle . \end{aligned} \tag{32}$$

Also the product is non-degenerate. Any linear transformation A in $\mathfrak{M}$ determines an adjoint (or transpose) transformation A^* in $\mathfrak{M}^*$ such that

$$\langle xA, y^* \rangle = \langle x, y^*A^* \rangle . \tag{33}$$

The mapping $A \rightarrow A^*$ is an associative anti-homomorphism of $\mathfrak{E}(\mathfrak{M})$ into $\mathfrak{E}(\mathfrak{M}^*)$. Now consider the mapping $A \rightarrow -A^*$. This is linear and

$$\begin{aligned} [-A^*, -B^*] = [A^*, B^*] &= A^*B^* - B^*A^* \\ &= (BA)^* - (AB)^* = [BA]^* \\ &= -[AB]^* . \end{aligned} \tag{34}$$

Hence $A \rightarrow -A^*$ is homomorphism of $\mathfrak{E}(\mathfrak{M})_L$ into $\mathfrak{E}(\mathfrak{M}^*)_L$. If we now take the resultant of the representation $l \to l^R$ determined by $\mathfrak{M}$ with $A \to -A^*$ we obtain a new representation $l \to -(l^R)^*$ of $\mathfrak{L}$. For the corresponding module $\mathfrak{M}^*$ we have

$$\begin{aligned}\langle x, y^*l\rangle &= \langle x, -y^*(l^R)^*\rangle \\ &= -\langle xl^R, y^*\rangle \\ &= -\langle xl, y^*\rangle .\end{aligned}$$

Hence the characteristic property relating the two modules is

$$\langle x, y^*l\rangle + \langle xl, y^*\rangle = 0 . \tag{35}$$

We call the module $\mathfrak{M}^*$ defined in this way the *contragredient module* $\mathfrak{M}^*$ of $\mathfrak{M}$ and denote the corresponding representation by R^*.

We recall that if $\mathfrak{M}$ is a finite-dimensional space, then there is a natural isomorphism of $\mathfrak{M} \otimes \mathfrak{M}^*$ onto the vector space $\mathfrak{E}(\mathfrak{M})$. If $\sum_i x_i \otimes y_i^* \in \mathfrak{M} \otimes \mathfrak{M}^*$, then the corresponding linear transformation in $\mathfrak{M}$ is $x \to \sum_i \langle x, y_i^*\rangle x_i$. If $\mathfrak{M}$ is a module for $\mathfrak{L}$, then $\mathfrak{M}^*$ and $\mathfrak{M} \otimes \mathfrak{M}^*$ are modules. By definition,

$$(\sum x_i \otimes y_i^*)l = \sum x_i l \otimes y_i^* + \sum x_i \otimes y_i^* l .$$

If we denote $x \to xl$ by l^R then $x^* \to x^*l$ is $-(l^R)^*$. Then in the representation in $\mathfrak{M} \otimes \mathfrak{M}^*$ the mapping corresponding to l sends $\sum x_i \otimes y_i^*$ into $\sum x_i l^R \otimes y_i^* - \sum x_i \otimes y_i^*(l^R)^*$. The elements of $\mathfrak{E}(\mathfrak{M})$ associated with these two transformations are

$$\begin{aligned}A&: x \to \sum \langle x, y_i^*\rangle x_i . \\ B&: x \to \sum \langle x, y_i^*\rangle x_i l^R - \sum \langle x, y_i^*(l^R)^*\rangle x_i \\ &\qquad = \sum \langle x, y_i^*\rangle x_i l^R - \sum \langle x l^R, y_i^*\rangle x_i .\end{aligned}$$

It is clear from these formulas that $B = [A, l^R]$. We can interpret this result as follows: Consider an arbitrary $\mathfrak{L}$-module $\mathfrak{M}$ and the algebra $\mathfrak{E}(\mathfrak{M})$. If l^R is the representing transformation of l, then the mapping $X \to [X, l^R]$ is a representation of $\mathfrak{L}$ acting in $\mathfrak{E}$. The result we have shown is that this representation is equivalent to the representation in $\mathfrak{M} \otimes \mathfrak{M}^*$; that is, the module $\mathfrak{E}$ is isomorphic to $\mathfrak{M} \otimes \mathfrak{M}^*$.

The result just indicated can be generalized to a pair of vector spaces $\mathfrak{M}_1, \mathfrak{M}_2$. We recall that the set of linear transformations $\mathfrak{E}(\mathfrak{M}_2, \mathfrak{M}_1)$ of $\mathfrak{M}_2$ into $\mathfrak{M}_1$ is a vector space under usual compositions of addition and scalar multiplication. If the spaces are finite-dimensional, then there is a natural isomorphisms of $\mathfrak{M}_1 \otimes \mathfrak{M}_2^*$

onto $\mathfrak{E}(\mathfrak{M}_2, \mathfrak{M}_1)$ mapping the element $\sum x_i \otimes y_i^*$, $x_i \in \mathfrak{M}_1$, $y_i^* \in \mathfrak{M}_2^*$, into the linear mapping $y \to \sum \langle y, y_i^* \rangle x_i$ of $\mathfrak{M}_2$ into $\mathfrak{M}_1$. If $\mathfrak{M}_1$ and $\mathfrak{M}_2$ are $\mathfrak{L}$-modules, then $\mathfrak{E}(\mathfrak{M}_2, \mathfrak{M}_1)$ is an $\mathfrak{L}$-module relative to the composition $Xl \equiv Xl^{R_1} - l^{R_2}X$. As for a single space, this module is isomorphic to $\mathfrak{M}_1 \otimes \mathfrak{M}_2^*$ under the space isomorphism we defined.

7. *Ideals, solvability, nilpotency*

If $\mathfrak{B}_1$ and $\mathfrak{B}_2$ are subspaces of a Lie algebra $\mathfrak{L}$ then we write $\mathfrak{B}_1 \cap \mathfrak{B}_2$, $\mathfrak{B}_1 + \mathfrak{B}_2$, respectively, for the intersection and space spanned by $\mathfrak{B}_1$ and $\mathfrak{B}_2$. The latter is just the collection of elements of the form $b_1 + b_2$, b_i in $\mathfrak{B}_i$. We now define $[\mathfrak{B}_1\mathfrak{B}_2]$ to be the subspace spanned by all products $[b_1b_2]$, $b_i \in \mathfrak{B}_i$. It is immediate that this is the set of sums $\sum_j [b_{1j}b_{2j}]$, $b_{ij} \in \mathfrak{B}_i$. We assume the reader is familiar with the (lattice) properties of the set of subspaces relative to the compositions $\cap$ and $+$ and we proceed to state the main properties of the composition $[\mathfrak{B}_1\mathfrak{B}_2]$. We list these as follows and leave the veification to the reader.

1. $[\mathfrak{B}_1\mathfrak{B}_2] = [\mathfrak{B}_2\mathfrak{B}_1]$,
2. $[\mathfrak{B}_1 + \mathfrak{B}_2, \mathfrak{B}_3] = [\mathfrak{B}_1\mathfrak{B}_3] + [\mathfrak{B}_2\mathfrak{B}_3]$,
3. $[[\mathfrak{B}_1\mathfrak{B}_2]\mathfrak{B}_3] \subseteq [[\mathfrak{B}_1\mathfrak{B}_3]\mathfrak{B}_2] + [[\mathfrak{B}_2\mathfrak{B}_3]\mathfrak{B}_1]$,
4. $[\mathfrak{B}_1 \cap \mathfrak{B}_2, \mathfrak{B}_3] \subseteq [\mathfrak{B}_1\mathfrak{B}_3] \cap [\mathfrak{B}_2\mathfrak{B}_3]$.

A subspace $\mathfrak{B}$ is an ideal if and only if $[\mathfrak{B}\mathfrak{L}] \subseteq \mathfrak{B}$. The intersection and sum of ideals is an ideal and 3 for $\mathfrak{B}_3 = \mathfrak{L}$ shows that the same is true of the (Lie) product of ideals. In particular, it is clear that the terms of the *derived series*

$$
\begin{aligned}
\mathfrak{L} &\supseteq \mathfrak{L}' = [\mathfrak{L}\mathfrak{L}] \\
&\supseteq \mathfrak{L}'' = [\mathfrak{L}'\mathfrak{L}'] \supseteq \cdots \\
&\supseteq \mathfrak{L}^{(k)} = [\mathfrak{L}^{(k-1)}\mathfrak{L}^{(k-1)}] \supseteq \cdots
\end{aligned} \tag{36}
$$

are ideals. The same is true of the terms of the *lower central series*

$$
\begin{aligned}
\mathfrak{L} &\supseteq \mathfrak{L}^2 = \mathfrak{L}' \\
&\supseteq \mathfrak{L}^3 = [\mathfrak{L}^2\mathfrak{L}] \supseteq \cdots \\
&\supseteq \mathfrak{L}^k = [\mathfrak{L}^{k-1}\mathfrak{L}] \supseteq \cdots .
\end{aligned} \tag{37}
$$

These series are analogous to the derived series and the lower central series of a group. More generally, if $\mathfrak{B}$ is an ideal in $\mathfrak{L}$, then the derived algebras $\mathfrak{B}^{(i)}$ and the powers $\mathfrak{B}^i$ are ideals. We note also that if η is a homomorphism of $\mathfrak{L}$ into a second Lie algebra then $(\mathfrak{L}^{(i)})\eta = (\mathfrak{L}\eta)^{(i)}$ and $(\mathfrak{L}^i)\eta = (\mathfrak{L}\eta)^i$. These are readily proved by induction on i.

A Lie algebra is said to be *solvable* if $\mathfrak{L}^{(h)} = 0$ for some positive integer h. Any abelian algebra is solvable, and it is immediate from the list of Lie algebras of dimensions $\leqq 3$ that all these are solvable with the exception of those such that $\dim \mathfrak{L} = 3 = \dim \mathfrak{L}'$. The algebra of triangular matrices is another example of a solvable Lie algebra (Exercise 12 below).

LEMMA *Every subalgebra and every homomorphic image of a solvable Lie algebra is solvable. If $\mathfrak{L}$ contains a solvable ideal $\mathfrak{B}$ such that $\mathfrak{L}/\mathfrak{B}$ is solvable, then $\mathfrak{L}$ is solvable.*

Proof: The first two statements are clear. If $\mathfrak{B}$ is an ideal such that $\mathfrak{L}/\mathfrak{B}$ is solvable, then $\mathfrak{L}^{(h)} \subseteqq \mathfrak{B}$ for some positive integer h. Thus let η be the canonical homomorphism $l \rightarrow l + \mathfrak{B}$ of $\mathfrak{L}$ onto $\mathfrak{L}/\mathfrak{B}$. Then $(\mathfrak{L}^{(h)})\eta = (\mathfrak{L}\eta)^{(h)} = (\mathfrak{L}/\mathfrak{B})^{(h)} = 0$ if h is sufficiently large. Hence $\mathfrak{L}^{(h)} \subseteqq \mathfrak{B}$. If $\mathfrak{B}$ is solvable we have $\mathfrak{B}^{(k)} = 0$ for some k. Hence $\mathfrak{L}^{(h)} \subseteqq \mathfrak{B}$ implies $\mathfrak{L}^{(h+k)} \subseteqq \mathfrak{B}^{(k)} = 0$ and $\mathfrak{L}$ is solvable.

PROPOSITION 4. *The sum of any two solvable ideals is a solvable ideal.*

Proof: Let $\mathfrak{B}_1$ and $\mathfrak{B}_2$ be solvable ideals. By one of the standard isomorphism theorems $\mathfrak{B}_1 \cap \mathfrak{B}_2$ is an ideal in $\mathfrak{B}_1$ and $(\mathfrak{B}_1 + \mathfrak{B}_2)/\mathfrak{B}_2 \cong \mathfrak{B}_1/(\mathfrak{B}_1 \cap \mathfrak{B}_2)$. This is solvable since it is a homomorphic image of the solvable algebra $\mathfrak{B}_1$. Since $\mathfrak{B}_2$ is solvable the lemma applies to prove $\mathfrak{B}_1 + \mathfrak{B}_2$ solvable.

Now suppose $\mathfrak{L}$ is finite-dimensional and let $\mathfrak{S}$ be a solvable ideal of maximum dimensionality of $\mathfrak{L}$. Then Proposition 4 implies that if $\mathfrak{B}$ is any solvable ideal $\mathfrak{S} + \mathfrak{B}$ is solvable and an ideal. Hence $\mathfrak{S} + \mathfrak{B} = \mathfrak{S}$ since dim $\mathfrak{S}$ is maximal. Consequently, $\mathfrak{S} \supseteqq \mathfrak{B}$. We therefore have the existence of a solvable ideal $\mathfrak{S}$ which contains every solvable ideal. We call $\mathfrak{S}$ the *radical* of $\mathfrak{L}$. If $\mathfrak{S} = 0$, that is, $\mathfrak{L}$ has no non-zero solvable ideal, then $\mathfrak{L}$ is called *semi-simple*. If $\mathfrak{L}$ has no ideals $\neq 0$ and $\mathfrak{L}$, and if $\mathfrak{L}' \neq 0$, then $\mathfrak{L}$ is called *simple*. If $\mathfrak{L}$ is simple and $\mathfrak{S}$ is its radical, then we must have $\mathfrak{L} = \mathfrak{S}$ or $\mathfrak{S} = 0$. But if $\mathfrak{L} = \mathfrak{S}$ then $\mathfrak{S}' \subset \mathfrak{S}$ and $\mathfrak{S}'$ is an ideal so $\mathfrak{S}' = \mathfrak{L}' = 0$ contrary to the definition. Hence $\mathfrak{S} = 0$; so simplicity implies semi-simplicity. If $\mathfrak{S}$ is the radical, then any solvable ideal of $\mathfrak{L}/\mathfrak{S}$ has the form $\mathfrak{B}/\mathfrak{S}$, $\mathfrak{B}$ an ideal in $\mathfrak{L}$. But $\mathfrak{B}$ is solvable by the lemma. Hence $\mathfrak{B} \subseteq \mathfrak{S}$ and $\mathfrak{B}/\mathfrak{S} = 0$. Thus $\mathfrak{L}/\mathfrak{S}$ is semi-simple. If $\mathfrak{B}$ is a non-zero solvable ideal in $\mathfrak{L}$ and $\mathfrak{B}^{(h-1)} \neq 0$, $\mathfrak{B}^{(h)} = 0$, then $\mathfrak{B}^{(h-1)}$ is an abelian ideal $\neq 0$ in $\mathfrak{L}$. Hence $\mathfrak{L}$ is semi-simple if it has no non-zero abelian ideals.

The three-dimensional Lie algebras $\mathfrak{L}$ with dim $\mathfrak{L}' = 3$ (or $\mathfrak{L}' = \mathfrak{L}$) are simple. For, if $\mathfrak{B}$ is an ideal in $\mathfrak{L}$ such that $0 \neq \mathfrak{B} \neq \mathfrak{L}$, then $\mathfrak{B}$ and $\mathfrak{L}/\mathfrak{B}$ are both one or two dimensional; hence solvable. Then $\mathfrak{L}$ is solvable contrary to $\mathfrak{L}' = \mathfrak{L}$.

A Lie algebra $\mathfrak{L}$ is called *nilpotent* if $\mathfrak{L}^k = 0$ for some positive integer k.

PROPOSITION 5. $[\mathfrak{L}^i\mathfrak{L}^j] \subseteq \mathfrak{L}^{i+j}$.

Proof: By definition $[\mathfrak{L}^i\mathfrak{L}] = \mathfrak{L}^{i+1}$. We assume $[\mathfrak{L}^i\mathfrak{L}^j] \subseteq \mathfrak{L}^{i+j}$ for all i. Then

$$[\mathfrak{L}^i\mathfrak{L}^{j+1}] = [\mathfrak{L}^i[\mathfrak{L}^j\mathfrak{L}]] \subseteq [[\mathfrak{L}^i\mathfrak{L}]\mathfrak{L}^j] + [[\mathfrak{L}^i\mathfrak{L}^j]\mathfrak{L}] \subseteq [\mathfrak{L}^{i+1}\mathfrak{L}^j] + [\mathfrak{L}^{i+j}\mathfrak{L}] \subseteq \mathfrak{L}^{i+j+1}.$$

This result implies that any product of k factors $\mathfrak{L}$ in any association is contained in $\mathfrak{L}^k$. Since $\mathfrak{L}^{(k)}$ is a product of 2^k such factors it follows that $\mathfrak{L}^{(k)} \subseteq \mathfrak{L}^{2^k}$. Hence if $\mathfrak{L}$ is nilpotent, say $\mathfrak{L}^{2^k} = 0$, then $\mathfrak{L}^{(k)} = 0$ and $\mathfrak{L}$ is solvable. The converse does not hold since the two-dimensional non-abelian Lie algebra is solvable but not nilpotent. The set of *nil triangular matrices*, that is, the triangular matrices with 0's on the diagonal is a nilpotent subalgebra of Φ_{nL} where Φ_n is the algebra of $n \times n$ matrices.

PROPOSITION 6. *The sum of nilpotent ideals is nilpotent.*

Proof: We note first that if $\mathfrak{B}$ is an ideal, then any product $[\cdots[[\mathfrak{A}_1\mathfrak{A}_2]\mathfrak{A}_3]\cdots\mathfrak{A}_k]$ in which h of the $\mathfrak{A}_i = \mathfrak{B}$ and the remaining $\mathfrak{A}_i = \mathfrak{L}$ is contained in $\mathfrak{B}^h(\mathfrak{B}^0 = \mathfrak{L})$. A simple induction on k establishes this result. Now consider two ideals $\mathfrak{B}_1$ and $\mathfrak{B}_2$ and the ideal $\mathfrak{B}_1 + \mathfrak{B}_2$. Then $(\mathfrak{B}_1 + \mathfrak{B}_2)^m$ is contained in a sum of terms $[\cdots[[\mathfrak{A}_1\mathfrak{A}_2]\mathfrak{A}_3]\cdots\mathfrak{A}_m]$ where $\mathfrak{A}_i = \mathfrak{B}_1$ or $\mathfrak{B}_2$. Any such term contains $[m/2]$ $\mathfrak{B}_1$'s or $[m/2]$ $\mathfrak{B}_2$'s, where $[m/2]$ is the integral part of $m/2$; hence it is contained in $\mathfrak{B}_1^{[m/2]}$ or in $\mathfrak{B}_2^{[m/2]}$. Consequently,

$$(\mathfrak{B}_1 + \mathfrak{B}_2)^m \subseteq \mathfrak{B}_1^{[m/2]} + \mathfrak{B}_2^{[m/2]}\,.$$

It follows that if $\mathfrak{B}_1$ and $\mathfrak{B}_2$ are nilpotent, then $\mathfrak{B}_1 + \mathfrak{B}_2$ is nilpotent.

As for solvability, we can now conclude that if $\mathfrak{L}$ is finite-dimensional, then $\mathfrak{L}$ contains a nilpotent ideal $\mathfrak{N}$ which contains every nilpotent ideal of $\mathfrak{L}$. We call $\mathfrak{N}$ the *nil radical* of $\mathfrak{L}$. This is contained in the radical $\mathfrak{S}$. For the two-dimensional non-abelian algebra $\Phi e + \Phi f$, $[ef] = e$, $\mathfrak{S} = \mathfrak{L}$ and $\mathfrak{N} = \Phi e$. Also $\mathfrak{L}/\mathfrak{N}$ is abelian, hence nilpotent. Thus we may have $\mathfrak{S} \supset \mathfrak{N}$ and $\mathfrak{L}/\mathfrak{N}$ may have a non-zero nil radical.

The theory of nilpotent ideals and radicals has a parallel for associative algebras. If $\mathfrak{B}_1$ and $\mathfrak{B}_2$ are subspaces of an associative algebra $\mathfrak{A}$, then $\mathfrak{B}_1\mathfrak{B}_2$ denotes the subspace spanned by all products b_1b_2, $b_i \in \mathfrak{B}_i$. $\mathfrak{A}$ is nilpotent if there exists a positive integer k such that $\mathfrak{A}^k = 0$ ($\mathfrak{A}^1 = \mathfrak{A}, \mathfrak{A}^k = \mathfrak{A}^{k-1}\mathfrak{A}$). This is equivalent to saying that every product of k elements of $\mathfrak{A}$ is 0. If $\mathfrak{N}_1$ and $\mathfrak{N}_2$ are nilpotent ideals of $\mathfrak{A}$, then it is easy to see that $\mathfrak{N}_1 + \mathfrak{N}_2$ is a nilpotent ideal. Hence if $\mathfrak{A}$ is finite-dimensional, then $\mathfrak{A}$ contains a maximal nilpotent ideal $\mathfrak{R}$ which contains every nilpotent ideal. The ideal $\mathfrak{R}$ is called the *radical* of $\mathfrak{A}$. The algebra $\mathfrak{A}/\mathfrak{R}$ is *semi-simple* in the sense that it has no nilpotent ideals $\neq 0$. The proofs of these statements are similar to the corresponding ones for Lie algebras and will be left as exercises.

8. Extension of the base field

We are assuming that the reader is familiar with the basic definitions and results on tensor products and extension of the field of operators of vector spaces and non-associative algebras. We now recall without proofs the main properties which will be needed in the sequel.

If $\mathfrak{A}$ and $\mathfrak{B}$ are arbitrary non-associative algebras over Φ, then the vector space $\mathfrak{A} \otimes \mathfrak{B}$ ($\mathfrak{A} \otimes_\Phi \mathfrak{B}$) can be made into a non-associative algebra by defining $(\sum_i a_i \otimes b_i)(\sum_j a_j' \otimes b_j') = \sum a_i a_j' \otimes b_i b_j'$ for $a_i, a_j' \in \mathfrak{A}$, $b_i, b_j' \in \mathfrak{B}$. If $\mathfrak{A}$ and $\mathfrak{B}$ are associative, then $\mathfrak{A} \otimes \mathfrak{B}$ is associative also. If P is a field extension of Φ and $\mathfrak{A}$ is arbitrary, then the Φ-algebra $P \otimes \mathfrak{A}$ can be considered as a non-associative algebra over P by defining $\rho(\sum \rho_i \otimes a_i) = \sum \rho\rho_i \otimes a_i$, $\rho, \rho_i \in P$, $a_i \in \mathfrak{A}$. We denote this "extension" of $\mathfrak{A}$ as $\mathfrak{A}_P$. Such extensions for Lie algebras will play an important role from time to time in the sequel.

We recall some of the main properties of $\mathfrak{A}_P$ and of $\mathfrak{M}_P$ where $\mathfrak{M}$ is any vector space over Φ and $\mathfrak{M}_P$ is $P \otimes \mathfrak{M}$ considered as vector space over P relative to $\rho(\sum \rho_i \otimes x_i) = \sum \rho\rho_i \otimes x_i$, $\rho, \rho_i \in P$, $x_i \in \mathfrak{M}$, If $\{e_i \mid i \in I\}$ is a basis for $\mathfrak{M}$ over Φ, then $\{1 \otimes e_i\}$ is a basis for $\mathfrak{M}_P$ over P. The set of Φ-linear combinations of these elements coincides with the subset $\{1 \otimes x \mid x \in \mathfrak{M}\}$ of $\mathfrak{M}_P$. This set is a Φ-subspace of $\mathfrak{M}_P$ which is isomorphic to $\mathfrak{M}$. We may identify $1 \otimes x$ with x and the set $\{1 \otimes x\}$ with $\mathfrak{M}$. In this way $\mathfrak{M}$ becomes a Φ-subspace of $\mathfrak{M}_P$ which has the following two properties: (1) the P-space spanned by $\mathfrak{M}$ is $\mathfrak{M}_P$, (2) any subset of $\mathfrak{M}$ which is linearly independent over Φ is linearly independent over P. These imply that any basis for $\mathfrak{M}$ over Φ is a basis for $\mathfrak{M}$ over P. If $\mathfrak{A}$ is a non-associative algebra, then the identification just made permits us to consider $\mathfrak{A}$ as a Φ-subalgebra of $\mathfrak{A}_P$. The properties (1) and (2) are characteristic. Thus, let $\tilde{\mathfrak{M}}$ be any vector space over P, Φ a subfield of P, so that $\tilde{\mathfrak{M}}$ can be considered also as space over Φ. Suppose $\mathfrak{M}$ is a subspace of $\tilde{\mathfrak{M}}$ over Φ satisfying (1) and (2). Then the mapping $\sum \rho_i \otimes x_i \rightarrow \sum \rho_i x_i$, $\rho_i \in P$, $x_i \in \mathfrak{M}$, is an isomorphism of $\mathfrak{M}_P$ onto $\tilde{\mathfrak{M}}$. Similarly, if $\tilde{\mathfrak{A}}$ is a non-associative algebra over P and $\mathfrak{A}$ is a Φ-subalgebra satisfying (1) and (2), then we have the isomorphism indicated of $\mathfrak{A}_P$ onto $\tilde{\mathfrak{A}}$.

If $\mathfrak{A}$ is associative, then $(aa')a'' = a(a'a'')$ in $\mathfrak{A}$ implies that $\mathfrak{A}_P$ is associative. Similarly, if $\mathfrak{A}$ is a Lie algebra, then $[aa] = 0$, $[aa'] = -[a'a]$, $[[aa']a''] + [[a'a'']a] + [[a''a]a'] = 0$ imply that $\mathfrak{A}_P$ is a Lie algebra.

If $\mathfrak{N}$ is a subspace of $\mathfrak{M}$, then the P-subspace $P\mathfrak{N}$ generated by $\mathfrak{N}$ may be identified with $\mathfrak{N}_P$. If $\mathfrak{B}$ is a subalgebra (ideal) of $\mathfrak{A}$, then $\mathfrak{B}_P (= P\mathfrak{B})$ is a subalgebra (ideal) of $\mathfrak{A}_P$. The ideal $(\mathfrak{A}_P)^2$ of $\mathfrak{A}_P$ is the set of P-linear combinations of the elements aa', $a, a' \in \mathfrak{A}$. Hence $(\mathfrak{A}_P)^2 = \mathfrak{A}^2_P$. If $\mathfrak{L}$ is a Lie algebra, $\mathfrak{L}^r$ is the set of linear combinations of the products $[\cdots[a_1a_2]\cdots a_r]$, $a_i \in \mathfrak{L}$. It follows that $(\mathfrak{L}^r)_P = (\mathfrak{L}_P)^r$. Similarly, the derived algebra $\mathfrak{L}''$ is the set of linear combinations of the products of the form $[[a_1a_2][a_3a_4]]$, $\mathfrak{L}'''$ the set of linear combinations of products

$$[[[a_1a_2][a_3a_4]], [[a_5a_6], [a_7a_8]]] , \qquad a_i \in \mathfrak{L} ,$$

etc., for $\mathfrak{L}^{(4)} \cdots$. It follows that $(\mathfrak{L}_P)^{(r)} = (\mathfrak{L}^{(r)})_P$. These observations imply that a Lie algebra $\mathfrak{L}$ is commutative, nilpotent, or solvable if and only if $\mathfrak{L}_P$ is, respectively, commutative, nilpotent, or solvable.

If Σ is an extension field of the field P, then we can form $\mathfrak{M}_\Sigma$ and $(\mathfrak{M}_P)_\Sigma$. The first of these is $\Sigma \otimes_\Phi \mathfrak{M}$ while the second is $\Sigma \otimes_P (P \otimes_\Phi \mathfrak{M})$. It is well known that there is a natural isomorphism of $(\mathfrak{M}_P)_\Sigma$ onto $\mathfrak{M}_\Sigma$ mapping $\sigma \otimes (\rho \otimes x) \rightarrow \sigma\rho \otimes x$, $\sigma \in \Sigma$, $\rho \in P$, $x \in \mathfrak{M}$. Hence we can identify $(\mathfrak{M}_P)_\Sigma$ with $\mathfrak{M}_\Sigma$ by means of this isomorphism.

If A is a linear transformation of $\mathfrak{M}$ into a second vector space $\mathfrak{N}$ over Φ, then A has a unique extension, which we shall usually denote by A again, to a linear transformation of $\mathfrak{M}_P$ into $\mathfrak{N}_P$. We have $(\sum \rho_i x_i)A = \sum \rho_i(x_i A)$, $\rho_i \in P$, $x_i \in \mathfrak{M}$. The image $\mathfrak{M}_P A = (\mathfrak{M}A)_P$ and the kernel of the extension A is $\mathfrak{K}_P$, $\mathfrak{K}$ the kernel of A in $\mathfrak{M}$. Hence A is surjective (onto) if and only if its extension is surjective and A is $1:1$ if and only if its extension is $1:1$. If $\mathfrak{A}$ is a non-associative algebra and A is a homomorphism (anti-homomorphism, derivation), then its extension A in $\mathfrak{A}_P$ is a homomorphism (anti-homomorphism, derivation).

Now let $\mathfrak{L}$ be a Lie algebra, $\mathfrak{M}$ an $\mathfrak{L}$-module and R the associated representation. If $a \in \mathfrak{L}$, a^R has a unique extension a^R which is a linear transformation in $\mathfrak{M}_P$. We have $(a+b)^R = a^R + b^R$, $(\alpha a)^R = \alpha a^R$, $[ab]^R = [a^R, b^R]$, $\alpha \in \Phi$, for these extensions. This implies that for $\rho_i \in P$, $a_i \in \mathfrak{L}$ the mapping $\sum \rho_i a_i \to \sum \rho_i a_i^R$ is a homomorphism R of $\mathfrak{L}_P$ into the Lie algebra $\mathfrak{E}(\mathfrak{M}_P)_L$ of linear transformations in $\mathfrak{M}_P$. Thus R has an extension R which is a representation of $\mathfrak{L}_P$ acting in $\mathfrak{M}_P$. For the module $\mathfrak{M}_P$ which is determined in this way we have $x(\sum \rho_i a_i) = \sum \rho_i(x a_i)$, $\rho_i \in P$, $a_i \in \mathfrak{L}$, $x \in \mathfrak{M}_P$ and if $x = \sum \rho_j' x_j$, $x_j \in \mathfrak{M}$, $\rho_j' \in P$, then $xa = \sum \rho_j'(x_j a)$, $a \in \mathfrak{L}$. In a similar fashion a representation R of an associative algebra $\mathfrak{A}$ defines a representation R of the extension $\mathfrak{A}_P$ and a right module $\mathfrak{M}$ for $\mathfrak{A}$ defines a right module $\mathfrak{M}_P$ for $\mathfrak{A}_P$.

Exercises

1. Let $\mathfrak{A}$ and $\mathfrak{B}$ be associative algebras. Show that if θ is a homomorphism of $\mathfrak{A}$ into $\mathfrak{B}$, then θ is a homomorphism of $\mathfrak{A}_L$ into $\mathfrak{B}_L$ and if θ is an anti-homomorphism of $\mathfrak{A}$ into $\mathfrak{B}$, then $-\theta$ is a homomorphism of $\mathfrak{A}_L$ into $\mathfrak{B}_L$. Show that if θ is an anti-homomorphism of $\mathfrak{A}$ into $\mathfrak{A}$, then the subset $\mathfrak{S}(\mathfrak{A}, \theta)$ of θ-skew elements $(a^\theta = -a)$ is a subalgebra of $\mathfrak{A}_L$. Show that if D is a derivation of $\mathfrak{A}$, then D is a derivation of $\mathfrak{A}_L$. Give examples of $\mathfrak{A}$ and $\mathfrak{B}$ which are neither isomorphic nor anti-isomorphic but $\mathfrak{A}_L \cong \mathfrak{B}_L$ ($\cong$ indicates isomorphism). (*Hint*: Take $\mathfrak{A}$, $\mathfrak{B}$ to be commutative.) Give an example of a derivation in $\mathfrak{A}_L$ which is not a derivation of $\mathfrak{A}$.

2. If S is a subset of a Lie algebra $\mathfrak{L}$ the *centralizer* $\mathfrak{C}(S)$ is the set of elements c such that $[sc] = 0$ for all $s \in S$. Show that $\mathfrak{C}(S)$ is a subalgebra. If $\mathfrak{B}$ is a subspace of $\mathfrak{L}$, then the *normalizer* of $\mathfrak{B}$ is the set of $l \in \mathfrak{L}$ such that $[bl] \in \mathfrak{B}$ for every $b \in \mathfrak{B}$. Show that the normalizer of $\mathfrak{B}$ is a subalgebra.

3. Let D be a derivation in a non-associatiative algebra $\mathfrak{A}$. Show that the set of elements z of $\mathfrak{A}$ satisfying $zD = 0$ is a subalgebra. (Such elements are called *D-constants*.) Show that the set of elements z satisfying $zD^i = 0$

for some i is a subalgebra.

4. Show that if D is a derivation of a Lie algebra such that D commutes with every inner derivation, then the image $\mathfrak{L}D \subseteq \mathfrak{C}$, $\mathfrak{C}$ the center. (If $\mathfrak{C} = 0$ this implies that the center of the derivation algebra $\mathfrak{C}(\mathfrak{D}) = 0$. Hence also $\mathfrak{C}(\mathfrak{D}(\mathfrak{D}(\mathfrak{L})) = 0$, etc.)

5. Show that any three-dimensional simple Lie algebra of characteristic not two is complete.

6. Prove that any four-dimensional nilpotent Lie algebra has a three-dimensional ideal. Use this to classify the four-dimensional nilpotent Lie algebras.

7. Verify that if $\mathfrak{L}$ has basis $(e_1, e_2, \cdots, e_8)$ with the multiplication table

$$[e_1e_2] = e_5 \quad [e_1e_3] = e_6 \quad [e_1e_4] = e_7 \quad [e_1e_5] = -e_8$$
$$[e_2e_3] = e_8 \quad [e_2e_4] = e_6 \quad [e_2e_6] = -e_7 \quad [e_3e_4] = -e_5$$
$$[e_3e_5] = -e_7 \quad [e_4e_6] = -e_8\,, \quad \text{all other } [e_ie_j] = 0$$
$$\text{for } i < j\,, \quad [e_ie_i] = 0\,, \quad [e_ie_j] = -[e_je_i]\,,$$

$\mathfrak{L}$ is a nilpotent Lie algebra.

8. A subalgebra $\mathfrak{B}$ of $\mathfrak{L}$ is called *subinvariant* in $\mathfrak{L}$ if there exists a chain of subalgebras $\mathfrak{L} = \mathfrak{L}_1 \supseteq \mathfrak{L}_2 \supseteq \cdots \supseteq \mathfrak{L}_s = \mathfrak{B}$ such that $\mathfrak{L}_i$ is an ideal in $\mathfrak{L}_{i-1}$. Show that if the chain is as indicated, then $[\cdots[\mathfrak{L}\mathfrak{B}]\mathfrak{B}\cdots\mathfrak{B}] \subseteq \mathfrak{B}$ if there are s $\mathfrak{B}$'s in the term on the left-hand side.

9. (Schenkman). Show that if $\mathfrak{A}$ and $\mathfrak{B}$ are subspaces of a Lie algebra then $[\mathfrak{A}\mathfrak{B}^n] \subseteq [\cdots[\mathfrak{A}\mathfrak{B}]\mathfrak{B}\cdots\mathfrak{B}]$ (n $\mathfrak{B}$'s). Use this to prove that if $\mathfrak{B}$ is subinvariant in $\mathfrak{L}$, then $\mathfrak{B}^\omega \equiv \cap_{i=1}^{\infty}\mathfrak{B}^i$ is an ideal in $\mathfrak{L}$.

10. Let $\mathfrak{L}$ be a nilpotent Lie algebra. Show that a subset S of $\mathfrak{L}$ generates $\mathfrak{L}$ if and only if the cosets $s + \mathfrak{L}^2$, $s \in S$, generate $\mathfrak{L}/\mathfrak{L}^2$. Hence show that for finite-dimensional $\mathfrak{L}$ the minimum number of generators of $\mathfrak{L}$ is $\dim \mathfrak{L}/\mathfrak{L}^2$.

11. Show that if $\mathfrak{L}$ is a nilpotent Lie algebra, then every subalgebra of $\mathfrak{L}$ is subinvariant. Show also that if $\mathfrak{B}$ is a non-zero ideal in $\mathfrak{L}$, then $\mathfrak{B} \cap \mathfrak{C} \neq 0$ for $\mathfrak{C}$ the center of $\mathfrak{L}$.

12. Let $\mathfrak{T}$ be the Lie algebra of triangular matrices of n rows and columns. Determine the derived series and the lower central series for $\mathfrak{T}$.

13. Prove that a finite-dimensional Lie algebra is solvable if and only if there exists a chain $\mathfrak{L} = \mathfrak{L}_0 \supset \mathfrak{L}_1 \supset \mathfrak{L}_2 \supset \cdots \supset \mathfrak{L}_n = 0$ where $\dim \mathfrak{L}_i = n - i$ and $\mathfrak{L}_i$ is an ideal in $\mathfrak{L}_{i-1}$.

14. If $\mathfrak{L}$ is a Lie algebra the *upper central series* $0 \subseteq \mathfrak{C}_1 \subseteq \mathfrak{C}_2 \subseteq \cdots$ is defined as follows: $\mathfrak{C}_1$ is the center of $\mathfrak{L}$ and $\mathfrak{C}_i$ is the ideal in $\mathfrak{L}$ such that $\mathfrak{C}_i/\mathfrak{C}_{i-1}$ is the center of $\mathfrak{L}/\mathfrak{C}_{i-1}$. Show that the upper central series leads to $\mathfrak{L}$, that is, there is an s such that $\mathfrak{C}_s = \mathfrak{L}$ if and only if $\mathfrak{L}$ is nilpotent. Show that the minimum s satisfying this is the same as the minimum t such that $\mathfrak{L}^{t+1} = 0$.

15. Prove that every nilpotent Lie algebra has an outer (= non-inner) derivation. (*Hint*: Write $\mathfrak{L} = \mathfrak{M} \oplus \Phi e$ where $\mathfrak{M}$ is an ideal. If $z \in \mathfrak{Z} = \mathfrak{C}(\mathfrak{M})$ the linear mapping such that $\mathfrak{M} \to 0$, $e \to z$ is a derivation. If z is chosen in $\mathfrak{Z}$ but not in $\mathfrak{L}^{n+1}$ where n satisfies $\mathfrak{Z} \subseteq \mathfrak{L}^n$, $\mathfrak{Z} \not\subseteq \mathfrak{L}^{n+1}$, then the derivation defined by z is outer.)

16. Let A be a linear transformation in an n dimensional space. Suppose A has n distinct characteristic roots $\xi_1, \xi_2, \cdots, \xi_n$. Show that ad A acting in $\mathfrak{E}$ has the n^2 characteristic roots $\xi_i - \xi_j$, $i, j = 1, 2, \cdots, n$.

17. Let $\mathfrak{M}$ be a finite-dimensional module, $\mathfrak{M}^*$ the contragredient module. Show that if $\mathfrak{N}$ is a submodule of $\mathfrak{M}$, then $\mathfrak{N}^{\perp} = \{z^* \mid < y, z^* > = 0, y \in \mathfrak{N}\}$ is a submodule of $\mathfrak{M}^*$. Hence show that $\mathfrak{M}$ is irreducible if and only if $\mathfrak{M}^*$ is irreducible.

18. Let $\mathfrak{M}$ and $\mathfrak{M}^*$ be as in Exercise 17. Show that $\mathfrak{M} \otimes \mathfrak{M}^*$ contains a $u \neq 0$ such that $ul = 0$ for all l. Assume $\mathfrak{M}$ irreducible and suppose $\mathfrak{N}$ is any module such that $\mathfrak{M} \otimes \mathfrak{N}$ contains a $u \neq 0$ such that $ul = 0$, $l \in \mathfrak{L}$. Show that $\mathfrak{N}$ contains a submodule isomorphic to $\mathfrak{M}^*$.

19. Let $\mathfrak{A}$ be a non-associative algebra such that $\mathfrak{A} = \mathfrak{A}_1 \oplus \mathfrak{A}_2 \oplus \cdots \oplus \mathfrak{A}_r$ where the $\mathfrak{A}_i$ are ideals satisfying $\mathfrak{A}_i^2 = \mathfrak{A}_i$. Show that the derivation algebra $\mathfrak{D} = \mathfrak{D}_1 \oplus \mathfrak{D}_2 \oplus \cdots \oplus \mathfrak{D}_r$ where $\mathfrak{D}_i$ is an ideal in $\mathfrak{D}$ and is isomorphic to the derivation algebra of $\mathfrak{A}_i$.

20. Show that the derived algebra Φ'_{nL} of the Lie algebra Φ_{nL} is the set of matrices of trace 0. Show that the center of Φ_{nL} is the set $\Phi 1$ of multiples of 1 and that the only ideals in Φ_{nL} are Φ'_{nL} and $\Phi 1$ unless $n = 2$ and the characteristic is 2.

21. Give an example of a Lie algebra over the field C of complex numbers which is not of the form $\mathfrak{L}_C$ where $\mathfrak{L}$ is a Lie algebra over the field R of real numbers. (*Hint*: Consider the Lie algebras satisfying dim $\mathfrak{L} = 3$, dim $\mathfrak{L}' = 2$.)

22. Let $\mathfrak{B}$ be an ideal in a non-associative algebra $\mathfrak{A}$, D a derivation in $\mathfrak{A}$. Show that $\mathfrak{B} + \mathfrak{B}D$ is an ideal. Show that if $\mathfrak{A}$ is finite-dimensional associative of characteristic zero with radical $\mathfrak{R}$, then $\mathfrak{R}D \subseteq \mathfrak{R}$ for every derivation D of $\mathfrak{A}$. (This fails for characteristic p; see p. 75.) Prove the same result for Lie algebras.

23. Show that if $\mathfrak{C}$ is a commutative, associative algebra (with 1) and $\mathfrak{L}$ is a Lie algebra, then $\mathfrak{C} \otimes \mathfrak{L}$ is a Lie algebra. Give an example to show that the tensor product of an associative algebra and a Lie algebra need not be a Lie algebra and an example to show that the tensor product of two Lie algebras need not be a Lie algebra. (*Hint*: For the first of these, take the associative algebra to be Φ_2 and note that $\Phi_n \otimes \mathfrak{B} \cong \mathfrak{B}_n$ where $\mathfrak{B}$ is any non-associative algebra and $\mathfrak{B}_n$ is the algebra of $n \times n$ matrices with entries in $\mathfrak{B}$.)

CHAPTER II

Solvable and Nilpotent Lie Algebras

The main theme of the last chapter has been the analogy between Lie algebras and groups. In this chapter we pursue another idea, namely, relations between Lie algebras and associative algebras. We consider an associative algebra $\mathfrak{A}$—usually the algebra of linear transformations in a finite dimensional vector space—and a subalgebra $\mathfrak{L}$ of $\mathfrak{A}_L$. We are interested in studying relations between the structure of $\mathfrak{L}$ and of the subalgebra $\mathfrak{L}^*$ of $\mathfrak{A}$ generated by $\mathfrak{L}$. We study this particularly for $\mathfrak{L}$ solvable or nilpotent. The results we obtain in this way include the classical theorems of Lie and Engel on solvable Lie algebras of linear transformations and a criterion for complete reducibility of a Lie algebra of linear transformations. We introduce the notion of weight spaces and we establish a decomposition into weight spaces for the vector space of a "split" nilpotent Lie algebra of linear transformations. These results will play an important role in the structure theory of the next chapter.

1. *Weakly closed subsets of an associative algebra*

Our first results can be established for subsets of an associative algebra which are more general than Lie algebras. It is not much more difficult to treat these more general systems. Moreover, occasionally these are useful in the study of Lie algebras themselves.

DEFINITIONS 1. A subset $\mathfrak{W}$ of an associative algebra $\mathfrak{A}$ over a field Φ is called *weakly closed* if for every ordered pair (a, b), $a, b \in \mathfrak{W}$, there is defined an element $\gamma(a, b) \in \Phi$ such that $ab + \gamma(a, b)ba \in \mathfrak{W}$. We assume the mapping $(a, b) \rightarrow \gamma(a, b)$ is fixed and write $a \times b = ab + \gamma(a, b)ba$. A subset $\mathfrak{U}$ of $\mathfrak{W}$ is called a *subsystem* if $c \times d \in \mathfrak{U}$ for every $c, d \in \mathfrak{U}$ and $\mathfrak{U}$ is a *left ideal* (*ideal*) if $a \times c \in \mathfrak{U}$ ($a \times c$ and $c \times a \in \mathfrak{U}$) for every $a \in \mathfrak{W}$, $c \in \mathfrak{U}$.

Examples

(1) Any subalgebra $\mathfrak{L}$ of $\mathfrak{A}_L$ is weakly closed in $\mathfrak{A}$ relative to $\gamma(a, b) \equiv -1$.

(2) If $\mathfrak{L}(\subseteqq \mathfrak{A})$ is the Lie algebra with basis (e, f, h) such that $[ef] = h$, $[eh] = 2e$, $[fh] = -2f$ then $\mathfrak{W} = \Phi e \cup \Phi f \cup \Phi h$ is a subsystem of $\mathfrak{L}$.

(3) The set of symmetric matrices is weakly closed in the algebra Φ_n of $n \times n$ matrices if we take $\gamma(a, b) = 1$.

(4) Let $\mathfrak{W} = \mathfrak{H} \cup \mathfrak{S}$ where $\mathfrak{H}$ is the set of symmetric matrices and $\mathfrak{S}$ is the set of skew matrices. Define $\gamma(a, b) = 1$ if a and b are symmetric and $\gamma(a, b) = -1$ otherwise. Then $\mathfrak{W}$ is weakly closed and $\mathfrak{H}$ is an ideal in $\mathfrak{W}$.

(5) If $\gamma(a, b) \equiv 0$ we have a multiplicative semigroup in $\mathfrak{A}$.

Definition 2. If $\mathfrak{S}$ is a subset of an associative algebra $\mathfrak{A}$ (an algebra = associative algebra with 1) we denote by $\mathfrak{S}^*(\mathfrak{S}^\dagger)$ the subalgebra of $\mathfrak{A}$ (subalgebra containing 1) generated by $\mathfrak{S}$. We call $\mathfrak{S}^*(\mathfrak{S}^\dagger)$ the *enveloping associative algebra* (*enveloping algebra*) *of* $\mathfrak{S}$ (*in* $\mathfrak{A}$).

We shall now note some properties of weakly closed systems which will be needed in the proof of our main theorem on such sets.

I. *If W is an element of a weakly closed system $\mathfrak{W}$, then $\mathfrak{Z} \equiv \{W\}^* \cap \mathfrak{W}$ is a subsystem of $\mathfrak{W}$ such that $\mathfrak{Z}^* = \{W\}^*$.*

Proof: The enveloping associative algebra $\{W\}^*$ is the algebra of polynomials in W with constant terms 0. If W_1 and W_2 are two such polynomials then $W_1 \times W_2$ is a polynomial. Hence $\mathfrak{Z} = \{W\}^* \cap \mathfrak{W}$ is a subsystem. Since $\mathfrak{Z} \ni W$, $\mathfrak{Z}^* \supseteqq \{W\}^*$. Since $\mathfrak{Z} \subseteqq \{W\}^*$ and the latter is a subalgebra, $\mathfrak{Z}^* \subseteqq \{W\}^*$. Hence $\mathfrak{Z}^* = \{W\}^*$.

II. *If $\mathfrak{B}$ is a subsystem of $\mathfrak{W}$ and W is an element of $\mathfrak{W}$ such that $B \times W \in \mathfrak{B}^*$ for every $B \in \mathfrak{B}$ then*

$$\mathfrak{B}^* W \subseteqq W\mathfrak{B}^* + \mathfrak{B}^*. \tag{1}$$

Proof: The elements of $\mathfrak{B}^*$ are linear combinations of monomials $B_1 B_2 \cdots B_r$, $B_i \in \mathfrak{B}$. If $B \in \mathfrak{B}$ then $BW = -\gamma(B, W)WB + B \times W \in W\mathfrak{B}^* + \mathfrak{B}^*$. Induction on r now shows that if $B_i \in \mathfrak{B}$ then $B_1 \cdots B_r W \in W\mathfrak{B}^* + \mathfrak{B}^*$. This proves (1).

III. *Let $\mathfrak{B}$ be a subsystem of $\mathfrak{W}$ such that $\mathfrak{B}^*$ is nilpotent and $\mathfrak{B}^* \neq \mathfrak{W}^*$. Then there exists a $W \in \mathfrak{W}$ such that $W \notin \mathfrak{B}^*$ but*

$B \times W \in \mathfrak{B}^*$ for every B in $\mathfrak{B}$.

Proof: The assumption $\mathfrak{B}^* \neq \mathfrak{W}^*$ implies that there exists a $W_1 \in \mathfrak{W}$, $W_1 \notin \mathfrak{B}^*$. If $B \times W_1 \in \mathfrak{B}^*$ for all $B \in \mathfrak{B}$ then we take $W = W_1$. Otherwise we have a $W_2 = B_1 \times W_1 \in \mathfrak{W}$, $\notin \mathfrak{B}^*$. We repeat the argument with W_2 in place of W_1. Either this can be taken to be the W of the statement or we obtain $W_3 = B_2 \times W_2 = B_2 \times (B_1 \times W_1) \in \mathfrak{W}$, $\notin \mathfrak{B}^*$. This procedure leads in a finite number of steps to the required element W or else we obtain an infinite sequence $W_1, W_2, \cdots, W_i = B_{i-1} \times W_{i-1}$, where $W_i \in \mathfrak{W}$ but $W_i \notin \mathfrak{B}^*$. We shall show that this last possibility cannot occur and this will complete the proof. We note that W_k is a linear combination of products of $k-1$ B's belonging to $\mathfrak{B}$ and W_1. Since $\mathfrak{B}^*$ is nilpotent there exists a positive integer n such that any product of n elements of $\mathfrak{B}$ is 0. Now W_{2n} is a linear combination of terms $C_1 \cdots C_j W_1 D_1 \cdots D_k$ where the C_r and $D_s \in \mathfrak{B}$ and $j + k = 2n - 1$. Either $j \geqq n$ or $k \geqq n$ so that we have $C_1 \cdots C_j W_1 D_1 \cdots D_k = 0$. Hence $W_{2n} = 0$ and $W_{2n} \in \mathfrak{B}^*$ contrary to assumption

2. *Nil weakly closed sets*

The main result we shall obtain on weakly closed systems is the following

THEOREM 1. *Let $\mathfrak{W}$ be a weakly closed subset of the associative algebra $\mathfrak{E}$ of linear transformations of a finite-dimensional vector space $\mathfrak{M}$ over Φ. Assume every $W \in \mathfrak{W}$ is associative nilpotent, that is, $W^k = 0$ for some positive integer k. Then the enveloping associative algebra $\mathfrak{W}^*$ of $\mathfrak{W}$ is nilpotent.*

Proof: We shall prove the result by induction on $\dim \mathfrak{M}$. The result is clear if $\dim \mathfrak{M} = 0$ or if $\mathfrak{W} = \{0\}$. Hence we assume $\dim \mathfrak{M} > 0$, $\mathfrak{W} \neq \{0\}$. Let Ω be the collection of subsystems $\mathfrak{B}$ of $\mathfrak{W}$ such that $\mathfrak{B}^*$ is nilpotent and let $\mathfrak{B}$ be an element of Ω such that $\dim \mathfrak{B}^*$ is maximal (for the elements of Ω). We shall show that $\mathfrak{B}^* = \mathfrak{W}^*$ and this will imply the theorem. We note first that $\mathfrak{B}^* \neq 0$. Thus let W be a non-zero element of $\mathfrak{W}$. Then by I, $\mathfrak{Z} = \mathfrak{W} \cap \{W\}^*$ is a subsystem and $\mathfrak{Z}^* = \{W\}^*$. Since $\{W\}^*$ is the set of polynomials in W with constant term 0, $\{W\}^*$ is nilpotent. Hence $\mathfrak{Z} \in \Omega$. Since $\mathfrak{Z} \neq 0$ it follows that $\mathfrak{B}^* \neq 0$. This implies that the subspace $\mathfrak{N}$ spanned by all the vectors xB^*, $x \in \mathfrak{M}$, $B^* \in \mathfrak{B}^*$, is not 0. Also $\mathfrak{N} \neq \mathfrak{M}$. For otherwise, any $x = \sum x_i B_i^*$, $x_i \in \mathfrak{M}$,

$B_i^* \in \mathfrak{B}^*$. If we use similar expressions for the x_i we obtain $x = \sum y_j B_j^* C_j^*$, B_j^*, C_j^* in $\mathfrak{B}^*$. A repetition of this process gives $x = \sum z_k B_{k1}^* B_{k2}^* \cdots B_{kr}^*$, B_{kj}^* in $\mathfrak{B}^*$. Since $\mathfrak{B}^*$ is nilpotent this implies $x = 0$, and $\mathfrak{M} = 0$ contrary to assumption. Hence we have $\mathfrak{M} \supset \mathfrak{N} \supset 0$. Let $\mathfrak{C}$ be the subset of $\mathfrak{W}$ of elements C such that $\mathfrak{N}C \subseteq \mathfrak{N}$. Then it is clear that $\mathfrak{C}$ is a subsystem containing $\mathfrak{B}$. Moreover $\mathfrak{C}$ induces weakly closed systems of nilpotent linear transformations in $\mathfrak{N}$ and in the factor space $\mathfrak{M}/\mathfrak{N}$. Since $\dim \mathfrak{N}$, $\dim \mathfrak{M}/\mathfrak{N} < \dim \mathfrak{M}$ we may assume that the induced systems have nilpotent enveloping associative algebras. The nilpotency of the algebra in $\mathfrak{M}/\mathfrak{N}$ implies that there exists a positive integer p such that if x is any element of $\mathfrak{M}$ and $C_1, C_2, \cdots, C_p$ are any C_i in $\mathfrak{C}$ then $xC_1C_2 \cdots C_p \in \mathfrak{N}$. Also there exists an integer q such that if $D_1, D_2, \cdots, D_q \in \mathfrak{C}$ and $y \in \mathfrak{N}$ then $yD_1 \cdots D_q = 0$. This implies that if $C_1, \cdots, C_{p+q} \in \mathfrak{C}$ then $C_1C_2 \cdots C_{p+q} = 0$. Thus $\mathfrak{C}^*$ is nilpotent and $\mathfrak{C} \in \Omega$. We can now conclude that $\mathfrak{W}^* = \mathfrak{B}^*$. Otherwise, by III., there exists $W \in \mathfrak{W}$, $\notin \mathfrak{B}^*$ such that $B \times W \in \mathfrak{B}^*$ for all B in $\mathfrak{B}$. By II., we have for any $x \in \mathfrak{M}$, $B^* \in \mathfrak{B}^*$, $xB^*W = x(WB_1^* + B_2^*)$, $B_i^* \in \mathfrak{B}^*$. Hence $\mathfrak{N}W \subseteq \mathfrak{N}$ so $W \in \mathfrak{C}$. Since $W \notin \mathfrak{B}^*$, $\dim \mathfrak{C}^* > \dim \mathfrak{B}^*$ and since $\mathfrak{C} \in \Omega$ this contradicts the choice of $\mathfrak{B}$. Hence $\mathfrak{B}^* = \mathfrak{W}^*$ is nilpotent.

If $\mathfrak{W}$ is a set of nil triangular matrices, that is, triangular matrices with 0 diagonal elements, then $\mathfrak{W}^*$ is contained in the associative algebra of nil triangular matrices. The latter is nilpotent; hence $\mathfrak{W}^*$ is nilpotent. The following result is therefore essentially just a slightly different formulation of Theorem 1.

THEOREM 1$'$. *Let $\mathfrak{M}$ be as in Theorem 1. Then there exists a basis for $\mathfrak{M}$ such that the matrices of all the $W \in \mathfrak{W}$ have nil triangular form relative to this basis.*

Proof: We may assume $\mathfrak{M} \neq 0$. Then the proof of Theorem 1 shows that $\mathfrak{M} \supset \mathfrak{M}\mathfrak{W}^*$ where $\mathfrak{M}\mathfrak{W}^*$ is the space spanned by the vectors xW^*, $x \in \mathfrak{M}$, $W^* \in \mathfrak{W}^*$. In general, if $\mathfrak{N}$ is a subspace and $\mathfrak{S}$ is a set of linear transformations, then we shall write $\mathfrak{N}\mathfrak{S}$ for the subspace spanned by all the vectors yS, $y \in \mathfrak{N}$, $S \in \mathfrak{S}$. Then it is immediate that $\mathfrak{M}(\mathfrak{W}^*)^2 = (\mathfrak{M}\mathfrak{W}^*)\mathfrak{W}^*$. Also if $\mathfrak{M}\mathfrak{W}^* \neq 0$ the argument used before shows that $\mathfrak{M}\mathfrak{W}^* \supset \mathfrak{M}(\mathfrak{W}^*)^2$. Hence we have a chain $\mathfrak{M} \supset \mathfrak{M}\mathfrak{W}^* \supset \mathfrak{M}(\mathfrak{W}^*)^2 \supset \mathfrak{M}(\mathfrak{W}^*)^3 \supset \cdots \supset \mathfrak{M}(\mathfrak{W}^*)^{N-1} \supset \mathfrak{M}(\mathfrak{W}^*)^N = 0$, if $(\mathfrak{W}^*)^N = 0$ and $(\mathfrak{W}^*)^{N-1} \neq 0$. We now choose a basis $(e_1, \cdots, e_n)$ for $\mathfrak{M}$ such that $(e_1, \cdots e_{n_1})$ is a basis for $\mathfrak{M}(\mathfrak{W}^*)^{N-1}$, $(e_1, \cdots, e_{n_1}, \cdots, e_{n_1+n_2})$ is a basis for $\mathfrak{M}(\mathfrak{W}^*)^{N-2}, \cdots, (e_1, \cdots, e_{n_1+n_2+\cdots+n_k})$ is a basis for

$\mathfrak{M}(\mathfrak{W}^*)^{N-k}$ Then the relations $\mathfrak{M}(\mathfrak{W}^*)^{N-k}\mathfrak{W} \subseteqq \mathfrak{M}(\mathfrak{W}^*)^{N-k+1}$ imply that the matrix of any $W \in \mathfrak{W}$ has the form

$$\begin{bmatrix} \boxed{0}_{\,n_1}^{\,n_1} & & & \\ * & \boxed{0}_{\,n_2}^{\,n_2} & & \\ \vdots & & \ddots & \\ * & \cdots & * & \boxed{0}_{\,n_N}^{\,n_N} \end{bmatrix}.$$

Theorem 1 can be generalized to a theorem about ideals in weakly closed sets. For this purpose we shall derive another general property of enveloping associative algebras of weakly closed sets:

IV. *Let $\mathfrak{W}$ be a weakly closed subset of an associative algebra and let $\mathfrak{B}$ be an ideal in $\mathfrak{W}$, $\mathfrak{B}^*$, $\mathfrak{W}^*$ the enveloping associative algebras of $\mathfrak{B}$ and $\mathfrak{W}$, respectively. Then*

$$\begin{aligned} \mathfrak{W}^*(\mathfrak{B}^*)^k &\subseteqq (\mathfrak{B}^*)^k\mathfrak{W}^* + (\mathfrak{B}^*)^k\,, \\ (\mathfrak{B}^*)^k\mathfrak{W}^* &\subseteqq \mathfrak{W}^*(\mathfrak{B}^*)^k + (\mathfrak{B}^*)^k\,, \end{aligned} \tag{2}$$

$$(\mathfrak{W}^*\mathfrak{B}^*)^k \subseteqq \mathfrak{W}^*(\mathfrak{B}^*)^k\,, \qquad (\mathfrak{B}^*\mathfrak{W}^*)^k \subseteqq (\mathfrak{B}^*)^k\mathfrak{W}^*\,. \tag{3}$$

Proof: By II. of § 1, the condition that $B \times W \in \mathfrak{B}$ for $B \in \mathfrak{B}$, $W \in \mathfrak{W}$, implies that $\mathfrak{B}^* W \subseteqq W\mathfrak{B}^* + \mathfrak{B}^*$. It follows that if $W_1, W_2, \cdots, W_r \in \mathfrak{W}$, then $\mathfrak{B}^* W_1 W_2 \cdots W_r \subseteqq \mathfrak{W}^*\mathfrak{B}^* + \mathfrak{B}^*$. Hence $\mathfrak{B}^*\mathfrak{W}^* \subseteqq \mathfrak{W}^*\mathfrak{B}^* + \mathfrak{B}^*$. Induction now proves $(\mathfrak{B}^*)^k\mathfrak{W}^* \subseteqq \mathfrak{W}^*(\mathfrak{B}^*)^k + (\mathfrak{B}^*)^k$. Similarly the condition that $\mathfrak{B}$ is a left ideal implies that $\mathfrak{W}^*(\mathfrak{B}^*)^k \subseteqq (\mathfrak{B}^*)^k\mathfrak{W}^* + (\mathfrak{B}^*)^k$. Hence (2) holds. Now (3) is clear for $k = 1$ and if it holds for some k, then

$$\begin{aligned} (\mathfrak{W}^*\mathfrak{B}^*)^{k+1} = (\mathfrak{W}^*\mathfrak{B}^*)^k\mathfrak{W}^*\mathfrak{B}^* &\subseteqq \mathfrak{W}^*(\mathfrak{B}^*)^k\mathfrak{W}^*\mathfrak{B}^* \\ &\subseteqq \mathfrak{W}^*(\mathfrak{W}^*\mathfrak{B}^{*k} + \mathfrak{B}^{*k})\mathfrak{B}^* \subseteqq \mathfrak{W}^*(\mathfrak{B}^*)^{k+1}\,. \end{aligned}$$

Similarly, the second part of (3) holds.

We can now prove

Theorem 2. *Let $\mathfrak{W}$ be a weakly closed set of linear transformations acting in a finite-dimensional vector space and let $\mathfrak{B}$ be an ideal in $\mathfrak{W}$ such that every element of $\mathfrak{B}$ is nilpotent. Then $\mathfrak{B}^*$(hence $\mathfrak{B}$) is contained in the radical $\mathfrak{R}$ of $\mathfrak{W}^*$.*

Proof: $\mathfrak{W}^*\mathfrak{B}^* + \mathfrak{B}^*$ is an ideal since $\mathfrak{W}^*(\mathfrak{W}^*\mathfrak{B}^* + \mathfrak{B}^*) \subseteqq \mathfrak{W}^*\mathfrak{B}^*$ and

$(\mathfrak{W}^*\mathfrak{B}^* + \mathfrak{B}^*)\mathfrak{W}^* \subseteq \mathfrak{W}^*(\mathfrak{W}^*\mathfrak{B}^* + \mathfrak{B}^*) + \mathfrak{W}^*\mathfrak{B}^* + \mathfrak{B}^* = \mathfrak{W}^*\mathfrak{B}^* + \mathfrak{B}^*$. Also $(\mathfrak{W}^*\mathfrak{B}^* + \mathfrak{B}^*)^k \subseteq \mathfrak{W}^*\mathfrak{B}^* + (\mathfrak{B}^*)^k$. By Theorem 1, $\mathfrak{B}^*$ is nilpotent. Hence k can be chosen so that $(\mathfrak{W}^*\mathfrak{B}^* + \mathfrak{B}^*)^k \subseteq \mathfrak{W}^*\mathfrak{B}^*$. On the other hand $(\mathfrak{W}^*\mathfrak{B}^*)^l \subseteq \mathfrak{W}^*(\mathfrak{B}^*)^l$ and l can be taken so that $(\mathfrak{W}^*\mathfrak{B}^*)^l = 0$. Thus $\mathfrak{W}^*\mathfrak{B}^* + \mathfrak{B}^*$ is a nilpotent ideal and so it is contained in $\mathfrak{R}$. It follows that $\mathfrak{B}^*$ and $\mathfrak{B}$ are contained in $\mathfrak{R}$,

3. *Engel's theorem*

Theorem 1 applies in particular to Lie algebras. In this case it is known as *Engel's theorem on Lie algebras of linear transformations*: *If $\mathfrak{L}$ is a Lie algebra of linear transformations in a finite-dimensional vector space and every $L \in \mathfrak{L}$ is nilpotent, then $\mathfrak{L}^*$ is nilpotent*. The conclusion can also be stated that a basis exists for the underlying space so that all the matrices are nil triangular. These results can be applied via the adjoint representation to arbitrary finite-dimensional Lie algebras. Thus we have

Engel's theorem on abstract Lie algebras. If $\mathfrak{L}$ is a finite-dimensional Lie algebra, then $\mathfrak{L}$ is nilpotent if and only if ad a *is nilpotent for every $a \in \mathfrak{L}$.*

Proof: A Lie algebra $\mathfrak{L}$ is nilpotent if and only if there exists an integer N such that $[\cdots[a_1a_2]a_3]\cdots a_N] = 0$ for $a_i \in \mathfrak{L}$. This implies that $[\cdots[xa]a]\cdots a] = 0$ if the product contains $N-1$ a's. Hence $(\mathrm{ad}\, a)^{N-1} = 0$. Conversely, let $\mathfrak{L}$ be finite-dimensional and assume that $\mathrm{ad}\, a$ is nilpotent for every $a \in \mathfrak{L}$. The set $\mathrm{ad}\, \mathfrak{L}$ of linear transformations $\mathrm{ad}\, a$ (acting in $\mathfrak{L}$) is a Lie algebra of nilpotent linear transformations. Hence $(\mathrm{ad}\, \mathfrak{L})^*$ is nilpotent. This means that there is an integer N such that $\mathrm{ad}\, a_1\, \mathrm{ad}\, a_2 \cdots \mathrm{ad}\, a_N = 0$ for $a_i \in \mathfrak{L}$. Thus $[\cdots[aa_1]a_2]\cdots a_N] = 0$ and so $\mathfrak{L}^{N+1} = 0$.

We can apply Theorem 2 of the last section to obtain two characterizations of the nil radical of a Lie algebra.

THEOREM 3. *Let $\mathfrak{L}$ be a finite-dimensional Lie algebra. Then the nil radical $\mathfrak{N}$ of $\mathfrak{L}$ can be characterized in the following two ways*: (1) *For every $a \in \mathfrak{N}$,* $\mathrm{ad}\, a$ *(acting in $\mathfrak{L}$) is nilpotent and if $\mathfrak{B}$ is any ideal such that* $\mathrm{ad}\, b$ *(in $\mathfrak{L}$) is nilpotent for every $b \in \mathfrak{B}$, then $\mathfrak{B} \subseteq \mathfrak{N}$.* (2) *$\mathfrak{N}$ is the set of elements $b \in \mathfrak{L}$ such that* $\mathrm{ad}\, b \in$ *the radical $\mathfrak{R}$ of* $(\mathrm{ad}\, \mathfrak{L})^*$.

Proof:

(1) If $b \in \mathfrak{N}$ and $a \in \mathfrak{L}$, then $[ab] \in \mathfrak{N}$ and $[\cdots[ab]b]\cdots b] = 0$ for

enough b's. Hence ad b is nilpotent. Next let $\mathfrak{B}$ be an ideal such that ad b is nilpotent for every $b \in \mathfrak{B}$. Then the restriction $\mathrm{ad}_{\mathfrak{B}} b$ of ad b to $\mathfrak{B}$ is nilpotent and $\mathfrak{B}$ is nilpotent by Engel's theorem. Hence $\mathfrak{B} \subseteq \mathfrak{N}$.

(2) The image ad $\mathfrak{N}$ of $\mathfrak{N}$ under the adjoint mapping is an ideal in ad $\mathfrak{L}$ consisting of nilpotent transformations. Hence, by Theorem 2, ad $\mathfrak{N} \subseteq \mathfrak{R}$. It is clear that $\mathfrak{R} \cap$ ad $\mathfrak{L}$ is an ideal in ad $\mathfrak{L}$; hence its inverse image $\mathfrak{N}_1$ under the adjoint representation is an ideal in $\mathfrak{L}$. But $\mathfrak{N}_1$ is the set of elements b such that ad $b \in \mathfrak{R}$. Every $\mathrm{ad}_{\mathfrak{N}_1} b$, $b \in \mathfrak{N}_1$, is nilpotent. Hence $\mathfrak{N}_1$ is nilpotent and $\mathfrak{N}_1 \subseteq \mathfrak{N}$. We saw before that ad $\mathfrak{N} \subseteq \mathfrak{R}$ so that $\mathfrak{N} \subseteq \mathfrak{N}_1$. Hence $\mathfrak{N} = \mathfrak{N}_1$.

4. *Primary components. Weight spaces*

For a linear transformation A in a finite-dimensional vector space $\mathfrak{M}$ the well-known *Fitting's lemma* asserts that $\mathfrak{M} = \mathfrak{M}_{0A} \oplus \mathfrak{M}_{1A}$ where the $\mathfrak{M}_{iA}$ are invariant relative to A and the induced transformation in $\mathfrak{M}_{0A}$ is nilpotent and in $\mathfrak{M}_{1A}$ is an isomorphism. We shall call $\mathfrak{M}_{0A}$ and $\mathfrak{M}_{1A}$, respectively, the *Fitting null* and *one component* of $\mathfrak{M}$ relative to A. The space $\mathfrak{M}_{1A} = \bigcap_{i=1}^{\infty} \mathfrak{M}A^i$ and $\mathfrak{M}_{0A} = \{z \mid zA^i = 0 \text{ for some } i\}$. The proof of the decomposition runs as follows. We have $\mathfrak{M} \supseteq \mathfrak{M}A \supseteq \mathfrak{M}A^2 \supseteq \cdots$. Hence there exists an r such that $\mathfrak{M}A^r = \mathfrak{M}A^{r+1} = \cdots = \mathfrak{M}_{1A}$. Let $\mathfrak{Z}_i = \{z_i \mid z_i A^i = 0\}$. Then $\mathfrak{Z}_1 \subseteq \mathfrak{Z}_2 \subseteq \cdots$, so we have an s such that $\mathfrak{Z}_s = \mathfrak{Z}_{s+1} = \cdots = \mathfrak{M}_{0A}$. Let $t = \max(r, s)$. Then $\mathfrak{M}_{0A} = \mathfrak{Z}_t$ and $\mathfrak{M}_{1A} = \mathfrak{M}A^t$. Let $x \in \mathfrak{M}$. Then $xA^t = yA^{2t}$ for some y since $\mathfrak{M}A^t = \mathfrak{M}A^{2t}$. Thus $x = (x - yA^t) + yA^t$ and $yA^t \in \mathfrak{M}_{1A}$, while $(x - yA^t)A^t = 0$ so $x - yA^t \in \mathfrak{M}_{0A}$. Hence $\mathfrak{M} = \mathfrak{M}_{0A} + \mathfrak{M}_{1A}$. Let $z \in \mathfrak{M}_{0A} \cap \mathfrak{M}_{1A}$. Then $z = wA^t$ and $0 = zA^t = wA^{2t}$. Since $wA^{2t} = 0$, $w \in \mathfrak{M}_{0A} = \mathfrak{Z}_t$ and $wA^t = 0$. Hence $z = 0$ and $\mathfrak{M}_{0A} \cap \mathfrak{M}_{1A} = 0$; hence $\mathfrak{M} = \mathfrak{M}_{0A} \oplus \mathfrak{M}_{1A}$. Since $\mathfrak{M}_{0A} = \mathfrak{Z}_t$, $A^t = 0$ in $\mathfrak{M}_{0A}$. Since $\mathfrak{M}_{1A} = \mathfrak{M}A^r = \mathfrak{M}A^{r+1} = \mathfrak{M}_{1A}A$, A is surjective in $\mathfrak{M}_{1A}$. Hence A is an isomorphism of $\mathfrak{M}_{1A}$.

We recall also another type of decomposition of $\mathfrak{M}$ relative to A, namely, the decomposition into *primary components*. Here $\mathfrak{M} = \mathfrak{M}_{\pi_1 A} \oplus \mathfrak{M}_{\pi_2 A} \oplus \cdots \oplus \mathfrak{M}_{\pi_r A}$ where $\pi_i = \pi_i(\lambda)$, $\{\pi_1(\lambda), \cdots, \pi_r(\lambda)\}$ are the irreducible factors with leading coefficients 1 of the minimum (or of the characteristic) polynomial of A, and if $\mu(\lambda)$ is any polynomial, then

$$\mathfrak{M}_{\mu A} = \{z \mid z\mu(A)^r = 0 \quad \text{for some } r\} \tag{4}$$

(cf. Jacobson, [2], vol. II, p. 130).

If $\mu(\lambda) = \pi(\lambda)$ is irreducible with leading coefficient one, then $\mathfrak{M}_{\pi A} = 0$ unless $\pi = \pi_i$ for some i. The space $\mathfrak{M}_{\mu A}$ is evidently invariant relative to A. The characteristic polynomial of the restriction of A to $\mathfrak{M}_{\pi_i A}$ has the form $\pi_i(\lambda)^{e_i}$ and $\dim \mathfrak{M}_{\pi_i A} = e_i r_i$ where $r_i = \deg \pi_i(\lambda)$. If $\pi_i(\lambda) = \lambda$, then the characteristic polynomial of A in $\mathfrak{M}_{\lambda A}$ is λ^e, so $A^r = 0$ in $\mathfrak{M}_{\lambda A}$ and $\mathfrak{M}_{\lambda A} \subseteq \mathfrak{M}_{0A}$. If $\pi_i(\lambda) \neq \lambda$, the characteristic polynomial of the restriction of A to $\mathfrak{M}_{\pi_i A}$ is not divisible by λ, so this restriction is an isomorphism. Hence $\mathfrak{M}_{\pi_i A} = \mathfrak{M}_{\pi_i A} A = \cdots$, so $\mathfrak{M}_{\pi_i A} \subseteq \mathfrak{M}_{1A}$. Thus $\sum_{\pi_i \neq \lambda} \mathfrak{M}_{\pi_i A} \subseteq \mathfrak{M}_{1A}$. Since $\mathfrak{M} = \mathfrak{M}_{0A} \oplus \mathfrak{M}_{1A} = \mathfrak{M}_{\lambda A} \oplus \sum_{\pi_i \neq \lambda} \mathfrak{M}_{\pi_i A}$, it follows that $\mathfrak{M}_{0A} = \mathfrak{M}_{\lambda A}$, $\mathfrak{M}_{1A} = \sum_{\pi_i \neq \lambda} \mathfrak{M}_{\pi_i A}$. In particular, we see that the Fitting null component is the characteristic space of the characteristic root 0 of A and $\dim \mathfrak{M}_{0A}$ is the multiplicity of the root 0 in the characteristic polynomial of A.

We shall now extend these results to nilpotent Lie algebras of linear transformations. It would suffice to consider the primary decomposition since the result on the Fitting decomposition could be deduced from it. However, the Fitting decomposition is applicable in other situations (e.g., vector spaces over division rings), so it seems to be of interest to treat this separately. We shall need two important commutation formulas. Let $\mathfrak{A}$ be an associative algebra, a an element of $\mathfrak{A}$ and consider the inner derivation D_a: $x \rightarrow x' \equiv [xa]$ in $\mathfrak{A}$. We write $x^{(k)} = (x^{(k-1)})'$, $x^{(0)} = x$. Then one proves directly by induction the following formulas:

$$xa^k = a^k x + \binom{k}{1} a^{k-1} x' + \binom{k}{2} a^{k-2} x'' + \cdots + x^{(k)}. \tag{5}$$

$$a^k x = x a^k - \binom{k}{1} x' a^{k-1} + \binom{k}{2} x'' a^{k-2} - \cdots \pm x^{(k)}. \tag{6}$$

We apply this to linear transformations to prove

LEMMA 1. *Let A and B be linear transformations in a finite-dimensional vector space. Suppose there exists a positive integer N such that $[\cdots[[B\overbrace{A]A]\cdots A}^{N}] = 0$. Then the Fitting components $\mathfrak{M}_{0A}$, $\mathfrak{M}_{1A}$ of $\mathfrak{M}$ relative to A are invariant under B.*

Proof: Suppose $xA^m = 0$. Then for $k = N + m - 1$

$$xBA^k = x\left(A^k B + \binom{k}{1} A^{k-1} B' + \cdots + \binom{k}{N-1} A^{k-N+1} B^{(N-1)}\right) = 0.$$

Hence $xB \in \mathfrak{M}_{0A}$. Next let $x \in \mathfrak{M}_{1A}$. If t is the integer used in the proof of Fitting's lemma, then we can write $x = yA^{t+N-1}$. Then

$$xB = yA^{t+N-1}B = y\left(BA^{t+N-1} - \binom{t+N-1}{1}B'A^{t+N-2} + \cdots \pm \binom{t+N-1}{N-1}B^{(N-1)}A^t\right) \in \mathfrak{M}A^t = \mathfrak{M}_{1A}\,.$$

Hence $\mathfrak{M}_{1A}B \subseteq \mathfrak{M}_{1A}$.

THEOREM 4. *Let $\mathfrak{L}$ be a nilpotent Lie algebra of linear transformations in a finite-dimensional vector space $\mathfrak{M}$ and let $\mathfrak{M}_0 = \bigcap_{A\in\mathfrak{L}}\mathfrak{M}_{0A}$, $\mathfrak{M}_1 = \bigcap_{i=1}^{\infty}\mathfrak{M}(\mathfrak{L}^*)^i$. Then the $\mathfrak{M}_i$ are invariant under $\mathfrak{L}$ (that is, under every $B \in \mathfrak{L}$) and $\mathfrak{M} = \mathfrak{M}_0 \oplus \mathfrak{M}_1$. Moreover, $\mathfrak{M}_1 = \sum_{A\in\mathfrak{L}}\mathfrak{M}_{1A}$.*

Proof: Suppose first that $\mathfrak{M} = \mathfrak{M}_{0A}$ for every $A \in \mathfrak{L}$. Then $\mathfrak{M}_{1A} = 0$ for all A and $\sum \mathfrak{M}_{1A} = 0$. Also by Engel's theorem $(\mathfrak{L}^*)^N = 0$ for some N. Hence $\mathfrak{M}_1 = \bigcap_{i=1}^{\infty}\mathfrak{M}(\mathfrak{L}^*)^i = 0$. Thus the result holds in this case. We shall now use induction on $\dim \mathfrak{M}$ and we may assume $\mathfrak{M}_{0A} \neq \mathfrak{M}$ for some A. By Lemma 1, the $B \in \mathfrak{L}$ map $\mathfrak{M}_{0A}$ into itself; hence $\mathfrak{L}$ induces a nilpotent Lie algebra of linear transformations in $\mathfrak{N} = \mathfrak{M}_{0A} \subset \mathfrak{M}$. We can write $\mathfrak{N} = \mathfrak{N}_0 \oplus \mathfrak{N}_1$ where $\mathfrak{N}_0 = \bigcap_{B\in\mathfrak{L}}\mathfrak{N}_{0B}$, $\mathfrak{N}_1 = \bigcap_i \mathfrak{N}(\mathfrak{L}^*)^i = \sum_{B\in\mathfrak{L}}\mathfrak{N}_{1B}$ and $\mathfrak{N}_0$ and $\mathfrak{N}_1$ are invariant under $\mathfrak{L}$. Then $\mathfrak{M} = \mathfrak{N}_0 \oplus \mathfrak{N}_1 \oplus \mathfrak{M}_{1A}$. It is clear from the definitions that $\mathfrak{N}_0 = \mathfrak{M}_0 = \bigcap_{B\in\mathfrak{L}}\mathfrak{M}_{0B}$ and $\mathfrak{N}_1 + \mathfrak{M}_{1A} \subseteq \sum_{B\in\mathfrak{L}}\mathfrak{M}_{1B} \subseteq \bigcap_i \mathfrak{M}(\mathfrak{L}^*)^i$. On the other hand, by Engel's theorem the algebra induced by $\mathfrak{L}^*$ in $\mathfrak{M}_0$ is nilpotent so that we have an N such that $\mathfrak{M}_0(\mathfrak{L}^*)^N = 0$. Then

$$\begin{aligned}\mathfrak{M}(\mathfrak{L}^*)^N &= \mathfrak{M}_0(\mathfrak{L}^*)^N + \mathfrak{N}_1(\mathfrak{L}^*)^N + \mathfrak{M}_{1A}(\mathfrak{L}^*)^N \\ &\subseteq \mathfrak{N}_1(\mathfrak{L}^*)^N + \mathfrak{M}_{1A}(\mathfrak{L}^*) \subseteq \mathfrak{N}_1 + \mathfrak{M}_{1A}\,.\end{aligned}$$

Hence $\mathfrak{N}_1 + \mathfrak{M}_{1A} = \sum_{B\in\mathfrak{L}}\mathfrak{M}_{1B} = \bigcap_i \mathfrak{M}(\mathfrak{L}^*)^i$ and the theorem is proved.

Our discussion of the primary decomposition will be based on a criterion for multiple factors of polynomials. Let $\Phi[\lambda]$ denote the polynomial ring in the indeterminate λ with coefficients in Φ. We define a sequence of linear mappings $D_0 = 1, D_1, D_2, \cdots$ in $\Phi[\lambda]$ as follows: D_i is the linear mapping in $\Phi[\lambda]$ whose effect on the element λ^j in the basis $(1, \lambda, \lambda^2, \cdots)$ for $\Phi[\lambda]$ over Φ is given by

$$\lambda^j D_i = \binom{j}{i}\lambda^{j-i} \tag{7}$$

where we take $\binom{j}{i} = 0$ if $i > j$. Thus we have $\lambda^j D_1 = j\lambda^{j-1}$, so that D_1 is the usual formal derivation in $\Phi[\lambda]$. Also $\lambda^j D_1^i = j(j-1)\cdots(j-i+1)\lambda^{j-i} = i!\binom{j}{i}\lambda^{j-i} = i!\,\lambda^j D_i$. Hence if the characteristic is zero, then $D_i = (1/i!)D_1^i$. If $\phi(\lambda) \in \Phi[\lambda]$ we shall write $\phi_i \equiv \phi D_i$. Then we have the Leibniz rule:

$$(\phi\psi)_k = \sum_{i=0}^{k} \phi_i \psi_{k-i} \,. \tag{8}$$

It suffices to check this for $\phi = \lambda^j$, $\psi = \lambda^{j'}$ in the basis $(1, \lambda, \lambda^2, \cdots)$. Then

$$\sum_{i=0}^{k} \phi_i\psi_{k-i} = \sum_{i=0}^{k}\binom{j}{i}\binom{j'}{k-i}\lambda^{j-i}\lambda^{j'-k+i}$$
$$= \left(\sum_{i=0}^{k}\binom{j}{i}\binom{j'}{k-i}\right)\lambda^{j+j'-k} \,.$$

Since $\sum_{i=0}^{k}\binom{j}{i}\binom{j'}{k-i} = \binom{j+j'}{k}$ the above reduces to $\binom{j+j'}{k}\lambda^{j+j'-k} = (\lambda^{j+j'})_k$. Hence (8) is valid.

We can now prove

LEMMA 2. *If* $\eta(\lambda)^{k+1} \mid \mu(\lambda)$, *then* $\eta(\lambda) \mid \mu_i(\lambda)$, $i = 0, 1, 2, \cdots, k$.

Proof: This is true for $k = 0$ since $\mu_0(\lambda) = \mu(\lambda)$. Write $\mu(\lambda) = \phi(\lambda)\psi(\lambda)$ where $\eta^k \mid \phi$ and $\eta \mid \psi$. Then we may assume that $\eta \mid \mu_j$, $\eta \mid \phi_j$, $j = 0, 1, 2, \cdots, k-1$, and $\eta \mid \psi_0$. Then $\eta \mid (\phi\psi)_k = \mu_k$ by (8).

Let $\psi(\lambda) = \sum_{k=0}^{r}\alpha_k\lambda^k$ and multiply (5) by α_k and sum on k. This gives

$$x\psi(a) = \psi(a)x + \psi_1(a)x' + \psi_2(a)x'' + \cdots + \alpha_r x^{(r)} \,, \tag{9}$$

which we shall use to establish

LEMMA 3. *Let A, B be linear transformations in a finite-dimensional vector space satisfying* $B^{(N)} = \overbrace{[\cdots[BA]A]\cdots A]}^{N} = 0$ *for some N. Let $\mu(\lambda)$ be a polynomial and let* $\mathfrak{M}_{\mu A} = \{y \mid y\mu(A)^m = 0$ *for some* $m\}$. *Then $\mathfrak{M}_{\mu A}$ is invariant under B.*

Proof: Let $y \in \mathfrak{M}_{\mu A}$ and suppose $y\mu(A)^m = 0$. Set $\eta(\lambda) = \mu(\lambda)^m$, $\psi(\lambda) = \eta(\lambda)^N = \mu(\lambda)^{mN}$. Then, by (9),

$$B\psi(A) = \psi(A)B + \psi_1(A)B' + \cdots + \psi_{N-1}(A)B^{(N-1)} \,.$$

By Lemma 2, $\eta(\lambda) \mid \psi_j(\lambda)$, $j \leqq N-1$. Hence $y\psi_j(A) = 0$, $j \leqq N-1$, and $yB\psi(A) = 0$. Hence $yB \in \mathfrak{M}_{\mu A}$.

THEOREM 5. *Let $\mathfrak{L}$ be a nilpotent Lie algebra of linear transformations in a finite-dimensional vector space $\mathfrak{M}$. Then we can* *decompose* $\mathfrak{M} = \mathfrak{M}_1 \oplus \mathfrak{M}_2 \oplus \cdots \oplus \mathfrak{M}_r$ *where* $\mathfrak{M}_i$ *is invariant under* $\mathfrak{L}$ *and the minimum polynomial of the restriction of every* $A \in \mathfrak{L}$ *to* $\mathfrak{M}_i$ *is a prime power.*

Proof: If the minimum polynomial of every $A \in \mathfrak{L}$ is a prime power, then there is nothing to prove. Otherwise, we have an $A \in \mathfrak{L}$ such that $\mathfrak{M} = \mathfrak{M}_{\pi_1 A} \oplus \cdots \oplus \mathfrak{M}_{\pi_s A}$, $\pi_i = \pi_i(\lambda)$ irreducible, $\mathfrak{M}_{\pi_i A} \neq 0$, $s > 1$. By Lemma 3, $\mathfrak{M}_{\pi_i A} B \subseteq \mathfrak{M}_{\pi_i A}$ for every $B \in \mathfrak{L}$. This permits us to complete the proof by induction on $\dim \mathfrak{M}$.

If the base field is algebraically closed, then the minimum polynomials in the subspaces $\mathfrak{M}_i$ are of the form $(\lambda - \alpha(A))^{k_{iA}}$, $A \in \mathfrak{L}$. Setting $Z_i(A) = A - \alpha(A)$ for the space $\mathfrak{M}_i$ we see that $Z_i(A)$ is a nilpotent transformation in $\mathfrak{M}_i$. Hence every $A \in \mathfrak{L}$ is a scalar plus a nilpotent in $\mathfrak{M}_i$. We therefore have the following

Corollary (*Zassenhaus*). *If* $\mathfrak{L}$ *is a nilpotent Lie algebra of linear transformations in a finite-dimensional vector space over an algebraically closed field, then the space* $\mathfrak{M}$ *can be decomposed as* $\mathfrak{M}_1 \oplus \mathfrak{M}_2 \oplus \cdots \oplus \mathfrak{M}_r$ *where the* $\mathfrak{M}_i$ *are invariant subspaces such that the restriction of every* $A \in \mathfrak{L}$ *to* $\mathfrak{M}_i$ *is a scalar plus a nilpotent linear transformation.*

Consider a decomposition of $\mathfrak{M}$ as in Theorem 5. For each i and each A we may associate the prime polynomial $\pi_{iA}(\lambda)$ such that $\pi_{iA}(\lambda)^{k_{iA}}$ is the minimum polynomial of A restricted to $\mathfrak{M}_i$. Then the mapping π_i: $A \rightarrow \pi_{iA}(\lambda)$ is a primary function for $\mathfrak{L}$ in the sense of the following

Definition 3. Let $\mathfrak{L}$ be a Lie algebra of linear transformations in $\mathfrak{M}$. Then a mapping π: $A \rightarrow \pi_A(\lambda)$, A in $\mathfrak{L}$, $\pi_A(\lambda)$, a prime polynomial with leading coefficient one, is called a *primary function* of $\mathfrak{M}$ for $\mathfrak{L}$ if there exists a non-zero vector x such that $x\pi_A(A)^{m_{x,A}} = 0$ for every $A \in \mathfrak{L}$. The set of vectors (zero included) satisfying this condition is a subspace called the *primary component* $\mathfrak{M}_\pi$ corresponding to π.

Using this terminology we may say that the space $\mathfrak{M}_i$ in Theorem 5 is contained in the primary component $\mathfrak{M}_{\pi_i}$. By adding together certain of the $\mathfrak{M}_i$, if necessary, we may suppose that if $i \neq j$, then π_i: $A \rightarrow \pi_{iA}(\lambda)$ and π_j: $A \rightarrow \pi_{jA}(\lambda)$ determined by $\mathfrak{M}_i$ and $\mathfrak{M}_j$ are distinct. We shall now show that in this case $\mathfrak{M}_i$ coincides with $\mathfrak{M}_{\pi_i}$ and the π_i are the only primary functions. Thus let π: $A \rightarrow \pi_A(\lambda)$ be a primary function and let $x \in \mathfrak{M}_\pi$. Assume $\pi \neq \pi_i$, so that there exists $A_i \in \mathfrak{L}$ such that $\pi_{A_i}(\lambda) \neq \pi_{iA_i}(\lambda)$. Write $x = x_1 + \cdots + x_r$, $x_i \in \mathfrak{M}_i$. Since $x \in \mathfrak{M}_\pi$ there exists a positive integer m such that

$$0 = x\pi_{A_i}(A_i)^m = x_1\pi_{A_i}(A_i)^m + \cdots + x_r\pi_{A_i}(A_i)^m .$$

Since the decomposition $\mathfrak{M} = \mathfrak{M}_1 \oplus \cdots \oplus \mathfrak{M}_r$ is direct, we have $x_i\pi_{A_i}(A_i)^m = 0$ and since $\pi_{A_i}(\lambda) \neq \pi_{iA_i}(\lambda)$ this implies $x_i = 0$. It follows that if $\pi \neq \pi_i$ for $i = 1, 2, \cdots, r$, then $x = 0$ and this contradicts the definition of a primary function. Thus the π_i are the only primary functions. The same argument shows that if $x \in \mathfrak{M}_{\pi_i}$, then $x \in \mathfrak{M}_i$; hence $\mathfrak{M}_i = \mathfrak{M}_{\pi_i}$.

The argument just given was based on the following two properties of the decomposition: $\mathfrak{M}_i \subseteq \mathfrak{M}_{\pi_i}$ where π_i is a primary function and $\pi_i \neq \pi_j$ if $i \neq j$. The existence of such a decomposition is a consequence of Theorem 5. Hence we have the following

THEOREM 6. *If $\mathfrak{L}$ is a nilpotent Lie algebra of linear transformations in a finite-dimensional vector space $\mathfrak{M}$, then $\mathfrak{L}$ has only a finite number of primary functions. The corresponding primary components are submodules and $\mathfrak{M}$ is a direct sum of these. Moreover, if $\mathfrak{M} = \mathfrak{M}_1 \oplus \mathfrak{M}_2 \oplus \cdots \oplus \mathfrak{M}_r$ is any decomposition of $\mathfrak{M}$ as a direct sum of subspaces $\mathfrak{M}_i \neq 0$ invariant under $\mathfrak{L}$ such that* (1) *for each i the minimum polynomial of the restriction of every $A \in \mathfrak{L}$ to $\mathfrak{M}_i$ is a prime power $\pi_{iA}(\lambda)^{m_{iA}}$ and* (2) *if $i \neq j$, there exists an A such that $\pi_{iA}(\lambda) \neq \pi_{jA}(\lambda)$, then the mappings $A \to \pi_{iA}(\lambda)$, $i = 1, 2, \cdots, r$, are the primary functions and the $\mathfrak{M}_i$ are the corresponding primary components.*

It is easy to establish the relation between the primary decomposition and the Fitting decomposition: $\mathfrak{M}_0$ is the primary component $\mathfrak{M}_\lambda$ if $A \to \lambda$ is a primary function and is 0 otherwise, $\mathfrak{M}_1$ is the sum of the primary components $\mathfrak{M}_\pi$, $\pi \neq \lambda$. We leave it to the reader to verify this.

We shall now assume that the nilpotent Lie algebra $\mathfrak{L}$ of linear transformations has the property that the characteristic roots of every $A \in \mathfrak{L}$ are in the base field. A Lie algebra of linear transformations having this property will be called *split*. Evidently, if the base field Φ is algebraically closed, then any Lie algebra of linear transformations in a finite-dimensional vector space over Φ is split. Now let $\mathfrak{L}$ be nilpotent and split. It is clear that the characteristic polynomial of the restriction of A to the primary component corresponding to $A \to \pi_A(\lambda)$ is of the form $\pi_A(\lambda)^r$. Since this is a factor of the characteristic polynomial of A and $\pi_A(\lambda)$ is irreducible, our hypothesis implies that $\pi_A(\lambda) = \lambda - \alpha(A)$, $\alpha(A) \in \Phi$. Under these circumstances it is natural to replace the mapping $A \to \pi_A(\lambda)$ by the mapping $A \to \alpha(A)$ of $\mathfrak{L}$ into Φ and to formulate the following

Definition 4. Let $\mathfrak{L}$ be a Lie algebra of linear transformations in $\mathfrak{M}$. Then a mapping α: $A \to \alpha(A)$ of $\mathfrak{L}$ into the base field Φ is called a *weight* of $\mathfrak{M}$ for $\mathfrak{L}$ if there exists a non-zero vector x such that $x(A - \alpha(A)1)^{m_{x,A}} = 0$ for all $A \in \mathfrak{L}$. The set of vectors (zero included) satisfying this condition is a subspace $\mathfrak{M}_\alpha$ called the *weight space* of $\mathfrak{M}$ corresponding to the weight α.

Theorem 6 specializes to the following result on weights and weight spaces.

Theorem 7. *Let $\mathfrak{L}$ be a split nilpotent Lie algebra of linear transformations in a finite-dimensional vector space. Then $\mathfrak{L}$ has only a finite number of distinct weights, the weight spaces are submodules, and $\mathfrak{M}$ is a direct sum of these. Moreover, let $\mathfrak{M} = \mathfrak{M}_1 \oplus \mathfrak{M}_2 \oplus \cdots \oplus \mathfrak{M}_r$ be any decomposition of $\mathfrak{M}$ into subspaces $\mathfrak{M}_i \neq 0$ invariant under $\mathfrak{L}$ such that* (1) *for each i, the restriction of any $A \in \mathfrak{L}$ has only one characteristic root $\alpha_i(A)$ (with a certain multiplicity) in $\mathfrak{M}_i$ and* (2) *if $i \neq j$, then there exists an $A \in \mathfrak{L}$ such that $\alpha_i(A) \neq \alpha_j(A)$. Then the mappings $A \to \alpha_i(A)$ are the weights and the spaces $\mathfrak{M}_i$ are the weight spaces.*

5. *Lie algebras with semi-simple enveloping associative algebras*

Our main result in this section gives the structure of a Lie algebra $\mathfrak{L}$ of linear transformations whose enveloping associative algebra $\mathfrak{L}^*$ is semi-simple. In the next section the proof of Lie's theorems will also be based on this result. For all of this we shall have to assume that the characteristic is 0. We recall that the trace of a linear transformation A in a finite-dimensional vector space, which is defined to be $\sum_1^n \alpha_{ii}$ for any matrix (α_{ij}) of A, is the sum of the characteristic roots ρ_i, $i = 1, \cdots, n$, of A. Also $\operatorname{tr} A^k = \sum_{i=1}^n \rho_i^k$. If A is nilpotent all the ρ_i are 0, so $\operatorname{tr} A^k = 0$, $k = 1, 2, \cdots$. If the characteristic of Φ is 0, the converse holds: If $\operatorname{tr} A^k = 0$, $k = 1, 2, \cdots$, then A is nilpotent. Thus we have $\sum \rho_i^k = 0$, $k = 1, 2, \cdots$. The formulas (Newton's identities; cf. Jacobson, [2], vol. I, p. 110) expressing $\sum \rho_i^k$ in terms of the elementary symmetric functions of the ρ_i show that if the characteristic is 0 our conditions imply that all the elementary symmetric functions of the ρ_i are 0, so all the ρ_i are 0. Hence A is nilpotent. We use this result in the following

LEMMA 4. *Let $C \in \mathfrak{E}$, the algebra of linear transformations in a finite-dimensional vector space over a field of characteristic 0 and suppose $C = \sum_1^r [A_iB_i]$, A_i, $B_i \in \mathfrak{E}$ and $[CA_i] = 0$, $i = 1, 2, \cdots, r$. Then C is nilpotent.*

Proof: We have $[C^{k-1}A_i] = 0$, $k = 1, 2, \cdots (C^0 = 1)$. Hence $C^k = \sum_1^r C^{k-1}(A_iB_i - B_iA_i) = \sum_i (A_i(C^{k-1}B_i) - (C^{k-1}B_i)A_i) = \sum [A_i, C^{k-1}B_i]$. Since the trace of any commutator is 0, this gives $\operatorname{tr} C^k = 0$ for $k = 1, 2, \cdots$. Hence C is nilpotent.

The key result for the present considerations is the following

THEOREM 8. *Let $\mathfrak{L}$ be a Lie algebra of linear transformations in a finite-dimensional vector space over a field of characteristic 0. Assume that the enveloping associative algebra $\mathfrak{L}^*$ is semi-simple. Then $\mathfrak{L} = \mathfrak{L}_1 \oplus \mathfrak{C}$ where $\mathfrak{C}$ is the center of $\mathfrak{L}$ and $\mathfrak{L}_1$ is an ideal of $\mathfrak{L}$ which is semi-simple (as a Lie algebra).*

Proof: Let $\mathfrak{S}$ be the radical of $\mathfrak{L}$. We show first that $\mathfrak{S} = \mathfrak{C}$ the center of $\mathfrak{L}$. Otherwise, $\mathfrak{S}_1 = [\mathfrak{L}\mathfrak{S}]$ is a non-zero solvable ideal. Suppose $\mathfrak{S}_1^{(h)} = 0$, $\mathfrak{S}_1^{(h-1)} \neq 0$ and set $\mathfrak{S}_2 = \mathfrak{S}_1^{(h-1)}$, $\mathfrak{S}_3 = [\mathfrak{S}_2\mathfrak{L}]$. If $C \in \mathfrak{S}_3$, $C = \sum [A_iB_i]$, $A_i \in \mathfrak{S}_2$, $B_i \in \mathfrak{L}$, and $[CA_i] = 0$ since $\mathfrak{S}_3 \subseteq \mathfrak{S}_2$ and $\mathfrak{S}_2$ is abelian. Hence, by Lemma 4, C is nilpotent. Thus every element of the ideal $\mathfrak{S}_3$ of $\mathfrak{L}$ is nilpotent; hence, by Theorem 2, $\mathfrak{S}_3$ is in the radical of $\mathfrak{L}^*$. Since $\mathfrak{L}^*$ is semi-simple, $\mathfrak{S}_3 = 0$. Since $\mathfrak{S}_3 = [\mathfrak{S}_2\mathfrak{L}]$ this shows that $\mathfrak{S}_2 \subseteq \mathfrak{C}$. Since $\mathfrak{S}_2 \subseteq [\mathfrak{L}\mathfrak{S}] \subseteq \mathfrak{L}'$ every element C of $\mathfrak{S}_2$ has the form $\sum [A_iB_i]$, A_i, $B_i \in \mathfrak{L}$, and we have $[CA_i] = 0$, since $\mathfrak{S}_2 \subseteq \mathfrak{C}$. The argument just used implies that $\mathfrak{S}_2 = 0$, which contradicts our original assumption that $\mathfrak{S}_1 = [\mathfrak{L}\mathfrak{S}] \neq 0$. Hence we have $[\mathfrak{L}\mathfrak{S}] = 0$, and $\mathfrak{S} = \mathfrak{C}$. The argument we have used twice can be applied to conclude also that $\mathfrak{C} \cap \mathfrak{L}' = 0$. Hence we can find a subspace $\mathfrak{L}_1 \supseteq \mathfrak{L}'$ such that $\mathfrak{L} = \mathfrak{L}_1 \oplus \mathfrak{C}$. Since $\mathfrak{L}_1$ contains $\mathfrak{L}'$, $\mathfrak{L}_1$ is an ideal. Also $\mathfrak{L}_1 \cong \mathfrak{L}/\mathfrak{C} = \mathfrak{L}/\mathfrak{S}$ is semi-simple. This concludes the proof.

Remark: In the next chapter we shall show that $\mathfrak{L}_1 = \mathfrak{L}'$, so we shall have $\mathfrak{L} = \mathfrak{L}' \oplus \mathfrak{C}$, $\mathfrak{L}'$ semi-simple.

COROLLARY 1. *Let $\mathfrak{L}$ be as in the theorem, $\mathfrak{L}^*$ semi-simple. Then $\mathfrak{L}$ is solvable if and only if $\mathfrak{L}$ is abelian. More generally, if $\mathfrak{L}$ is* *solvable and* $\mathfrak{R}$ *is the radical of* $\mathfrak{L}^*$, *then* $\mathfrak{L}^*/\mathfrak{R}$ *is a commutative algebra.*

Proof: If $\mathfrak{L}$ is abelian it is solvable, and if $\mathfrak{L}$ is solvable and $\mathfrak{L}^*$ is semi-simple, then $\mathfrak{L} = \mathfrak{C} \oplus \mathfrak{L}_1$ where $\mathfrak{L}_1$ is semi-simple. Since the only Lie algebra which is solvable and semi-simple is 0 we have $\mathfrak{L}_1 = 0$ and $\mathfrak{L} = \mathfrak{C}$ is abelian. To prove the second statement it is convenient to change the point of view slightly and consider $\mathfrak{L}$ as a subalgebra of $\mathfrak{A}_L$, $\mathfrak{A}$ a finite-dimensional associative algebra over a field of characteristic 0. Since any such $\mathfrak{A}$ can be considered as a subalgebra of an associative algebra $\mathfrak{E}$ of linear transformations in a finite-dimensional vector space, Theorem 8 is applicable to $\mathfrak{L}$ and $\mathfrak{A}$. We now assume $\mathfrak{L}$ solvable, so $(\mathfrak{L} + \mathfrak{R})/\mathfrak{R}$, which is a homomorphic image of $\mathfrak{L}$, is solvable. Moreover, the enveloping associative algebra of this Lie algebra is the semi-simple associative algebra $\mathfrak{L}^*/\mathfrak{R}$. Hence $(\mathfrak{L} + \mathfrak{R})/\mathfrak{R}$ is abelian. This implies that $(a + \mathfrak{R})(b + \mathfrak{R}) = (b + \mathfrak{R})(a + \mathfrak{R})$ for any $a, b \in \mathfrak{L}$. Since the cosets $a + \mathfrak{R}$ generate $\mathfrak{L}^*/\mathfrak{R}$, it follows that $\mathfrak{L}^*/\mathfrak{R}$ is commutative.

COROLLARY 2. *Let* $\mathfrak{L}$ *be a Lie algebra of linear transformations in a finite-dimensional vector space over a field of characteristic* 0, *let* $\mathfrak{S}$ *be the radical of* $\mathfrak{L}$ *and* $\mathfrak{R}$ *the radical of* $\mathfrak{L}^*$. *Then* $\mathfrak{L} \cap \mathfrak{R}$ *is the totality of nilpotent elements of* $\mathfrak{S}$ *and* $[\mathfrak{S}\mathfrak{L}] \subseteq \mathfrak{R}$.

Proof: Since $\mathfrak{R}$ is associative nilpotent it is Lie nilpotent. Hence $\mathfrak{L} \cap \mathfrak{R} \subseteq \mathfrak{S}$. Moreover, the elements of $\mathfrak{R}$ are nilpotent so $\mathfrak{L} \cap \mathfrak{R} \subseteq \mathfrak{S}_0$ the set of nilpotent elements of $\mathfrak{S}$. If $\mathfrak{R}_0$ denotes the radical of the enveloping associative algebra $\mathfrak{S}^*$ then, by Corollary 1, $\mathfrak{S}^*/\mathfrak{R}_0$ is commutative. Now any nilpotent element of a commutative algebra generates a nilpotent ideal and so belongs to the radical. Since $\mathfrak{S}^*/\mathfrak{R}_0$ is semi-simple, it has no non-zero nilpotent elements. It follows that $\mathfrak{R}_0$ is the set of nilpotent elements of $\mathfrak{S}^*$ and so $\mathfrak{S}_0 = \mathfrak{R}_0 \cap \mathfrak{S}$ is a subspace of $\mathfrak{L}$. Next consider $(\mathfrak{L} + \mathfrak{R})/\mathfrak{R}$. The enveloping associative algebra of this Lie algebra is $\mathfrak{L}^*/\mathfrak{R}$, which is semi-simple. Hence the radical of $(\mathfrak{L} + \mathfrak{R})/\mathfrak{R}$ is contained in its center. Since $(\mathfrak{S} + \mathfrak{R})/\mathfrak{R}$ is a solvable ideal in $(\mathfrak{L} + \mathfrak{R})/\mathfrak{R}$ we must have $[(\mathfrak{S} + \mathfrak{R})/\mathfrak{R}, (\mathfrak{L} + \mathfrak{R})/\mathfrak{R}] = 0$, which means that $[\mathfrak{S}\mathfrak{L}] \subseteq \mathfrak{R}$. Hence $[\mathfrak{S}\mathfrak{L}] \subseteq \mathfrak{L} \cap \mathfrak{R} \subseteq \mathfrak{S}_0$. Then $[\mathfrak{S}_0\mathfrak{L}] \subseteq \mathfrak{S}_0$ and $\mathfrak{S}_0$ is an ideal in $\mathfrak{L}$. Since its elements are nilpotent, $\mathfrak{S}_0 \subseteq \mathfrak{R}$ by Theorem 2. Hence $\mathfrak{S}_0 \subseteq \mathfrak{L} \cap \mathfrak{R}$ and $\mathfrak{S}_0 = \mathfrak{L} \cap \mathfrak{R}$, which completes the proof.

There is a more useful formulation of Theorem 8 in which the hypothesis on the structure of $\mathfrak{L}^*$ is replaced by one on the action of $\mathfrak{L}$ on $\mathfrak{M}$. In order to give this we need to recall some standard notions on sets of linear transformations. Let Σ be a set of linear transformations in a finite-dimensional vector space $\mathfrak{M}$ over a field Φ. We recall that the collection $L(\Sigma)$ of subspaces which are invariant under Σ ($\mathfrak{N}A \subseteq \mathfrak{N}$, $A \in \Sigma$) is a sublattice of the lattice of subspaces of $\mathfrak{M}$. We refer to the elements of $L(\Sigma)$ as *Σ-subspaces* of $\mathfrak{M}$. If $\mathfrak{N}$ is a subspace then the collection of $A \in \mathfrak{E}$ such that $\mathfrak{N}A \subseteq \mathfrak{N}$ is a subalgebra of $\mathfrak{E}$. It follows that if $\mathfrak{N} \in L(\Sigma)$, then $\mathfrak{N}$ is invariant under every element of the enveloping associative algebra Σ^* of Σ and under every element of the enveloping algebra $\Sigma^\dagger$. Thus we see that $L(\Sigma) = L(\Sigma^*) = L(\Sigma^\dagger)$.

The set Σ is called an *irreducible* set of linear transformations and $\mathfrak{M}$ is called Σ-irreducible if $L(\Sigma) = \{\mathfrak{M}, 0\}$ and $\mathfrak{M} \neq 0$. Σ is *indecomposable* and $\mathfrak{M}$ is Σ-indecomposable if there exists no decomposition $\mathfrak{M} = \mathfrak{M}_1 \oplus \mathfrak{M}_2$ where the $\mathfrak{M}_i$ are non-zero elements of $L(\Sigma)$. Of course, irreducibility implies indecomposability. Σ (and $\mathfrak{M}$ relative to Σ) is called *completely reducible* if $\mathfrak{M} = \sum \mathfrak{M}_\alpha$, $\mathfrak{M}_\alpha \in L(\Sigma)$ and $\mathfrak{M}_\alpha$ irreducible. We recall the following well-known result.

THEOREM 9. *Σ is completely reducible if and only if $L(\Sigma)$ is complemented, that is, for every $\mathfrak{N} \in L(\Sigma)$ there exists an $\mathfrak{N}' \in L(\Sigma)$ such that $\mathfrak{M} = \mathfrak{N} \oplus \mathfrak{N}'$. If the condition holds then $\mathfrak{M} = \mathfrak{M}_1 \oplus \mathfrak{M}_2 \oplus \cdots \oplus \mathfrak{M}_r$ where the $\mathfrak{M}_i \in L(\Sigma)$ and are irreducible.*

Proof: Assume $\mathfrak{M} = \sum \mathfrak{M}_\alpha$, $\mathfrak{M}_\alpha$ irreducible in $L(\Sigma)$ and let $\mathfrak{N} \in L(\Sigma)$. If $\dim \mathfrak{N} = \dim \mathfrak{M}$, $\mathfrak{N} = \mathfrak{M}$ and $\mathfrak{M} = \mathfrak{N} \oplus 0$. We now suppose $\dim \mathfrak{N} < \dim \mathfrak{M}$ and we may assume the theorem for subspaces $\mathfrak{N}_1$ such that $\dim \mathfrak{N}_1 > \dim \mathfrak{N}$. Since $\mathfrak{N} \subset \mathfrak{M} = \sum \mathfrak{M}_\alpha$ there is an $\mathfrak{M}_\alpha$ such that $\mathfrak{M}_\alpha \not\subseteq \mathfrak{N}$. Consider the subspace $\mathfrak{M}_\alpha \cap \mathfrak{N}$. This is a Σ-subspace of the irreducible Σ-space $\mathfrak{M}_\alpha$. Hence either $\mathfrak{M}_\alpha \cap \mathfrak{N} = \mathfrak{M}_\alpha$ or $\mathfrak{M}_\alpha \cap \mathfrak{N} = 0$. If $\mathfrak{M}_\alpha \cap \mathfrak{N} = \mathfrak{M}_\alpha$, $\mathfrak{N} \supseteq \mathfrak{M}_\alpha$ contrary to assumption. Hence $\mathfrak{M}_\alpha \cap \mathfrak{N} = 0$ and $\mathfrak{N}_1 \equiv \mathfrak{N} + \mathfrak{M}_\alpha = \mathfrak{N} \oplus \mathfrak{M}_\alpha$. We can now apply the induction hypothesis to conclude that $\mathfrak{M} = \mathfrak{N}_1 \oplus \mathfrak{N}_1'$, $\mathfrak{N}_1' \in L(\Sigma)$. Then $\mathfrak{M} = \mathfrak{N} \oplus \mathfrak{M}_\alpha \oplus \mathfrak{N}_1' = \mathfrak{N} \oplus \mathfrak{N}'$ where $\mathfrak{N}' = \mathfrak{M}_\alpha \oplus \mathfrak{N}_1' \in L(\Sigma)$. Conversely, assume $L(\Sigma)$ is complemented. Let $\mathfrak{M}_1$ be a minimal element $\neq 0$ of $L(\Sigma)$. (Such elements exist since $\dim \mathfrak{M}$ is finite.) Then we have $\mathfrak{M} = \mathfrak{M}_1 \oplus \mathfrak{N}$ where $\mathfrak{N} \in L(\Sigma)$. We note now that the condition assumed for $\mathfrak{M}$ carries over to $\mathfrak{N}$. Thus let $\mathfrak{P}$ be a Σ-subspace of $\mathfrak{N}$. Then we can write $\mathfrak{M} = \mathfrak{P} \oplus \mathfrak{P}'$, $\mathfrak{P}' \in L(\Sigma)$. Then, by Dedekind's modular law, $\mathfrak{N} = \mathfrak{M} \cap \mathfrak{N} = \mathfrak{P} + (\mathfrak{P}' \cap \mathfrak{N})$ since $\mathfrak{N} \supseteq \mathfrak{P}$. If we set $\mathfrak{P}'' = \mathfrak{P}' \cap \mathfrak{N}$, then $\mathfrak{P}'' \cap \mathfrak{P} =$

$\mathfrak{P} \cap \mathfrak{P}' \cap \mathfrak{N} = 0$. Hence $\mathfrak{N} = \mathfrak{P} \oplus \mathfrak{P}''$ where $\mathfrak{P}'' \in L(\Sigma)$ and $\mathfrak{P}'' \subseteqq \mathfrak{N}$. We can now repeat for $\mathfrak{N}$ the step taken for $\mathfrak{M}$; that is, we can write $\mathfrak{N} = \mathfrak{M}_2 \oplus \mathfrak{P}$ where $\mathfrak{M}_2$, $\mathfrak{P} \in L(\Sigma)$ and $\mathfrak{M}_2$ is irreducible. Continuing in this way we obtain, because of the finiteness of dimensionality, that $\mathfrak{M} = \mathfrak{M}_1 \oplus \mathfrak{M}_2 \oplus \cdots \oplus \mathfrak{M}_r$, $\mathfrak{M}_i$ irreducible, $\mathfrak{M}_i \in L(\Sigma)$. This completes the proof.

We suppose next that $\Sigma = \mathfrak{A}$ is a subalgebra of the associative algebra $\mathfrak{E}$ (possibly not containing 1), and we obtain the following necessary condition for complete reducibility.

THEOREM 10. *If $\mathfrak{A}$ is an associative algebra of linear transformations in a finite-dimensional vector space, then $\mathfrak{A}$ completely reducible implies that $\mathfrak{A}$ is semi-simple.*

Proof: Let $\mathfrak{R}$ be the radical of $\mathfrak{A}$ and suppose $\mathfrak{M} = \sum \mathfrak{M}_\alpha$, $\mathfrak{M}_\alpha \in L(\mathfrak{A})$ and irreducible. Consider the subspace $\mathfrak{M}_\alpha\mathfrak{R}$ spanned by the vectors of the form yN, $y \in \mathfrak{M}_\alpha$, $N \in \mathfrak{R}$. This is an $\mathfrak{A}$-subspace contained in $\mathfrak{M}_\alpha$. Since $\mathfrak{R}^k = 0$ for some k we must have $\mathfrak{M}_\alpha\mathfrak{R} \subset \mathfrak{M}_\alpha$ (cf. the proof of Theorem 1). Since $\mathfrak{M}_\alpha$ is irreducible we can conclude that $\mathfrak{M}_\alpha\mathfrak{R} = 0$ for every $\mathfrak{M}_\alpha$. Since $\mathfrak{M} = \sum \mathfrak{M}_\alpha$ this implies that $\mathfrak{M}\mathfrak{R} = 0$, that is, $\mathfrak{R} = 0$. Hence $\mathfrak{A}$ is semi-simple.

Since Σ is completely reducible if and only if Σ^* and $\Sigma^\dagger$ are completely reducible, we have the

COROLLARY. *If Σ is completely reducible, then Σ^* and $\Sigma^\dagger$ are semi-simple.*

We shall say that a single linear transformation A is *semi-simple* if the minimum polynomial $\mu(\lambda)$ of A is a product of distinct prime polynomials. This condition is equivalent to the condition that $\{A\}^\dagger$ has no nilpotent elements $\neq 0$. Thus if $\mu(\lambda) = \pi_1(\lambda)^{e_1}\pi_2(\lambda)^{e_2} \cdots \pi_r(\lambda)^{e_r}$ then $Z = \pi_1(A)\pi_2(A) \cdots \pi_r(A)$ is nilpotent and $Z \neq 0$ if some $e_i > 1$. Conversely, suppose the π_i are distinct primes and all $e_i = 1$. Let $Z = \phi(A)$ be nilpotent. If $Z^r = 0$, $\phi(A)^r = 0$ and $\phi(\lambda)^r$ is divisible by $\mu(\lambda)$. Hence $\phi(\lambda)$ is divisible by $\mu(\lambda)$ and $Z = \phi(A) = 0$. We can now prove

THEOREM 11. *Let $\mathfrak{L}$ be a completely reducible Lie algebra of linear transformations in a finite-dimensional vector space over a field of characteristic 0. Then $\mathfrak{L} = \mathfrak{C} \oplus \mathfrak{L}_1$ where $\mathfrak{C}$ is the center and $\mathfrak{L}_1$ is a semi-simple ideal. Moreover, the elements of $\mathfrak{C}$ are semi-simple.*

Proof: If $\mathfrak{L}$ is completely reducible, $\mathfrak{L}^*$ is semi-simple. Hence the statement about $\mathfrak{L} = \mathfrak{C} \oplus \mathfrak{L}_1$ follows from Theorem 8. Now suppose that $\mathfrak{C}$ contains a C which is not semi-simple. Then $\{C\}^\dagger$ contains a non-zero nilpotent N. Moreover, $\{C\}^\dagger$ is the set of polynomials in C, so that N is a polynomial in C. Hence N is in the center of $\mathfrak{L}^\dagger$. Since this is the case, $N\mathfrak{L}^\dagger = \mathfrak{L}^\dagger N$ is an ideal in $\mathfrak{L}^\dagger$. Moreover, $(N\mathfrak{L}^\dagger)^k \subseteq N^k\mathfrak{L}^\dagger = 0$ if k is sufficiently large. Since $N \in N\mathfrak{L}^\dagger$, $N\mathfrak{L}^\dagger$ is a non-zero nilpotent ideal in $\mathfrak{L}^\dagger$ contrary to the semi-simplicity of $\mathfrak{L}^\dagger$.

We shall show later (Theorem 3.10) that the converse of this result holds, so the conditions given here are necessary and sufficient for complete reducibility of a Lie algebra of linear transformations in the characteristic zero case. Nothing like this can hold for characteristic $p \neq 0$ (cf. § 6.3). We remark also that the converse of Theorem 10 is valid too. For the associative case one therefore has a simple necessary and sufficient condition for complete reducibility, valid for any characteristic of the field. The converse of Theorem 10 is considerably deeper than the theorem itself. This will not play an essential role in the sequel.

6. *Lie's theorems*

We need to recall the notion of a composition series for a set Σ of linear transformations and its meaning in terms of matrices. We recall that a chain $\mathfrak{M} = \mathfrak{M}_1 \supset \mathfrak{M}_2 \supset \cdots \supset \mathfrak{M}_s \supset \mathfrak{M}_{s+1} = 0$ of Σ-subspaces is a *composition series* for $\mathfrak{M}$ relative to Σ if for every i there exists no $\mathfrak{M}' \in L(\Sigma)$ such that $\mathfrak{M}_i \supset \mathfrak{M}' \supset \mathfrak{M}_{i+1}$. If $\mathfrak{N}$ is a Σ-subspace of $\mathfrak{M}$, then Σ induces a set $\bar{\Sigma}$ of linear transformations in $\mathfrak{M}/\mathfrak{N}$. As is well known (and easy to see) for groups with operators the $\bar{\Sigma}$-subspaces of $\bar{\mathfrak{M}} = \mathfrak{M}/\mathfrak{N}$ have the form $\mathfrak{P}/\mathfrak{N}$ where $\mathfrak{P}$ is a Σ-subspace of $\mathfrak{M}$ containing $\mathfrak{N}$. It follows that there exists no $\mathfrak{P} \in L(\Sigma)$ with $\mathfrak{M} \supset \mathfrak{P} \supset \mathfrak{N}$ if and only if $\mathfrak{M}/\mathfrak{N}$ is $\bar{\Sigma}$-irreducible. Hence $\mathfrak{M} = \mathfrak{M}_1 \supset \mathfrak{M}_2 \supset \cdots \supset \mathfrak{M}_{s+1} = 0$, $\mathfrak{M}_i \in L(\Sigma)$, is a composition series if and only if every $\mathfrak{M}_i/\mathfrak{M}_{i+1}$ is $\bar{\Sigma}_i$-irreducible, $\bar{\Sigma}_i$ the set of induced transformations in $\mathfrak{M}_i/\mathfrak{M}_{i+1}$ determined by the $A \in \Sigma$.

The finiteness of dimensionality of $\mathfrak{M}$ assures the existence of a composition series. Thus let $\mathfrak{M}_2$ be a maximal Σ-subspace properly contained in $\mathfrak{M}_1 = \mathfrak{M}$. Then $\mathfrak{M}_1 \supset \mathfrak{M}_2$ and $\mathfrak{M}_1/\mathfrak{M}_2$ is irreducible. Next let $\mathfrak{M}_3$ be a maximal invariant subspace of $\mathfrak{M}_2$, $\neq \mathfrak{M}_2$, etc. This leads to a composition series $\mathfrak{M}_1 \supset \mathfrak{M}_2 \supset \mathfrak{M}_3 \supset \cdots \supset \mathfrak{M}_s \supset \mathfrak{M}_{s+1} = 0$ for $\mathfrak{M}$.

Let $\mathfrak{M} = \mathfrak{M}_1 \supset \mathfrak{M}_2 \supset \mathfrak{M}_3 \supset \cdots \supset \mathfrak{M}_{s+1} = 0$ be a descending chain of Σ-subspaces and let $(e_1, \cdots, e_n)$ be a basis for $\mathfrak{M}$ such that $(e_1, \cdots, e^{n_1})$ is a basis for $\mathfrak{M}_s$, $(e_1, \cdots, e_{n_1+n_2})$ is a basis for $\mathfrak{M}_{s-1}$, etc. Then it is clear that if $A \in \Sigma$, the matrix of A relative to (e_i) is of the form

$$(10) \qquad M = \begin{bmatrix} M_1 & & & & & 0 \\ * & M_2 & & & & \\ \cdot & & \cdot & & & \\ \cdot & & & \cdot & & \\ \cdot & & & & \cdot & \\ * & \cdot & \cdot & \cdot & * & M_s \end{bmatrix}.$$

Thus we have

$$(11) \qquad e_{n_1+\cdots+n_{j-1}+k}A = \sum_l \mu_{kl}^{(j)} e_{n_1+\cdots+n_{j-1}+l} + x\,, \qquad k, l = 1, \cdots, n_j\,,$$

where x is a vector in $\mathfrak{M}_{s-j+2}$. Hence we have (10) with $M_j = (\mu_{kl}^{(j)})$. The cosets $\bar{e}_{n_1+\cdots+n_{j-1}+k} \equiv e_{n_1+\cdots+n_{j-1}+k} + \mathfrak{M}_{s-j+2}$, $k = 1, \cdots, n_j$ form a basis for the factor space $\mathfrak{M}_{s-j+1}/\mathfrak{M}_{s-j+2}$ and (11) gives

$$(12) \qquad \bar{e}_{n_1+\cdots+n_{j-1}+k}A = \sum_l \mu_{kl}^{(j)} \bar{e}_{n_1+\cdots+n_{j-1}+l}\,;$$

which shows that the matrix of the linear transformation A induced in $\mathfrak{M}_{s-j+1}/\mathfrak{M}_{s-j+2}$ is M_j.

We shall now see what can be said about solvable Lie algebras of linear transformations in a finite-dimensional vector space over an algebraically closed field of characteristic zero. For this we need the following

LEMMA 5. *Let $\mathfrak{L}$ be an abelian Lie algebra of linear transformations in the finite-dimensional space $\mathfrak{M}$ over Φ, Φ algebraically closed. Suppose $\mathfrak{M}$ is $\mathfrak{L}$-irreducible. Then $\mathfrak{M}$ is one-dimensional.*

Proof: If $A \in \mathfrak{L}$, A has a non-zero characteristic vector x. Thus $xA = \alpha x$, $\alpha \in \Phi$. Now let $\mathfrak{M}_\alpha$ be the set of vectors y satisfying this equation. Then if $B \in \mathfrak{L}$ and $y \in \mathfrak{M}_\alpha$, $(yB)A = yAB = \alpha yB$. Hence $yB \in \mathfrak{M}_\alpha$. This shows that $\mathfrak{M}_\alpha$ is invariant under every $B \in \mathfrak{L}$. Since $\mathfrak{M}$ is irreducible, $\mathfrak{M} = \mathfrak{M}_\alpha$, which means that $A = \alpha 1$ in $\mathfrak{M}$. Now this holds for every $A \in \mathfrak{L}$. It follows that any subspace of $\mathfrak{M}$ is an $\mathfrak{L}$-subspace. Since $\mathfrak{M}$ is $\mathfrak{L}$-irreducible, $\mathfrak{M}$ has no subspaces other than itself and 0. Hence $\mathfrak{M}$ is one-dimensional.

We can now prove

Lie's theorem. If $\mathfrak{L}$ is a solvable Lie algebra of linear transformations in a finite-dimensional vector space $\mathfrak{M}$ over an algebraically closed field of characteristic 0, *then the matrices of $\mathfrak{L}$ can be taken in simultaneous triangular form.*

Proof: Let $\mathfrak{M} = \mathfrak{M}_1 \supset \mathfrak{M}_2 \supset \cdots \supset \mathfrak{M}_{s+1} = 0$ be a composition series for $\mathfrak{M}$ relative to $\mathfrak{L}$. Let $\bar{\mathfrak{L}}_i$ denote the set of induced linear transformations in the irreducible space $\mathfrak{M}_i/\mathfrak{M}_{i+1}$. Then $\bar{\mathfrak{L}}_i$ is a solvable Lie algebra of linear transformations since it is a homomorphic image of $\mathfrak{L}$. Also $\bar{\mathfrak{L}}_i$ is irreducible, hence completely reducible in $\mathfrak{M}_i/\mathfrak{M}_{i+1}$. Hence $\bar{\mathfrak{L}}_i$ is abelian by Theorem 11. It follows from the lemma that $\dim \mathfrak{M}_i/\mathfrak{M}_{i+1} = 1$. This means that if we use a basis corresponding to the composition series then the matrices M_i in (10) are one-rowed. Hence every M corresponding to the $A \in \mathfrak{L}$ is triangular.

If $\mathfrak{L}$ is nilpotent, it is solvable, so Lie's theorem is applicable. We observe also that Lemma 5 and, consequently, Lie's theorem are valid for any field of characteristic 0, provided the characteristic roots of the linear transformations belong to the field. We have called such Lie algebras of linear transformations split. If we combine this extension of Lie's theorem with Theorem 7 we obtain

THEOREM 12. *Let $\mathfrak{L}$ be a split nilpotent Lie algebra of linear transformations in a finite-dimensional vector space $\mathfrak{M}$ over a field of characteristic* 0. *Then $\mathfrak{M}$ is a direct sum of its weight spaces $\mathfrak{M}_\alpha$ and the matrices in the weight space $\mathfrak{M}_\alpha$ can be taken simultaneously in the form*

$$A_\alpha = \begin{bmatrix} \alpha(A) & & & & 0 \\ & \alpha(A) & & & \\ & & \cdot & & \\ & * & & \cdot & \\ & & & & \cdot \\ & & & & \alpha(A) \end{bmatrix}. \tag{13}$$

Proof: The fact that $\mathfrak{M}$ is a direct sum of the weight spaces $\mathfrak{M}_\alpha$ has been proved before. We have $x_\alpha(A - \alpha(A)1)^m = 0$ for every x_α in $\mathfrak{M}_\alpha$, which implies that the only characteristic root of A in $\mathfrak{M}_\alpha$ is $\alpha(A)$. Since the diagonal elements in (10) are characteristic roots it follows that the M_i in (10) for A are $\alpha(A)$. Hence we have the form (13) for the matrix of A in $\mathfrak{M}_\alpha$.

The proof of this result (or the form of (13)) shows that in $\mathfrak{M}_\alpha$ there exists a non-zero vector x such that $xA = \alpha(A)x$, $A \in \mathfrak{L}$. If $A, B \in \mathfrak{L}$ we have $x(A + B) = (\alpha(A) + \alpha(B))x = \alpha(A + B)x$, $x(\rho A) = \rho\alpha(A)x = \alpha(\rho(A))x$, $\rho \in \Phi$, and $x[AB] = (\alpha(A)\alpha(B) - \alpha(B)\alpha(A))x = \alpha([AB])x$. The first two of these equations imply that α is a linear function on $\mathfrak{L}$. Since every element of the derived algebra $\mathfrak{L}'$ is a linear combination of elements $[AB]$, the linearity and the last condition imply that $\alpha(C) = 0$, $C \in \mathfrak{L}'$. We therefore have the following important consequence of Theorem 12.

Corollary. *Under the same assumptions as in Theorem* 12, *the weights* α: $A \to \alpha(A)$ *are linear functions on* $\mathfrak{L}$ *which vanish on* $\mathfrak{L}'$.

7. *Applications to abstract Lie algebras. Some counter examples*

As usual, the results we have derived for Lie algebras of linear transformations apply to abstract Lie algebras on considering the set of linear transformations ad $\mathfrak{L}$. We state two of the results which can be obtained in this way.

Theorem 13. *Let* $\mathfrak{L}$ *be a finite-dimensional Lie algebra over a field of characteristic* 0, $\mathfrak{S}$ *the radical, and* $\mathfrak{N}$ *the nil radical. Then* $[\mathfrak{L}\mathfrak{S}] \subseteq \mathfrak{N}$.

Proof: By Corollary 2 to Theorem 8, $[\text{ad}\,\mathfrak{L}, \text{ad}\,\mathfrak{S}]$ is contained in the radical of $(\text{ad}\,\mathfrak{L})^*$. This implies that there exists an integer N such that for any N transformations of the form $[\text{ad}\,a_i, \text{ad}\,s_i]$, $a_i \in \mathfrak{L}$, $s_i \in \mathfrak{S}$, we have $[\text{ad}\,a_1, \text{ad}\,s_1][\text{ad}\,a_2, \text{ad}\,s_2]\cdots[\text{ad}\,a_N, \text{ad}\,s_N] = 0$. Hence $\text{ad}\,[a_1s_1]\,\text{ad}\,[a_2s_2]\cdots\text{ad}\,[a_Ns_N] = 0$. Thus for any $x \in \mathfrak{L}$ we have $[\cdots[[x[a_1s_1]][a_2s_2]]\cdots,[a_Ns_N]] = 0$. This implies that $[\mathfrak{S}\mathfrak{L}]^{N+1} = 0$, so $[\mathfrak{S}\mathfrak{L}]$ is nilpotent. Since this is an ideal, $[\mathfrak{S}\mathfrak{L}] \subseteq \mathfrak{N}$.

Corollary 1. *The derived algebra of any finite-dimensional solvable Lie algebra of characteristic* 0 *is nilpotent.*

Proof: Take $\mathfrak{L} = \mathfrak{S}$ in the theorem.

Corollary 2. *Let* $\mathfrak{S}$ *be solvable finite-dimensional of characteristic* 0, $\mathfrak{N}$ *the nil radical of* $\mathfrak{S}$. *Then* $\mathfrak{S}D \subseteq \mathfrak{N}$ *for any derivation* D *of* $\mathfrak{S}$.

Proof: Let $\mathfrak{L} = \mathfrak{S} \oplus \Phi D$ the split extension of ΦD by $\mathfrak{S}$ (cf. § 1.5). Here $\mathfrak{S}$ is an ideal and $[s, D] = sD$ for $s \in \mathfrak{S}$. Since $\mathfrak{L}/\mathfrak{S}$ is one-dimensional, it is solvable. Hence $\mathfrak{L}'$ is a nilpotent ideal in $\mathfrak{L}$. On the other hand, $\mathfrak{L}' = \mathfrak{S}' + \mathfrak{S}D$, so $\mathfrak{S}' + \mathfrak{S}D$ is a nilpotent ideal in $\mathfrak{S}$ and $\mathfrak{S}' + \mathfrak{S}D \subseteq \mathfrak{N}$. Hence $\mathfrak{S}D \subseteq \mathfrak{N}$.

THEOREM 14 (Lie). *Let $\mathfrak{L}$ be a finite-dimensional solvable Lie algebra over an algebraically closed field of characteristic* 0. *Then there exists a chain of ideals $\mathfrak{L} = \mathfrak{L}_n \supset \mathfrak{L}_{n-1} \supset \cdots \supset \mathfrak{L}_1 \supset 0$ such that* $\dim \mathfrak{L}_i = i$.

Proof: Since $\mathfrak{L}$ is solvable, $\operatorname{ad}\mathfrak{L}$ is a solvable Lie algebra of linear transformations acting in the vector space $\mathfrak{L}$. Let $\mathfrak{L} = \mathfrak{L}_1 \supset \mathfrak{L}_2 \supset \cdots \supset \mathfrak{L}_{s+1} = 0$ be a composition series of $\mathfrak{L}$ relative to $\operatorname{ad}\mathfrak{L}$. Then $\mathfrak{L}_i \operatorname{ad}\mathfrak{L} \subseteq \mathfrak{L}_i$ is equivalent to the statement that $\mathfrak{L}_i$ is an ideal. By the proof of Lie's theorem we have that $\dim \mathfrak{L}_i/\mathfrak{L}_{i+1} = 1$. Hence the composition series provides a chain of ideals of the type required.

We shall now show that the assumption that the characteristic is 0 is essential in the results of §§ 5 and 6. We begin our construction of counter examples in the characteristic $p \neq 0$ case with a p-dimensional vector space $\mathfrak{M}$ over Φ of characteristic p. Let E and F be the linear transformations in $\mathfrak{M}$ whose effect on a basis $(e_1, e_2, \cdots, e_p)$ is given by

$$\begin{aligned} e_iE &= e_{i+1}\,, \qquad i \leqq p-1\,, \\ e_pE &= e_1\,, \\ e_iF &= (i-1)e_i\,. \end{aligned} \tag{14}$$

Then $e_iEF = ie_{i+1}$, $i \leqq p-1$, $e_pEF = 0$ and $e_iFE = (i-1)e_{i+1}$, $e_pFE = -e_1$. Hence

$$e_i[EF] = e_{i+1}\,, \qquad e_p[EF] = e_1\,,$$

so that we have $[EF] = E$. Hence $\mathfrak{L} = \Phi E + \Phi F$ is a two-dimensional solvable Lie algebra. We assert that $\mathfrak{L}$ acts irreducibly in $\mathfrak{M}$. Thus let $\mathfrak{N}$ be an $\mathfrak{L}$-subspace $\neq 0$ and let $x = \sum \xi_i e_i \neq 0$ be in $\mathfrak{N}$. Then

$$\begin{aligned} x &= \xi_1e_1 + \xi_2e_2 + \cdots + \xi_pe_p \in \mathfrak{N} \\ xF &= 1\xi_2e_2 + \cdots + (p-1)\xi_pe_p \in \mathfrak{N} \\ xF^2 &= 1^2\xi_2e_2 + \cdots + (p-1)^2\xi_pe_p \in \mathfrak{N} \\ &\cdots\cdots\cdots\cdots\cdots\cdots \\ xF^{p-1} &= 1^{p-1}\xi_2e_2 + \cdots + (p-1)^{p-1}\xi_pe_p \in \mathfrak{N}\,. \end{aligned} \tag{15}$$

The Vandermonde determinant

$$(16) \qquad V = \begin{vmatrix} 1 & 1 & \cdots & 1 \\ 0 & 1 & \cdots & p-1 \\ \cdots & \cdots & \cdots & \cdots \\ 0 & 1^{p-1} & \cdots & (p-1)^{p-1} \end{vmatrix} \neq 0 .$$

If we multiply xF^{j-1} by the cofactor of the (j, i) element of V and sum on j we obtain that $V\xi_i e_i \in \mathfrak{N}$ and so $\xi_i e_i \in \mathfrak{N}$. It follows from this that $\mathfrak{N}$ contains one of the e_i. If we operate on this e_i by the powers of E we obtain every e_j. Hence $\mathfrak{N} = \mathfrak{M}$, which proves irreducibility and shows that Theorem 11 fails for characteristic p. Also since complete reducibility implies semi-simplicity (actually $\mathfrak{L}^* = \mathfrak{E}$ is simple), Theorem 8 also fails for characteristic p.

Next let $\mathfrak{S}$ denote the two-dimensional non-abelian Lie algebra $\Phi e + \Phi f$, $[ef] = e$, Φ of characteristic p. Then $e \rightarrow E$, $f \rightarrow F$, given by (14), defines a representation of $\mathfrak{S}$ acting in $\mathfrak{M}$ and $\mathfrak{M}$ is an $\mathfrak{S}$-module with basis $(e_1, e_2, \cdots, e_p)$. Let $\mathfrak{K}$ be the split extension $\mathfrak{K} = \mathfrak{S} \oplus \mathfrak{M}$. Then $\mathfrak{M}$ is an abelian ideal in $\mathfrak{K}$ and $\mathfrak{K}/\mathfrak{M} \cong \mathfrak{S}$ is solvable. Hence $\mathfrak{K}$ is solvable. The derived algebra $\mathfrak{K}' = \Phi e + [\mathfrak{M}\mathfrak{S}] = \Phi e + \mathfrak{M}$. Since $[\mathfrak{M}e] = \mathfrak{M}E = \mathfrak{M}$, $(\mathfrak{K}')^2 = (\mathfrak{K}')^3 = \cdots = \mathfrak{M}$. Hence $\mathfrak{K}'$ is not nilpotent. This shows that also Theorem 13 fails for characteristic p.

It is still conceivable that Lie's theorem might hold for characteristic p if one replaces the word "solvable" by "nilpotent." This would imply that we have the nice canonical form (13) for nilpotent Lie algebras. However, this is not the case either. Thus let E and F be as before. Then $EF - FE = E$ implies that $E(FE^{-1}) - (FE^{-1})E = 1$. Set $G = FE^{-1}$, $H = 1$. Then we have

$$(17) \qquad [EG] = H , \qquad [EH] = 0 = [GH] .$$

Hence $\mathfrak{N} = \Phi E + \Phi G + \Phi H$ is a nilpotent Lie algebra. On the other hand, $\mathfrak{N}^* = \mathfrak{L}^*(\mathfrak{L} = \Phi E + \Phi F)$ is irreducible even if the base field is algebraically closed. Hence the matrices cannot be put in simultaneous triangular form. If $X = \xi E + \eta G + \zeta 1$ one can verify that $(X - (\xi + \zeta)1)^p = 0$ if $p \neq 2$. Hence the mapping $X \rightarrow \xi + \zeta$ is the only weight for $\mathfrak{N}$. Note that this is linear. (Cf. Exercise 24, Chap. V.)

Exercises

In these exercises all algebras and vector spaces are finite-dimensional.

1. Let Φ be algebraically closed of characteristic 0 and let $\mathfrak{B}$ be a solvable

subalgebra of m dimensions of the Lie algebra $\mathfrak{L}$ properly containing $\mathfrak{B}$. Show that $\mathfrak{B}$ is contained in a subalgebra of $m+1$ dimensions.

2. Φ as in Exercise 1, $\mathfrak{L}$ solvable. Show that $\mathfrak{L}$ has a basis $(e_1, e_2, \cdots, e_n)$ with a multiplication table $[e_ie_j] = \sum_{k=1}^{i} \gamma_{ijk}e_k$, $i < j$.

3. Φ as in Exercise 1. Let $\mathfrak{L}$ be a solvable subalgebra of Φ_{nL}. Prove that $\dim \mathfrak{L} \leqq n(n+1)/2$ and that $\mathfrak{L}$ can be imbedded in a solvable subalgebra of $n(n+1)/2$ dimensions. Show that if $\mathfrak{L}_1$ and $\mathfrak{L}_2$ are solvable subalgebras of $n(n+1)/2$ dimensions, then there exists a non-singular matrix A such that $\mathfrak{L}_2 = A^{-1}\mathfrak{L}_1A$.

4. Show that if $\mathfrak{L}$ is not nilpotent over an algebraically closed field then $\mathfrak{L}$ contains a two-dimensional non-abelian subalgebra.

5. Φ algebraically closed, any characteristic. If α is a characteristic root of a linear transformation A, the characteristic space $\mathfrak{M}_\alpha = \{x_\alpha \mid x_\alpha(A - \alpha 1)^k = 0$ for some $k\}$. We have $\mathfrak{M} = \sum \oplus \mathfrak{M}_\alpha$. Let A be an automorphism in a non-associative algebra $\mathfrak{A}$ and let $\mathfrak{A} = \sum \oplus \mathfrak{A}_\alpha$ be the decomposition of $\mathfrak{A}$ into characteristic spaces relative to A. Show that

$$\mathfrak{A}_\alpha\mathfrak{A}_\beta = \begin{cases} 0 \text{ if } \alpha\beta \text{ is not a characteristic root;} \\ \subseteq \mathfrak{A}_{\alpha\beta} \text{ if } \alpha\beta \text{ is a characteristic root.} \end{cases}$$

Let S be the linear transformation in $\mathfrak{A}$ such that $S = \alpha 1$ in $\mathfrak{A}_\alpha$ for every α. Show that S is an automorphism which commutes with A and that $U = S^{-1}A$ is an automorphism of the form $1 + Z$, Z nilpotent.

6. Φ as in Exercise 5, $\mathfrak{L}$ a Lie algebra over Φ, A an automorphism in $\mathfrak{L}$. Let $\mathfrak{L} = \sum \oplus \mathfrak{L}_\alpha$ be the decomposition of $\mathfrak{L}$ into characteristic subspaces relative to A. Let $\mathfrak{W} = \cup \operatorname{ad} \mathfrak{L}_\alpha$ where $\operatorname{ad} \mathfrak{L}_\alpha = \{\operatorname{ad} x_\alpha \mid x_\alpha \in \mathfrak{L}_\alpha\}$. Show that $\mathfrak{W}$ is weakly closed relative to $A \times B = [AB]$.

7. Φ, $\mathfrak{L}$, A, $\mathfrak{W}$ as in Exercise 6. Assume A is of prime order. Show that if the only $x \in \mathfrak{L}$ such that $xA = x$ is $x = 0$ then every element of $\mathfrak{W}$ is nilpotent. Hence prove that if a Lie algebra over an arbitrary field has an automorphism of prime order without fixed points except 0, then $\mathfrak{L}$ is nilpotent.

8. Let D be a derivation in a non-associative algebra $\mathfrak{A}$ and let $\mathfrak{A} = \sum \oplus \mathfrak{A}_\alpha$ be the decomposition of $\mathfrak{A}$ into characteristic spaces relative to D. Show that

$$\mathfrak{A}_\alpha\mathfrak{A}_\beta = \begin{cases} 0 \text{ if } \alpha + \beta \text{ is not a characteristic root;} \\ \subseteq \mathfrak{A}_{\alpha+\beta} \text{ if } \alpha + \beta \text{ is a characteristic root.} \end{cases}$$

Let S be the linear transformation in $\mathfrak{A}$ such that $S = \alpha 1$ in $\mathfrak{A}_\alpha$ for every α. Show that S is a derivation which commutes with D and that $D - S = Z$ is a nilpotent derivation.

9. Prove that if $\mathfrak{L}$ is a Lie algebra over a field of characteristic 0 and $\mathfrak{L}$ has a derivation without non-zero constants, then $\mathfrak{L}$ is nilpotent.

10. (Dixmier-Lister). Show that if $\mathfrak{L}$ is the nilpotent Lie algebra of Exercise 1.7 and the characteristic is not 2 or 3, then every derivation D satisfies $\mathfrak{L}D \subseteq \mathfrak{L}'$ and hence is nilpotent. Show that this Lie algebra cannot be the derived algebra of any Lie algebra.

11. Let $\mathfrak{L}$ be a Lie algebra of linear transformations such that every $A \in \mathfrak{L}$ has the form $\alpha(A)1 + Z$ where $\alpha(A) \in \Phi$ and Z is nilpotent. Prove that $\mathfrak{L}$ is nilpotent.

The next group of exercises is designed to prove the finiteness of the derivation tower of any Lie algebra with 0 center. The corresponding result for finite groups is due to Wielandt. The proof in the Lie algebra case is due to Schenkman [1]. It follows fairly closely Wielandt's proof of the group result but makes use of several results which are peculiar to the Lie algebra case. These effect a substantial simplification and give a final result which is sharper than that of the group case.

Let $\mathfrak{L}$ be a Lie algebra with 0 center. Then $\mathfrak{L}$ is isomorphic to the ideal $\mathfrak{L}_1$ of inner derivations in the derivation algebra $\mathfrak{L}_2 = \mathfrak{D}(\mathfrak{L})$. We can identify $\mathfrak{L}$ with $\mathfrak{L}_1$ and so consider $\mathfrak{L}$ as an ideal in $\mathfrak{L}_2$. We have seen (Exercise 1.4) that $\mathfrak{L}_2$ has 0 center. Hence the process can be repeated and we obtain $\mathfrak{L}_1 \subseteq \mathfrak{L}_2 \subseteq \mathfrak{L}_3$ where $\mathfrak{L}_3$ is the derivation algebra of $\mathfrak{D}(\mathfrak{L})$, $\mathfrak{D}(\mathfrak{D}(\mathfrak{L}))$, and $\mathfrak{L}_2$ is identified with the ideal of inner derivations of $\mathfrak{L}_2$. This process leads to a chain $\mathfrak{L} = \mathfrak{L}_1 \subseteq \mathfrak{L}_2 \subseteq \mathfrak{L}_3 \subseteq \cdots$, where each $\mathfrak{L}_i$ is invariant in $\mathfrak{L}_{i+1}$. Hence $\mathfrak{L}_1 = \mathfrak{L}$ is sub-invariant in $\mathfrak{L}_i$ (Exercise 1.8) and every $\mathfrak{L}_i$ has 0 center. The tower theorem states that from a certain point on we must have $\mathfrak{L}_r = \mathfrak{L}_{r+1} = \cdots$. This means that if $\mathfrak{L}$ has 0 center, then for a suitable K, $\overbrace{\mathfrak{D}(\mathfrak{D}(\cdots\mathfrak{D}}^{K}(\mathfrak{L})\cdots))$ is a complete Lie algebra.

Exercise 1.4, 1.8, and 1.9 will be needed in the present discussion. Besides these all that is required are elementary results and Engel's theorem that if $\operatorname{ad} a$ is nilpotent for every a, then $\mathfrak{L}$ is nilpotent.

12. Prove that every Lie algebra $\mathfrak{L}$ has a decomposition $\mathfrak{L} = \mathfrak{L}^{\omega} + \mathfrak{H}$ where $\mathfrak{L}^{\omega} = \bigcap_{i=1}^{\infty}\mathfrak{L}^{i}$ and $\mathfrak{H}$ is a nilpotent subalgebra.

13. Let $\mathfrak{L}$ be a Lie algebra and let $\mathfrak{Z}$ be the centralizer of $\mathfrak{L}^{\omega}$. Show that if $\mathfrak{L}$ has 0 center, then $\mathfrak{Z} \subseteq \mathfrak{L}^{\omega}$.

Sketch of proof: $\mathfrak{Z}$ is an ideal. Write $\mathfrak{L} = \mathfrak{L}^{\omega} + \mathfrak{H}$, $\mathfrak{H}$ a nilpotent subalgebra. Set $\mathfrak{L}_1 = \mathfrak{Z} + \mathfrak{H}$, which is a subalgebra since $\mathfrak{Z}$ is an ideal. Write $\mathfrak{L}_1 = \mathfrak{H}_1 + \mathfrak{L}_1^{\omega}$, $\mathfrak{H}_1$ a nilpotent subalgebra. Then $\mathfrak{L}_1^{\omega} \subseteq \mathfrak{Z}$ and $\mathfrak{L} = \mathfrak{L}^{\omega} + \mathfrak{H}_1$. If $\mathfrak{C}_1$ is the center of $\mathfrak{H}_1$, $\mathfrak{Z} \cap \mathfrak{C}_1$ is contained in the center of $\mathfrak{L}$ so that $\mathfrak{Z} \cap \mathfrak{C}_1 = 0$. This implies that $\mathfrak{H}_1 \cap \mathfrak{Z} = 0$ since $\mathfrak{H}_1 \cap \mathfrak{Z}$ is an ideal in $\mathfrak{H}_1$. Then $\mathfrak{Z} = \mathfrak{L}_1^{\omega} \subseteq \mathfrak{L}^{\omega}$.

14. Let $\mathfrak{A}$ be subinvariant in $\mathfrak{L}$ and assume that the centralizer of $\mathfrak{A}$ in $\mathfrak{L}$ is 0. Prove that the centralizer $\mathfrak{Z}$ of $\mathfrak{A}^{\omega}$ in $\mathfrak{L}$ is contained in $\mathfrak{A}^{\omega}$.

Sketch of proof: Assume $\mathfrak{Z} \not\subseteq \mathfrak{A}^{\omega}$. Then, by Exercise 13, $\mathfrak{Z} \not\subseteq \mathfrak{A}$. $\mathfrak{Z}$ is an ideal and $\mathfrak{K} = \mathfrak{Z} + \mathfrak{A}$ is a subalgebra; $\mathfrak{K} \supset \mathfrak{A}$. The normalizer of $\mathfrak{A}$ in $\mathfrak{K}$ properly contains $\mathfrak{A}$ and contains a $z \in \mathfrak{Z}$, $\notin \mathfrak{A}$. Then $\mathfrak{B} = \Phi z + \mathfrak{A}$ is a subalgebra such that $\mathfrak{B}^{\omega} = \mathfrak{A}^{\omega}$. The centralizer of $\mathfrak{B}^{\omega}$ in $\mathfrak{B}$ contains $z \notin \mathfrak{B}^{\omega}$. Hence, by Exercise 13, $\mathfrak{B}$ has a non-zero center and so $\mathfrak{A}$ has non-zero centralizer in $\mathfrak{L}$ contrary to assumption.

15. Let $\mathfrak{L}$ be a Lie algebra, $\mathfrak{B}$ a subinvariant subalgebra, $\mathfrak{A}$ an ideal in $\mathfrak{B}$. Show that if the centralizer of $\mathfrak{A}$ in $\mathfrak{B}$ is 0 and the centralizer of $\mathfrak{B}$ in $\mathfrak{L}$ is 0, then the centralizer of $\mathfrak{A}$ in $\mathfrak{L}$ is 0.

16. Let $\mathfrak{L} = \mathfrak{L}_1 \subseteqq \mathfrak{L}_2 \subseteqq \mathfrak{L}_3 \subseteqq \cdots$, $\mathfrak{L}_i = \mathfrak{D}(\mathfrak{L}_{i-1})$ be the tower of derivation algebras for a Lie algebra with 0 center. Show that the centralizer of $\mathfrak{L}_1$ in $\mathfrak{L}_i$ is 0 and that the centralizer of $\mathfrak{L}_1^{\omega}$ in $\mathfrak{L}_i$ is contained in $\mathfrak{L}_1^{\omega}$. Let $\mathfrak{C}(\mathfrak{L}_1^{\omega})$ be the center of $\mathfrak{L}_1^{\omega}$ and $\mathfrak{D}(\mathfrak{L}_1^{\omega})$ the derivation algebra of $\mathfrak{L}_1^{\omega}$. Prove that

$$\dim \mathfrak{L}_i \leqq \dim \mathfrak{C}(\mathfrak{L}_1^{\omega}) + \dim \mathfrak{D}(\mathfrak{L}_1^{\omega})$$

Hence prove that there exists an m such that $\mathfrak{L}_m = \mathfrak{L}_{m+1} = \cdots$. This is *Schenkman's derivation tower theorem.*

17. Let $\mathfrak{L}$ be a Lie algebra of linear transformations in a vector space over a field of characteristic 0. Let $\mathfrak{K}$ be a subinvariant subalgebra of $\mathfrak{L}$ such that every $K \in \mathfrak{K}$ is nilpotent. Prove that $\mathfrak{K}$ is contained in the radical of $\mathfrak{L}^*$ (*Hint*: Use Exercise 1.22.)

18. Let $\mathfrak{L}$ be a nilpotent Lie algebra of linear transformations in a vector space $\mathfrak{M}$ such that $\mathfrak{M}$ is a sum of finite-dimensional subspaces invariant under $\mathfrak{L}$. Show that $\mathfrak{M} = \sum \oplus \mathfrak{M}_\pi$ where the $\mathfrak{M}_\pi$ are the primary components corresponding to the primary functions π: $A \rightarrow \pi(A)$. Show that the $\mathfrak{M}_\pi$ are invariant. Show that if $\mathfrak{N}$ is any invariant subspace, then $\mathfrak{N} = \sum \oplus \mathfrak{N}_\pi$ where $\mathfrak{N}_\pi = \mathfrak{N} \cap \mathfrak{M}_\pi$.

CHAPTER III

Cartan's Criterion and Its Consequences

In this chapter we shall get at the heart of the structure theory of finite-dimensional Lie algebras of characteristic zero. We shall obtain the structure of semi-simple algebras of this type, prove complete reducibility of finite-dimensional representations and prove the Levi radical splitting theorem. All these results are consequences of certain trace criteria for solvability and semi-simplicity. One of these is Cartan's criterion that a finite-dimensional Lie algebra of characteristic 0 is solvable if and only if $\mathrm{tr}\,(\mathrm{ad}\, a)^2 = 0$ for every a in $\mathfrak{L}'$. Clearly, this is a weakening of Engel's condition—that $\mathrm{ad}\, a$ is nilpotent for every $a \in \mathfrak{L}'$. The method which we employ to establish this result and others of the same type is classical and is based on the study of certain nilpotent subalgebras, called Cartan subalgebras. We shall pursue this method further in the next chapter to obtain the classification of simple Lie algebras over algebraically closed fields of characteristic 0.

1. Cartan subalgebras

If $\mathfrak{B}$ is a subalgebra of a Lie algebra $\mathfrak{L}$ then the *normalizer* $\mathfrak{N}$ of $\mathfrak{B}$ is the set of $x \in \mathfrak{L}$ such that $[x\mathfrak{B}] \subseteq \mathfrak{B}$, that is, $[xb] \in \mathfrak{B}$ for every $b \in \mathfrak{B}$. It is immediate that $\mathfrak{N}$ is a subalgebra containing $\mathfrak{B}$, and $\mathfrak{B}$ is an ideal in $\mathfrak{N}$. In fact, as in group theory, $\mathfrak{N}$ is the largest subalgebra in which $\mathfrak{B}$ is contained as an ideal. We now give the following

Definition 1. A subalgebra $\mathfrak{H}$ of a Lie algebra $\mathfrak{L}$ is called a *Cartan subalgebra* if (1) $\mathfrak{H}$ is nilpotent and (2) $\mathfrak{H}$ is its own normalizer in $\mathfrak{L}$.

Let $\mathfrak{H}$ be a nilpotent subalgebra of a finite-dimensional Lie algebra $\mathfrak{L}$ and let $\mathfrak{L} = \mathfrak{L}_0 \oplus \mathfrak{L}_1$ be the Fitting decomposition of $\mathfrak{L}$ relative to $\mathrm{ad}_{\mathfrak{L}}\mathfrak{H}$. We recall that $\mathfrak{L}_0 = \{x \mid x(\mathrm{ad}\, h)^k = 0, h \in \mathfrak{H}$, for some integer $k\}$. We can now establish the following criterion.

Proposition 1. *Let $\mathfrak{H}$ be a nilpotent subalgebra of a finite-dimen-*

sional Lie algebra $\mathfrak{L}$. *Then* $\mathfrak{H}$ *is a Cartan subalgebra if and only if* $\mathfrak{H}$ *coincides with the Fitting component* $\mathfrak{L}_0$ *of* $\mathfrak{L}$ *relative to* ad $\mathfrak{H}$.

Proof: We note first that $\mathfrak{L}_0 \supseteq \mathfrak{N}$, the normalizer of $\mathfrak{H}$. Thus if $x \in \mathfrak{N}$ then $[xh] \in \mathfrak{H}$ for any $h \in \mathfrak{H}$. Since $\mathfrak{H}$ is nilpotent,

$$[\cdots[xh]h]\cdots h] = x(\mathrm{ad}\, h)^k = 0$$

for some k. Hence $x \in \mathfrak{L}_0$. Hence if $\mathfrak{N} \supset \mathfrak{H}$, then $\mathfrak{L}_0 \supset \mathfrak{H}$. Next assume $\mathfrak{L}_0 \supset \mathfrak{H}$. Now $\mathfrak{L}_0$ is invariant under ad $\mathfrak{H}$ and every restriction $\mathrm{ad}_{\mathfrak{L}_0} h$, $h \in \mathfrak{H}$, is nilpotent. Also $\mathfrak{H}$ is an invariant subspace of $\mathfrak{L}_0$ relative to ad $\mathfrak{H}$. Hence we obtain an induced Lie algebra $\bar{\mathfrak{H}}$ of linear transformations acting in the non-zero space $\mathfrak{L}_0/\mathfrak{H}$. Since these transformations are nilpotent, one of the versions of Engel's theorem implies that there exists a non-zero vector $x + \mathfrak{H}$ such that $(x + \mathfrak{H})\bar{\mathfrak{H}} = 0$. This means that we have $[xh] \in \mathfrak{H}$ for every h; hence $x \in \mathfrak{N}$ and $x \notin \mathfrak{H}$ so that $\mathfrak{N} \supset \mathfrak{H}$. Thus $\mathfrak{N} \supset \mathfrak{H}$ if and only if $\mathfrak{L}_0 \supset \mathfrak{H}$, which is what we wished to prove

PROPOSITION 2. *Let* $\mathfrak{H}$ *be a nilpotent subalgebra of the finite-dimensional Lie algebra* $\mathfrak{L}$ *and let* $\mathfrak{L} = \mathfrak{L}_0 \oplus \mathfrak{L}_1$ *be the Fitting decomposition of* $\mathfrak{L}$ *relative to* ad $\mathfrak{H}$. *Then* $\mathfrak{L}_0$ *is a subalgebra and* $[\mathfrak{L}_1\mathfrak{L}_0] \subseteq \mathfrak{L}_1$.

Proof: Let $h \in \mathfrak{H}$ and $a \in \mathfrak{L}_0$. Then $[\cdots[[a\overbrace{h]h]\cdots h}^{k}] = 0$ for some k. Hence

$$[\cdots[\mathrm{ad}\, a\, \mathrm{ad}\, h]\mathrm{ad}\, h]\cdots \mathrm{ad}\, h] = 0\ .$$

This relation and Lemma 1 of § 2.4 implies that the Fitting spaces $\mathfrak{L}_{0\,\mathrm{ad}\,h}$ and $\mathfrak{L}_{1\,\mathrm{ad}\,h}$ of $\mathfrak{L}$ relative to ad h are invariant under ad a. Since $\mathfrak{L}_0 = \bigcap_{h\in\mathfrak{H}}\mathfrak{L}_{0\,\mathrm{ad}\,h}$ and $\mathfrak{L}_1 = \sum_{h\in\mathfrak{H}}\mathfrak{L}_{1\,\mathrm{ad}\,h}$, it follows that $\mathfrak{L}_0\, \mathrm{ad}\, a \subseteq \mathfrak{L}_0$ and $\mathfrak{L}_1\, \mathrm{ad}\, a \subseteq \mathfrak{L}_1$. Since a is any element of $\mathfrak{L}_0$, $[\mathfrak{L}_0\mathfrak{L}_0] \subseteq \mathfrak{L}_0$ and $[\mathfrak{L}_1\mathfrak{L}_0] \subseteq \mathfrak{L}_1$.

DEFINITION 2. An element $h \in \mathfrak{L}$ is called *regular* if the dimensionality of the Fitting null component of $\mathfrak{L}$ relative to ad h is minimal. If this dimensionality is l, then $n - l$, where $n = \dim \mathfrak{L}$, is called the *rank* of $\mathfrak{L}$.

We have seen that the dimensionality of the Fitting null component of a linear transformation A is the multiplicity of the root 0 of the characteristic polynomial $f(\lambda) = \det(\lambda 1 - A)$ of A. Hence h is regular if and only if the multiplicity of the characteristic root 0 of ad h is minimal. Since $[hh] = 0$ for every h it is clear that ad h is singular for every h. Hence the number l of the above definition is > 0. We remark also that $\mathfrak{L}$ is of rank 0 if and only if every $\operatorname{ad} h$ is nilpotent. By Engel's theorem this holds if and only if $\mathfrak{L}$ is a nilpotent Lie algebra. Regular elements can be used to construct Cartan subalgebras, for we have the following

THEOREM 1. *If $\mathfrak{L}$ is a finite-dimensional Lie algebra over an infinite field Φ and a is a regular element of $\mathfrak{L}$, then the Fitting null component $\mathfrak{H}$ of $\mathfrak{L}$ relative to $\operatorname{ad} a$ is a Cartan subalgebra.*

Proof: Let $\mathfrak{L} = \mathfrak{H} \oplus \mathfrak{K}$ be the Fitting decomposition of $\mathfrak{L}$ relative to $\operatorname{ad} a$. Then, by Proposition 2, $\mathfrak{H}$ is a subalgebra and $[\mathfrak{H}\mathfrak{K}] \subseteq \mathfrak{K}$. We assert that every $\operatorname{ad}_{\mathfrak{H}} b$, $b \in \mathfrak{H}$, is nilpotent. Otherwise let b be an element of $\mathfrak{H}$ such that $\operatorname{ad}_{\mathfrak{H}} b$ is not nilpotent. We choose a basis for $\mathfrak{L}$ which consists of a basis for $\mathfrak{H}$ and a basis for $\mathfrak{K}$. Then the matrix of any $\operatorname{ad} h$, $h \in \mathfrak{H}$, relative to this basis has the form

$$(1) \qquad \begin{pmatrix} (\rho_1) & 0 \\ 0 & (\rho_2) \end{pmatrix}$$

where (ρ_1) is a matrix of $\operatorname{ad}_{\mathfrak{H}} h$ and (ρ_2) is a matrix of $\operatorname{ad}_{\mathfrak{K}} h$. Let

$$(2) \qquad A = \begin{pmatrix} (\alpha_1) & 0 \\ 0 & (\alpha_2) \end{pmatrix}, \qquad B = \begin{pmatrix} (\beta_1) & 0 \\ 0 & (\beta_2) \end{pmatrix},$$

respectively be the matrices for $\operatorname{ad} a$ and $\operatorname{ad} b$. Then we know that (α_2) is non-singular; hence $\det(\alpha_2) \neq 0$. Also, by assumption, (β_1) is not nilpotent. Hence if $n - l$ is the rank then $\dim \mathfrak{H} = l$ and the characteristic polynomial of (β_1) is not divisible by λ^l. Now let λ, μ, ν be (algebraically independent) indeterminates and let $F(\lambda, \mu, \nu)$ be the characteristic polynomial $F(\lambda, \mu, \nu) = \det(\lambda 1 - \mu A - \nu B)$. We have $F(\lambda, \mu, \nu) = F_1(\lambda, \mu, \nu)F_2(\lambda, \mu, \nu)$ where

$$F_i(\lambda, \mu, \nu) = \det(\lambda 1 - \mu(\alpha_i) - \nu(\beta_i)) .$$

We have seen that $F_2(\lambda, 1, 0) = \det(\lambda 1 - (\alpha_2))$ is not divisible by λ and $F_1(\lambda, 0, 1) = \det(\lambda 1 - (\beta_1))$ is not divisible by λ^l. Hence the highest power of λ dividing $F(\lambda, \mu, \nu)$ is $\lambda^{l'}$, $l' < l$. Since Φ is infinite we can choose μ_0, ν_0 in Φ such that $F(\lambda, \mu_0, \nu_0)$ is not divisible by $\lambda^{l'+1}$. Set $c = \mu_0 a + \nu_0 b$. Then the characteristic polynomial $\det(\lambda 1 - \operatorname{ad} c) = \det(\lambda 1 - \mu_0 A - \nu_0 B) = F(\lambda, \mu_0, \nu_0)$ is not divisible by $\lambda^{l'+1}$. Hence the multiplicity of the characteristic root 0 of $\operatorname{ad} c$ is $l' < l$. This contradicts the regularity of a. We have therefore proved that for every $b \in \mathfrak{H}$, $\operatorname{ad}_{\mathfrak{H}} b$ is nilpotent. Consequently, by Engel's theorem, $\mathfrak{H}$ is a nilpotent Lie algebra. Let $\mathfrak{L}_0$ be the Fitting null component of $\mathfrak{L}$ relative to ad $\mathfrak{H}$. Then $\mathfrak{L}_0 \subseteq \mathfrak{H}$ since the latter is the Fitting null component of ad a and $a \in \mathfrak{H}$. On the other hand, we always have that $\mathfrak{L}_0 \supseteq \mathfrak{H}$ for a nilpotent subalgebra. Hence $\mathfrak{L}_0 = \mathfrak{H}$, and $\mathfrak{H}$ is a Cartan subalgebra by Proposition 1.

Another useful remark about regular elements and Cartan subalgebras is that if a Cartan subalgebra $\mathfrak{H}$ contains a regular element a, then $\mathfrak{H}$ is uniquely determined by a as the Fitting null component of ad a. Thus if $\mathfrak{N}$ is this component then it is clear that $\mathfrak{N} \supseteq \mathfrak{H}$ since $\mathfrak{H}$ is nilpotent. On the other hand, we have just seen that $\mathfrak{N}$ is nilpotent so that if $\mathfrak{N} \supset \mathfrak{H}$, then $\mathfrak{N}$ contains an element $z \notin \mathfrak{H}$ such that $[z\mathfrak{H}] \subseteq \mathfrak{H}$ (cf. Exercise 1). This contradicts the assumption that $\mathfrak{H}$ is a Cartan subalgebra. An immediate consequence of our result is that if two Cartan subalgebras have a regular element in common then they coincide. We shall see later (Chapter IX) that if Φ is algebraically closed of characteristic zero, then every Cartan subalgebra contains a regular element. We now indicate a fairly concrete way of determining the regular elements assuming again that Φ is infinite. For this purpose we need to introduce the notion of a generic element and the characteristic polynomial of a Lie algebra.

Let $\mathfrak{L}$ be a Lie algebra with basis $(e_1, e_2, \cdots, e_n)$ over the field Φ. Let $\xi_1, \xi_2, \cdots, \xi_n$ be indeterminates and let $P = \Phi(\xi_1, \xi_2 \cdots, \xi_n)$, the field of rational expressions in the ξ_i. We form the extension $\mathfrak{L}_P = Pe_1 + Pe_2 + \cdots + Pe_n$. The element $x = \sum_1^n \xi_i e_i$ of $\mathfrak{L}_P$ is called a *generic element* of $\mathfrak{L}$ and the characteristic polynomial $f_x(\lambda)$ of ad x (in $\mathfrak{L}_P$) is called the *characteristic polynomial of the Lie algebra* $\mathfrak{L}$. If we use the basis $(e_1, e_2, \cdots, e_n)$ for $\mathfrak{L}_P$, then we can write

$$(3) \qquad [e_i x] = \sum_{j=1}^n \rho_{ij} e_j\,, \qquad i = 1, 2, \cdots, n\,,$$

where the ρ_{ij} are homogeneous expressions of degree one in the ξ_k. It follows that

$$(4) \qquad \begin{aligned} f_x(\lambda) &= \det(\lambda 1 - (\rho)) \\ &= \lambda^n - \tau_1(\xi)\lambda^{n-1} + \tau_2(\xi)\lambda^{n-2} - \cdots + (-1)^l \tau_{n-l}(\xi)\lambda^l\,, \end{aligned}$$

where τ_i is a homogeneous polynomial of degree i in the ξ's and $\tau_{n-l}(\xi) \neq 0$ but $\tau_{n-l+k} = 0$, if $k > 0$. Since $x \operatorname{ad} x = 0$ and $x \neq 0$, $\det(\rho) = 0$ and $l > 0$. The characteristic polynomial of any $a = \sum \alpha_i e_i \in \mathfrak{L}$ is obtained by specializing $\xi_i = \alpha_i$, $i = 1, 2, \cdots, n$, in (4). Hence it is clear that the multiplicity of the root 0 for the characteristic polynomial of ad a is at least l. On the other hand, if Φ is an infinite field then, since the polynomial $\tau_{n-l}(\xi) \neq 0$ in the polynomial algebra $\Phi[\xi_1, \xi_2, \cdots, \xi_n]$, we can choose $\xi_i = \alpha_i$ so that $\tau_{n-l}(\alpha) \neq 0$. Then ad a for $a = \sum \alpha_i e_i$ has exactly l characteristic roots 0, and so a is regular. Thus we see that for an infinite field, a is regular if and only if

$$\tau_{n-l}(\alpha) \neq 0 \,. \tag{5}$$

In this sense "almost all" the elements of $\mathfrak{L}$ are regular. (In the sense of algebraic geometry the regular elements form an open set.) It is also clear that $n - l$ is the rank of $\mathfrak{L}$.

All of this depends on the choice of the basis (e). However, it is easy to see what happens if we change to another basis $(f_1, f_2, \cdots, f_n)$ where $f_i = \sum \mu_{ij} e_j$. Thus if $\eta_1, \eta_2, \cdots, \eta_n$ are indeterminates, then $y = \sum \eta_i f_i = \sum \eta_i \mu_{ij} e_j$. Hence the characteristic polynomial $f_y(\lambda)$ is obtained from $f_x(\lambda)$ by the substitutions $\xi_j \to \sum_i \eta_i \mu_{ij}$ in its coefficients (of the powers of λ).

If Ω is any extension field of Φ, then (e) is a basis for $\mathfrak{L}_\Omega$ over Ω. Hence $x = \sum \xi_i e_i$ can be considered also as a generic element of $\mathfrak{L}_\Omega$ and the characteristic polynomial $f_x(\lambda)$ is unchanged on extending the base field Φ to Ω. It is clear from this also that if Φ is infinite, $a \in \mathfrak{L}$ is regular in $\mathfrak{L}$ if and only if a is regular as an element in $\mathfrak{L}_\Omega$. (In either case, (5) is the condition for regularity.) We have seen that the Fitting null component $\mathfrak{H}$ of ad a, a regular is a Cartan subalgebra. The dimensionality of $\mathfrak{H}$ is l, which is the multiplicity of the characteristic root 0. It follows that the Cartan subalgebra determined by a in $\mathfrak{L}_\Omega$ is $\mathfrak{H}_\Omega$.

2. *Products of weight spaces*

It is convenient to carry over the notion of weights and weight spaces for a Lie algebra of linear transformations to an abstract Lie algebra $\mathfrak{L}$ and a representation R of $\mathfrak{L}$. Let $\mathfrak{M}$ be the module for $\mathfrak{L}$. A mapping $a \to \alpha(a)$ of $\mathfrak{L}$ into Φ is called a *weight* of $\mathfrak{M}$ if there exists a non-zero vector x in $\mathfrak{M}$ such that

$$x(a^R - \alpha(a)1)^k = 0 \tag{6}$$

for a suitable k. The set of vectors satisfying this condition together with 0 is a subspace $\mathfrak{M}_\alpha$ called the *weight space* corresponding to the weight α. If $\mathfrak{L}$ is nilpotent, then Lemma 2.1 shows that $\mathfrak{M}_\alpha$ is a submodule. If $\mathfrak{M} = \mathfrak{M}_\alpha$, then we shall say that $\mathfrak{M}$ is a *weight module* for $\mathfrak{L}$ corresponding to the weight α.

Let $\mathfrak{L}$ be a Lie algebra and let $\mathfrak{M}$ be a finite-dimensional weight module for $\mathfrak{L}$ relative to the weight α. Then for any $x \in \mathfrak{M}$, $x(a^R - \alpha(a)1)^k = 0$ if k is sufficiently large. Moreover, if $\dim \mathfrak{M} = n$, then $(\lambda - \alpha(a))^n$ is the characteristic polynomial of a^R. Hence we have $x(a^R - \alpha(a)1)^n = 0$ for all $x \in \mathfrak{M}$. We consider the contragredient module $\mathfrak{M}^*$ which carries the representation R^* satisfying

$$\langle xa^R, y^* \rangle + \langle x, ya^{R^*} \rangle = 0 \,, \tag{7}$$

$x \in \mathfrak{M}$, $y^* \in \mathfrak{M}^*$. We have

$$-\langle \alpha(a)x, y^* \rangle + \langle x, \alpha(a)y^* \rangle = 0 \,, \tag{8}$$

which we can add to (7) to obtain

$$\langle x(a^R - \alpha(a)1), y^* \rangle + \langle x, y^*(a^{R^*} + \alpha(a)1) \rangle = 0 \,. \tag{9}$$

Iteration of this gives

$$\langle x(a^R - \alpha(a)1)^k, y^* \rangle + \langle x, y^*(a^{R^*} + (-1)^{k-1}\alpha(a)1)^k \rangle = 0 \,. \tag{10}$$

If $k = n$, $x(a^R - \alpha(a)1)^n = 0$ for all x and consequently, by (10), $\langle x, y^*(a^{R^*} + \alpha(a)1)^n \rangle = 0$. Hence $y^*(a^{R^*} + \alpha(a)1)^n = 0$ for all $y^* \in \mathfrak{M}^*$. This shows that $\mathfrak{M}^*$ is a weight module with the weight $-\alpha$.

PROPOSITION 3. *If $\mathfrak{M}$ is a finite-dimensional weight module for $\mathfrak{L}$ with the weight α, then the contragredient module $\mathfrak{M}^*$ is a weight module with the weight $-\alpha$.*

We consider next what happens if we take the tensor product of two weight spaces. Thus let $\mathfrak{M}, \mathfrak{N}$ be weight modules of $\mathfrak{L}$ relative to the weights α and β. Let R and S denote the representations in $\mathfrak{M}$ and $\mathfrak{N}$, respectively. Then any $x \in \mathfrak{M}$ satisfies

$$x(a^R - \alpha(a)1)^k = 0 \tag{11}$$

for some positive integer k, and every $y \in \mathfrak{N}$ satisfies

$$y(a^S - \beta(a)1)^{k'} = 0 \tag{12}$$

for some positive integer k'. Let $\mathfrak{P} = \mathfrak{M} \otimes \mathfrak{N}$ and denote the representation of $\mathfrak{L}$ in $\mathfrak{P}$ by T. Then we have

$$(x \otimes y)a^T = xa^R \otimes y + x \otimes ya^S \tag{13}$$

or $a^T = a^R \otimes 1 + 1 \otimes a^S$. Hence

$$(14)\qquad a^T - (\alpha(a) + \beta(a))1 \\ = (a^R \otimes 1 - \alpha(a)1 \otimes 1) + (1 \otimes a^S - 1 \otimes \beta(a)1)\,.$$

Since the two transformations in the parentheses commute we can apply the binomial theorem to obtain

$$(15)\qquad (a^T - (\alpha(a) + \beta(a))1)^m \\ = \sum_{i=0}^{m} \binom{m}{i} (a^R \otimes 1 - \alpha(a) \otimes 1)^i (1 \otimes a^S - 1 \otimes \beta(a))^{m-i}\,.$$

If we apply this to $x \otimes y$, we obtain

$$(16)\qquad (x \otimes y)(a^T - (\alpha(a) + \beta(a))1)^m \\ = \sum_{i=0}^{m} \binom{m}{i} x(a^R - \alpha(a)1)^i \otimes y(a^S - \beta(a)1)^{m-i}\,.$$

If we take $m = k + k' - 1$, then for every i, either $x(a^R - \alpha(a)1)^i = 0$ or $y(a^S - \beta(a)1)^{m-i} = 0$. Hence $(x \otimes y)(a^T - (\alpha(a) + \beta(a))1)^m = 0$ and we have the following

PROPOSITION 4. *If $\mathfrak{M}$ and $\mathfrak{N}$ are weight modules for $\mathfrak{L}$ for the weights α and β, respectively, then $\mathfrak{P} = \mathfrak{M} \otimes \mathfrak{N}$ is a weight module with the weight $\alpha + \beta$.*

We suppose now that $\mathfrak{L}$ is a finite-dimensional Lie algebra, $\mathfrak{H}$ a nilpotent subalgebra, and $\mathfrak{M}$ is a finite-dimensional module for $\mathfrak{L}$, hence for $\mathfrak{H}$. If R denotes the representation in $\mathfrak{M}$ and $\mathrm{ad}_{\mathfrak{L}}$ the adjoint representation in $\mathfrak{L}$, then we assume that $\mathfrak{H}^R$ and $\mathrm{ad}_{\mathfrak{L}}\mathfrak{H}$ are split Lie algebras of linear transformations, that is, the characteristic roots of h^R and $\mathrm{ad}_{\mathfrak{L}} h$, $h \in \mathfrak{H}$, are in the base field Φ. This will be automatically satisfied if Φ is algebraically closed. If $h^R \rightarrow \alpha(h^R)$ is a weight on $\mathfrak{H}^R$, then $h \rightarrow \alpha(h) \equiv \alpha(h^R)$ is a weight for $\mathfrak{H}$ in the module $\mathfrak{M}$. The result on weight spaces for a split nilpotent Lie algebra of linear transformations (Theorem 2.7) implies that $\mathfrak{M}$ is a direct sum of weight modules $\mathfrak{M}_\rho$. Similarly, we have a decomposition of $\mathfrak{L}$ into weight modules $\mathfrak{L}_\alpha$. Thus we have

$$(17)\qquad \mathfrak{M} = \mathfrak{M}_\rho \oplus \mathfrak{M}_\sigma \oplus \cdots \oplus \mathfrak{M}_\tau$$

and

$$(18)\qquad \mathfrak{L} = \mathfrak{L}_\alpha \oplus \mathfrak{L}_\beta \oplus \cdots \oplus \mathfrak{L}_\delta$$

where $\rho, \sigma, \cdots,$ are mappings of $\mathfrak{H}$ into Φ such that if $x_\rho \in \mathfrak{M}_\rho$ then $x_\rho(h^R - \rho(h)1)^m = 0$ for some m, etc. The weights $\alpha, \beta, \cdots,$ associated with $\mathrm{ad}_{\mathfrak{L}}\mathfrak{H}$ will be called the *roots* of $\mathfrak{H}$ in $\mathfrak{L}$. Since $h \to h^R$ and $h \to \mathrm{ad}_{\mathfrak{L}} h$ are linear, it is clear that in the characteristic 0 case the weights $\rho, \cdots, \tau$ and the roots $\alpha, \cdots, \delta$ are linear functions on $\mathfrak{H}$ which vanish on $\mathfrak{H}'$.

The following important result relates the decompositions (17) and (18) relative to $\mathfrak{H}$.

PROPOSITION 5. *$\mathfrak{M}_\rho\mathfrak{L}_\alpha \subseteq \mathfrak{M}_{\rho+\alpha}$ if $\rho + \alpha$ is a weight of $\mathfrak{M}$ relative to $\mathfrak{H}$; otherwise $\mathfrak{M}_\rho\mathfrak{L}_\alpha = 0$.*

Proof: The elements of $\mathfrak{M}_\rho\mathfrak{L}_\alpha$ are of the form $\sum_i x_\rho^{(i)} a_\alpha^{(i)}$, $x_\rho^{(i)} \in \mathfrak{M}_\rho$, $a_\alpha^{(i)} \in \mathfrak{L}_\alpha$. The characteristic property of the tensor product of two spaces shows that we have a linear mapping π of $\mathfrak{M}_\rho \otimes \mathfrak{L}_\alpha$ onto $\mathfrak{M}_\rho\mathfrak{L}_\alpha$ such that

$$(\sum_i x_\rho^{(i)} \otimes a_\alpha^{(i)})\pi = \sum_i x_\rho^{(i)} a_\alpha^{(i)} .$$

We shall now show that π is in fact a homomorphism for the $\mathfrak{H}$-modules. Thus let $h \in \mathfrak{H}$. Then we have

$$(x_\rho \otimes a_\alpha)h = x_\rho h \otimes a_\alpha + x_\rho \otimes [a_\alpha h] \xrightarrow[\pi]{} (x_\rho h)a_\alpha + x_\rho[a_\alpha h] = (x_\rho a_\alpha)h .$$

On the other hand, the image of $x_\rho \otimes a_\alpha$ under π is $x_\rho a_\alpha$. Following this with the module product by h we again obtain $(x_\rho a_\alpha)h$. We have therefore proved that $\mathfrak{M}_\rho\mathfrak{L}_\alpha$ is a homomorphic image of $\mathfrak{M}_\rho \otimes \mathfrak{L}_\alpha$. Moreover, the latter is a weight module for the weight $\rho + \alpha$. Now it is clear from the definition that the homomorphic image of a weight module with weight β is either 0 or it is a weight module with weight β. The result stated follows from this.

If we apply the last result to the case in which $\mathfrak{M} = \mathfrak{L}$ and the representation is the adjoint representation of $\mathfrak{L}$, we obtain the

COROLLARY. *$[\mathfrak{L}_\alpha\mathfrak{L}_\beta] \subseteq \mathfrak{L}_{\alpha+\beta}$ if $\alpha + \beta$ is a root, and $[\mathfrak{L}_\alpha\mathfrak{L}_\beta] = 0$ otherwise.*

3. *An example*

Before plunging into the structure theory it will be well to look at an example. Let $\mathfrak{M}$ be an n-dimensional vector space over an algebraically closed field Φ, $\mathfrak{E}$, the associative algebra of linear transformations in $\mathfrak{M}$, $\mathfrak{L} = \mathfrak{E}_L$, the corresponding Lie algebra. We wish to determine the regular elements of $\mathfrak{L}$ and the corresponding Cartan subalgebras.

Let $A \in \mathfrak{L}$ and let $\mathfrak{M} = \mathfrak{M}_{\alpha_1} \oplus \mathfrak{M}_{\alpha_2} \oplus \cdots \oplus \mathfrak{M}_{\alpha_r}$ be the decomposition of $\mathfrak{M}$ into the weight spaces relative to ΦA. The α_i are distinct elements of Φ and these are just the characteristic roots of the linear transformation A and the $\mathfrak{M}_{\alpha_i}$ are the corresponding characteristic spaces. It is well known that a decomposition of a space as a direct sum leads to a decomposition of the dual space into a direct sum of subspaces which can be considered as the conjugate spaces of the components. Hence we have

$$\mathfrak{M}^* = \mathfrak{M}_1^* \oplus \mathfrak{M}_2^* \oplus \cdots \oplus \mathfrak{M}_r^*$$

where $\mathfrak{M}_i^*$ can be identified with the conjugate space of $\mathfrak{M}_{\alpha_i}$. $\mathfrak{M}_i^*$ is invariant under $-A^*$ and Proposition 3 shows that this is a weight module for the Lie algebra ΦA with the weight $-\alpha_i$. Accordingly, we write $\mathfrak{M}_{-\alpha_i}^*$ for $\mathfrak{M}_i^*$ and we have

$$\mathfrak{M}^* = \mathfrak{M}_{-\alpha_1}^* \oplus \mathfrak{M}_{-\alpha_2}^* \oplus \cdots \oplus \mathfrak{M}_{-\alpha_r}^* .$$

We now consider the module $\mathfrak{M} \otimes \mathfrak{M}^*$ relative to ΦA. As is well known, we have $\mathfrak{M} \otimes \mathfrak{M}^* = \sum_{i,j=1}^{r} \mathfrak{M}_{\alpha_i} \otimes \mathfrak{M}_{-\alpha_j}^*$. By Proposition 4, $\mathfrak{M}_{\alpha_i} \otimes \mathfrak{M}_{-\alpha_j}^*$ is a weight module for ΦA for the weight $\alpha_i - \alpha_j$. If A_{ij} denotes the linear transformation in $\mathfrak{M}_{\alpha_i} \otimes \mathfrak{M}_{-\alpha_j}^*$ corresponding to A, then $(A_{ij} - (\alpha_i - \alpha_j)1)^{k_{ij}} = 0$ for suitable k_{ij}. Hence A_{ij} is non-singular if $\alpha_i - \alpha_j \neq 0$ and A_{ij} is nilpotent if $\alpha_i = \alpha_j$. This implies that the Fitting null component of $\mathfrak{M} \otimes \mathfrak{M}^*$ relative to the linear transformation $\tilde{A}$ corresponding to A ($A \to \tilde{A}$ is the representation of $\mathfrak{L}$ in $\mathfrak{M} \otimes \mathfrak{M}^*$) is $\sum_{i=1}^{r} \mathfrak{M}_{\alpha_i} \otimes \mathfrak{M}_{-\alpha_i}^*$. If $\dim \mathfrak{M}_{\alpha_i} = n_i$, then the dimensionality of this Fitting space is $\sum_1^r n_i^2$. Since $\sum n_i = n$, $\sum n_i^2$ is minimal if and only if every $n_i = 1$, which is equivalent to saying that A has n distinct characteristic roots. Now, we recall that the module $\mathfrak{M} \otimes \mathfrak{M}^*$ is isomorphic to the module $\mathfrak{L} = \mathfrak{E}_L$ relative to the adjoint mapping $X \to [XA]$. Hence we see that *the dimensionality of the Fitting null component of* $\mathfrak{E}$ *relative to* ad A *is minimal if and only if* A *has* n *distinct characteristic roots*. Thus A is regular in $\mathfrak{L} = \mathfrak{E}_L$ if and only if A has n distinct characteristic roots. The corresponding Cartan subalgebra $\mathfrak{H}$ is the Fitting null component of $\mathfrak{L}$ relative to ad A and $\dim \mathfrak{H} = n$.

Since A has n distinct characteristic roots we can choose a basis for $\mathfrak{M}$ such that the matrix of A relative to this basis is diagonal. Let $\mathfrak{H}_1$ be the set of linear transformations whose matrices relative to this basis are diagonal. Then $\dim \mathfrak{H}_1 = n$ and $[HA] = 0$ for every $H \in \mathfrak{H}_1$. Hence $\mathfrak{H}_1 \subseteq \mathfrak{H}$. On the other hand, $\dim \mathfrak{H} = n$. Hence

$\mathfrak{H}_1 = \mathfrak{H}$ and the Cartan subalgebra determined by A is just the centralizer of the linear transformation A.

Let $(e_1, e_2, \cdots, e_n)$ be a basis such that $e_iA = \alpha_ie_i$. Then we have seen that the Cartan subalgebra $\mathfrak{H}$ is the set of linear transformations which have diagonal matrices relative to the basis (e_i). Hence if $H \in \mathfrak{H}$ then we have $e_iH = \lambda_i(H)e_i$, $\lambda_i(H) \in \Phi$, and we may assume the notation is such that $\lambda_i(A) = \alpha_i$. The spaces Φe_i $(= \mathfrak{M}_{\alpha_i})$ are weight spaces and Φe_i corresponds to the weight $\lambda_i(H)$. We shall therefore write $\mathfrak{M}_{\lambda_i}$ for Φe_i. As for the single linear transformation A we can now write $\mathfrak{M}^* = \mathfrak{M}^*_{-\lambda_1} \oplus \mathfrak{M}^*_{-\lambda_2} \oplus \cdots \oplus \mathfrak{M}^*_{-\lambda_n}$ and $\mathfrak{M} \otimes \mathfrak{M}^* = \sum_{i,j} \oplus (\mathfrak{M}_{\lambda_i} \otimes \mathfrak{M}^*_{-\lambda_j})$. $\mathfrak{M}_{\lambda_i} \otimes \mathfrak{M}^*_{-\lambda_j}$ is a weight space for the Cartan subalgebra $\mathfrak{H}$ corresponding to the weight $\lambda_i - \lambda_j$. We can summarize our results in the following

THEOREM 2. *Let $\mathfrak{L} = \mathfrak{E}_L$, the Lie algebra of linear transformations in an n-dimensional vector space $\mathfrak{M}$ over an algebraically closed field Φ. Then $A \in \mathfrak{L}$ is a regular element if and only if the characteristic polynomial det $(\lambda 1 - A)$ has n distinct roots. The Cartan subalgebra $\mathfrak{H}$ determined by A is the set of linear transformations H such that $[HA] = 0$. If the weights of H acting in $\mathfrak{M}$ are $\lambda_1(H), \cdots, \lambda_n(H)$ then the roots (weights of the adjoint representation in $\mathfrak{L}$) are $\lambda_i(H) - \lambda_j(H)$, $i, j = 1, \cdots, n$.*

If Φ is not algebraically closed but is infinite then the extension field argument of § 1 shows that again A is regular in $\mathfrak{E}_L$ if and only if A has n distinct characteristic roots (in the algebraic closure Ω of Φ). It is well known that the centralizer of such an A is the algebra $\Phi[A]$ of polynomials in A and dim $\Phi[A] = n$. Since the dimensionality of the Cartan subalgebra determined by A is n, it follows that this subalgebra is $\Phi[A]$.

If $\mathfrak{L}$ is the orthogonal or symplectic Lie algebra in a $2l$-dimensional space over an infinite field one can show that the rank is $N - l$, $N = \dim \mathfrak{L}$ (cf. Exercises 4 and 5). It is not difficult to determine Cartan subalgebras for the other important examples of Lie algebras which we have encountered (cf. Exercise 3).

4. *Cartan's criteria*

We now consider a finite-dimensional Lie algebra over an algebraically closed field Φ, a nilpotent subalgebra $\mathfrak{H}$ of $\mathfrak{L}$ and $\mathfrak{M}$ a module for $\mathfrak{L}$, which is finite-dimensional over Φ. Let

$$\mathfrak{M} = \mathfrak{M}_\rho \oplus \mathfrak{M}_\sigma \oplus \cdots \oplus \mathfrak{M}_\tau\,, \tag{19}$$
$$\mathfrak{L} = \mathfrak{L}_\alpha \oplus \mathfrak{L}_\beta \oplus \cdots \oplus \mathfrak{L}_\gamma$$

be the decompositions of $\mathfrak{M}$ and $\mathfrak{L}$ into weight modules relative to $\mathfrak{H}$. We have seen that

$$\mathfrak{M}_\rho \mathfrak{L}_\alpha = \begin{cases} 0 \text{ if } \rho + \alpha \text{ is not a weight of } \mathfrak{M} \\ \subseteq \mathfrak{M}_{\rho+\alpha} \text{ if } \rho + \alpha \text{ is a weight} \end{cases} \tag{20}$$
$$[\mathfrak{L}_\alpha \mathfrak{L}_\beta] = \begin{cases} 0 \text{ if } \alpha + \beta \text{ is not a root} \\ \subseteq \mathfrak{L}_{\alpha+\beta} \text{ if } \alpha + \beta \text{ is a root}\,. \end{cases}$$

Now suppose $\mathfrak{H}$ is a Cartan subalgebra. Then $\mathfrak{H} = \mathfrak{L}_0$, the root module corresponding to the root 0. Also we have $\mathfrak{L}' = [\mathfrak{L}\mathfrak{L}] = \sum[\mathfrak{L}_\alpha \mathfrak{L}_\beta]$ where the sum is taken over all the roots α, β. The formula for $[\mathfrak{L}_\alpha \mathfrak{L}_\beta]$ shows that

$$\mathfrak{H} \cap \mathfrak{L}' = \sum_\alpha [\mathfrak{L}_\alpha \mathfrak{L}_{-\alpha}]\,, \tag{21}$$

where the summation is taken over all the α such that $-\alpha$ is also a root (e.g., $\alpha = 0$).

We now prove the following

Lemma 1. *Let Φ be algebraically closed of characteristic 0, $\mathfrak{H}$ a Cartan subalgebra of $\mathfrak{L}$ over Φ, $\mathfrak{M}$ a module for $\mathfrak{L}$. Suppose α is a root such that $-\alpha$ is also a root. Let $e_\alpha \in \mathfrak{L}_\alpha$, $e_{-\alpha} \in \mathfrak{L}_{-\alpha}$, $h_\alpha = [e_\alpha e_{-\alpha}]$. Then $\rho(h_\alpha)$ is a rational multiple of $\alpha(h_\alpha)$ for every weight ρ of $\mathfrak{H}$ in $\mathfrak{M}$.*

Proof: Consider the functions of the form $\rho(h) + i\alpha(h)$, $i = 0, \pm 1, \pm 2, \cdots$, which are weights, and form the subspace $\mathfrak{N} = \sum_i \mathfrak{M}_{\rho+i\alpha}$ summed over the corresponding weight spaces. This space is invariant relative to $\mathfrak{H}$ and, by (20), it is also invariant relative to the linear transformations e_α^R, $e_{-\alpha}^R$, where R is the representation of $\mathfrak{L}$ determined by $\mathfrak{M}$. Hence, if $\mathrm{tr}_{\mathfrak{N}}$ denotes the trace of an induced mapping in $\mathfrak{N}$, then $\mathrm{tr}_{\mathfrak{N}} h_\alpha^R = \mathrm{tr}_{\mathfrak{N}}[e_\alpha^R, e_{-\alpha}^R] = 0$. On the other hand, the restriction of h_α^R to $\mathfrak{M}_\sigma$ has the single characteristic root $\sigma(h_\alpha)$. Hence

$$0 = \mathrm{tr}_{\mathfrak{N}} h_\alpha^R = \sum_i n_{\rho+i\alpha}(\rho + i\alpha)(h_\alpha)$$

where, in general, $n_\sigma = \dim \mathfrak{M}_\sigma$. Thus we have

$$(\sum_i n_{\rho+i\alpha})\rho(h_\alpha) + (\sum_i n_{\rho+i\alpha} i)\alpha(h_\alpha) = 0\,.$$

Since $\sum_i n_{\rho+i\alpha}$ is a positive integer this shows that $\rho(h_\alpha)$ is a rational multiple of $\alpha(h_\alpha)$.

We can now prove

Cartan's criterion for solvable Lie algebras. Let $\mathfrak{L}$ be a finite-dimensional Lie algebra over a field of characteristic 0. *Suppose $\mathfrak{L}$ has a finite-dimensional module $\mathfrak{M}$ such that* (1) *the kernel $\mathfrak{K}$ of the associated representation R is solvable and* (2) $\operatorname{tr}(a^R)^2 = 0$ *for every $a \in \mathfrak{L}'$. Then $\mathfrak{L}$ is solvable.*

Proof: Assume first that the base field Φ is algebraically closed. It suffices to prove that $\mathfrak{L}' \subset \mathfrak{L}$; for, conditions (1) and (2) carry over to $\mathfrak{L}'$ and $\mathfrak{M}$ as $\mathfrak{L}'$-module. Hence we shall have that

$$\mathfrak{L} \supset \mathfrak{L}' \supset \mathfrak{L}'' \supset \cdots \supset \mathfrak{L}^{(k)} = 0 .$$

We therefore suppose that $\mathfrak{L}' = \mathfrak{L}$. Let $\mathfrak{H}$ be a Cartan subalgebra and let the decomposition of $\mathfrak{M}$ and $\mathfrak{L}$ relative to $\mathfrak{H}$ be as in (19). Then (21) implies that $\mathfrak{H} = \sum[\mathfrak{L}_\alpha \mathfrak{L}_{-\alpha}]$ summed on the α such that $-\alpha$ is also a root. Choose such an α, let $e_\alpha \in \mathfrak{L}_\alpha$, $e_{-\alpha} \in \mathfrak{L}_{-\alpha}$, and consider the element $h_\alpha = [e_\alpha e_{-\alpha}]$. The formula $\mathfrak{H} = \sum[\mathfrak{L}_\alpha \mathfrak{L}_{-\alpha}]$ implies that every element of $\mathfrak{H}$ is a sum of terms of the form $[e_\alpha e_{-\alpha}]$. The restriction of h_α^R to $\mathfrak{M}_\rho$ has the single characteristic root $\rho(h_\alpha)$. Hence the restriction of $(h_\alpha^R)^2$ has the single characteristic root $\rho(h_\alpha)^2$, and if n_ρ is the dimensionality of $\mathfrak{M}_\rho$, then we have

$$0 = \operatorname{tr}(h_\alpha^R)^2 = \sum n_\rho \rho(h_\alpha)^2 .$$

By the lemma, $\rho(h_\alpha) = r_\rho \alpha(h_\alpha)$, r_ρ rational. Hence $\alpha(h_\alpha)^2(\sum n_\rho r_\rho^2) = 0$. Since the n_ρ are positive integers, this implies that $\alpha(h_\alpha) = 0$ and hence $\rho(h_\alpha) = 0$. Since the ρ are linear functions and every $h \in \mathfrak{H}$ is a sum of elements of the form $h_\alpha, h_\beta, \cdots$, etc., we see that $\rho(h) = 0$. Thus 0 is the only weight for $\mathfrak{M}$; that is, we have $\mathfrak{M} = \mathfrak{M}_0$. If α is a root then (20) now implies that $\mathfrak{M}\mathfrak{L}_\alpha = 0$ for every $\alpha \neq 0$. This means the kernel $\mathfrak{K}$ of R contains all the $\mathfrak{L}_\alpha$, $\alpha \neq 0$. Hence $\mathfrak{L}^R \cong \mathfrak{L}/\mathfrak{K}$ is a homomorphic image of $\mathfrak{H}$. Thus $\mathfrak{L}^R$ is nilpotent and $\mathfrak{L}$ is solvable contrary to $\mathfrak{L}' = \mathfrak{L}$.

If the base field Φ is not algebraically closed, then let Ω be its algebraic closure. Then $\mathfrak{M}_\Omega$ is a module for $\mathfrak{L}_\Omega$ and $\mathfrak{K}_\Omega$ is the kernel of the corresponding representation. Since $\mathfrak{K}$ is solvable, $\mathfrak{K}_\Omega$ is solvable. Next we note that the condition $\operatorname{tr}(a^R)^2 = 0$ and $\operatorname{tr} a^R b^R = \operatorname{tr} b^R a^R$ imply that

$$\operatorname{tr} a^R b^R = \frac{1}{2} \operatorname{tr}(a^R b^R + b^R a^R)$$
$$= \frac{1}{2} [\operatorname{tr}(a^R + b^R)^2 - \operatorname{tr}(a^R)^2 - \operatorname{tr}(b^R)^2] = 0 .$$

Hence if the $a_i \in \mathfrak{L}$ and $\omega_i \in \Omega$, then $\operatorname{tr}(\sum \omega_i a_i^R)^2 = \sum \omega_i \omega_j \operatorname{tr} a_i^R a_j^R = 0$. Hence the condition (2) holds also in $\mathfrak{L}_\Omega$. The first part of the proof therefore implies that $\mathfrak{L}_\Omega$ is solvable. Hence $\mathfrak{L}$ is solvable and the proof is complete.

COROLLARY. *If Φ is of characteristic* 0 *then $\mathfrak{L}$ is solvable if and only if* $\operatorname{tr}(\operatorname{ad} a)^2 = 0$ *for every $a \in \mathfrak{L}'$.*

Proof: The sufficiency of the condition is a consequence of Cartan's criterion since the kernel of the adjoint representation is the center. Conversely, assume $\mathfrak{L}$ solvable. Then, by Corollary 2 to Theorem 2.8, applied to $\operatorname{ad} \mathfrak{L}$, the elements $\operatorname{ad}_{\mathfrak{L}} a$, $a \in \mathfrak{L}'$, are in the radical of $(\operatorname{ad} \mathfrak{L})^*$. Hence $\operatorname{ad}_{\mathfrak{L}} a$ is nilpotent and $\operatorname{tr}(\operatorname{ad}_{\mathfrak{L}} a)^2 = 0$.

Let R be a representation of a Lie algebra in a finite-dimensional space $\mathfrak{M}$. Then the function

(22) $$f(a, b) \equiv \operatorname{tr} a^R b^R$$

is evidently a symmetric bilinear form on $\mathfrak{M}$ with values in Φ. Such a form will be called a *trace form* for $\mathfrak{L}$. In particular, if $\mathfrak{L}$ is finite-dimensional, then we have the trace form $\operatorname{tr}(\operatorname{ad} a)(\operatorname{ad} b)$, which we shall call *the Killing form* of $\mathfrak{L}$. If f is the trace form determined by the representation R then

$$\begin{aligned} f([ac], b) + f(a, [bc]) &= \operatorname{tr}([ac]^R b^R + a^R [bc]^R) \\ &= \operatorname{tr}([a^R c^R] b^R + a^R [b^R c^R]) \\ &= \operatorname{tr}[a^R b^R, c^R] = 0 . \end{aligned}$$

A bilinear form $f(a, b)$ on $\mathfrak{L}$ which satisfies this condition

(23) $$f([ac], b) + f(a, [bc]) = 0$$

is called an *invariant* form on $\mathfrak{L}$. Hence we have verified that trace forms are invariant. We note next that if $f(a, b)$ is any symmetric invariant form on $\mathfrak{L}$, then the radical $\mathfrak{L}^\perp$ of the form; that is, the set of elements z such that $f(a, z) = 0$ for all $a \in \mathfrak{L}$, is an ideal. This is clear since $f(a, [zb]) = -f([ab], z) = 0$.

We can now derive

Cartan's criterion for semi-simplicity. If $\mathfrak{L}$ is a finite-dimensional *semi-simple Lie algebra over a field of characteristic* 0, *then the trace form of any* 1:1 *representation of* $\mathfrak{L}$ *is non-degenerate. If the Killing form is non-degenerate, then* $\mathfrak{L}$ *is semi-simple.*

Proof: Let R be a 1 : 1 representation of $\mathfrak{L}$ in a finite-dimensional space $\mathfrak{M}$ and let $f(a, b)$ be the associated trace form. Then $\mathfrak{L}^{\perp}$ is an ideal of $\mathfrak{L}$ and $f(a, a) = \mathrm{tr}(a^R)^2 = 0$ for every $a \in \mathfrak{L}^{\perp}$. Hence $\mathfrak{L}^{\perp}$ is solvable by the first Cartan criterion. Since $\mathfrak{L}$ is semi-simple, $\mathfrak{L}^{\perp} = 0$ and $f(a, b)$ is non-degenerate. Next suppose that $\mathfrak{L}$ is not semi-simple. Then $\mathfrak{L}$ has an abelian ideal $\mathfrak{B} \neq 0$. If we choose a basis for $\mathfrak{L}$ such that the first vectors form a basis for $\mathfrak{B}$, then the matrices of $\mathrm{ad}\, a$, $a \in \mathfrak{L}$, and $\mathrm{ad}\, b$, $b \in \mathfrak{B}$, are, respectively, of the forms

$$\begin{pmatrix} * & 0 \\ * & * \end{pmatrix}, \quad \begin{pmatrix} 0 & 0 \\ * & 0 \end{pmatrix}.$$

This implies that $\mathrm{tr}(\mathrm{ad}\, b)(\mathrm{ad}\, a) = 0$. Hence $\mathfrak{B} \subseteq \mathfrak{L}^{\perp}$ and the Killing form is degenerate.

If $f(a, b)$ is a symmetric bilinear form in a finite-dimensional space and $(e_1, e_2, \cdots, e_n)$ is a basis for the space, then it is well known that f is non-degenerate if and only if $\det(f(e_i, e_j)) \neq 0$. If $\mathfrak{L}$ is a finite-dimensional Lie algebra of characteristic zero with basis $(e_1, e_2, \cdots, e_n)$ and we set $\beta_{ij} = \mathrm{tr}\, \mathrm{ad}\, e_i\, \mathrm{ad}\, e_j$, then $\mathfrak{L}$ is semi-simple if and only if $\det(\beta_{ij}) \neq 0$. This is the determinant form of Cartan's criterion which we have just proved. If Ω is an extension of the base field of $\mathfrak{L}$ then $(e_1, e_2, \cdots, e_n)$ is a basis for $\mathfrak{L}_\Omega$ over Ω. Hence it is clear that we have the following consequence of our criterion.

COROLLARY. *A finite-dimensional Lie algebra* $\mathfrak{L}$ *over a field* Φ *of characteristic zero is semi-simple if and only if* $\mathfrak{L}_\Omega$ *is semi-simple for every extension field* Ω *of* Φ.

5. Structure of semi-simple algebras

We are now in a position to obtain the main structure theorem on semi-simple Lie algebras. The proof of this result which we shall give is a simplification, due to Dieudonné, of Cartan's original proof. The argument is actually applicable to arbitrary non-associative algebras and we shall give it in this general form.

Let $\mathfrak{A}$ be a non-associative algebra over a field Φ. A bilinear form $f(a, b)$ on $\mathfrak{A}$ (to Φ) is called *associative* if

$$f(ac, b) = f(a, cb) . \tag{24}$$

If f is an invariant form in a Lie algebra then $f([ac], b) + f(a, [bc]) = 0$. Hence $f([ac], b) - f(a, [cb]) = 0$ and f is associative. Now let $f(a, b)$ be a symmetric associative bilinear form on $\mathfrak{A}$ and let $\mathfrak{B}$ be an ideal in $\mathfrak{A}$. Let $a \in \mathfrak{B}^\perp$ so that $f(a, b) = 0$ for all $b \in \mathfrak{B}$. Then for any c in $\mathfrak{A}$, $f(ac, b) = f(a, cb) = 0$ since $cb \in \mathfrak{B}$. Also $f(ca, b) = f(b, ca) = f(bc, a) = 0$ since $bc \in \mathfrak{B}$. Hence $\mathfrak{B}^\perp$ is an ideal.

The importance of associative forms is indicated in the following result.

THEOREM 3. *Let $\mathfrak{A}$ be a finite-dimensional non-associative algebra over a field Φ such that* (1) *$\mathfrak{A}$ has a non-degenerate symmetric associative form f and* (2) *$\mathfrak{A}$ has no ideals $\mathfrak{B}$ with $\mathfrak{B}^2 = 0$. Then $\mathfrak{A}$ is a direct sum of ideals which are simple algebras.*

(We recall that $\mathfrak{A}$ simple means that $\mathfrak{A}$ has no ideals $\neq 0$, $\mathfrak{A}$, and $\mathfrak{A}^2 \neq 0$.)

Proof: Let $\mathfrak{B}$ be a minimal ideal ($\neq 0$) in $\mathfrak{A}$. Then $\mathfrak{B} \cap \mathfrak{B}^\perp$ is an ideal contained in $\mathfrak{B}$. Hence either $\mathfrak{B} \cap \mathfrak{B}^\perp = \mathfrak{B}$ or $\mathfrak{B} \cap \mathfrak{B}^\perp = 0$. Suppose the first case holds and let $b_1, b_2 \in \mathfrak{B}$, $a \in \mathfrak{A}$. Then $f(b_1b_2, a) = f(b_1, b_2a) = 0$. Since f is non-degenerate $b_1b_2 = 0$ and $\mathfrak{B}^2 = 0$ contrary to hypothesis. Hence $\mathfrak{B} \cap \mathfrak{B}^\perp = 0$. It is well known that this implies that $\mathfrak{A} = \mathfrak{B} \oplus \mathfrak{B}^\perp$ and $\mathfrak{B}^\perp$ is an ideal. This decomposition implies that $\mathfrak{B}\mathfrak{B}^\perp = 0 = \mathfrak{B}^\perp\mathfrak{B}$; hence every $\mathfrak{B}$-ideal is an ideal. Consequently, $\mathfrak{B}$ is simple. Moreover, $\mathfrak{B}^\perp$ satisfies the same conditions as $\mathfrak{A}$ since the restriction of f to $\mathfrak{B}^\perp$ is non-degenerate and any $\mathfrak{B}^\perp$-ideal is an ideal. Hence, induction on $\dim \mathfrak{A}$ implies that $\mathfrak{B}^\perp = \mathfrak{A}_2 \oplus \cdots \oplus \mathfrak{A}_r$ where the $\mathfrak{A}_i$ are ideals and are simple algebras. Then for $\mathfrak{A}_1 = \mathfrak{B}$ we have $\mathfrak{A} = \mathfrak{A}_1 \oplus \mathfrak{A}_2 \oplus \cdots \oplus \mathfrak{A}_r$, $\mathfrak{A}_i$ simple and ideals.

This result and the non-degeneracy of the Killing form for a semi-simple Lie algebra of characteristic zero imply the difficult half of the fundamental

Structure theorem. A finite-dimensional Lie algebra over a field of characteristic 0 *is semi-simple if and only if $\mathfrak{L} = \mathfrak{L}_1 \oplus \mathfrak{L}_2 \oplus \cdots \oplus \mathfrak{L}_r$ where the $\mathfrak{L}_i$ are ideals which are simple algebras.*

Proof: If $\mathfrak{L}$ is semi-simple, then $\mathfrak{L}$ has the structure indicated. Conversely, suppose $\mathfrak{L} = \mathfrak{L}_1 \oplus \mathfrak{L}_2 \oplus \cdots \oplus \mathfrak{L}_r$, $\mathfrak{L}_i$ ideals and simple. We consider the set of linear transformations $\operatorname{ad} \mathfrak{L} = \{\operatorname{ad} a \mid a \in \mathfrak{L}\}$

acting in $\mathfrak{L}$. The invariant subspaces relative to this set are the ideals of $\mathfrak{L}$. Since $\mathfrak{L} = \mathfrak{L}_1 \oplus \mathfrak{L}_2 \oplus \cdots \oplus \mathfrak{L}_r$ where the $\mathfrak{L}_i$ are irreducible, we see that the set ad $\mathfrak{L}$ is completely reducible. Hence if $\mathfrak{B}$ is any ideal $\neq 0$ in $\mathfrak{L}$, then $\mathfrak{L} = \mathfrak{B} \oplus \mathfrak{D}$ where $\mathfrak{D}$ is an ideal (Theorem 2.9). Moreover, the proof of the theorem referred to shows that we can take $\mathfrak{D}$ to have the form $\mathfrak{D} = \mathfrak{L}_{i_1} \oplus \mathfrak{L}_{i_2} \oplus \cdots \mathfrak{L}_{i_k}$ for a subset $\{\mathfrak{L}_{i_u}\}$ of the $\mathfrak{L}_i$. Then $\mathfrak{B} \cong \mathfrak{L}/(\mathfrak{L}_{i_1} \oplus \cdots \oplus \mathfrak{L}_{i_k}) \cong \mathfrak{L}_{j_1} \oplus \mathfrak{L}_{j_2} \oplus \cdots \oplus \mathfrak{L}_{j_l}$ where the $\mathfrak{L}_{j_v}$ are the remaining $\mathfrak{L}_i$. Since $\mathfrak{L}_i$ is simple, $\mathfrak{L}_i^2 = \mathfrak{L}_i$. Hence $(\mathfrak{L}_{j_1} \oplus \mathfrak{L}_{j_2} \oplus \cdots \oplus \mathfrak{L}_{j_l})^2 = \mathfrak{L}_{j_1} \oplus \mathfrak{L}_{j_2} \oplus \cdots \oplus \mathfrak{L}_{j_l}$ and consequently $\mathfrak{B}^2 = \mathfrak{B}$. Thus $\mathfrak{B}$ is not solvable. We have therefore proved that $\mathfrak{L}$ has no non-zero solvable ideals; so $\mathfrak{L}$ is semi-simple.

The argument just given that $\mathfrak{B} \cong \mathfrak{L}_{j_1} \oplus \mathfrak{L}_{j_2} \oplus \cdots \oplus \mathfrak{L}_{j_l}$ has the following consequence.

COROLLARY 1. *Any ideal in a semi-simple Lie algebra of characteristic* 0 *is semi-simple.*

If $\mathfrak{L}_i$ is simple then the derived algebra $\mathfrak{L}_i' = \mathfrak{L}_i$; hence the structure theorem implies the following

COROLLARY 2. *If* $\mathfrak{L}$ *is semi-simple of characteristic* 0, *then* $\mathfrak{L}' = \mathfrak{L}$.

Remark. We have proved in Chapter II that if $\mathfrak{L}$ is a completely reducible Lie algebra of linear transformations in a finite-dimensional vector space over a field of characteristic 0, then $\mathfrak{L} = \mathfrak{C} \oplus \mathfrak{L}_1$ where $\mathfrak{C}$ is the center and $\mathfrak{L}_1$ is a semi-simple ideal. Then $\mathfrak{L}' = \mathfrak{L}_1' = \mathfrak{L}_1$. Hence $\mathfrak{L} = \mathfrak{C} \oplus \mathfrak{L}'$, $\mathfrak{L}'$ semi-simple.

We prove next the following general uniqueness theorem.

THEOREM 4. *If* $\mathfrak{A}$ *is a non-associative algebra and*

$$\mathfrak{A} = \mathfrak{A}_1 \oplus \mathfrak{A}_2 \oplus \cdots \oplus \mathfrak{A}_r = \mathfrak{B}_1 \oplus \mathfrak{B}_2 \oplus \cdots \oplus \mathfrak{B}_s$$

where the $\mathfrak{A}_i$ *and* $\mathfrak{B}_j$ *are ideals and are simple, then* $r = s$ *and the* $\mathfrak{B}_j$'s *coincide with the* $\mathfrak{A}_i$*'s (except for order).*

Proof: Consider $\mathfrak{A}_1 \cap \mathfrak{B}_j$, $j = 1, 2, \cdots, s$. This is an ideal contained in $\mathfrak{A}_1$ and $\mathfrak{B}_j$. Hence if $\mathfrak{A}_1 \cap \mathfrak{B}_j \neq 0$, then $\mathfrak{A}_1 = \mathfrak{A}_1 \cap \mathfrak{B}_j = \mathfrak{B}_j$ since $\mathfrak{A}_1$ and $\mathfrak{B}_j$ are simple. It follows that $\mathfrak{A}_1 \cap \mathfrak{B}_j \neq 0$ for at most one $\mathfrak{B}_j$. On the other hand, if $\mathfrak{A}_1 \cap \mathfrak{B}_j = 0$ for all j then $\mathfrak{A}_1\mathfrak{B}_j \subseteq \mathfrak{A}_1 \cap \mathfrak{B}_j = 0$ for all j. Since $\mathfrak{A} = \mathfrak{B}_1 \oplus \mathfrak{B}_2 \oplus \cdots \oplus \mathfrak{B}_s$, this implies that $\mathfrak{A}_1\mathfrak{A} = 0$ contrary to the assumption that $\mathfrak{A}_1^2 \neq 0$. Hence there is a j such that $\mathfrak{A}_1 = \mathfrak{B}_j$. Similarly, we have that every $\mathfrak{A}_i$ coincides with one of the $\mathfrak{B}_j$ and every $\mathfrak{B}_j$ coincides with one of the $\mathfrak{A}_i$.

The result follows from this.

It is easy to see also that if $\mathfrak{A}$ is as in Theorem 4, then $\mathfrak{A}$ has just 2^r ideals, namely, the ideals $\mathfrak{A}_{i_1} \oplus \cdots \oplus \mathfrak{A}_{i_k}$, $\{i_1, \cdots, i_k\}$ a subset of $\{1, 2, \cdots, r\}$. We omit the proof.

The main structure theorem fails if the characteristic is $p \neq 0$. To obtain a counter example we consider the Lie algebra $\mathfrak{E}_L$ of linear transformations in a vector space $\mathfrak{M}$ whose dimensionality n is divisible by p. It is easy to prove (Exercise 1.20) that the only ideals in $\mathfrak{E}_L$ are $\mathfrak{E}_L'$ and $\Phi 1$, the set of multiples of 1. Since $\mathfrak{E}_L'$ is the set of linear transformations of trace 0 and $\text{tr}\, 1 = n = 0$, $\Phi 1 \subseteq \mathfrak{E}_L'$. Hence $\mathfrak{L} = \mathfrak{E}_L/\Phi 1$ has only one ideal, namely, $\mathfrak{E}_L'/\Phi 1$, and the latter is simple. This implies that $\mathfrak{E}_L/\Phi 1$ is semi-simple, but since $\mathfrak{E}_L'/\Phi 1$ is the only ideal in $\mathfrak{E}_L/\Phi 1$, $\mathfrak{E}_L/\Phi 1$ is not a direct sum of simple ideals. This and Theorem 3 imply that $\mathfrak{E}_L/\Phi 1$ possesses no non-degenerate symmetric associative bilinear form.

We conclude this section with the following characterization of the radical in the characteristic 0 case.

THEOREM 5. *If $\mathfrak{L}$ is a finite-dimensional Lie algebra over a field of characteristic 0, then the radical $\mathfrak{S}$ of $\mathfrak{L}$ is the orthogonal complement $\mathfrak{L}'^{\perp}$ of $\mathfrak{L}'$ relative to the Killing form $f(a, b)$.*

Proof: $\mathfrak{B} = \mathfrak{L}'^{\perp}$ is an ideal and if $b \in \mathfrak{B}'$, then $\text{tr}(\text{ad}_{\mathfrak{L}} b)^2 = f(b, b) = 0$. The kernel of the representation $a \to \text{ad}_{\mathfrak{L}} a$, $a \in \mathfrak{B}$, is abelian. Hence $\mathfrak{B}$ is solvable, by Cartan's criterion, and $\mathfrak{B} \subseteq \mathfrak{S}$. Next let $s \in \mathfrak{S}$, $a, b \in \mathfrak{L}$. Then $f(s, [ab]) = f([sa], b)$. We have seen (Corollary 2 to Theorem 2.8) that $\text{ad}\,[sa]$ is contained in the radical of the enveloping associative algebra $(\text{ad}\,\mathfrak{L})^*$. Consequently, $\text{ad}\,[sa]\,\text{ad}\, b$ is nilpotent for every b and hence $f([sa], b) = 0$. Thus $f(s, [ab]) = 0$ and $s \in \mathfrak{L}'^{\perp}$. Thus $\mathfrak{S} \subseteq \mathfrak{L}'^{\perp}$ and so $\mathfrak{S} = \mathfrak{L}'^{\perp}$.

6. *Derivations*

We recall that $\text{ad}\, a$ is a derivation called *inner* and the set $\text{ad}\,\mathfrak{L}$ of these derivations is an ideal in the derivation algebra $\mathfrak{D}(\mathfrak{L})$. In fact, we have the formula $[\text{ad}\, a, D] = \text{ad}\, aD$ for D a derivation. Hence

$$[\text{ad}\, a\, \text{ad}\, b, D] = \text{ad}\, a\, \text{ad}\,(bD) + \text{ad}\,(aD)\, \text{ad}\, b$$

which implies that

$$0 = \mathrm{tr}\,[\mathrm{ad}\, a\, \mathrm{ad}\, b, D] = \mathrm{tr}\, \mathrm{ad}\, a\, \mathrm{ad}\,(bD) + \mathrm{tr}\, \mathrm{ad}\,(aD)\, \mathrm{ad}\, b\,.$$

Thus, for the Killing form $f(a, b) = \mathrm{tr}\, \mathrm{ad}\, a\, \mathrm{ad}\, b$ we have

$$f(a, bD) + f(aD, b) = 0\,; \tag{25}$$

that is, every derivation is a skew-symmetric transformation relative to the Killing form.

We prove next the following theorem which is due to Zassenhaus.

THEOREM 6. *If $\mathfrak{L}$ is a finite-dimensional Lie algebra which has a non-degenerate Killing form, then every derivation D of $\mathfrak{L}$ is inner.*

Proof: The mapping $x \to \mathrm{tr}\,(\mathrm{ad}\, x)D$ is a linear mapping of $\mathfrak{L}$ into Φ; that is, it is an element of the conjugate space $\mathfrak{L}^*$ of $\mathfrak{L}$. Since $f(a, b)$ is non-degenerate it follows that there exists an element $d \in \mathfrak{L}$ such that $f(d, x) = \mathrm{tr}\,(\mathrm{ad}\, x)D$ for all $x \in \mathfrak{L}$.

Let E be the derivation $D - \mathrm{ad}\, d$. Then

$$\mathrm{tr}\,(\mathrm{ad}\, x)E = \mathrm{tr}\,(\mathrm{ad}\, x)D - \mathrm{tr}\,(\mathrm{ad}\, x)(\mathrm{ad}\, d) = \mathrm{tr}\,(\mathrm{ad}\, x)D - f(d, x) = 0\,.$$

Thus

$$\mathrm{tr}\,(\mathrm{ad}\, x)E = 0\,. \tag{26}$$

Now consider

$$\begin{aligned} f(xE, y) &= \mathrm{tr}\,(\mathrm{ad}\, xE)\, \mathrm{ad}\, y \\ &= \mathrm{tr}\,[\mathrm{ad}\, x, E]\, \mathrm{ad}\, y \\ &= \mathrm{tr}\,((\mathrm{ad}\, x)E\, \mathrm{ad}\, y - E\, \mathrm{ad}\, x\, \mathrm{ad}\, y) \\ &= \mathrm{tr}\,(E\, \mathrm{ad}\, y\, \mathrm{ad}\, x - E\, \mathrm{ad}\, x\, \mathrm{ad}\, y) \\ &= \mathrm{tr}\, E[\mathrm{ad}\, y, \mathrm{ad}\, x] \\ &= \mathrm{tr}\, E\, \mathrm{ad}\,[yx] = 0\,, \end{aligned}$$

by (26). Since f is non-degenerate, this implies that $E = 0$. Hence $D = \mathrm{ad}\, d$ is inner.

This result implies that the derivations of any finite dimensional semi-simple Lie algebra over a field of characteristic zero are all inner. We recall also that if $\mathfrak{S}$ is solvable, finite-dimensional of characteristic 0, then $\mathfrak{S}$ is mapped into the nil radical $\mathfrak{N}$ by every derivation of $\mathfrak{S}$ (Corollary 2 to Theorem 2.13). We can now prove

THEOREM 7. (1) *Let $\mathfrak{L}$ be a finite-dimensional Lie algebra over a field of characteristic* 0, *$\mathfrak{S}$ the radical, $\mathfrak{N}$ the nil radical. Then any derivation D of $\mathfrak{L}$ maps $\mathfrak{S}$ into $\mathfrak{N}$.* (2) *Let $\mathfrak{L}$ be an ideal in a finite-dimensional algebra $\mathfrak{L}_1$, $\mathfrak{S}_1$, $\mathfrak{N}_1$ the radical and nil radical of $\mathfrak{L}_1$.*

Then $\mathfrak{S} = \mathfrak{L} \cap \mathfrak{S}_1$, $\mathfrak{N} = \mathfrak{L} \cap \mathfrak{N}_1$.

Proof: We first prove (2) for the radical. Thus it is clear that $\mathfrak{L} \cap \mathfrak{S}_1$ is a solvable ideal in $\mathfrak{L}$, hence $\mathfrak{L} \cap \mathfrak{S}_1 \subseteq \mathfrak{S}$. Then $\mathfrak{S}/(\mathfrak{L} \cap \mathfrak{S}_1)$ is a solvable ideal in $\mathfrak{L}/(\mathfrak{L} \cap \mathfrak{S}_1)$. On the other hand, $\mathfrak{L}/(\mathfrak{L} \cap \mathfrak{S}_1) \cong (\mathfrak{L} + \mathfrak{S}_1)/\mathfrak{S}_1$, which is an ideal in $\mathfrak{L}_1/\mathfrak{S}_1$. Hence $(\mathfrak{L} + \mathfrak{S}_1)/\mathfrak{S}_1$ and $\mathfrak{L}/(\mathfrak{L} \cap \mathfrak{S}_1)$ are semi-simple. Hence $\mathfrak{S}/(\mathfrak{L} \cap \mathfrak{S}_1) = 0$ and $\mathfrak{S} = \mathfrak{L} \cap \mathfrak{S}_1$. Now let $\mathfrak{L}_1$ be the holomorph of $\mathfrak{L}$, $\mathfrak{S}_1$, $\mathfrak{N}_1$ the radical and nil radical of $\mathfrak{L}_1$. Then we know that $[\mathfrak{L}_1\mathfrak{S}_1] \subseteq \mathfrak{N}_1$ (Theorem 2.13). Since $\mathfrak{S} \subseteq \mathfrak{S}_1$ by the first part of the argument, $[\mathfrak{L}_1\mathfrak{S}] \subseteq (\mathfrak{N}_1 \cap \mathfrak{L}) \subseteq \mathfrak{N}$. This implies that every derivation of $\mathfrak{L}$ maps $\mathfrak{S}$ into $\mathfrak{N}$, which proves (1). Now let $\mathfrak{L}_1$ be any finite-dimensional Lie algebra containing $\mathfrak{L}$ as an ideal. If $a_1 \in \mathfrak{L}_1$ then $\operatorname{ad} a_1$ induces a derivation in $\mathfrak{L}$. Hence $\mathfrak{N} \operatorname{ad} a_1 \subseteq \mathfrak{N}$ by (1). This means that $\mathfrak{N}$ is an ideal in $\mathfrak{L}_1$ so that $\mathfrak{N} \subseteq \mathfrak{N}_1 \cap \mathfrak{L}$, $\mathfrak{N}_1$ the nil radical of $\mathfrak{L}_1$. Since the reverse inequality is clear, $\mathfrak{N} = \mathfrak{N}_1 \cap \mathfrak{L}$.

This result fails for characteristic $p \neq 0$. To construct a counter example we consider first the commutative associative algebra $\mathfrak{Z}$ with the basis $(1, z, z^2, \cdots, z^{p-1})$ with $z^p = 0$. The radical $\mathfrak{R}$ of $\mathfrak{Z}$ has the basis $(z, z^2, \cdots, z^{p-1})$ and $\mathfrak{Z}/\mathfrak{R} \cong \Phi 1$. It is easy to prove that if w is any element of $\mathfrak{Z}$, then there exists a derivation of $\mathfrak{Z}$ mapping z into w. In particular, there is a derivation D such that $zD = 1$. Now let $\mathfrak{B}$ be any simple Lie algebra and let $\mathfrak{L}$ be the Lie algebra $\mathfrak{B} \otimes \mathfrak{Z}$. The elements of this algebra have the form $\sum b_i \otimes z_i$, $b_i \in \mathfrak{B}$, $z_i \in \mathfrak{Z}$, and $[(b \otimes z)(b' \otimes z')] = [bb'] \otimes zz'$. Then $\mathfrak{L}$ is a Lie algebra (Exercise 1.23) and $\mathfrak{B} \otimes \mathfrak{R}$ is a nilpotent ideal in $\mathfrak{L}$. Moreover, $\mathfrak{L}/(\mathfrak{B} \otimes \mathfrak{R}) \cong \mathfrak{B} \otimes \Phi 1 \cong \mathfrak{B}$ is simple. Hence $\mathfrak{B} \otimes \mathfrak{R}$ is the radical and the nil radical of $\mathfrak{L}$. If D is any derivation in the associative algebra $\mathfrak{Z}$, then the mapping $\sum b_i \otimes z_i \rightarrow \sum b_i \otimes z_i D$ is a derivation in $\mathfrak{L}$. If we take D so that $zD = 1$ and let $b \neq 0$ in $\mathfrak{B}$ then $b \otimes z \rightarrow b \otimes 1 \notin \mathfrak{B} \otimes \mathfrak{R}$. Hence we have a derivation which does not leave the radical invariant.

7. *Complete reducibility of the representations of semi-simple algebras*

In this section we shall prove the main structure theorem for modules of a semi-simple Lie algebra of characteristic zero and we shall obtain its most important consequences. The main theorem is due to Weyl and was proved by him by transcendental methods based on the connection between Lie algebras and compact groups.

The first algebraic proof of the result was given by Casimir and van der Waerden. The proof we shall give is in essence due to Whitehead. It should be mentioned that Whitehead's proof was one of the stepping stones to the cohomology theory of Lie algebras which we shall consider in § 10. We note also that in the characteristic p case there appears to be little connection between the structure of a Lie algebra and the structure of its modules since, as will be shown later, every finite-dimensional Lie algebra of characteristic $p \neq 0$ has faithful representations which are not completely reducible and also faithful representations which are completely reducible.

We obtain first a criterion that a set Σ of linear transformations in a finite-dimensional vector space $\mathfrak{M}$ be completely reducible. We have seen (Theorem 2.9) that Σ is completely reducible if and only if every invariant subspace $\mathfrak{N}$ of $\mathfrak{M}$ has a complement $\mathfrak{P}$ which is invariant relative to Σ. Now let $\mathfrak{N}'$ be any complementary subspace to $\mathfrak{N}$: $\mathfrak{M} = \mathfrak{N} \oplus \mathfrak{N}'$. Such a decomposition is associated with a projection E of $\mathfrak{M}$ onto $\mathfrak{N}$. Thus if $x \in \mathfrak{M}$, then we can write x in one and only one way as $x = y + y'$, $y \in \mathfrak{N}$, $y' \in \mathfrak{N}'$, and E is the linear mapping $x \to y$. Conversely, if E is any idempotent linear mapping such that $\mathfrak{N} = \mathfrak{M}E$ then $\mathfrak{N}' = \mathfrak{M}(1 - E)$ is a complement of $\mathfrak{N}$ in $\mathfrak{M}$. Now let $A \in \Sigma$ and consider the linear transformation $[AE] \equiv AE - EA$. If $x \in \mathfrak{M}$, $xAE \in \mathfrak{N}$ and $xEA \in \mathfrak{N}$; hence $[AE]$ maps $\mathfrak{M}$ into $\mathfrak{N}$. If $y \in \mathfrak{N}$ then $yAE = yA$. Hence $[AE]$ maps $\mathfrak{N}$ into 0. Then if $\mathfrak{X}$ denotes the set of linear tranformations of $\mathfrak{M}$ which map $\mathfrak{M}$ into $\mathfrak{N}$ and $\mathfrak{N}$ into 0, $[AE] \in \mathfrak{X}$. It is clear that $\mathfrak{X}$ is a subspace of the space $\mathfrak{E}$ of linear transformations in $\mathfrak{M}$. We now prove the following

LEMMA 2. *$\mathfrak{N}$ has a complement $\mathfrak{P}$ which is invariant if and only if there exists a $D \in \mathfrak{X}$ such that $[AE] = [AD]$ for all $A \in \Sigma$. Here E is any projection onto $\mathfrak{N}$.*

Proof: Let $\mathfrak{P}$ be a complement of $\mathfrak{N}$ which is invariant relative to Σ and let F be the projection of $\mathfrak{M}$ onto $\mathfrak{N}$ determined by the decomposition $\mathfrak{M} = \mathfrak{N} \oplus \mathfrak{P}$. Since $\mathfrak{P}$ is invariant, F commutes with every $A \in \Sigma$, that is, $[AF] = 0$. Hence $[AD] = [AE]$ for $D = E - F$. Also, since E and F are projections on $\mathfrak{N}$, $E - F$ maps $\mathfrak{M}$ into $\mathfrak{N}$ and $\mathfrak{N}$ into 0. Hence $D \in \mathfrak{X}$ as required. Conversely, suppose there exists a $D \in \mathfrak{X}$ such that $[AE] = [AD]$. Then $F = E - D$ commutes with every $A \in \Sigma$. If $x \in \mathfrak{M}$ then $xF = x(E - D) \in \mathfrak{N}$

and if $y \in \mathfrak{N}$ then $yF = yE = y$. Hence $F^2 = F$ and $\mathfrak{N} = \mathfrak{M}F$. Then $\mathfrak{P} = \mathfrak{M}(1 - F)$ is a complement of $\mathfrak{N}$ and $\mathfrak{P}$ is invariant under Σ since F commutes with every $A \in \Sigma$.

Suppose now that $\mathfrak{L}$ is a Lie algebra and $\mathfrak{M}$ is an $\mathfrak{L}$-module, $\mathfrak{N}$ a submodule. We can apply our considerations to the set $\mathfrak{L}^R$ of representing transformations a^R determined by $\mathfrak{M}$. Let E be a projection of $\mathfrak{M}$ onto $\mathfrak{N}$. If $a \in \mathfrak{L}$, set $f(a) = [a^R, E]$. Then $a \to f(a)$ is a linear mapping of $\mathfrak{L}$ into the space $\mathfrak{X}$ of linear transformations of $\mathfrak{M}$ which map $\mathfrak{M}$ into $\mathfrak{N}$ and $\mathfrak{N}$ into 0. If $X \in \mathfrak{X}$ and $a \in \mathfrak{L}$ then $[Xa^R] \in \mathfrak{X}$. Thus if $x \in \mathfrak{M}$ then $x[Xa^R] \in \mathfrak{N}$ and if $y \in \mathfrak{N}$ then $yXa^R = 0$ and $ya^RX = 0$. We denote the mapping $X \to [Xa^R]$ by $\tilde{a}^R$. It is immediate that $a \to \tilde{a}^R$ is a representation of $\mathfrak{L}$ whose associated module is the space $\mathfrak{X}$. We have

$$\begin{aligned} f([ab]) &= [[ab]^R, E] = [[a^Rb^R]E] \\ &= [[a^RE]b^R] + [a^R[b^RE]] \\ &= [f(a), b^R] + [a^R, f(b)] \\ &= f(a)\tilde{b}^R - f(b)\tilde{a}^R \,. \end{aligned}$$

We are now led to consider the following situation: We have a module $\mathfrak{X}$ for $\mathfrak{L}$ and a linear mapping $a \to f(a)$ of $\mathfrak{L}$ into $\mathfrak{X}$ such that

$$f([ab]) = f(a)b - f(b)a \,. \tag{27}$$

A "trivial" example of such a mapping is obtained by taking $f(a) = da$, where d is an element of $\mathfrak{X}$. For, we have

$$f([ab]) = d[ab] = (da)b - (db)a = f(a)b - f(b)a \,.$$

The key result for the proof of complete reducibility of the modules for a semi-simple Lie algebra of characteristic zero is the following

Lemma 3 (Whitehead). *Let $\mathfrak{L}$ be finite-dimensional semi-simple of characteristic zero and let $\mathfrak{X}$ be a finite-dimensional module for $\mathfrak{L}$ and $a \to f(a)$ a linear mapping of $\mathfrak{L}$ into $\mathfrak{X}$ satisfying* (27). *Then there exists a $d \in \mathfrak{X}$ such that $f(a) = da$.*

Proof: The proof will be based on the important notion of a *Casimir operator*. First, suppose that $\mathfrak{L}$ is a Lie algebra and $\mathfrak{B}_1$ and $\mathfrak{B}_2$ are ideals in $\mathfrak{L}$ such that the representations of $\mathfrak{L}$ in $\mathfrak{B}_1$ and $\mathfrak{B}_2$ are contragredient. Thus we are assuming that the spaces $\mathfrak{B}_1$ and $\mathfrak{B}_2$ are connected by a bilinear form (b_1, b_2), $b_i \in \mathfrak{B}_i$, $(b_1, b_2) \in \Phi$, which is non-degenerate and that for any $a \in \mathfrak{L}$, we have

(28) $$([b_1a], b_2) + (b_1, [b_2a]) = 0 .$$

If $(u_1, \cdots, u_m)$ is a basis for $\mathfrak{B}_1$ then we can choose a complementary or dual basis $(u^1, u^2 \cdots, u^m)$ for $\mathfrak{B}_2$ satisfying $(u_i, u^j) = \delta_{ij}$. Let $[u_ia] = \sum_j \alpha_{ij}u_j$ and $[u^ka] = \sum_l \beta_{kl}u^l$. Then $([u_ia], u^k) = (\sum \alpha_{ij}u_j, u^k) = \sum \alpha_{ij}\delta_{jk} = \alpha_{ik}$ and $(u_i, [u^ka]) = (u_i, \sum \beta_{kl}u^l) = \beta_{ki}$. Hence (28) implies that $\alpha_{ik} = -\beta_{ki}$, that is, the matrices (α) and (β) determined by dual bases satisfy $(\beta) = -(\alpha)'$ ($(\alpha)'$ the transpose of (α)). Now let R be a representation of $\mathfrak{L}$. Then the element

$$\Gamma = \sum_{i=1}^{m} u_i^R u^{iR}$$

is called a *Casimir operator* of R. We have

$$\begin{aligned}
[\Gamma, a^R] &= \sum_i [u_i^R a^R]u^{iR} + \sum u_i^R[u^{iR}a^R] \\
&= \sum_{i,j} \alpha_{ij}u_j^R u^{iR} + \sum_{i,j} \beta_{ij}u_i^R u^{jR} \\
&= \sum_{i,j} \alpha_{ij}u_j^R u^{iR} - \sum_{i,j} \alpha_{ji}u_i^R u^{jR} \\
&= \sum \alpha_{ij}u_j^R u^{iR} - \sum_{i,j} \alpha_{ij}u_j^R u^{iR} \\
&= 0.
\end{aligned}$$

Hence we have the important property that Γ commutes with all the representing transformations a^R.

Now let $\mathfrak{L}$ satisfy the hypotheses of the lemma. Let $\mathfrak{K}$ be the kernel of the representation R determined by $\mathfrak{X}$. Then we can write $\mathfrak{L} = \mathfrak{K} \oplus \mathfrak{L}_1$ where $\mathfrak{L}_1$ is an ideal. Then the restriction of R to $\mathfrak{L}_1$ is $1:1$ and $\mathfrak{L}_1$ is semi-simple. Hence the trace form $(b_1, b_2) \equiv \operatorname{tr} b_1^R b_2^R$, b_i in $\mathfrak{L}_1$, is non-degenerate on $\mathfrak{L}_1$. Also we know that the trace form of a representation is invariant. Hence the equation (28) holds for $b_i \in \mathfrak{L}_1$ and $a \in \mathfrak{L}$. Thus the representation of $\mathfrak{L}$ in $\mathfrak{L}_1$ coincides with its contragredient and if $(u_1, \cdots, u_m)$, $(u^1, \cdots, u^m)$ are bases for $\mathfrak{L}_1$ satisfying $(u_i, u^j) = \delta_{ij}$ then $\Gamma = \sum_{i=1}^m u_i^R u^{iR}$ is a Casimir operator which commutes with every a^R. We note also that $\operatorname{tr} \Gamma = \sum_i \operatorname{tr} u_i^R u^{iR} = \sum (u_i, u^i) = m = \dim \mathfrak{L}_1$.

We now decompose $\mathfrak{X}$ into its Fitting components $\mathfrak{X}_0$ and $\mathfrak{X}_1$ relative to Γ so that Γ induces a nilpotent linear transformation in $\mathfrak{X}_0$ and a non-singular one in $\mathfrak{X}_1$. Since $\Gamma a^R = a^R\Gamma$, $\mathfrak{X}_j a^R \subseteq \mathfrak{X}_j$ so that the $\mathfrak{X}_j$ are submodules. We can write $f(a) = f_0(a) + f_1(a)$ where $f_j(a) \in \mathfrak{X}_j$ and it is immediate that $a \rightarrow f_j(a)$ is a linear mapping of

$\mathfrak{L}$ into $\mathfrak{X}_j$ satisfying (27). Now if both spaces $\mathfrak{X}_j$ are $\neq 0$, then $\dim \mathfrak{X}_j < \dim \mathfrak{X}$ for $j = 0, 1$. Hence we can use induction on $\dim \mathfrak{X}$ to conclude that there is a $d_j \in \mathfrak{X}_j$ such that $f_j(a) = d_j a$. Then $d = d_0 + d_1$ satisfies $f(a) = da$ as required. Thus it remains to consider the following two cases: $\mathfrak{X} = \mathfrak{X}_0$ and $\mathfrak{X} = \mathfrak{X}_1$.

$\mathfrak{X} = \mathfrak{X}_0$: In this case Γ is nilpotent. Hence $m = \operatorname{tr} \Gamma = 0$. This means that the kernel of R is the whole of $\mathfrak{L}$, that is, $a^R = 0$ for all a. Then the condition (27) is that $f([ab]) = 0$, $a, b \in \mathfrak{L}$. Thus $f(a') = 0$ for all $a' \in \mathfrak{L}'$. Since $\mathfrak{L}' = \mathfrak{L}$, this implies that $f(a) = 0$ so that $d = 0$ satisfies the condition.

$\mathfrak{X} = \mathfrak{X}_1$: Set $y = \sum_{1=1}^m f(u_i)u^i$ where the (u_i) and (u^i) are dual bases for $\mathfrak{L}_1$ as before. Then

$$
\begin{aligned}
ya &= \sum_i (f(u_i)u^i)a \\
&= \sum_i (f(u_i)a)u^i + \sum_i f(u_i)[u^i a] \\
&= \sum_i (f(u_i)a)u^i + \sum_{i,j} \beta_{ij} f(u_i)u^j \\
&= \sum_i (f(u_i)a)u^i - \sum_{i,j} \alpha_{ji} f(u_i)u^j \\
&= \sum_i (f(u_i)a)u^i - \sum_j (f[u_j a])u^j \\
&= \sum_i (f(u_i)a)u^i - \sum_i (f(u_i)a)u^i + \sum_i (f(a)u_i)u^i \\
&= f(a)\Gamma
\end{aligned}
$$

Since Γ is non-singular, $d = y\Gamma^{-1}$ satisfies the required condition $f(a) = da$. This completes the proof of Whitehead's lemma.

We can now prove the following fundamental theorem:

THEOREM 8. *If $\mathfrak{L}$ is finite-dimensional semi-simple of characteristic* 0, *then every finite-dimensional module for $\mathfrak{L}$ is completely reducible.*

Proof: Let $\mathfrak{M}$ be a finite-dimensional $\mathfrak{L}$-module, $\mathfrak{N}$ a submodule. Let $\mathfrak{X}$ be the space of linear transformations of $\mathfrak{M}$ which map $\mathfrak{M}$ into $\mathfrak{N}$, $\mathfrak{N}$ into 0, and consider $\mathfrak{X}$ as $\mathfrak{L}$-module relative to the composition $Xa \equiv [X, a^R]$, R the representation of $\mathfrak{M}$. Let E be any projection of $\mathfrak{M}$ onto $\mathfrak{N}$ and set $f(a) = [a^R, E]$. Then $f(a)$ satisfies the conditions of Whitehead's lemma. Hence there exists a $D \in \mathfrak{X}$ such that $f(a) = Da = [D, a^R]$. As we saw before, this implies that $\mathfrak{N}$ has a complementary subspace which is invariant under $\mathfrak{L}$. Since this applies to every submodule $\mathfrak{N}$, $\mathfrak{M}$ is completely reducible.

If $\mathfrak{L}$ is a subalgebra of a Lie algebra $\mathfrak{B}$, then a *derivation D of*

$\mathfrak{L}$ *into* $\mathfrak{B}$ is a linear mapping of $\mathfrak{L}$ into $\mathfrak{B}$ such that

$$[l_1l_2]D = [l_1D, l_2] + [l_1, l_2D] \tag{29}$$

for every $l_1, l_2 \in \mathfrak{L}$. It is immediate that the set $\mathfrak{D}(\mathfrak{L}, \mathfrak{B})$ of derivations of $\mathfrak{L}$ into $\mathfrak{B}$ is a subspace of the space of linear transformations of $\mathfrak{L}$ into $\mathfrak{B}$. Whitehead's lemma to the theorem on complete reducibility has the following important consequence on derivations:

THEOREM 9. *Let* $\mathfrak{B}$ *be a finite-dimensional Lie algebra of characteristic* 0 *and let* $\mathfrak{L}$ *be a semi-simple subalgebra of* $\mathfrak{B}$. *Then every derivation of* $\mathfrak{L}$ *into* $\mathfrak{B}$ *can be extended to an inner derivation of* $\mathfrak{B}$.

Proof: Consider $\mathfrak{B}$ as $\mathfrak{L}$-module relative to the multiplication $[bl]$, $l \in \mathfrak{L}$, $b \in \mathfrak{B}$. Then a derivation D of $\mathfrak{L}$ into $\mathfrak{B}$ defines $f(l) \equiv lD$ satisfying the condition

$$\begin{aligned} f([l_1l_2]) &= [l_1l_2]D = [l_1, l_2D] + [l_1D, l_2] \\ &= [f(l_1), l_2] - [f(l_2), l_1] \end{aligned}$$

of Whitehead's lemma. Hence there exists a $d \in \mathfrak{B}$ such that $lD = f(l) = [d, l]$. Then D can be extended to the inner derivation determined by the element $-d$.

We recall that we have shown in Chapter II (Theorem 2.11) that if $\mathfrak{L}$ is a completely reducible Lie algebra of linear transformations in a finite-dimensional vector space $\mathfrak{M}$ over a field of characteristic 0, then $\mathfrak{L} = \mathfrak{L}_1 \oplus \mathfrak{C}$ where $\mathfrak{L}_1$ is a semi-simple ideal and $\mathfrak{C}$ is the center. Moreover, the elements $C \in \mathfrak{C}$ are semi-simple in the sense that their minimum polynomials are products of distinct irreducible polynomials. We are now in a position to establish the converse of this result. Our proof will be based on a field extension argument of the following type: Suppose we have a set Σ of linear transformations in $\mathfrak{M}$ over Φ. If Ω is an extension field of Φ, every $A \in \Sigma$ has a unique extension to a linear transformation, denoted again by A, in $\mathfrak{M}_\Omega$. In this way we get a set $\Sigma = \{A\}$ of linear transformations in $\mathfrak{M}_\Omega$ over Ω. We shall now prove the following

LEMMA 4. *Let* Σ *be a set of linear transformations in a finite-dimensional vector space* $\mathfrak{M}$ *over* Φ *and let* Σ *be the set of extensions of these transformations to* $\mathfrak{M}_\Omega$ *over* Ω, Ω *an extension field of* Φ. *Suppose the set* Σ *in* $\mathfrak{M}_\Omega$ *is completely reducible. Then the original* Σ *is completely reducible in* $\mathfrak{M}$.

Proof: Let $\mathfrak{N}$ be a subspace of $\mathfrak{M}$ which is invariant under Σ

and let E be a projection on $\mathfrak{N}$. Then our criterion for complementation (Lemma 2) shows that $\mathfrak{N}$ will have a complement which is invariant relative to Σ if and only if there exists a linear transformation D of $\mathfrak{M}$ mapping $\mathfrak{M}$ into $\mathfrak{N}$, $\mathfrak{N}$ into 0 such that $[AE] = [AD]$ for all $A \in \Sigma$. If $A_1, A_2, \cdots, A_k$ is a maximal set of linearly independent elements in Σ, and we set $B_i = [A_iE]$, then it suffices to find a D such that $[A_iD] = B_i$, $i = 1, 2, \cdots, k$. This is a system of k linear equations for D in the finite-dimensional space $\mathfrak{X}$ of linear transformations of $\mathfrak{M}$ mapping $\mathfrak{M}$ into $\mathfrak{N}$, $\mathfrak{N}$ into 0. Thus if we have the basis $(U_1, U_2, \cdots, U_r)$ for $\mathfrak{X}$ we can write $B_i = \sum_{s=1}^{r} \beta_{is} U_s$, $[A_i U_h] = \sum_s \gamma_{ihs} U_s$, $D = \sum_1^r \delta_j U_j$, then our equations are equivalent to the ordinary system: $\sum_h \gamma_{ihs} \delta_h = \beta_{is}$, $i = 1, 2, \cdots, k$, $s = 1, 2, \cdots, r$, for the δ_j in Φ. Hence $\mathfrak{N}$ has a Σ-invariant complement if and only if this system has a solution. We now pass to $\mathfrak{M}_\Omega$ and the invariant subspace $\mathfrak{N}_\Omega$ relative to the set Σ of extensions of the $A \in \Sigma$. Then our hypothesis is that $\mathfrak{N}_\Omega$ has a Σ-invariant complement in $\mathfrak{M}_\Omega$. Now the extension E of E is a projection of $\mathfrak{M}_\Omega$ onto $\mathfrak{N}_\Omega$. Hence we have a linear mapping $\tilde{D}$ of $\mathfrak{M}_\Omega$ mapping $\mathfrak{M}_\Omega$ into $\mathfrak{N}_\Omega$, $\mathfrak{N}_\Omega$ into 0, such that $[A_i\tilde{D}] = B_i \equiv [A_iE]$, $i = 1, 2, \cdots, k$. The extensions $U_1, U_2, \cdots, U_r$ form a basis for the space of linear transformations of $\mathfrak{M}_\Omega$ mapping $\mathfrak{M}_\Omega$ into $\mathfrak{N}_\Omega$, $\mathfrak{N}_\Omega$ into 0. Hence if $\tilde{D} = \sum_1^r \delta_j U_j$ then the $\tilde{\delta}$ satisfy the system $\sum_h \gamma_{ihs} \tilde{\delta}_h = \beta_{is}$. Since the γ_{ihs} and β_{is} belong to Φ, it follows that this system has a solution $(\delta_1, \cdots, \delta_r)$, δ's in Φ. Hence there exists a $D \in \mathfrak{X}$ such that $[AD] = [AE]$, $A \in \Sigma$, and so $\mathfrak{N}$ has a Σ-invariant complement in $\mathfrak{M}$.

We can now prove the following

THEOREM 10. *Let $\mathfrak{L}$ be a Lie algebra of linear transformations in a finite-dimensional vector space $\mathfrak{M}$ over a field of characteristic zero. Then $\mathfrak{L}$ is completely reducible in $\mathfrak{M}$ if and only if the following conditions hold*: (1) $\mathfrak{L} = \mathfrak{L}_1 \oplus \mathfrak{C}$, *$\mathfrak{L}_1$ a semi-simple ideal and $\mathfrak{C}$ the center and* (2) *the elements of $\mathfrak{C}$ are semi-simple.*

Proof: The necessity has been proved before. Now assume (1) and (2) and let Ω be the algebraic closure of the base field. Then the lemma shows that it suffices to prove that the set $\mathfrak{L}$ of extensions of the elements of $\mathfrak{L}$ is completely reducible in $\mathfrak{M}_\Omega$. The set of Ω-linear combinations of the elements of $\mathfrak{L}$ can be identified with $\mathfrak{L}_\Omega$ and similar statements hold for $\mathfrak{L}_1$ and $\mathfrak{C}$. Now let $C \in \mathfrak{C}$. Since the minimum polynomial of C in $\mathfrak{M}$ has distinct irreducible factors and since the field is of characteristic 0, the minimum polynomial of C in $\mathfrak{M}_\Omega$ has distinct linear factors in Ω. Consequently, we can decompose $\mathfrak{M}_\Omega$ as $\mathfrak{M}_{\alpha_1} \oplus \cdots \oplus \mathfrak{M}_{\alpha_k}$ where

$$\mathfrak{M}_{\alpha_i} = \{x_i \mid x_i \in \mathfrak{M}_\Omega, x_iC = \alpha_i x_i\}$$

and $\alpha_1, \alpha_2, \cdots, \alpha_r$ are the different characteristic roots of C. Since $AC = CA$ for $A \in \mathfrak{L}$, $\mathfrak{M}_{\alpha_i}A \subseteq \mathfrak{M}_{\alpha_i}$. We can apply the same procedure to the $\mathfrak{M}_{\alpha_i}$ relative to any other $D \in \mathfrak{C}$. This leads to a decomposition of $\mathfrak{M}_\Omega = \mathfrak{M}_1 \oplus \mathfrak{M}_2 \oplus \cdots \oplus \mathfrak{M}_r$ into $\mathfrak{L}$-invariant subspaces such that the transformation induced in the $\mathfrak{M}_i$ by every $C \in \mathfrak{C}$ is a scalar multiplication. To prove $\mathfrak{L}$ completely reducible in $\mathfrak{M}_\Omega$ it suffices to show that the sets of induced transformations in the $\mathfrak{M}_i$ are completely reducible and since the elements of $\mathfrak{C}$ are scalars in $\mathfrak{M}_i$ it suffices to show that $\mathfrak{L}_1$ is completely reducible in every $\mathfrak{M}_i$. The invariant subspaces of $\mathfrak{M}_i$ relative to $\mathfrak{L}_1$ are invariant relative to $\Omega\mathfrak{L}_1$, the set of Ω-linear combinations of the elements of $\mathfrak{L}_1$. Now $\Omega\mathfrak{L}_1$ is a homomorphic (actually isomorphic) image of the extension algebra $\mathfrak{L}_{1\Omega}$, which is semi-simple. Hence $\Omega\mathfrak{L}_1$ is semi-simple and consequently this Lie algebra of linear transformations is completely reducible by Theorem 8. Thus we have proved that $\mathfrak{L}$ is completely reducible in $\mathfrak{M}_\Omega$ and hence in $\mathfrak{M}$.

We now shift our point of view and consider a finite dimensional Lie algebra $\mathfrak{L}$ of characteristic 0 and two finite-dimensional completely reducible modules $\mathfrak{M}$ and $\mathfrak{N}$ for $\mathfrak{L}$. We shall show that $\mathfrak{M} \otimes \mathfrak{N}$ is completely reducible. Now the space $\mathfrak{P} = \mathfrak{M} \oplus \mathfrak{N}$ is a module relative to the product $(x + y)l = xl + yl$, $x \in \mathfrak{M}$, $y \in \mathfrak{N}$. Evidently $\mathfrak{P}$ is completely reducible and $\mathfrak{M} \otimes \mathfrak{N}$ is a submodule of $\mathfrak{P} \otimes \mathfrak{P}$. Hence it suffices to prove that $\mathfrak{P} \otimes \mathfrak{P}$ is completely reducible. If we replace $\mathfrak{L}$ by $\mathfrak{L}/\mathfrak{K}$ where $\mathfrak{K}$ is the kernel of the representation in $\mathfrak{P}$, then we may assume that the associated representation R in $\mathfrak{P}$ is $1:1$. Then we know that $\mathfrak{L} = \mathfrak{L}_1 \oplus \mathfrak{C}$ where $\mathfrak{L}_1$ is a semi-simple ideal and $\mathfrak{C}$ is the center. Moreover, the elements C^R, $C \in \mathfrak{C}$, are semi-simple. Now, in general, if R is a faithful representation of a Lie algebra $\mathfrak{L}$, then the representation $R \otimes R$ in $\mathfrak{P} \otimes \mathfrak{P}$ is also faithful. Thus, if $a \in \mathfrak{L}$ and a^R is not a scalar multiplication, then, since the algebra of linear transformations in $\mathfrak{P} \otimes \mathfrak{P}$ is the tensor product of the algebras of linear transformations in $\mathfrak{P}$, $a^R \otimes a^R$, $a^R \otimes 1$, $1 \otimes a^R$ and $1 \otimes 1$ are linearly independent, so $a^R \otimes 1 + 1 \otimes a^R \neq 0$. Hence if $a^{R\otimes R} = 0$, a^R must be a scalar, say $a^R = \alpha$. Then $a^{R\otimes R} = 2\alpha$ (in $\mathfrak{P} \otimes \mathfrak{P}$) and $\alpha = 0$. Since R is $1:1$ this implies that $a = 0$. We can now conclude that

$\mathfrak{L}^{R\otimes R} = \mathfrak{L}_1^{R\otimes R} + \mathfrak{C}^{R\otimes R}$ where $\mathfrak{L}_1^{R\otimes R} \cong \mathfrak{L}_1$ is semi-simple and $\mathfrak{C}^{R\otimes R}$ is the center. Our result will therefore follow from the criterion of Th. 10 provided that we can prove that every

$$C^{R\otimes R} = C^R \otimes 1 + 1 \otimes C^R ,$$

$C \in \mathfrak{C}$, is semi-simple.

Let Ω be the algebraic closure of the base field and let $\alpha_1, \alpha_2, \cdots, \alpha_k$ be the different characteristic roots of C^R. Then the proof of Theorem 10 shows that $\mathfrak{P}_\Omega = \mathfrak{P}_{\alpha_1} \oplus \mathfrak{P}_{\alpha_2} \oplus \cdots \oplus \mathfrak{P}_{\alpha_k}$ where $x_{\alpha_i} C^R = \alpha_i x_{\alpha_i}$ for $x_{\alpha_i} \in \mathfrak{P}_{\alpha_i}$. Hence $(\mathfrak{P} \otimes_\Phi \mathfrak{P})_\Omega = \mathfrak{P}_\Omega \otimes_\Omega \mathfrak{P}_\Omega = \sum \mathfrak{P}_{\alpha_i} \otimes \mathfrak{P}_{\alpha_j}$ and $yC^{R\otimes R} = (\alpha_i + \alpha_j)y$ for every $y \in \mathfrak{P}_{\alpha_i} \otimes \mathfrak{P}_{\alpha_j}$. It follows that the minimum polynomial of $C^{R\otimes R}$ has distinct roots in $(\mathfrak{P} \otimes \mathfrak{P})_\Omega$. Since this is also the minimum polynomial of $C^{R\otimes R}$ in $\mathfrak{P} \otimes \mathfrak{P}$, it follows that this polynomial is a product of distinct irreducible factors. Thus $C^{R\otimes R}$ is semi-simple and we have proved

THEOREM 11. *Let $\mathfrak{L}$ be a finite-dimensional Lie algebra over a field of characteristic zero and let $\mathfrak{M}$ and $\mathfrak{N}$ be finite-dimensional completely reducible modules for $\mathfrak{L}$. Then $\mathfrak{M} \otimes \mathfrak{N}$ is completely reducible.*

8. *Representations of the split three-dimensional simple Lie algebra*

In § 1.4 we called a three-dimensional simple Lie algebra $\mathfrak{K}$ split if $\mathfrak{K}$ contains an element h such that ad h has a non-zero characteristic root ρ belonging to the base field. We showed that any such algebra has a basis (e, f, h) with the multiplication table

$$[eh] = 2e, \qquad [fh] = -2f, \qquad [ef] = h . \tag{30}$$

The representation theory of this algebra is the key for unlocking the deeper parts of the structure and representation theory of semi-simple Lie algebras (Chapters IV, VII, and VIII). We consider this now for the case of a field Φ of characteristic 0. We suppose first that Φ is algebraically closed and that $\mathfrak{M}$ is a finite-dimensional module for $\mathfrak{K}$. The representation in $\mathfrak{M}$ is determined by the images E, F, H of the base elements e, f, h and we have

$$[E, H] = 2E , \qquad [F, H] = -2F , \qquad [E, F] = H . \tag{31}$$

Conversely, any three linear transformations E, F, H satisfying these relations determine a representation of $\mathfrak{K}$ and hence a $\mathfrak{K}$-module. Let α be a characteristic root of H and x a corresponding characteristic vector: $x \neq 0$, $xH = \alpha x$. Then

$$(32) \qquad (xE)H = x(HE + 2E) = (xE)(\alpha + 2) .$$

If $xE \neq 0$ then (32) shows that $\alpha + 2$ is a characteristic root for H and xE a corresponding characteristic vector. We can replace x by xE and repeat the process. This leads to a sequence of non-zero vectors $x, xE, xE^2, \cdots$, belonging to the characteristic roots $\alpha, \alpha + 2, \alpha + 4, \cdots$, respectively, for H. Now H has only a finite number of distinct characteristic roots; hence, our sequence breaks off and this means that we obtain a k such that $xE^k \neq 0$ and $xE^{k+1} = 0$.

If we replace x by xE^k we may suppose at the start that $x \neq 0$ and

$$(33) \qquad xH = \alpha x , \qquad xE = 0 .$$

Now set $x_0 = x$ and let $x_i = x_{i-1}F$. Then, analogous to (32), we obtain

$$34) \qquad x_iH = (\alpha - 2i)x_i ,$$

and the argument used for the vectors xE^i shows that there exists a non-negative integer m such that $x_0, x_1, \cdots, x_m$ are $\neq 0$ but $x_{m+1} = 0$. Thus $xF^{m+1} = 0$, $xF^m \neq 0$.

Then x_i, $0 \leqq i \leqq m$, is a characteristic vector of H belonging to the characteristic root $\alpha - 2i$. Since $\alpha, \alpha - 2, \alpha - 4, \cdots, \alpha - 2m$ are all different it follows that the x_i are linearly independent. Let $\mathfrak{N} = \sum_{i=0}^{m} \Phi x_i$ so that $\mathfrak{N}$ is an $(m + 1)$-dimensional subspace of $\mathfrak{M}$. We shall now show that $\mathfrak{N}$ is invariant and irreducible relative to $\mathfrak{K}$. We first establish the formula

$$(35) \qquad x_iE = (-i\alpha + i(i - 1))x_{i-1} .$$

Thus we have $x_0E = 0$ as given in (35). Assume (35) for $i - 1$. Then

$$\begin{aligned} x_i E &= x_{i-1}FE = x_{i-1}(EF - H) \\ &= (-(i - 1)\alpha + (i - 1)(i - 2))x_{i-2}F \\ &\qquad\qquad - (\alpha - 2(i - 1))x_{i-1} \\ &= (-i\alpha + i(i - 1))x_{i-1} \end{aligned}$$

as required. It is now clear from (34), (35), and $x_iF = x_{i+1}$ that $\mathfrak{N}$ is a $\mathfrak{K}$-subspace of $\mathfrak{M}$. Since $H = [EF]$ we must have $\mathrm{tr}_{\mathfrak{N}} H = 0$. This, using (34), gives $(m+1)\alpha - m(m+1) = 0$. Hence we obtain the result that $\alpha = m$. Our formulas now read

$$
\begin{aligned}
& x_iH = (m - 2i)x_i\,, \qquad i = 0, \cdots, m \\
(36) \qquad & x_iF = x_{i+1}\,, \qquad i = 0, \cdots, m-1\,, \qquad x_mF = 0 \\
& x_0E = 0\,, \qquad x_iE = (-mi + i(i-1))x_{i-1}\,, \qquad i = 1, \cdots, m
\end{aligned}
$$

and we note that in the last equation

$$- mi + i(i-1) \neq 0 \,.$$

Now let $\mathfrak{N}_1$ be a non-zero invariant subspace of $\mathfrak{N}$ and let

$$y = \beta_i x_i + \beta_{i+1}x_{i+1} + \cdots + \beta_m x_m \,,$$

$\beta_i \neq 0$, be in $\mathfrak{N}_1$. Then $x_m = \beta_i^{-1}yF^{m-i} \in \mathfrak{N}_1$. Hence by the last equation of (36) every $x_i \in \mathfrak{N}_1$ and $\mathfrak{N}_1 = \mathfrak{N}$. Hence if $\mathfrak{M}$ is $\mathfrak{K}$-irreducible to begin with, then $\mathfrak{M} = \mathfrak{N}$. In general, the theorem on complete reducibility shows that $\mathfrak{M}$ is a direct sum of irreducible invariant subspaces which are like the space $\mathfrak{N}$.

We can now drop the hypothesis that Φ is algebraically closed, assuming only that Φ is of characteristic 0. We note first the following

LEMMA 5. *Let $\mathfrak{K}$ be the split three-dimensional simple Lie algebra over a field Φ of characteristic zero and let $e \to E$, $f \to F$, $h \to H$ define a finite-dimensional representation of $\mathfrak{K}$. Then the characteristic roots of H are integers.*

Proof: If $\mathfrak{M}$ is the module of the representation and Ω is the algebraic closure of Φ then $\mathfrak{M}_\Omega$ is a module for $\mathfrak{K}_\Omega$ which satisfies the same conditions over Ω as $\mathfrak{K}$ over Φ. Then $\mathfrak{M}_\Omega$ is a direct sum of irreducible subspaces $\mathfrak{N}$ with bases $(x_0, x_1, \cdots, x_m)$ satisfying (36). Hence if we choose a suitable basis for $\mathfrak{M}_\Omega$ then the matrix of H relative to this is a diagonal matrix with integral entries. Hence the characteristic roots of H in $\mathfrak{M}_\Omega$ are integers. These are also the characteristic roots of H in $\mathfrak{M}$.

We can now prove the following

THEOREM 12. *Let $\mathfrak{K}$ be the split three-dimensional simple Lie algebra over a field of characteristic 0. Then for each integer $m = 0, 1, 2, \cdots$ there exists one and, in the sense of isomorphism, only* *one irreducible* $\mathfrak{K}$*-module* $\mathfrak{M}$ *of dimension* $m+1$. $\mathfrak{M}$ *has a basis* $(x_0, x_1, \cdots, x_m)$ *such that the representing transformations* (E, F, H) *corresponding to the canonical basis* (e, f, h) *are given by* (36).

Proof: Let $\mathfrak{M}$ be a finite-dimensional irreducible module for $\mathfrak{K}$. Then the characteristic roots of H are integers. Hence we can find an integer α and a vector $x \neq 0$ in $\mathfrak{M}$ such that $xH = \alpha x$. As before we may suppose $xE = 0$. Then we obtain that $\alpha = m$ and that $\mathfrak{M}$ has a basis $(x_0, x_1, \cdots, x_m)$ such that (36) hold. These formulas are completely determined by the dimensionality $m+1$ of $\mathfrak{M}$. Hence any two $(m+1)$-dimensional irreducible modules for $\mathfrak{K}$ are isomorphic. It remains to show that there is an irreducible $(m+1)$-dimensional module for $\mathfrak{K}$ for every $m = 0, 1, \cdots$. To see this we let $\mathfrak{M}$ be a space with the basis $(x_0, x_1, \cdots, x_m)$ and we define the linear transformations E, F, H by (36). Then we have

$$\begin{aligned} x_i(EH-HE) &= (-mi + i(i-1))(m - 2(i-1))x_{i-1} \\ &\quad - (m-2i)(-mi + i(i-1))x_{i-1} \\ &= 2(-mi + i(i-1))x_{i-1} \\ &= 2x_iE\,, \\ x_i(FH-HF) &= (m - 2(i+1))x_{i+1} - (m-2i)x_{i+1} \\ &= -2x_{i+1} \\ &= -2x_iF\,, \\ x_i(EF-FE) &= (-mi + i(i-1))x_i + (m(i+1) - (i+1)i)x_i \\ &= (m-2i)x_i \\ &= x_iH\,. \end{aligned}$$

Hence E, F, and H satisfy the required commutation relations and so they define a representation of $\mathfrak{K}$. As before, $\mathfrak{M}$ is $\mathfrak{K}$-irreducible.

The theorem of complete reducibility applies here also and together with the foregoing result gives the structure of any finite-dimensional $\mathfrak{K}$-module.

9. *The theorems of Levi and Malcev-Harish-Chandra*

The "radical splitting" theorem of Levi asserts that if $\mathfrak{B}$ is a finite-dimensional Lie algebra of characteristic 0 with solvable radical $\mathfrak{S}$, then $\mathfrak{B}$ contains a semi-simple subalgebra $\mathfrak{L}$ such that $\mathfrak{B} = \mathfrak{L} + \mathfrak{S}$. It will follow that $\mathfrak{L} \cap \mathfrak{S} = 0$ so that $\mathfrak{B} = \mathfrak{L} \oplus \mathfrak{S}$ and

$\mathfrak{L} \cong \mathfrak{B}/\mathfrak{S}$. Thus the subalgebra $\mathfrak{L}$ is isomorphic to the difference algebra of $\mathfrak{B}$ modulo its radical. Conversely, if $\mathfrak{B}$ contains a subalgebra $\mathfrak{L}$ isomorphic to $\mathfrak{B}/\mathfrak{S}$, then $\mathfrak{L}$ is semi-simple. Hence $\mathfrak{L} \cap \mathfrak{S} = 0$ and since $\dim \mathfrak{B} = \dim \mathfrak{S} + \dim \mathfrak{B}/\mathfrak{S} = \dim \mathfrak{S} + \dim \mathfrak{L}$, $\mathfrak{B} = \mathfrak{L} + \mathfrak{S}$.

We note next that it suffices to prove the theorem for the case $\mathfrak{S}^2 = 0$, that is, $\mathfrak{S}$ is abelian. Thus suppose $\mathfrak{S}^2 \neq 0$. Then if $\bar{\mathfrak{B}} = \mathfrak{B}/\mathfrak{S}^2$, $\dim \bar{\mathfrak{B}} < \dim \mathfrak{B}$. Hence if we use induction on the dimensionality we may assume the result for $\bar{\mathfrak{B}}$. Now $\bar{\mathfrak{S}} = \mathfrak{S}/\mathfrak{S}^2$ is the radical of $\bar{\mathfrak{B}}$ and $\bar{\mathfrak{B}}/\bar{\mathfrak{S}} \cong \mathfrak{B}/\mathfrak{S}$. Hence $\bar{\mathfrak{B}}$ contains a subalgebra $\bar{\mathfrak{L}} \cong \mathfrak{B}/\mathfrak{S}$. As subalgebra of $\bar{\mathfrak{B}}$, $\bar{\mathfrak{L}}$ has the form $\mathfrak{B}_1/\mathfrak{S}^2$ where $\mathfrak{B}_1$ is a subalgebra of $\mathfrak{B}$ containing $\mathfrak{S}^2$. Now $\mathfrak{S}^2$ is the radical of $\mathfrak{B}_1$ and $\mathfrak{B}_1/\mathfrak{S}^2 \cong \mathfrak{B}/\mathfrak{S}$ so that $\dim \mathfrak{B}_1 < \dim \mathfrak{B}$. The induction hypothesis can therefore be used to conclude that $\mathfrak{B}_1$ contains a subalgebra $\mathfrak{L} \cong \mathfrak{B}/\mathfrak{S}$, and this completes the proof for $\mathfrak{B}$.

We now assume that $\mathfrak{S}^2 = 0$ and for the moment we drop the assumption that $\bar{\mathfrak{B}} = \mathfrak{B}/\mathfrak{S}$ is semi-simple. Now $\mathfrak{S}$ is a submodule of $\mathfrak{B}$ for $\mathfrak{B}$ (adjoint representation). Since $\mathfrak{S}^2 = 0$, $\mathfrak{S}$ is in the kernel of the representation of $\mathfrak{B}$ determined by the module $\mathfrak{S}$. Hence we have an induced representation for $\bar{\mathfrak{B}} = \mathfrak{B}/\mathfrak{S}$. For the corresponding module we have $s\bar{b} = [s, b]$, $s \in \mathfrak{S}$, $b \in \mathfrak{B}$.

We can find a $1:1$ linear mapping $\sigma: \bar{b} \to \bar{b}^\sigma$ of $\bar{\mathfrak{B}}$ into $\mathfrak{B}$ such that $\overline{\bar{b}^\sigma} = \bar{b}$. Such a mapping is obtained by writing $\mathfrak{B} = \mathfrak{S} \oplus \mathfrak{G}$ where $\mathfrak{G}$ is a subspace. Then we have a projection of $\mathfrak{B}$ onto $\mathfrak{G}$ defined by this decomposition. Since $\mathfrak{S}$ is the kernel, we have an induced linear isomorphism σ of $\bar{\mathfrak{B}}$ onto $\mathfrak{G}$; hence into $\mathfrak{B}$. If $b = s + g$, $s \in \mathfrak{S}$, $g \in \mathfrak{G}$, then by definition $\bar{b}^\sigma = g$ and $\bar{b} = \bar{g}$ so that $\overline{\bar{b}^\sigma} = \bar{g} = \bar{b}$ as required. Conversely, let $\bar{b} \to \bar{b}^\sigma$ be any $1:1$ linear mapping of $\bar{\mathfrak{B}}$ into $\mathfrak{B}$ such that $\overline{\bar{b}^\sigma} = \bar{b}$. Then $\mathfrak{G} = \bar{\mathfrak{B}}^\sigma$ is a complement of $\mathfrak{S}$ in $\mathfrak{B}$. If $s \in \mathfrak{S}$ and $\bar{b} \in \bar{\mathfrak{B}}$, then $s\bar{b} = s\bar{g} = [sg] = [s\bar{b}^\sigma]$ holds for the module multiplications in $\mathfrak{S}$.

Let $\bar{b}_1, \bar{b}_2 \in \bar{\mathfrak{B}}$ and consider the element

$$[\bar{b}_1^\sigma, \bar{b}_2^\sigma] - [\bar{b}_1\bar{b}_2]^\sigma \in \mathfrak{B}\,. \tag{37}$$

If we apply the algebra homomorphism $b \to \bar{b}$ of $\mathfrak{B}$ onto $\bar{\mathfrak{B}}$ and make use of the property $\overline{\bar{b}^\sigma} = \bar{b}$ we obtain $\overline{[\bar{b}_1^\sigma \bar{b}_2^\sigma]} = [\overline{\bar{b}_1^\sigma}, \overline{\bar{b}_2^\sigma}] = [\bar{b}_1\bar{b}_2]$ and $\overline{[\bar{b}_1\bar{b}_2]^\sigma} = [\overline{b_1 b_2}] = [\bar{b}_1\bar{b}_2]$. Hence we see that

$$g(\bar{b}_1, \bar{b}_2) \equiv [\bar{b}_1^\sigma, \bar{b}_2^\sigma] - [\bar{b}_1\bar{b}_2]^\sigma \in \mathfrak{S}\,.$$

One verifies immediately that $(\bar{b}_1, \bar{b}_2) \to g(\bar{b}_1, \bar{b}_2)$ is a bilinear mapping of $\bar{\mathfrak{B}} \times \bar{\mathfrak{B}}$ into $\mathfrak{S}$.

Now suppose $\mathfrak{G} = \bar{\mathfrak{B}}^\sigma$ is a subalgebra of $\mathfrak{B}$. Then $[\bar{b}_1^\sigma, \bar{b}_2^\sigma] \in \mathfrak{G}$; hence $g(\bar{b}_1, \bar{b}_2) \in \mathfrak{G} \cap \mathfrak{S} = 0$ so that we must have $g(\bar{b}_1, \bar{b}_2) = 0$ for all $\bar{b}_1, \bar{b}_2$. The converse is also clear since $g(\bar{b}_1, \bar{b}_2) = 0$ implies that $[\bar{b}_1^\sigma, \bar{b}_2^\sigma] = [\bar{b}_1\bar{b}_2]^\sigma \in \mathfrak{G}$. Hence $\mathfrak{G}$ is a subalgebra if and only if the bilinear mapping g is 0.

If $\bar{\mathfrak{B}}^\sigma$ is not a subalgebra, then we seek to modify σ to obtain a second mapping τ of $\bar{\mathfrak{B}}$ so that $\bar{\mathfrak{B}}^\tau$ is a subalgebra. Suppose this is possible. Then we have a $1:1$ linear mapping τ of $\bar{\mathfrak{B}}$ into $\mathfrak{B}$ such that $\overline{\bar{b}^\tau} = \bar{b}$ and $[\bar{b}_1^\tau \bar{b}_2^\tau] - [\bar{b}_1\bar{b}_2]^\tau = 0$ for all $\bar{b}_1, \bar{b}_2$. Now let $\rho = \sigma - \tau$. Then ρ is a linear mapping of $\bar{\mathfrak{B}}$ into $\mathfrak{B}$ such that

$$\overline{\bar{b}^\rho} = \overline{\bar{b}^\sigma} - \overline{\bar{b}^\tau} = \bar{b} - \bar{b} = 0 \,.$$

Hence $\bar{b}^\rho \in \mathfrak{S}$, and we can consider ρ as a linear mapping of $\bar{\mathfrak{B}}$ into $\mathfrak{S}$. Also we have

$$\begin{aligned} g(\bar{b}_1, \bar{b}_2) &= [\bar{b}_1^\sigma, \bar{b}_2^\sigma] - [\bar{b}_1\bar{b}_2]^\sigma \\ &= [\bar{b}_1^\rho + \bar{b}_1^\tau, \bar{b}_2^\rho + \bar{b}_2^\tau] - [\bar{b}_1\bar{b}_2]^\rho - [\bar{b}_1\bar{b}_2]^\tau \\ &= [\bar{b}_1^\rho \bar{b}_2^\tau] - [\bar{b}_2^\rho, \bar{b}_1^\tau] - [\bar{b}_1\bar{b}_2]^\rho \,. \end{aligned}$$

If $s\bar{b}$ is defined as before, we have $s\bar{b} = [s\bar{b}^\tau]$. Thus, if we can somehow choose a complement of $\mathfrak{S}$ which is a subalgebra then the bilinear mapping $g(\bar{b}_1, \bar{b}_2)$ of $\bar{\mathfrak{B}} \times \bar{\mathfrak{B}}$ into $\mathfrak{S}$ can be expressed in terms of the linear mapping ρ of $\bar{\mathfrak{B}}$ into $\mathfrak{S}$ by the formula

$$g(\bar{b}_1, \bar{b}_2) = \bar{b}_1^\rho \bar{b}_2 - \bar{b}_2^\rho \bar{b}_1 - [\bar{b}_1\bar{b}_2]^\rho \,. \tag{38}$$

Conversely, suppose we have a linear mapping ρ of $\bar{\mathfrak{B}}$ into $\mathfrak{S}$ satisfying this condition. Then $\tau = \sigma - \rho$ is another $1:1$ linear mapping of $\bar{\mathfrak{B}}$ into $\mathfrak{B}$ such that $\overline{\bar{b}^\tau} = \bar{b}$ and one can re-trace the steps to show that $[\bar{b}_1^\tau, \bar{b}_2^\tau] = [\bar{b}_1\bar{b}_2]^\tau$ so that $\bar{\mathfrak{B}}^\tau$ is subalgebra.

Our results can be stated in the following way:

Criterion. Let $\mathfrak{B}$ be a Lie algebra, $\mathfrak{S}$ an ideal in $\mathfrak{B}$ such that $\mathfrak{S}^2 = 0$ and set $\bar{\mathfrak{B}} = \mathfrak{B}/\mathfrak{S}$. Then $\mathfrak{S}$ is a $\bar{\mathfrak{B}}$-module relative to the composition $s\bar{b} = [sb]$. Also there exist $1:1$ linear mappings σ of $\bar{\mathfrak{B}}$ into $\mathfrak{B}$ such that $\overline{\bar{b}^\sigma} = \bar{b}$, $\bar{b} \in \bar{\mathfrak{B}}$. If σ is such a mapping then

$$g(\bar{b}_1, \bar{b}_2) = [\bar{b}_1^\sigma \bar{b}_2^\sigma] - [\bar{b}_1\bar{b}_2]^\sigma \in \mathfrak{S} \,.$$

Moreover, $\mathfrak{S}$ has a complementary space which is a subalgebra if and only if there exists a linear mapping ρ of $\bar{\mathfrak{B}}$ into $\mathfrak{S}$ such that

$$g(\bar{b}_1, \bar{b}_2) = \bar{b}_1^\rho \bar{b}_2 - \bar{b}_2^\rho \bar{b}_1 - [\bar{b}_1\bar{b}_2]^\rho \,.$$

We observe next that the bilinear mapping g, which we shall call a $\bar{\mathfrak{B}}$ *factor set* in $\mathfrak{S}$, satisfies certain conditions which are consequences of the special properties of the multiplication in a Lie algebra. Thus it is clear that

$$g(\bar{b}, \bar{b}) = 0 \tag{39}$$

which implies $g(\bar{b}_1, \bar{b}_2) = -g(\bar{b}_2, \bar{b}_1)$. We next write

$$[\bar{b}_1^\sigma, \bar{b}_2^\sigma] = [\bar{b}_1\bar{b}_2]^\sigma + g(\bar{b}_1, \bar{b}_2)$$

and calculate

$$\begin{aligned}[][[\bar{b}_1^\sigma, \bar{b}_2^\sigma]\bar{b}_3^\sigma] &= [[\bar{b}_1\bar{b}_2]^\sigma\bar{b}_3^\sigma] + [g(\bar{b}_1, \bar{b}_2)\bar{b}_3^\sigma] \\ &= [[\bar{b}_1\bar{b}_2]\bar{b}_3]^\sigma + g([\bar{b}_1\bar{b}_2], \bar{b}_3) + [g(\bar{b}_1, \bar{b}_2), \bar{b}_3^\sigma] .\end{aligned}$$

If we permute $\bar{b}_1, \bar{b}_2, \bar{b}_3$ cyclically, add, and make use of the Jacobi identities in $\bar{\mathfrak{B}}$ and $\mathfrak{B}$, we obtain

$$\begin{aligned} &g([\bar{b}_1\bar{b}_2], \bar{b}_3) + [g(\bar{b}_1, \bar{b}_2), \bar{b}_3^\sigma] + g([\bar{b}_2\bar{b}_3], \bar{b}_1) \\ &\quad + [g(\bar{b}_2, \bar{b}_3), \bar{b}_1^\sigma] + g([\bar{b}_3\bar{b}_1], \bar{b}_2) + [g(\bar{b}_3, \bar{b}_1), \bar{b}_2^\sigma] = 0 .\end{aligned} \tag{40}$$

Our proof of Levi's theorem will be completed by proving the following lemma, which is due to Whitehead.

LEMMA 6. *Let $\mathfrak{L}$ be a finite-dimensional semi-simple Lie algebra of characteristic* 0, *$\mathfrak{M}$ a finite-dimensional $\mathfrak{L}$-module and $(l_1, l_2) \rightarrow g(l_1, l_2)$ a bilinear mapping of $\mathfrak{L} \times \mathfrak{L}$ into $\mathfrak{M}$ such that*

(i) $g(l, l) = 0$,

(ii) $g([l_1l_2], l_3) + g(l_1, l_2)l_3 + g([l_2l_3], l_1) + g(l_2, l_3)l_1 + g([l_3l_1], l_2) + g(l_3, l_1)l_2 = 0$.

Then there exists a linear mapping $l \rightarrow l^\rho$ of $\mathfrak{L}$ into $\mathfrak{M}$ such that

(iii) $g(l_1, l_2) = l_1^\rho l_2 - l_2^\rho l_1 - [l_1l_2]^\rho$.

Proof: Let $\mathfrak{R}, \mathfrak{L}_1, u_i, u^i, \Gamma$ be as in the proof of Whitehead's first lemma: $\mathfrak{R}$ is the kernel of the representation, $\mathfrak{L}_1$ is an ideal such that $\mathfrak{L} = \mathfrak{R} \oplus \mathfrak{L}_1$, (u_i), and (u^i), $i = 1, \cdots, m$, are dual bases of $\mathfrak{L}_1$ relative to the trace form of the given representation, and Γ is the Casimir operator determined by the u_i and u^i. We recall that Γ is the mapping $x \rightarrow \sum_{i=1}^m (xu_i)u^i$ in $\mathfrak{M}$. Set $l_3 = u_i$ in (ii) and take the module product with u^i. Add for i. This gives

$$\begin{aligned}0 &= \sum_i g([l_1 l_2], u_i)u^i + g(l_1, l_2)\Gamma \\ &\quad + \sum_i g([l_2 u_i], l_1)u^i + \sum_i (g(l_2, u_i)l_1)u^i \\ &\quad + \sum_i g([u_i l_1], l_2)u^i + \sum_i (g(u_i, l_1)l_2)u^i \\ &= g(l_1, l_2)\Gamma + \sum_i g([l_1 l_2], u_i)u^i \\ &\quad + \sum_i g([l_2 u_i], l_1)u^i + \sum_i g(l_2, u_i)[l_1 u^i] \\ &\quad + \sum_i (g(l_2, u_i)u^i)l_1 + \sum_i g([u_i, l_1], l_2)u^i \\ &\quad + \sum_i g(u_i, l_1)[l_2 u^i] + \sum_i (g(u_i l_1)u^i)l_2 .\end{aligned}$$

If we make use of $[u_i l] = \sum \alpha_{ij} u_j$, $[u^i l] = \sum \beta_{ij} u^j$, and recall that $\beta_{ij} = -\alpha_{ji}$ (cf. (28)) we can verify that

$$\sum g(l_2, u_i)[l_1 u^i] = \sum g(l_2, [u_i l_1])u^i \tag{41}$$

$$\sum g(u_i, l_1)[l_2 u^i] = \sum g([u_i l_2], l_1)u^i . \tag{42}$$

These and the skew symmetry of g permit the cancellation of four terms in the foregoing equations. Hence we obtain

$$\begin{aligned}-g(l_1, l_2)\Gamma &= \sum g([l_1, l_2], u_i)u^i \\ &\quad + \sum (g(l_2, u_i)u^i)l_1 + \sum_i (g(u_i, l_1)u^i)l_2 .\end{aligned} \tag{43}$$

If Γ is non-singular we define

$$l^\rho = \sum_{i=1}^{m} g(l, u_i)u^i \Gamma^{-1} . \tag{44}$$

Then (43) gives the required relation (iii). If Γ is nilpotent, then, as in the proof of Whitehead's first lemma, $m = 0$, $\mathfrak{K} = \mathfrak{L}$, so that the representation is a zero representation. Then (ii) reduces to

$$g([l_1 l_2], l_3) + g([l_2 l_3], l_1) + g([l_3 l_1], l_2) = 0 . \tag{ii$'$}$$

Now let $\mathfrak{X}$ denote the vector space of linear mappings of $\mathfrak{L}$ into $\mathfrak{M}$. We make this into an $\mathfrak{L}$-module by defining for $A \in \mathfrak{X}$, $x, l \in \mathfrak{L}$, $x(Al) \equiv -[xl]A$, that is, $Al \equiv -(\operatorname{ad} l)A$. It is easy to see that this satisfies the module conditions (cf. § 1.6). For each $l \in \mathfrak{L}$ we define an element $A_l \in \mathfrak{X}$ as the mapping $x \to g(x, l) \in \mathfrak{M}$. Then $l \to A_l$ is a linear mapping of $\mathfrak{L}$ into $\mathfrak{X}$ and

$$xA_{[l_1l_2]} = g(x, [l_1l_2]),$$
$$xA_{l_1}l_2 = -g([xl_2], l_1),$$
$$xA_{l_2}l_1 = -g([xl_1], l_2).$$

Hence the skew symmetry of g and (ii′) imply that

$$A_{[l_1l_2]} = A_{l_1}l_2 - A_{l_2}l_1. \tag{45}$$

Thus the hypothesis of Whitehead's first lemma holds. The conclusion states that there exists a $\rho \in \mathfrak{X}$ such that $A_l = \rho l$. This means that we have a linear mapping ρ of $\mathfrak{L}$ into $\mathfrak{M}$ such that

$$g(x, l) = -[x, l]^\rho. \tag{46}$$

By definition of $\mathfrak{X}$ as module, this gives (iii). This proves the result for the case Γ nilpotent. If Γ is neither non-singular nor nilpotent, then we have the decomposition of $\mathfrak{M}$ as $\mathfrak{M}_0 \oplus \mathfrak{M}_1$ where the $\mathfrak{M}_i$ are the Fitting components of $\mathfrak{M}$ relative to Γ and these are $\neq 0$. These spaces are submodules and we can write $g(l_1, l_2) = g_0(l_1, l_2) + g_1(l_1, l_2)$, $g_i \in \mathfrak{M}_i$. Then the g_i satisfy the conditions imposed on g, so we can represent these in the form (iii), by virtue of an induction hypothesis on the dimensionality of $\mathfrak{M}$. This gives the result for $\mathfrak{M}$ by adding the linear transformations for the $\mathfrak{M}_i$.

As we have noted before, the lemma completes our proof of

Levi's theorem. If $\mathfrak{B}$ is a finite-dimensional Lie algebra of characteristic zero with radical $\mathfrak{S}$ then there exists a semi-simple subalgebra $\mathfrak{L}$ of $\mathfrak{B}$ such that $\mathfrak{B} = \mathfrak{L} \oplus \mathfrak{S}$.

A subalgebra $\mathfrak{L}$ satisfying these conditions is called a *Levi factor* of $\mathfrak{B}$. A first consequence of Levi's theorem is the following result:

COROLLARY 1. *Let $\mathfrak{B}$, $\mathfrak{S}$, and $\mathfrak{L}$ be as in the theorem. Then* $[\mathfrak{B}\mathfrak{S}] = \mathfrak{B}' \cap \mathfrak{S}$.

Proof: We have $\mathfrak{B} = \mathfrak{L} \oplus \mathfrak{S}$ so that $\mathfrak{B}' = [\mathfrak{L}\mathfrak{L}] + [\mathfrak{B}\mathfrak{S}]$. Since $[\mathfrak{B}\mathfrak{S}] \subseteq \mathfrak{S}$ have $\mathfrak{B}' \cap \mathfrak{S} = ([\mathfrak{L}\mathfrak{L}] \cap \mathfrak{S}) + [\mathfrak{B}\mathfrak{S}] = [\mathfrak{B}\mathfrak{S}]$.

We have seen that $[\mathfrak{S}\mathfrak{B}] \subseteq \mathfrak{N}$ the nil radical of $\mathfrak{B}$ (Theorem 2.13), so we can now state that $\mathfrak{B}' \cap \mathfrak{S} \subseteq \mathfrak{N}$. We know also that the radical of an ideal is the intersection of the ideal with the radical of the containing algebra. Hence $\mathfrak{B}' \cap \mathfrak{S}$ is the radical of $\mathfrak{B}'$. We therefore have the following

COROLLARY 2. *The radical of the derived algebra of a finite-dimensional Lie algebra of characteristic* 0 *is nilpotent.*

We take up next the question of uniqueness of the Levi factors. It will turn out that these are not usually unique; however, they are conjugate in a rather strong sense which we shall now define. We recall that if $z \in \mathfrak{N}$, the nil radical of $\mathfrak{B}$, then $\operatorname{ad} z$ is nilpotent. Since $\operatorname{ad} z$ is a derivation we know also that $A = \exp(\operatorname{ad} z)$ is an automorphism. Let $\mathfrak{A}$ denote the group of automorphisms generated by the elements $\exp(\operatorname{ad} z)$, $z \in \mathfrak{N}$. Then we have the following conjugacy

Theorem of Malcev-Harish-Chandra. Let $\mathfrak{B} = \mathfrak{S} \oplus \mathfrak{L}$ *where* $\mathfrak{S}$ *is a solvable ideal and* $\mathfrak{L}$ *is a semi-simple subalgebra and let* $\mathfrak{L}_1$ *be a semi-simple subalgebra of* $\mathfrak{B}$. *Assume* $\mathfrak{B}$ *finite-dimensional and of characteristic* 0. *Then there exists an automorphism* $A \in \mathfrak{A}$ *such that* $\mathfrak{L}_1^A \subseteq \mathfrak{L}$.

Proof: Any $l_1 \in \mathfrak{L}_1$ can be written in one and only one way as $l_1 = l_1^\lambda + l_1^\sigma$, where $l_1^\lambda \in \mathfrak{L}$ and $l_1^\sigma \in \mathfrak{S}$ so that we have the linear mappings λ and σ of $\mathfrak{L}_1$ into $\mathfrak{L}$ and $\mathfrak{S}$, respectively. Since $\mathfrak{L}_1$ is semi-simple, $\mathfrak{L}_1 \cap \mathfrak{S} = 0$; hence λ is $1:1$. If $l_2 \in \mathfrak{L}_1$ then

$$\begin{aligned}[l_1 l_2] &= [l_1 l_2]^\lambda + [l_1 l_2]^\sigma \\ &= [l_1^\lambda l_2^\lambda] + [l_1^\lambda l_2^\sigma] + [l_1^\sigma l_2^\lambda] + [l_1^\sigma l_2^\sigma]\,.\end{aligned} \tag{47}$$

Hence

$$\begin{aligned}[l_1 l_2]^\lambda &= [l_1^\lambda l_2^\lambda]\,, \\ [l_1 l_2]^\sigma &= [l_1^\sigma l_2^\lambda] - [l_2^\sigma l_1^\lambda] + [l_1^\sigma l_2^\sigma]\,.\end{aligned} \tag{48}$$

The second of these equations shows that $[l_1 l_2]^\sigma \in [\mathfrak{B}\mathfrak{S}] \subseteq \mathfrak{N}$, the nil radical of $\mathfrak{B}$. Since $\mathfrak{L}_1' = \mathfrak{L}_1$ this implies that $l_1^\sigma \in \mathfrak{N}$ for every $l_1 \in \mathfrak{L}_1$ and so $\mathfrak{L}_1 \subseteq \mathfrak{L} \oplus \mathfrak{N}$. We shall prove by induction that there exists an automorphism $A_i \in \mathfrak{A}$ $(A_1 = 1)$ such that $\mathfrak{L}_1^{A_i} \subseteq \mathfrak{L} + \mathfrak{N}^{(i)}$ where $\mathfrak{N}^{(i)}$ is the ith derived algebra of $\mathfrak{N}$. Since $\mathfrak{N}$ is solvable this will prove the result. Since we have proved that $\mathfrak{L}_1 \subseteq \mathfrak{L} + \mathfrak{N}$ it suffices to prove the inductive step and we may simplify the notation and assume that $\mathfrak{L}_1 \subseteq \mathfrak{L} + \mathfrak{N}^{(k)}$. Then we shall show that there exists $A \in \mathfrak{A}$ such that $\mathfrak{L}_1^A \subseteq \mathfrak{L} + \mathfrak{N}^{(k+1)}$. If we use the notation introduced before, $\mathfrak{L}_1 \subseteq \mathfrak{L} + \mathfrak{N}^{(k)}$ implies that $l_1^\sigma \in \mathfrak{N}^{(k)}$, $l_1 \in \mathfrak{L}_1$. The first equation in (48) implies that if we set $zl_1 = [z, l_1^\lambda]$, $z \in \mathfrak{N}^{(k)}$, $l_1 \in \mathfrak{L}_1$, then this makes $\mathfrak{N}^{(k)}$ into an $\mathfrak{L}_1$-module. Now $\mathfrak{N}^{(k+1)}$ is a submodule so that $\mathfrak{N}^{(k)}/\mathfrak{N}^{(k+1)}$ is an $\mathfrak{L}_1$-module relative to $\bar{z}l_1 = \overline{[zl_1^\lambda]}$ where $z \in \mathfrak{N}^{(k)}$ and $\bar{z} = z + \mathfrak{N}^{(k+1)}$. We now take the cosets relative to $\mathfrak{N}^{(k+1)}$ of the terms in the second equation of (48). Since $[l_1^\sigma l_2^\sigma] \in \mathfrak{N}^{(k+1)}$, we have

$$\overline{[l_1 l_2]^\sigma} = \overline{[l_1^\sigma l_2^\lambda]} - \overline{[l_2^\sigma l_1^\lambda]} = \overline{l_1^\sigma} l_2 - \overline{l_2^\sigma} l_1 .$$

Now set $f(l_1) = \overline{l_1^\sigma}$; then $l_1 \to f(l_1)$ is a linear mapping of $\mathfrak{L}_1$ into the $\mathfrak{L}_1$-module $\mathfrak{N}^{(k)}/\mathfrak{N}^{(k+1)}$ and the foregoing equation can be re-written as

(49) $$f([l_1 l_2]) = f(l_1)l_2 - f(l_2)l_1 .$$

Hence by Whitehead's first lemma there exists a $\bar{z} \in \mathfrak{N}^{(k)}/\mathfrak{N}^{(k+1)}$ such that $f(l_1) = \bar{z}l_1$, which means that

(50) $$\overline{l_1^\sigma} = \overline{[z, l_1^\lambda]} \text{ or } l_1^\sigma \equiv [z, l_1^\lambda] \,(\text{mod}\, \mathfrak{N}^{(k+1)}) .$$

Let $A = \exp(\text{ad}\, z)$. Then

(51) $$\begin{aligned} l_1^A &= l_1 + [l_1 z] + 1/2!\, [[l_1 z]z] + \cdots \\ &\equiv l_1 + [l_1 z] \,(\text{mod}\, \mathfrak{N}^{(k+1)}) \\ &\equiv l_1^\lambda + l_1^\sigma + [l_1^\lambda z] + [l_1^\sigma z] \,(\text{mod}\, \mathfrak{N}^{(k+1)}) \\ &\equiv l_1^\lambda \,(\text{mod}\, \mathfrak{N}^{(k+1)}) . \end{aligned}$$

Now $\mathfrak{L}_1^A \cong \mathfrak{L}_1$ and (51) shows that $\mathfrak{L}_1^A \subseteq \mathfrak{L} + \mathfrak{N}^{(k+1)}$. We can therefore prove the result by induction on k.

Corollary 1. *Any semi-simple subalgebra of a finite-dimensional Lie algebra of characteristic zero can be imbedded in a Levi factor.*

Proof: If A is as in the theorem, then $\mathfrak{L}_1$ is contained in the Levi factor $\mathfrak{L}^{A^{-1}}$.

Corollary 2. *If $\mathfrak{B} = \mathfrak{L}_1 \oplus \mathfrak{S} = \mathfrak{L}_2 \oplus \mathfrak{S}$ where $\mathfrak{L}_1$ and $\mathfrak{L}_2$ are semi-simple subalgebras then there exists an automorphism $A \in \mathfrak{A}$ such that $\mathfrak{L}_1^A = \mathfrak{L}_2$.*

This is an immediate consequence of the theorem.

10. Cohomology groups of a Lie algebra

The two lemmas of Whitehead can be formulated as theorems in the cohomology theory of Lie algebras. Historically, these constituted one of the clues which led to the discovery of this theory. Another impetus to the theory came from the study of the topology of Lie groups which was initiated by Cartan. In this section we give the definition of the cohomology groups which is concrete and we indicate an extension of the "Γ non-singular" case of Whitehead's lemmas to a general cohomology theorem. Later (Chapter

V) we shall give the definition of the cohomology groups which follows the general pattern of derived functors of Cartan-Eilenberg.

Let $\mathfrak{L}$ be a Lie algebra, $\mathfrak{M}$ an $\mathfrak{L}$-module. If $i \geqq 1$, an *i-dimensional $\mathfrak{M}$-cochain* for $\mathfrak{L}$ is a skew symmetric i-linear mapping of $\mathfrak{L} \times \mathfrak{L} \times \cdots \times \mathfrak{L}$ (i times) into $\mathfrak{M}$. Such a mapping f sends an i-tuple $(l_1, l_2, \cdots, l_i)$, $l_q \in \mathfrak{L}$, into $f(l_1, \cdots, l_i) \in \mathfrak{M}$ in such a way that for fixed values of $l_1, \cdots, l_{q-1}, l_{q+1}, \cdots, l_i$ the mapping $l_q \rightarrow f(l_1, \cdots, l_i)$ is a linear mapping of $\mathfrak{L}$ into $\mathfrak{M}$. The skew symmetry means that f is changed to $-f$ if any two of the l_i are interchanged (the remaining ones unchanged). If $i = 0$ one defines a *0-dimensional $\mathfrak{M}$-cochain* for $\mathfrak{L}$ as a "constant" function from $\mathfrak{L}$ to $\mathfrak{M}$, that is, a mapping $l \rightarrow u$, u a fixed element of $\mathfrak{M}$. If f is an i-dimensional cochain (or simply "an i-cochain"), $i \geqq 0$, f determines an $(i + 1)$-dimensional cochain $f\delta$, called the *coboundary* of f, defined by the formula

$$f\delta(l_1, \cdots, l_{i+1}) = \sum_{q=1}^{i+1}(-1)^{i+1-q} f(l_1, \cdots, \hat{l}_q, \cdots, l_{i+1})l_q \tag{52}$$
$$+ \sum_{q<r=1}^{i+1} (-1)^{r+q} f(l_1, \cdots, \hat{l}_q, \cdots, \hat{l}_r, \cdots, l_{i+1}, [l_q l_r]) \, .$$

Here the $\hat{}$ over an argument means that this argument is omitted (e.g., $f(l_1, \hat{l}_2, l_3) = f(l_1, l_3)$). For $i = 0$ this is to be interpreted as $(f\delta)(l) = ul$, if f is the mapping $x \rightarrow u \in \mathfrak{M}$.

The set $C^i(\mathfrak{L}, \mathfrak{M})$ of i-cochains for $\mathfrak{M}$ is a vector space relative to the usual definitions of addition and scalar multiplication of functions. Moreover $f \rightarrow f\delta$ is a linear mapping, the coboundary operator, of $C^i(\mathfrak{L}, \mathfrak{M})$ into $C^{i+1}(\mathfrak{L}, \mathfrak{M})$, $i \geqq 0$. Besides the case

$$f\delta(l) = ul \, , \qquad \text{if } f\colon x \rightarrow u \, , \tag{53}$$

we have

$$f\delta(l_1, l_2) = -f(l_2)l_1 + f(l_1)l_2 - f([l_1 l_2]) \, , \tag{54}$$

$$(f\delta)(l_1, l_2, l_3) = f(l_2, l_3)l_1 - f(l_1, l_3)l_2 + f(l_1, l_2)l_3 \tag{55}$$
$$- f(l_3, [l_1 l_2]) + f(l_2, [l_1 l_3]) - f(l_1, [l_2 l_3]) \, .$$

An i-cochain f is called a *cocycle* if $f\delta = 0$ and a *coboundary* if $f = g\delta$ for some $(i - 1)$-cochain g. The set $Z^i(\mathfrak{L}, \mathfrak{M})$ of i-cocycles is the kernel of the homomorphism δ of C^i into C^{i+1}, so Z^i is a subspace of C^i. Similarly, the set $B^i(\mathfrak{L}, \mathfrak{M})$ of i-coboundaries is a subspace of C^i since it is the image under δ of C^{i-1}. It can be proved fairly directly that $B^i \subseteq Z^i$, that is, coboundaries are cocycles.

This amounts to the fundamental property: $\delta^2=0$ of the coboundary operator. We shall not give the verification in the general case at this point since it will follow from the abstract point of view later on. At this point we shall be content to verify $f\delta^2=0$ for f a 0- or a 1-cochain. Thus if $f=u$, that is, f is the mapping $x\rightarrow u$, then $f\delta(l)=ul$ and $f\delta^2(l_1,l_2)=-ul_2l_1+ul_1l_2-u[l_1l_2]=0$ by the definition of a module. If f is a 1-cochain, $f\delta(l_1,l_2)$ is given by (54). Hence, by (55),

$$\begin{aligned} f\delta^2(l_1,l_2,l_3) = &-f(l_3)l_2l_1+f(l_2)l_3l_1-f([l_2l_3])l_1\\ &+f(l_3)l_1l_2-f(l_1)l_3l_2+f([l_1l_3])l_2-f(l_2)l_1l_3\\ &+f(l_1)l_2l_3-f([l_1l_2])l_3+f([l_1l_2])l_3-f(l_3)[l_1l_2]\\ &+f([l_3[l_1l_2]])-f([l_1l_3])l_2+f(l_2)[l_1l_3]\\ &-f([l_2[l_1l_3]])+f([l_2l_3])l_1-f(l_1)[l_2l_3]\\ &+f([l_1[l_2l_3]])\,. \end{aligned}$$

One checks that this sum is 0; hence $f\delta^2=0$ for any 1-cochain f.

Once the verification $\delta^2=0$ has been made, one can define the *i-dimensional cohomology group (space) of $\mathfrak{L}$ relative to the module $\mathfrak{M}$* as the factor space $H^i(\mathfrak{L},\mathfrak{M})\equiv Z^i(\mathfrak{L},\mathfrak{M})/B^i(\mathfrak{L},\mathfrak{M})$. If $i=0$ we agree to take $B^i=0$ since there are no $(i-1)$-cochains. Hence in this case it is understood that $H^0(\mathfrak{L},\mathfrak{M})=Z^0(\mathfrak{L},\mathfrak{M})$. This can be identified with the subspace $I(\mathfrak{M})$ of elements $u\in\mathfrak{M}$ such that $ul=0$ for all l. Such elements are called *invariants* of the module $\mathfrak{M}$. $H^i(\mathfrak{L},\mathfrak{M})=0$ means that $Z^i(\mathfrak{L},\mathfrak{M})=B^i(\mathfrak{L},\mathfrak{M})$, that is, every i-cocycle is a coboundary. For $i=1$ this states that if $l\rightarrow f(l)$ is a linear mapping of $\mathfrak{L}$ into $\mathfrak{M}$ such that $-f(l_2)l_1+f(l_1)l_2-f([l_1l_2])=0$, then there exists a u in $\mathfrak{M}$ such that $f(l)=ul$. This is just the type of statement which appears in Whitehead's first lemma. Similarly, Whitehead's second lemma is a statement about the second cohomology groups. In fact, these two results can now be stated in the following way.

THEOREM 13. *If $\mathfrak{L}$ is finite-dimensional semi-simple of characteristic 0, then $H^1(\mathfrak{L},\mathfrak{M})=0$ and $H^2(\mathfrak{L},\mathfrak{M})=0$ for every finite-dimensional module $\mathfrak{M}$ of $\mathfrak{L}$.*

It is easy to see that if $\mathfrak{M}=\mathfrak{M}_1\oplus\mathfrak{M}_2$ where the $\mathfrak{M}_i$ are submodules of $\mathfrak{M}$, then $H^i(\mathfrak{L},\mathfrak{M})=H^i(\mathfrak{L},\mathfrak{M}_1)\oplus H^i(\mathfrak{L},\mathfrak{M}_2)$. This and the theorem of complete reducibility permits the reduction of the $H^i(\mathfrak{L},\mathfrak{M})$ for finite-dimensional $\mathfrak{M}$ to the case $\mathfrak{M}$ irreducible and here one distinguishes two cases: (1) $\mathfrak{M}\mathfrak{L} \neq 0$ and (2) $\mathfrak{M}\mathfrak{L} = 0$. In the second case irreducibility implies $\dim \mathfrak{M} = 1$, so $\mathfrak{M}$ can be identified with the field Φ. Then an i-cochain is a skew symmetric i-linear function of $(l_1, \cdots, l_i)$ with values in Φ, and since the representation is a zero representation, the coboundary formula reduces to

$$(56) \quad f\delta(l_1, \cdots, l_{i+1}) = \sum_{q<r=1}^{i+1} (-1)^{r+q} f(l_1, \cdots, \hat{l}_q, \cdots, \hat{l}_r, \cdots, l_{i+1}, [l_q l_r])$$

It turns out that for semi-simple Lie algebras the cohomology groups with values in $\mathfrak{M} = \Phi$ are the really interesting ones, since these correspond to cohomology groups of Lie groups. On the other hand, the case $\mathfrak{M}\mathfrak{L} \neq 0$ is not very interesting (for semi-simple $\mathfrak{L}$, finite-dimensional irreducible $\mathfrak{M}$) except for its applications to the theorem of complete reducibility and the Levi theorem, since one has the following general result.

THEOREM 14 (*Whitehead*). *Let $\mathfrak{L}$ be a finite-dimensional semi-simple Lie algebra over a field of characteristic* 0 *and let $\mathfrak{M}$ be a finite-dimensional irreducible module such that $\mathfrak{M}\mathfrak{L} \neq 0$. Then $H^i(\mathfrak{L}, \mathfrak{M}) = 0$ for all $i \geqq 0$.*

If $i = 0$ the irreducibility and $\mathfrak{M}\mathfrak{L} \neq 0$ imply that $u\mathfrak{L} = 0$ holds only for $u = 0$. This means that $H^0(\mathfrak{L}, \mathfrak{M}) = 0$. The proof for $i > 0$ is similar to the proof of the case: Γ non-singular, in the two Whitehead lemmas. We leave the details to the reader.

11. More on complete reducibility

For our further study of this question we require a notion of a type of closure for Lie algebras of linear transformations and an imbedding theorem for nilpotent elements in three-dimensional split simple algebras. The first of these is based on a special case of a property of associative algebras (the so-called Wedderburn principal theorem), which is the analogue of Levi's theorem on Lie algebras. The result is the following

THEOREM 15. *Let $\mathfrak{A} = \Phi[x]$ be a finite-dimensional algebra (associative with identity* 1) *generated by a single element x over Φ of characteristic zero and let $\mathfrak{R}$ be the radical of $\mathfrak{A}$. Then $\mathfrak{A}$ contains a semi-simple subalgebra $\mathfrak{A}_1$ such that $\mathfrak{A} = \mathfrak{A}_1 \oplus \mathfrak{R}$.*

Proof: Let $f(\lambda)$ be the minimum polynomial of x and let

(57) $$f(\lambda) = \pi_1(\lambda)^{e_1}\pi_2(\lambda)^{e_2}\cdots\pi_r(\lambda)^{e_r}$$

be the factorization of $f(\lambda)$ into irreducible polynomials with the leading coefficients one such that $\pi_i(\lambda) \neq \pi_j(\lambda)$ if $i \neq j$ and $\deg \pi_i(\lambda) > 0$. We note first that if all the $e_i = 1$, $\mathfrak{A}$ has no non-zero nilpotent elements (cf. p. 47), so $\mathfrak{A}$ is semi-simple and there is nothing to prove. In any case, set

(58) $$f_1(\lambda) = \pi_1(\lambda)\pi_2(\lambda)\cdots\pi_r(\lambda)$$

and $z = f_1(x)$. Then if $e = \max(e_i)$, $z^e = (f_1(x))^e = \pi_1(x)^e\cdots\pi_r(x)^e = 0$, so that z is nilpotent. Since $\mathfrak{A}$ is commutative, the ideal (z) generated by z is nilpotent; hence $(z) \subseteq \mathfrak{R}$. On the other hand, $f_1(x) \equiv 0 \pmod{(z)}$. Hence the minimum polynomial of the coset $\bar{x} = x + (z)$ in $\mathfrak{A}/(z)$ is a product of distinct prime factors. Since $\bar{x}$ generates $\mathfrak{A}/(z)$, this means that $\mathfrak{A}/(z)$ is semi-simple. Hence $(z) = \mathfrak{R}$. It follows also easily that the minimum polynomial of $\bar{x} = x + \mathfrak{R}$ is $f_1(\lambda)$. Hence it suffices to prove that $\mathfrak{A}$ contains an element y whose minimum polynomial is $f_1(\lambda)$. We shall obtain such an element by a method of "successive approximations" beginning with $x_1 = x$. To begin with we have $f_1(x_1) \equiv 0 \pmod{\mathfrak{R}}$ and $x \equiv x_1 \pmod{\mathfrak{R}}$. Suppose we have already determined x_k such that $f_1(x_k) \equiv 0 \pmod{\mathfrak{R}^k}$ and $x \equiv x_k \pmod{\mathfrak{R}}$. Set $x_{k+1} = x_k + w$ where w is to be determined in $\mathfrak{R}^k$ so that $f_1(x_{k+1}) \equiv 0 \pmod{\mathfrak{R}^{k+1}}$. We have, by Taylor's theorem for polynomials,

$$f_1(x_{k+1}) = f_1(x_k + w) = f_1(x_k) + f_1'(x_k)w + \frac{f_1''(x_k)}{2!}w^2 + \cdots .$$

Since the base field is of characteristic 0, $f_1(\lambda)$ has distinct roots in the algebraic closure of Φ. Hence $f_1(\lambda)$ is prime to the derivative $f_1'(\lambda)$. It follows that $\bar{u} = \overline{f_1'(x_k)} = f_1'(x_k) + \mathfrak{R}$ has an inverse $\bar{v}$ in $\mathfrak{A}/\mathfrak{R}$. Set $w = -f_1(x_k)v$. Then $w \equiv 0 \pmod{\mathfrak{R}^k}$, so that

$$\begin{aligned} f_1(x_{k+1}) &\equiv f_1(x_k) + f_1'(x_k)w \pmod{\mathfrak{R}^{k+1}} \\ &\equiv f_1(x_k) - f_1'(x_k)vf_1(x_k) \pmod{\mathfrak{R}^{k+1}} \\ &\equiv f_1(x_k) - f_1(x_k) \pmod{\mathfrak{R}^{k+1}} \\ &\equiv 0 \pmod{\mathfrak{R}^{k+1}}. \end{aligned}$$

Thus we have determined x_{k+1} such that $f_1(x_{k+1}) \equiv 0 \pmod{\mathfrak{R}^{k+1}}$ and $x \equiv x_{k+1} \pmod{\mathfrak{R}}$. Since $\mathfrak{R}$ is nilpotent this process leads to a y such that $f_1(y) = 0$ and $y \equiv x \pmod{\mathfrak{R}}$. Hence $\mathfrak{A}_1 = \Phi[y]$ satisfies $\mathfrak{A} = \mathfrak{A}_1 \oplus \mathfrak{R}$. Since the minimum polynomial of $\bar{x}$ is $f_1(\lambda)$, it follows that the minimum polynomial of y is $f_1(\lambda)$ also.

We can now prove

THEOREM 16. *Let X be a linear transformation in a finite-dimensional vector space over a field of characteristic zero. Then we can write $X = Y + Z$ where Y and Z are polynomials in X such that Y is semi-simple and Z is nilpotent. Moreover, if $X = Y_1 + Z_1$ where Y_1 is semi-simple and Z_1 is nilpotent and Y_1 and Z_1 commute with X, then $Y_1 = Y$, $Z_1 = Z$.*

Proof: The existence of the decomposition $X = Y + Z$ is obtained by applying Theorem 15 to the algebra $\Phi[X]$. Now suppose $X = Y_1 + Z_1$, where Y_1 and Z_1 have the properties stated in the theorem. Since Y and Z are polynomials in X they commute with Y_1 and Z_1. We have $Y - Y_1 = Z_1 - Z$. Since Z and Z_1 are nilpotent and commute, $Z - Z_1$ is nilpotent. Since Y and Y_1 are semi-simple and the base field is of characteristic zero, the proof of Theorem 11 shows that $Y - Y_1$ is semi-simple. Since the only transformation which is both semi-simple and nilpotent is 0,

$$Y - Y_1 = 0 = Z - Z_1 .$$

Hence $Y = Y_1$, $Z = Z_1$.

We call the uniquely determined linear transformations Y and Z of Theorem 16 *the semi-simple and the nilpotent* components of X.

DEFINITION 3. A Lie algebra $\mathfrak{L}$ of linear transformations of a finite-dimensional vector space over a field of characteristic 0 is called *almost algebraic** if it contains the nilpotent and semi-simple components of every $X \in \mathfrak{L}$.

To prove our imbedding theorem we require the following two lemmas.

LEMMA 7 (Morozov). *Let $\mathfrak{L}$ be a finite-dimensional Lie algebra of characteristic 0 and suppose $\mathfrak{L}$ contains elements f, h such that $[fh] = -2f$ and $h \in [\mathfrak{L}f]$. Then there exists an element $e \in \mathfrak{L}$ such that*

* This concept is due to Malcev, who used the term *splittable*. We have changed the term to "almost algebraic" since this is somewhat weaker than Chevalley's notion of an algebraic Lie algebra of linear transformations. Moreover, we have preferred to use the term "split Lie algebra" in a connection which is totally unrelated to Malcev's notion.

(59) $$[eh] = 2e\,, \qquad [ef] = h \qquad ([fh] = -2f)\,.$$

Proof: There exists a $z \in \mathfrak{L}$ such that $h = [zf]$. Set $F = \operatorname{ad} f$, $H = \operatorname{ad} h$, $Z = \operatorname{ad} z$ so that we have

(60) $$[FH] = -2F\,, \qquad H = [ZF]\,.$$

The first of these relations implies that F is nilpotent (Lemma 2.4). Also

$$\begin{aligned}[][[zh] - 2z, f] &= [[zf]h] + [z[hf]] - 2[zf] \\ &= 0 + 2h - 2h = 0\,.\end{aligned}$$

Hence $[zh] = 2z + x_1$ where $x_1 \in \mathfrak{K}$, the subalgebra of elements x such that $[xf] = 0$. Since $[FH] = -2F$, if $b \in \mathfrak{K}$, then

$$bHF = b(FH + 2F) = 0\,.$$

Hence $bH \in \mathfrak{K}$ and so $\mathfrak{K}H \subseteq \mathfrak{K}$. Also we have

$$\begin{aligned}[][ZF^i] &= [ZF]F^{i-1} + F[ZF]F^{i-2} + \cdots + F^{i-1}[ZF] \\ &= HF^{i-1} + FHF^{i-2} + \cdots + F^{i-1}H\end{aligned}$$

and since $HF^k = F^kH + 2kF^k$, we have

(61) $$\begin{aligned}[][ZF^i] &= F^{i-1}(H + 2(i-1) + H + 2(i-2) + \cdots + H) \\ &= iF^{i-1}(H + (i-1))\,.\end{aligned}$$

Let $b \in \mathfrak{K} \cap \mathfrak{L}F^{i-1}$. Then $b = aF^{i-1}$ and $bF = aF^i = 0$. Hence

$$iaF^{i-1}(H + (i-1)) = a(ZF^i - F^iZ) = (aZ)F^i \in \mathfrak{L}F^i\,.$$

Hence $b(H + (i-1)) \in \mathfrak{K} \cap \mathfrak{L}F^i$. It follows from this relation and the nilpotency of F that if b is any element of $\mathfrak{K}$ then

(62) $$bH(H+1)(H+2)\cdots(H+m) = 0$$

for some positive integer m. Thus the characteristic roots of the restriction of H to $\mathfrak{K}$ are non-positive integers. Hence $H - 2$ induces a non-singular linear transformation in $\mathfrak{K}$ and consequently there exists a $y_1 \in \mathfrak{K}$ such that $y_1(H-2) = x_1$ where x_1 is the element such that $[zh] = 2z + x_1$. Then $[y_1h] = 2y_1 + x_1$. Hence if we set $e = z - y_1$ we have $[eh] = 2e$. Also $[ef] = [zf] = h$. Hence (59) holds.

LEMMA 8. *Let $\mathfrak{L}$ be a Lie algebra of linear transformations in a finite-dimensional vector space over a field of characteristic* 0. *Suppose every nilpotent element $F \neq 0$ of $\mathfrak{L}$ can be imbedded in a subalgebra* *with basis* (E, F, H) *such that* $[EH] = 2E$, $[FH] = -2F$, $[EF] = H$. *Let* $\mathfrak{K}$ *be any subalgebra of* $\mathfrak{L}$ *which has a complementary space* $\mathfrak{N}$ *in* $\mathfrak{L}$ *invariant under multiplication by* $\mathfrak{K}$: $\mathfrak{L} = \mathfrak{K} \oplus \mathfrak{N}$, $[\mathfrak{N}\mathfrak{K}] \subseteq \mathfrak{N}$. *Then* $\mathfrak{K}$ *has the property stated for* $\mathfrak{L}$.

Proof: Let F be a non-zero nilpotent of $\mathfrak{K}$. Then we can choose E and H in $\mathfrak{L}$ so that the indicated relations hold. Write $H = H_1 + H_2$, $H_1 \in \mathfrak{K}$, $H_2 \in \mathfrak{N}$, $E = E_1 + E_2$, $E_1 \in \mathfrak{K}$, $E_2 \in \mathfrak{N}$. Then we have $-2F = [FH] = [FH_1] + [FH_2]$ and $[FH_1] \in \mathfrak{K}$, $[FH_2] \in \mathfrak{N}$. Hence $-2F = [FH_1]$. Also $H = [EF] = [E_1F] + [E_2F]$. This implies that $H_1 = [E_1F] \in [\mathfrak{K}F]$. Thus H_1 satisfies for $F, \mathfrak{K}$ the conditions on H in Lemma 7. Hence there exists E', H' in $\mathfrak{K}$ such that $[FH'] = -2F$, $[E'H'] = 2E'$, $[E'F] = H'$. The subalgebra generated by F, E', H' is a homomorphic image of the split three-dimensional simple algebra. Since $F \neq 0$ we have an isomorphism, so that F, E', H' are linearly independent and satisfy the required conditions

We can now establish our second criterion for complete reducibility.

Theorem 17. *Let* $\mathfrak{L}$ *be a Lie algebra of linear transformations in a finite-dimensional vector space* $\mathfrak{M}$ *over a field of characteristic* 0. (1) *Assume* $\mathfrak{L}$ *completely reducible. Then every non-zero nilpotent element of* $\mathfrak{L}$ *can be imbedded in a three-dimensional split simple subalgebra of* $\mathfrak{L}$ *and* $\mathfrak{L}$ *is almost algebraic.* (2) *Assume that every non-zero nilpotent element of* $\mathfrak{L}$ *can be imbedded in a three-dimensional simple subalgebra of* $\mathfrak{L}$ *and that the center* $\mathfrak{C}$ *of* $\mathfrak{L}$ *is almost algebraic. Then* $\mathfrak{L}$ *is completely reducible.*

Proof: (1) Assume $\mathfrak{L}$ is completely reducible and let $\mathfrak{E}$ denote the complete algebra of linear transformations in $\mathfrak{M}$. Let F be a nilpotent linear transformation and let $\mathfrak{M} = \mathfrak{M}_1 \oplus \mathfrak{M}_2 \oplus \cdots \oplus \mathfrak{M}_r$ be a decomposition of $\mathfrak{M}$ into cyclic invariant subspaces relative to F. Thus in $\mathfrak{M}_i$ we have a basis $(x_0, x_1, \cdots, x_{m_i})$ such that $x_jF = x_{j+1}$, $x_{m_i}F = 0$. We define H and E to be the linear transformations leaving every $\mathfrak{M}_i$ invariant and satisfying $x_jH = (m_i - 2j)x_j$, $x_0E = 0$, $x_jE = (-m_ij + j(j-1))x_{j-1}$, $j > 0$ (cf. (36)). Then as in § 8, $[EH] = 2E$, $[FH] = -2F$, $[EF] = H$. This shows that F can be imbedded in a subalgebra $\Phi E + \Phi F + \Phi H$ of the type indicated. We shall show next that we can write $\mathfrak{E}_L = \mathfrak{L} \oplus \mathfrak{N}$ where $\mathfrak{N}$ is a subspace such that $[\mathfrak{N}\mathfrak{L}] \subseteq \mathfrak{N}$. It will then follow from Lemma 8 that every nilpotent element $\neq 0$ of $\mathfrak{L}$ can be imbedded in a split three-dimensional simple subalgebra of $\mathfrak{L}$. We recall that $\mathfrak{E}_L$ as module relative to $\mathfrak{L}$ (adjoint representation) is equivalent to $\mathfrak{M} \otimes \mathfrak{M}^*$, $\mathfrak{M}^*$

the contragredient module. It is also easy to see that $\mathfrak{M}^*$ is completely reducible. Hence, by Theorem 11, $\mathfrak{M} \otimes \mathfrak{M}^*$, and consequently $\mathfrak{E}_L$, is completely reducible relative to $\mathfrak{L}$. Since $\mathfrak{L}$ is a submodule of $\mathfrak{E}_L$ relative to $\mathfrak{L}$, there exists a complement $\mathfrak{N}$ such that $\mathfrak{E}_L = \mathfrak{L} \oplus \mathfrak{N}$, $[\mathfrak{N}\mathfrak{L}] \subseteq \mathfrak{N}$. This completes the proof of the first assertion in (1). Now let X be any element of $\mathfrak{L}$ and let Y and Z be the semi-simple and nilpotent components of X. Then

$$\mathrm{ad}_{\mathfrak{E}} X = \mathrm{ad}_{\mathfrak{E}} Y + \mathrm{ad}_{\mathfrak{E}} Z\,,$$

$[\mathrm{ad}_{\mathfrak{E}} Y \mathrm{ad}_{\mathfrak{E}} Z]=0$ and $\mathrm{ad}_{\mathfrak{E}} Z$ is nilpotent ($Z^m=0$ implies $(\mathrm{ad}_{\mathfrak{E}} Z)^{2m-1}=0$). Also the identification of $\mathfrak{E}$ with $\mathfrak{M} \otimes \mathfrak{M}^*$ and the proof of Theorem 11 show that $\mathrm{ad}_{\mathfrak{E}} Y$ is semi-simple. Hence $\mathrm{ad}_{\mathfrak{E}} Y$ and $\mathrm{ad}_{\mathfrak{E}} Z$ are the semi-simple and nilpotent components of $\mathrm{ad}_{\mathfrak{E}} X$ and so these are polynomials in $\mathrm{ad}_{\mathfrak{E}} X$. Since $\mathfrak{L}\,\mathrm{ad}_{\mathfrak{E}} X \subseteq \mathfrak{L}$ and $\mathrm{ad}_{\mathfrak{E}} Y$, $\mathrm{ad}_{\mathfrak{E}} Z$ are polynomials in $\mathrm{ad}_{\mathfrak{E}} X$, $\mathfrak{L}\,\mathrm{ad}_{\mathfrak{E}} Y \subseteq \mathfrak{L}$, $\mathfrak{L}\,\mathrm{ad}_{\mathfrak{E}} Z \subseteq \mathfrak{L}$. Thus $L \rightarrow [LY]$, $L \rightarrow [LZ]$ are derivations in $\mathfrak{L}$. We can write $\mathfrak{L} = \mathfrak{L}' \oplus \mathfrak{C}$ where $\mathfrak{L}'$ is semi-simple and $\mathfrak{C}$ is the center. Since the derivations of $\mathfrak{L}'$ are all inner it follows that any derivation of $\mathfrak{L}$ which maps $\mathfrak{C}$ into 0 is an inner derivation determined by an element of $\mathfrak{L}'$. Since Z is a polynomial in X, $[XC] = 0$ implies $[ZC] = 0$. This implies that the derivation $L \rightarrow [LZ]$ maps $\mathfrak{C}$ into 0. Hence there exists a $Z_1 \in \mathfrak{L}'$ such that $[LZ] = [LZ_1]$, $L \in \mathfrak{L}$. Since Z is nilpotent, $\mathrm{ad}_{\mathfrak{L}'} Z_1 = \mathrm{ad}_{\mathfrak{L}'} Z$ is nilpotent. Since $\mathfrak{L}'$ is semi-simple, the result just proved (applied to $\mathrm{ad}\,\mathfrak{L}'$) implies that there exists an element $U \in \mathfrak{L}'$ such that $[\mathrm{ad}_{\mathfrak{L}'} Z_1, \mathrm{ad}_{\mathfrak{L}'} U] = 2\,\mathrm{ad}_{\mathfrak{L}'} Z_1$. Then $[Z_1 U] = 2Z_1$, which implies that Z_1 is nilpotent. Since $[XZ] = 0$, $[XZ_1] = 0$ and since Z is a polynomial in X, $[ZZ_1] = 0$. It now follows that $Z - Z_1$ is nilpotent. Since $[L, Z - Z_1] = 0$, $L \in \mathfrak{L}$, and $Z - Z_1$ is in the enveloping associative algebra $\mathfrak{L}^*$, $Z - Z_1$ is in the center of $\mathfrak{L}^*$. Since $\mathfrak{L}^*$ is completely reducible and $Z - Z_1$ is nilpotent, this implies $Z - Z_1 = 0$ so $Z = Z_1 \in \mathfrak{L}$. Hence also $Y = X - Z \in \mathfrak{L}$. This completes the proof that $\mathfrak{L}$ is almost algebraic. (2) Assume $\mathfrak{L}$ has an almost algebraic center and has the property stated for nilpotent elements. Let $\mathfrak{S}$ be the radical of $\mathfrak{L}$ and let $F \in [\mathfrak{L}\mathfrak{S}]$. Then we know that F is nilpotent (Corollary 2 to Theorem 2.8). If F is not zero it can be imbedded in a three-dimensional simple subalgebra $\mathfrak{K}$. Since $\mathfrak{K} \cap \mathfrak{S} \neq 0$ and $\mathfrak{K}$ is simple, $\mathfrak{K} \subseteq \mathfrak{S}$ which is impossible because of the solvability of $\mathfrak{S}$. Hence $F = 0$ and $[\mathfrak{L}\mathfrak{S}] = 0$. This implies that $\mathfrak{S} = \mathfrak{C}$ the center. By Levi's theorem $\mathfrak{L} = \mathfrak{C} \oplus \mathfrak{L}_1$ where $\mathfrak{L}_1$ is a semi-simple subalgebra. Since $\mathfrak{C}$ is the center this implies that

$\mathfrak{L}_1$ is an ideal. We can now invoke Theorem 10 to prove that $\mathfrak{L}$ is completely reducible, provided that we can show that every $C \in \mathfrak{C}$ is semi-simple. Now we are assuming that $C = D + E$ where D is semi-simple, E nilpotent, and D and E are in $\mathfrak{C}$. If $E \neq 0$ we can imbed this in a three-dimensional simple subalgebra. Clearly this is impossible since E is in the center. This completes the proof of (2).

It is immediate that if $\mathfrak{L}$ is almost algebraic, then the center $\mathfrak{C}$ of $\mathfrak{L}$ is almost algebraic. Hence we can replace the assumption in (2) that $\mathfrak{C}$ is almost algebraic by the assumption that $\mathfrak{L}$ is almost algebraic. We recall that the centralizer $\mathfrak{C}_{\mathfrak{L}}(S)$ of a subset S is the set of elements $y \in \mathfrak{L}$ such that $[sy] = 0$ for all $s \in S$. This is a subalgebra of $\mathfrak{L}$. We shall now use the foregoing criterion to prove

THEOREM 18. *Let $\mathfrak{L}$ be a completely reducible Lie algebra of linear transformations in a finite-dimensional vector space $\mathfrak{M}$ of characteristic zero and let $\mathfrak{L}_1$ be a completely reducible subalgebra of $\mathfrak{L}$. Then the centralizer $\mathfrak{L}_2 = \mathfrak{C}_{\mathfrak{L}}(\mathfrak{L}_1)$ is completely reducible.*

Proof: Let $X \in \mathfrak{L}_2$. Then since $\mathfrak{L}$ is almost algebraic, the semi-simple and nilpotent parts Y, Z of X are in $\mathfrak{L}$. Since these are polynomials in X, $[CX] = 0$ for $C \in \mathfrak{L}_1$ implies $[CY] = 0 = [CZ]$. Hence $Y, Z \in \mathfrak{L}_2$ and $\mathfrak{L}_2$ is almost algebraic. We shall show next that $\mathfrak{L} = \mathfrak{L}_2 \oplus \mathfrak{L}_3$ where $\mathfrak{L}_3$ is a subspace of $\mathfrak{L}$ such that $[\mathfrak{L}_3\mathfrak{L}_2] \subseteq \mathfrak{L}_3$. It will then follow from Theorem 17 and Lemma 8 that every nilpotent element of $\mathfrak{L}_2$ can be imbedded in a three-dimensional split simple algebra. Then $\mathfrak{L}_2$ will be completely reducible by Theorem 17. Now we know that $\mathrm{ad}_{\mathfrak{E}}\mathfrak{L}_1$ is completely reducible (proof of Theorem 17). Since $\mathfrak{L}$ is a submodule of $\mathfrak{E}$ relative to $\mathfrak{L}_1$, $\mathfrak{L}$ is completely reducible relative to $\mathrm{ad}_{\mathfrak{L}}\mathfrak{L}_1$. Thus we may write

$$\mathfrak{L} = \mathfrak{M}_1 \oplus \mathfrak{M}_2 \oplus \cdots \oplus \mathfrak{M}_k$$

where $[\mathfrak{M}_i\mathfrak{L}_1] \subseteq \mathfrak{M}_i$, $i = 1, \cdots, k$, and $\mathfrak{M}_i$ is irreducible relative to $\mathrm{ad}_{\mathfrak{L}}\mathfrak{L}_1$. We assume the $\mathfrak{M}_i$ are ordered so that $[\mathfrak{M}_i\mathfrak{L}_1] = 0$, $i = 1, \cdots, h$, and $[\mathfrak{M}_j\mathfrak{L}_1] \neq 0$ if $j > h$. Since the subset $\mathfrak{Z}_i$ of elements z_i such that $[z_i\mathfrak{L}_1] = 0$ is a submodule of $\mathfrak{M}_i$, it is immediate that

$$\mathfrak{L}_2 = \mathfrak{M}_1 + \cdots + \mathfrak{M}_h .$$

Set $\mathfrak{L}_3 = \mathfrak{M}_{h+1} + \cdots + \mathfrak{M}_k$. Then $\mathfrak{L} = \mathfrak{L}_2 \oplus \mathfrak{L}_3$. If $i > h$, then $[\mathfrak{M}_i\mathfrak{L}_1] \neq 0$ and $[\mathfrak{M}_i\mathfrak{L}_1] + [[\mathfrak{M}_i\mathfrak{L}_1]\mathfrak{L}_1] + \cdots$ is an $\mathfrak{L}_1$-submodule $\neq 0$ of $\mathfrak{M}_i$. Hence $\mathfrak{M}_i = [\mathfrak{M}_i\mathfrak{L}_1] + [[\mathfrak{M}_i\mathfrak{L}_1]\mathfrak{L}_1] + \cdots$. This implies that

$$\mathfrak{L}_3 = [\mathfrak{L}_3\mathfrak{L}_1] + [[\mathfrak{L}_3\mathfrak{L}_1]\mathfrak{L}_1] + \cdots \subseteq [\mathfrak{L}\mathfrak{L}_1] .$$

On the other hand, $\mathfrak{L} = \mathfrak{L}_2 \oplus \mathfrak{L}_3$; hence, $[\mathfrak{L}\mathfrak{L}_1] = [\mathfrak{L}_3\mathfrak{L}_1] \subseteq \mathfrak{L}_3$. Hence $\mathfrak{L}_3 = [\mathfrak{L}\mathfrak{L}_1]$ and

$$[\mathfrak{L}_3\mathfrak{L}_2] = [[\mathfrak{L}\mathfrak{L}_1]\mathfrak{L}_2] \subseteq [\mathfrak{L}[\mathfrak{L}_1\mathfrak{L}_2]] + [[\mathfrak{L}\mathfrak{L}_2]\mathfrak{L}_1] \subseteq [\mathfrak{L}\mathfrak{L}_1] = \mathfrak{L}_3 .$$

This shows that $\mathfrak{L}_3$ is a complement of $\mathfrak{L}_2$ in $\mathfrak{L}$ such that $[\mathfrak{L}_3\mathfrak{L}_2] \subseteq \mathfrak{L}_3$, which is what we needed to prove.

Exercises

In all these exercises the characteristic of the base field will be zero, and unless the number is indicated with an asterisk the dimensionalities of the spaces will be finite.

1. Show that if $\mathfrak{H}$ is a Cartan subalgebra of $\mathfrak{L}$ then $\mathfrak{H}$ is a maximal nilpotent subalgebra of $\mathfrak{L}$. Show that the converse is false for Φ_{nL} $(n \geqq 2)$.

2. Let $\mathfrak{H}$ be a nilpotent Lie algebra of linear transformations in $\mathfrak{M}$ and let $\mathfrak{M} = \mathfrak{M}_0 \oplus \mathfrak{M}_1$ be the Fitting decomposition relative to $\mathfrak{H}$. Show that if Φ is infinite, then there exists an $A \in \mathfrak{H}$ such that $\mathfrak{M}_0 = \mathfrak{M}_{0A}$, $\mathfrak{M}_1 = \mathfrak{M}_{1A}$, $\mathfrak{M}_{iA}$, the Fitting components relative to A.

3. Show that the diagonal matrices of trace 0 form a Cartan subalgebra in the Lie algebra $\mathfrak{L}$ of triangular matrices of trace 0. Show that $\mathfrak{L}$ is complete

4. Let $\mathfrak{L}$ be the subalgebra of Φ_{2lL} of matrices A satisfying $S^{-1}A'S = -A$ where

$$S = \begin{pmatrix} 0 & 1_l \\ 1_l & 0 \end{pmatrix} .$$

(This is isomorphic to an orthogonal Lie algebra.) Show that the diagonal matrices

$$\text{diag}\,\{\lambda_1, \cdots, \lambda_l, -\lambda_1, \cdots, -\lambda_l\}$$

form a Cartan subalgebra of $\mathfrak{L}$.

5. Same as Exercise 4 but with S replaced by

$$Q = \begin{pmatrix} 0 & 1_l \\ -1_l & 0 \end{pmatrix} .$$

6. Generalize Exercise 2.9 to the following: Let $\mathfrak{L}$ be a Lie algebra, $\mathfrak{H}$ a nilpotent subalgebra of the derivation algebra of $\mathfrak{L}$. Suppose the only element $l \in \mathfrak{L}$ such that $lD = 0$ for all $D \in \mathfrak{H}$ is $l = 0$. Then prove that $\mathfrak{L}$ is nilpotent.

7. Show that if $\mathfrak{L}_1$ is a semi-simple ideal in $\mathfrak{L}$ then $\mathfrak{L} = \mathfrak{L}_1 \oplus \mathfrak{L}_2$ where $\mathfrak{L}_2$ is a second ideal.

8. Let $\mathfrak{R}$ be an ideal in $\mathfrak{L}$ such that $\mathfrak{L}/\mathfrak{R}$ is semi-simple. Show that there exists a subalgebra $\mathfrak{L}_1$ of $\mathfrak{L}$ such that $\mathfrak{L} = \mathfrak{R} \oplus \mathfrak{L}_1$.

9. Let $\mathfrak{L}$ be simple over an algebraically closed field Φ and let $f(a, b)$ be an invariant symmetric bilinear form on $\mathfrak{L}$. Show that f is a multiple of the Killing form. Generalize this to semi-simple $\mathfrak{L}$.

10. Let $\mathfrak{M}_n$ be the n-dimensional irreducible module for a split three-dimensional simple algebra $\mathfrak{R}$. Obtain a decomposition of $\mathfrak{M}_n \otimes \mathfrak{M}_r$ into irreducible submodules.

11*. Let e, h be elements of an associative algebra such that $[[eh]h] = 0$. Show that if h is *algebraic*, that is, there exists a non-zero polynomial $\phi(\lambda)$ such that $\phi(h) = 0$, then $[eh]$ is nilpotent.

12*. Let $\mathfrak{A}$ be an associative algebra with an identity element 1 and suppose $\mathfrak{A}$ contains elements e, f, h such that $[eh] = 2e$, $[fh] = -2f$, $[ef] = h$. Show that if $\phi(h) \in \mathfrak{A}$ is a polynomial in h then $e^i\phi(h) = \phi(h + 2i)e^i$, $i = 0, 1, 2,$ $\phi(h)f^i = f^i\phi(h + 2i)$. Also prove that if r and n are positive integers, $r \leqq n$, then

$$[\cdots[e^n \overbrace{f]f]\cdots f}^{r}] = \sum_{j=0}^{[r/2]} c_{nrj} \prod_{i=1}^{r-2j} (h+n-i)e^{n-r+j}$$

where $c_{nrj} = \binom{n}{j}\binom{n-j}{r-2j}r\,.$

13*. $\mathfrak{A}, h, e, f$ as in Exercise 12. Show that if $e^m = 0$, then

$$\prod_{i=1}^{2m-1} (h + m - i) = 0$$

14. Prove that if e is an element of a semi-simple Lie algebra $\mathfrak{L}$ of characteristic zero such that ad e is nilpotent, then e^R is nilpotent for every representation R of $\mathfrak{L}$.

15. Prove that if $\mathfrak{L}$ is semi-simple over an algebraically closed field then $\mathfrak{L}$ contains an $e \neq 0$ with ad e nilpotent.

16. Prove that every finite-dimensional Lie algebra $\neq 0$ over an algebraically closed field has indecomposable modules of arbitrarily high finite dimensionalities. (*Hint*: Show that there exists an $e \in \mathfrak{L}$ and a representation R such that e^R is nilpotent $\neq 0$. If $\mathfrak{M}$ is the corresponding module, then the dimensionalities of the indecomposable components of $\mathfrak{M}$, $\mathfrak{M}\otimes\mathfrak{M}$, $\mathfrak{M}\otimes\mathfrak{M}\otimes\mathfrak{M}, \cdots$ are not bounded.)

17. Prove that any semi-simple algebra has irreducible modules of arbitrarily high dimensionalities.

18. Show that the derivation algebra of any Lie algebra is algebraic. (*Hint*: Use Exercise 2.8.)

19. A Lie algebra $\mathfrak{L}$ is called *reductive* if ad $\mathfrak{L}$ is completely reducible. Show that $\mathfrak{L}$ is reductive if and only if $\mathfrak{L}$ has a $1:1$ completely reducible representation.

20. A subalgebra $\mathfrak{K}$ of $\mathfrak{L}$ is called *reductive in* $\mathfrak{L}$ if $\mathrm{ad}_{\mathfrak{L}}\mathfrak{K}$ is completely reducible. Prove that if $\mathfrak{L}$ is a completely reducible Lie algebra of linear transformations and $\mathfrak{K}$ is reductive in $\mathfrak{L}$ then $\mathfrak{K}$ is completely reducible.

21. Show that any reductive commutative subalgebra of a semi-simple Lie algebra can be imbedded in a Cartan subalgebra.

22. Show that any semi-simple Lie algebra contains commutative Cartan subalgebras.

23. Let A be an automorphism of a semi-simple Lie algebra $\mathfrak{L}$. Show that the subalgebra of elements y such that $y(A-1)^m = 0$ for some m is a reductive subalgebra. (*Hint*: Use Exercise 2.5.)

24. (Mostow-Taft). Let G be a finite group of automorphisms in a Lie algebra. Show that $\mathfrak{L}$ has a Levi factor which is invariant under G.

25. Let $f_a(\lambda) = \det(\lambda 1 - \mathrm{ad}\, a)$, the characteristic polynomial of ad a in a Lie algebra $\mathfrak{L}$, and let D be a derivation of $\mathfrak{L}$. Show that if t is an indeterminate, then

$$f_{a+taD}(\lambda) \equiv f_a(\lambda) \pmod{t^2} .$$

(*Hint*: Use the fact $f_{aA}(\lambda) = f_a(\lambda)$ if A is an automorphism and the fact that $\exp tD = 1 + tD + (t^2D^2/2!) + \cdots$ is a well-defined automorphism in $\mathfrak{L}_P$, P the field of power series in t with coefficients in Φ.)

26. Write $f_a(\lambda) = \lambda^n - \tau_1(a)\lambda^{n-1} + \tau_2(a)\lambda^{n-2} + \cdots + (-1)^i\tau_i(a)\lambda^{n-i}\cdots$ and let $\tau_i(a_1, \cdots, a_i)$ be the linearized form of τ_i defined by

$$\tau_i(a_1, \cdots, a_i) = \frac{1}{i!}[\tau_i(a_1 + \cdots + a_i)$$
$$-\sum \tau_i(a_1+\cdots+\hat{a}_j+\cdots+a_i) + \sum_{j<k} \tau_i(a_1+\cdots+\hat{a}_j+\cdots+\hat{a}_k+\cdots+a_i)$$
$$-\sum_{j<k<l} \tau_i(a_1+\cdots+\hat{a}_j+\cdots+\hat{a}_k+\cdots+\hat{a}_l+\cdots+a_i)+\cdots] .$$

Show that $\tau_i(a_1, \cdots, a_i)$ is a symmetric i-linear function and that

$$\tau_i(a_1D, a_2, \cdots, a_i) + \tau_i(a_1, a_2D, a_3, \cdots, a_i) + \cdots + \tau_i(a_1, \cdots, a_{i-1}, a_iD) = 0$$

for any derivation D.

CHAPTER IV

Split Semi-Simple Lie Algebras

In this chapter we shall obtain the classification of simple Lie algebras over an algebraically closed field of characteristc 0. This was given first by Killing, modulo some errors which were corrected by Cartan. Later simplifications are due to Weyl, van der Waerden, Coxeter, Witt, and Dynkin. In our discussion we shall follow Dynkin's method which is fairly close to Cartan's original method. However, we shall formulate everything in terms of "split" semi-simple Lie algebras over an arbitrary base field of characteristic 0. It is easy to see that the assumption of algebraic closure in the classical treatments is used only to ensure the existence of a decomposition of the algebra as $\mathfrak{L} = \mathfrak{H} \oplus \mathfrak{L}_\alpha \oplus \mathfrak{L}_\beta \oplus \cdots \oplus \mathfrak{L}_\delta$ where $\mathfrak{H}$ is a Cartan subalgebra and the $\mathfrak{L}_\alpha$ are the root spaces relative to $\mathfrak{H}$. This can be achieved by assuming the existence of a "splitting Cartan subalgebra" (cf. § 1). It appears to be clearer and more natural to employ this hypothesis in place of the stronger one of algebraic closure of the base field. For the benefit of a reader who has some familiarity with the associative theory it might be remarked that the split simple Lie algebras which are singled out in the classical theory are the analogues of the simple matrix algebras Φ_n of the associative theory.

A part of the results of this chapter (the isomorphism and existence theorems) will be derived again in Chapter VII in a more sophisticated way. In Chapter X we shall take up the problem of extending the classification from algebraically closed base fields—or from split Lie algebras—to simple Lie algebras over any field of characteristic 0. It should be noted that the classification we shall give is valid also for characteristic $p \neq 0$ under fairly simple hypotheses which are stronger than simplicity. This has been shown by Seligman and, in an improved form, by Mills and Seligman.

1. *Properties of roots and root spaces*

We shall call a Cartan subalgebra $\mathfrak{H}$ of a finite-dimensional Lie algebra $\mathfrak{L}$ a *splitting Cartan subalgebra* (abbreviated s.c.s.) if the characteristic roots of every $\mathrm{ad}_{\mathfrak{L}}h$, $h \in \mathfrak{H}$, are in the base field. We shall say that $\mathfrak{L}$ is *split* if it has a splitting Cartan subalgebra. If the base field Φ is algebraically closed, any finite-dimensional $\mathfrak{L}$ is split and any Cartan subalgebra is a s.c.s.

Example. Let $\mathfrak{L} = \Phi_{nL}$ and let $\mathfrak{H}$ denote the subalgebra of diagonal matrices. We have seen (§ 3.3) that $\mathfrak{H}$ is a s.c.s. in Φ_{nL}, so Φ_{nL} is split. Next let Φ be the field of real numbers and let A be a matrix whose characteristic roots ξ_i (in the complex field) are distinct but not every $\xi_i \in \Phi$. Then the polynomial algebra $\Phi[A]$ is a Cartan subalgebra of $\mathfrak{L}$. The characteristic polynomial of ad A is $\Pi_{i,j=1}^{n}(\lambda - (\xi_i - \xi_j))$ and some of its roots are not in Φ. Hence $\Phi[A]$ is not a s.c.s.

In the remainder of this chapter $\mathfrak{L}$ will denote a split finite-dimensional semi-simple Lie algebra over a field Φ of characteristic 0, $\mathfrak{H}$ will be a splitting Cartan subalgebra of $\mathfrak{L}$ and we shall write (a, b) for $f(a, b) = \mathrm{tr}\,\mathrm{ad}\,a\,\mathrm{ad}\,b$, the Killing form on $\mathfrak{L}$. We know that (a, b) is non-degenerate. Our assumption on $\mathfrak{H}$ is that $\mathrm{ad}_{\mathfrak{L}}\mathfrak{H}$ is a split algebra of linear transformations. Hence we know that we can decompose $\mathfrak{L}$ as

$$(1) \qquad \mathfrak{L} = \mathfrak{H} \oplus \mathfrak{L}_\alpha \oplus \mathfrak{L}_\beta \oplus \cdots \oplus \mathfrak{L}_\delta$$

where $\alpha, \beta, \cdots, \delta$ are the non-zero roots. These are linear functions on $\mathfrak{H}$ and $\mathfrak{L}_\alpha$ is the set of elements $x_\alpha \in \mathfrak{L}$ such that $x_\alpha(\mathrm{ad}\,h - \alpha(h))^r = 0$ for some $r = r(h)$, $h \in \mathfrak{H}$ (cf. § 2.4 and § 3.2). In the same way $\mathfrak{H} = \mathfrak{L}_0$, the Fitting null component of $\mathfrak{L}$ relative to ad $\mathfrak{H}$. We have $[\mathfrak{L}_\alpha\mathfrak{L}_\beta] \subseteq \mathfrak{L}_{\alpha+\beta}$ if $\alpha + \beta$ is a root while $[\mathfrak{L}_\alpha\mathfrak{L}_\beta] = 0$ if $\alpha + \beta$ is not a root. Our first task will be to obtain additional information on $\mathfrak{H}$, on the roots α and on the corresponding root spaces $\mathfrak{L}_\alpha$. We shall number these results by Roman numerals.

I. *If α and β are any two roots* (*including* 0) *and* $\beta \neq -\alpha$, *then $\mathfrak{L}_\alpha$ and $\mathfrak{L}_\beta$ are orthogonal relative to the Killing form.*

Proof: We show first that $\mathfrak{H} = \mathfrak{L}_0 \perp \mathfrak{L}_\alpha$ if $\alpha \neq 0$. Let $h \in \mathfrak{H}$, $e_\alpha \in \mathfrak{L}_\alpha$ and choose h' so that $\alpha(h') \neq 0$. Then the restriction of ad h' to $\mathfrak{L}_\alpha$ is a non-zero scalar plus a nilpotent and so this is non-singular. It follows that for any $k = 1, 2, \cdots$ we can find an $e_\alpha^{(k)} \in \mathfrak{L}_\alpha$ such that $e_\alpha = [\cdots [e_\alpha^{(k)} \overbrace{h']h'] \cdots h'}^{k}]$. Since the Killing form is invariant we have

$$(e_\alpha, h) = ([\cdots [e_\alpha^{(k)} \underbrace{h'] \cdots h'}_{k}], h)$$
$$= (e_\alpha^{(k)}, \overbrace{[h'[h' \cdots [h'}h] \cdots]) .$$

Since $\mathfrak{H}$ is nilpotent, k can be chosen so that $[h' \cdots [h'h] \cdots] = 0$. Then the above relation implies that $(e_\alpha, h) = 0$. Thus $\mathfrak{H} \perp \mathfrak{L}_\alpha$ for $\alpha \neq 0$. Now let $\beta \neq -\alpha$ and let $e_\beta \in \mathfrak{L}_\beta$. As before, write $e_\alpha = [e_\alpha^{(1)} h']$. Then $(e_\alpha, e_\beta) = ([e_\alpha^{(1)} h'], e_\beta) = -(h', [e_\alpha^{(1)} e_\beta])$. If $\alpha + \beta$ is a root it is non-zero and $[e_\alpha^{(1)} e_\beta] \in \mathfrak{L}_{\alpha+\beta}$. Hence $(h', [e_\alpha^{(1)} e_\beta]) = 0$. If $\alpha + \beta$ is not a root, $[e_\alpha^{(1)} e_\beta] = 0$ and again $(h', [e_\alpha^{(1)} e_\beta]) = 0$. Hence $(e_\alpha, e_\beta) = 0$ and $\mathfrak{L}_\alpha \perp \mathfrak{L}_\beta$.

II. *$\mathfrak{H}$ is a non-isotropic subspace of $\mathfrak{L}$ (relative to (a, b)). If α is a root, then $-\alpha$ is a root and $\mathfrak{L}_\alpha$ and $\mathfrak{L}_{-\alpha}$ are dual spaces relative to (a, b).*

Proof: If $z \in \mathfrak{H}$ and $z \perp \mathfrak{H}$, then $z \perp \mathfrak{L}$ since, by I, $z \perp \mathfrak{L}_\alpha$ for all $\alpha \neq 0$. Then $z = 0$ by the non-degeneracy of (a, b). If α is a root and $-\alpha$ is not a root, then $\mathfrak{L}_\alpha \perp \mathfrak{L}_\beta$ for every root β. Then $\mathfrak{L}_\alpha \perp \mathfrak{L}$ contrary to the non-degeneracy of the Killing form. Also the argument used for $\alpha = 0$ shows that if $z \neq 0$ is in $\mathfrak{L}_\alpha$, then there exists a $w \in \mathfrak{L}_{-\alpha}$ such that $(z, w) \neq 0$. Similarly, if $w \neq 0$ is in $\mathfrak{L}_{-\alpha}$, then there exists a $z \in \mathfrak{L}_\alpha$ such that $(z, w) \neq 0$. This shows that $\mathfrak{L}_\alpha$ and $\mathfrak{L}_{-\alpha}$ are dual spaces relative to (a, b).

We recall that the matrices of the restrictions of $\operatorname{ad} h$, $h \in \mathfrak{H}$, to $\mathfrak{L}_\alpha$ can be taken simultaneously in the form

$$(2) \qquad \begin{bmatrix} \alpha(h) & & & & 0 \\ & \alpha(h) & & & \\ & & \cdot & & \\ & & & \cdot & \\ & & & & \cdot \\ * & & & & \alpha(h) \end{bmatrix}$$

(§ 2.6). If $\dim \mathfrak{L}_\alpha = n_\alpha$ and $h, k \in \mathfrak{H}$, then this gives the formula

$$(3) \qquad (h, k) = \sum n_\alpha \alpha(h) \alpha(k) \qquad \alpha \text{ a root.}$$

We recall also that $\alpha(h') = 0$ for every $h' \in \mathfrak{H}'$. These results imply our next two results.

III. *There are l linearly independent roots where $l = n_0 = \dim \mathfrak{H}$.*

Proof: The roots are linear functions and so belong to the conjugate space $\mathfrak{H}^*$ of $\mathfrak{H}$. We have $\dim \mathfrak{H}^* = l$. Hence if the assertion is false, then the subspace of $\mathfrak{H}^*$ spanned by the roots has dimensionality $l' < l$. This implies that there exists a non-zero vector $h \in \mathfrak{H}$ such that $\alpha(h) = 0$ for every root α. Then (3) implies that $(h, k) = 0$ for every $k \in \mathfrak{H}$, contrary to II.

IV. *$\mathfrak{H}$ is abelian.*

Proof: If $h' \in \mathfrak{H}' = [\mathfrak{H}\mathfrak{H}]$, then $\alpha(h') = 0$ for all α. Then $(h', k) = 0$ for all k and $h' = 0$. Hence $\mathfrak{H}' = 0$ and $\mathfrak{H}$ is abelian.

The fact that the restriction of the Killing form to $\mathfrak{H}$ is non-degenerate implies that if $\rho(h)$ is any element of $\mathfrak{H}^*$, that is, any linear function, then there exists a unique vector $h_\rho \in \mathfrak{H}$ such that

$$(h, h_\rho) = \rho(h) . \tag{4}$$

The mapping $\rho \to h_\rho$ of $\mathfrak{H}^*$ into $\mathfrak{H}$ is surjective and 1 : 1. If $\rho, \sigma \in \mathfrak{H}^*$ we *define*

$$(\rho, \sigma) = (h_\rho, h_\sigma) . \tag{5}$$

Then

$$(\rho, \sigma) = \rho(h_\sigma) = \sigma(h_\rho) , \tag{6}$$

and it is immediate that (ρ, σ) is a non-degenerate symmetric bilinear form on $\mathfrak{H}^*$.

V. *Let e_α be an element of $\mathfrak{L}_\alpha$ such that $[e_\alpha h] = \alpha(h)e_\alpha$, $h \in \mathfrak{H}$, and let $e_{-\alpha}$ be any element of $\mathfrak{L}_{-\alpha}$. Then*

$$[e_{-\alpha}e_\alpha] = (e_{-\alpha}, e_\alpha)h_\alpha . \tag{7}$$

Proof: We have

$$\begin{aligned}([e_{-\alpha}e_\alpha], h) &= (e_{-\alpha}, [e_\alpha h]) = (e_{-\alpha}, \alpha(h)e_\alpha) \\ &= \alpha(h)(e_{-\alpha}, e_\alpha) \\ ((e_{-\alpha}, e_\alpha)h_\alpha, h) &= (e_{-\alpha}, e_\alpha)(h_\alpha, h) = (e_{-\alpha}, e_\alpha)\alpha(h) .\end{aligned}$$

Hence (7) follows from the non-degeneracy of (h, k) in $\mathfrak{H}$.

VI. *Every non-zero root α is non-isotropic relative to the bilinear form (ρ, σ) in $\mathfrak{H}^*$.*

Proof: We note first that $\operatorname{ad} e_\alpha$ is nilpotent if α is a non-zero root and $e_\alpha \in \mathfrak{L}_\alpha$. For this it suffices to show that if $x_\beta \in \mathfrak{L}_\beta$, then there exists a positive integer k such that $[\cdots[[x_\beta \overbrace{e_\alpha]e_\alpha]\cdots e_\alpha}^{k}] = 0$. Consider the sequence

$$[x_\beta e_\alpha], [[x_\beta e_\alpha]e_\alpha] , \quad [[[x_\beta e_\alpha]e_\alpha]e_\alpha], \cdots .$$

The vectors in this sequence are either 0 or belong respectively to the roots $\beta + \alpha$, $\beta + 2\alpha$, $\beta + 3\alpha, \cdots$. Since there are only a finite number of distinct roots for $\mathfrak{H}$ it follows that $[\cdots[x_\beta \overbrace{e_\alpha] \cdots e_\alpha]}^{k} =$ 0 for a suitable k. We can choose $e_\alpha \neq 0$ in $\mathfrak{L}_\alpha$ so that $[e_\alpha h] = \alpha(h)e_\alpha$ (cf. (2)) and $e_{-\alpha} \in \mathfrak{L}_{-\alpha}$ so that $(e_\alpha, e_{-\alpha}) = 1$. The latter choice can be made since $\mathfrak{L}_\alpha, \mathfrak{L}_{-\alpha}$ are dual spaces relative to the Killing form. If ρ is any element of $\mathfrak{H}^*$ we define $\mathfrak{L}_\rho = 0$ if ρ is not a root and $\mathfrak{L}_\rho$ is the root space corresponding to ρ if ρ is a root. Set

$$\mathfrak{K} = \Phi e_\alpha + \Phi h_\alpha + \sum_{k=1}^{\infty} \mathfrak{L}_{-k\alpha} \,. \tag{8}$$

It is clear from the rule on the product of root spaces that $\sum_{k=1}^{\infty}\mathfrak{L}_{-k\alpha}$ is a nilpotent subalgebra of $\mathfrak{L}$. Also $\Phi h_\alpha + \sum_{k=1}^{\infty}\mathfrak{L}_{-k\alpha}$ is a subalgebra containing $\sum_{k=1}^{\infty}\mathfrak{L}_{-k\alpha}$ as ideal. Hence $\Phi h_\alpha + \sum_{k=1}^{\infty}\mathfrak{L}_{-k\alpha}$ is solvable. If $x_{-\alpha} \in \mathfrak{L}_{-\alpha}$, then $[e_\alpha x_{-\alpha}]$ is a multiple of h_α, by (7). Also if $k > 1$, then $[e_\alpha \mathfrak{L}_{-k\alpha}] \subseteq \mathfrak{L}_{-(k-1)\alpha}$ and $[e_\alpha h_\alpha] = \alpha(h_\alpha)e_\alpha = (\alpha, \alpha)e_\alpha$. This shows that $\mathfrak{K}$ is a subalgebra of $\mathfrak{L}$ and that if $(\alpha, \alpha) = 0$, then $\Phi h_\alpha + \sum_{k=1}^{\infty}\mathfrak{L}_{-k\alpha}$ is an ideal in $\mathfrak{K}$. Since this ideal is solvable and has one-dimensional difference algebra, $(\alpha, \alpha) = 0$ implies that $\mathfrak{K}$ is solvable. Then $\mathrm{ad}_{\mathfrak{L}}\mathfrak{K}$ is a solvable algebra of linear transformations acting in $\mathfrak{L}$. Since $\mathrm{ad}\, e_\alpha$ is nilpotent, this element is in the radical of the enveloping associative algebra of $\mathrm{ad}_{\mathfrak{L}}\mathfrak{K}$ (Corollary 2 to Theorem 2.8). Hence the same is true of $(\mathrm{ad}\, e_\alpha)\,(\mathrm{ad}\, e_{-\alpha})$, which implies that $(e_\alpha, e_{-\alpha}) = \mathrm{tr}\, \mathrm{ad}\, e_\alpha\, \mathrm{ad}\, e_{-\alpha} = 0$, contrary to $(e_\alpha, e_{-\alpha}) =$ 1. Hence $(\alpha, \alpha) \neq 0$.

VII. *If α is a non-zero root, then $n_\alpha = \dim \mathfrak{L}_\alpha = 1$. Moreover, the only integral multiples $k\alpha$ of α which are roots are α, 0 and $-\alpha$.*

Proof: Let $e_\alpha, e_{-\alpha}, \mathfrak{K}$ be defined as in the proof of VI (cf. (8)). Then $\mathfrak{K}$ is an invariant subspace of $\mathfrak{L}$ relative to $\mathrm{ad}\, h$, $h \in \mathfrak{H}$, since $[e_\alpha h] = \alpha(h)e_\alpha$, $[h_\alpha h] = 0$ and $[\mathfrak{L}_{-k\alpha} h] \subseteq \mathfrak{L}_{-k\alpha}$. The restriction of $\mathrm{ad}\, h$ to $\mathfrak{L}_{-k\alpha}$ has the single characteristic root $-k\alpha(h)$. Hence we have

$$\mathrm{tr}_{\mathfrak{K}}\, \mathrm{ad}\, h = \alpha(h)\,(1 - n_{-\alpha} - 2n_{-2\alpha} - \cdots)$$

and, in particular,

$$\mathrm{tr}_{\mathfrak{K}}\, \mathrm{ad}\,(h_\alpha) = (\alpha, \alpha)\,(1 - n_{-\alpha} - 2n_{-2\alpha} - \cdots)\,. \tag{9}$$

Since $\mathfrak{K}$ is a subalgebra containing e_α and $e_{-\alpha}$ it is invariant under $\mathrm{ad}\, e_\alpha$ and $\mathrm{ad}\, e_{-\alpha}$ and since $[e_{-\alpha} e_\alpha] = h_\alpha$, $[\mathrm{ad}_{\mathfrak{K}} e_{-\alpha}, \mathrm{ad}_{\mathfrak{K}} e_\alpha] = \mathrm{ad}_{\mathfrak{K}} h_\alpha$. Hence $\mathrm{tr}\,(\mathrm{ad}_{\mathfrak{K}} h_\alpha) = 0$. Since $(\alpha, \alpha) \neq 0$ this and (9) imply that $1 - n_{-\alpha} - 2n_{-2\alpha} - \cdots = 0$. This occurs only if $n_{-\alpha} = 1$, $n_{-2\alpha} = n_{-3\alpha} = \cdots = 0$. Thus $-2\alpha, -3\alpha, \cdots$ are not roots and $n_{-\alpha} = 1$.

Since we can replace α by $-\alpha$ in the argument, we have also that $n_\alpha = 1$ and $2\alpha, 3\alpha, \cdots$ are not roots.

If α is a non-zero root, $\mathfrak{L}_\alpha = \Phi e_\alpha$ and $[e_\alpha h] = \alpha(h)e_\alpha$. Moreover, (7) shows that if we choose e_α and $e_{-\alpha}$ so that $(e_\alpha, e_{-\alpha}) = -1$ then $[e_\alpha e_{-\alpha}] = h_\alpha$. Then $[e_\alpha h_\alpha] = (\alpha, \alpha)e_\alpha$, $[e_{-\alpha}h_\alpha] = -(\alpha, \alpha)e_{-\alpha}$. If we set

$$(10) \qquad e'_\alpha = e_\alpha\,, \qquad e'_{-\alpha} = \frac{2e_{-\alpha}}{(\alpha, \alpha)}\,, \qquad h'_\alpha = \frac{2h_\alpha}{(\alpha, \alpha)}\,,$$

then

$$(11) \qquad [e'_\alpha h'_\alpha] = 2e'_\alpha\,, \qquad [e'_{-\alpha} h'_\alpha] = -2e'_{-\alpha}\,, \qquad [e'_\alpha e'_{-\alpha}] = h'_\alpha\,.$$

Thus $(e'_\alpha, e'_{-\alpha}, h'_\alpha)$ is the type of normalized basis we have considered before for a split three-dimensional simple algebra. This proves

VIII. *If α is a non-zero root then $\Phi h_\alpha + \mathfrak{L}_\alpha + \mathfrak{L}_{-\alpha}$ is a split three-dimensional simple subalgebra of $\mathfrak{L}$.*

At this point we have all the information we need on the multiplication in $\mathfrak{H}$, the products of elements of $\mathfrak{H}$ by elements of the $\mathfrak{L}_\alpha$ and the products $[e_\alpha e_{-\alpha}]$, $e_\alpha \in \mathfrak{L}_\alpha$, $e_{-\alpha} \in \mathfrak{L}_{-\alpha}$. It remains to investigate products of the form $[e_\alpha e_\beta]$ where α and β are non-zero roots and $\beta \neq -\alpha$. We shall obtain the results required after we have established in the next section a basic result on representations of semi-simple Lie algebras.

2. *A basic theorem on representations and its consequences for the structure theory*

We shall need only a part of the following theorem at this point. Later the full result will play an important role in the representation theory.

THEOREM 1. *Let $\mathfrak{L}$ be a finite-dimensional split semi-simple Lie algebra over a field of characteristic* 0, *$\mathfrak{H}$ a splitting Cartan subalgebra, $\mathfrak{L} = \mathfrak{H} + \sum \Phi e_\alpha$ the decomposition of $\mathfrak{L}$ into root spaces relative to $\mathfrak{H}$. Let h_α and (ρ, σ) for ρ, σ in the conjugate space $\mathfrak{H}^*$ be defined as before* (cf. (4), (5)) *and let e_α and $e_{-\alpha}$ be chosen so that $[e_\alpha e_{-\alpha}] = h_\alpha$. Let $\mathfrak{M}$ be a finite-dimensional module for $\mathfrak{L}$, R the representation. Then $\mathfrak{M}$ is a split module for $\mathfrak{H}$ and if $\mathfrak{M}_\Lambda$ is the weight module of $\mathfrak{H}$ corresponding to a weight Λ, then the linear transformation induced by h^R in $\mathfrak{M}_\Lambda$ is the scalar $\Lambda(h)1$. Let $\mathfrak{L}^{(\alpha)} = \mathfrak{H} + \Phi e_\alpha + \Phi e_{-\alpha}$. Then $\mathfrak{L}^{(\alpha)}$ is a subalgebra and $\mathfrak{M}$ is a completely reducible $\mathfrak{L}^{(\alpha)}$-module. Any irreducible $\mathfrak{L}^{(\alpha)}$-submodule $\mathfrak{N}$ of $\mathfrak{M}$ has* *a basis* $(y_0, y_1, \cdots, y_m)$ *such that*

$$
\begin{aligned}
&y_i h = (M - i\alpha)(h) y_i\,, \qquad i = 0, 1, \cdots, m \\
(12)\quad &y_i e_{-\alpha} = y_{i+1}\,, \qquad y_m e_{-\alpha} = 0\,, \qquad i = 0, 1, \cdots, m-1 \\
&y_0 e_\alpha = 0\,, \qquad y_i e_\alpha = \frac{i(i-1-m)}{2}(\alpha, \alpha) y_{i-1}\,, \qquad i = 1, 2, \cdots, m
\end{aligned}
$$

where M *is a weight and* $2(M, \alpha)/(\alpha, \alpha) = m$. *Moreover, if* y *is any non-zero vector such that* $yh = M(h)y$ *and* $ye_\alpha = 0$, *then* y *generates an irreducible* $\mathfrak{L}^{(\alpha)}$*-submodule of* $\mathfrak{M}$. *Let* Λ *be a weight of* $\mathfrak{H}$ *in* $\mathfrak{M}$ *and let* Σ *be the collection of weights of the form* $\Lambda + i\alpha$, i *an integer,* α *a fixed non-zero root. Then* Σ *is an arithmetic progression with first term* $\Lambda - r\alpha$, *difference* α, *and last term* $\Lambda + q\alpha$ *and we have*

$$
(13)\qquad \frac{2(\Lambda, \alpha)}{(\alpha, \alpha)} = (r - q)1\,.
$$

If x *is a non-zero vector such that* $xh = (\Lambda + q\alpha)(h)x$, *then* x *generates an irreducible* $\mathfrak{L}^{(\alpha)}$*-submodule and every weight belonging to* Σ *occurs as weight of* $\mathfrak{H}$ *in this submodule. If* Λ *is a weight, then*

$$
(14)\qquad \Lambda' = \Lambda - \frac{2(\Lambda, \alpha)}{(\alpha, \alpha)}\alpha
$$

is also a weight and

$$
(15)\qquad \dim \mathfrak{M}_\Lambda = \dim \mathfrak{M}_{\Lambda'}\,.
$$

Proof: By III we can find a basis for $\mathfrak{H}^*$ consisting of l roots $\alpha_1, \alpha_2, \cdots, \alpha_l$. The corresponding elements $h_{\alpha_1}, h_{\alpha_2}, \cdots, h_{\alpha_l}$ form a basis for $\mathfrak{H}$. Also $\mathfrak{L}_i = \Phi h_{\alpha_i} + \Phi e_{\alpha_i} + \Phi e_{-\alpha_i}$ is a split three-dimensional simple Lie algebra and, by (10), $e'_{\alpha_i}, e'_{-\alpha_i} = 2e_{-\alpha_i}/(\alpha_i, \alpha_i)$, $h'_{\alpha_i} = 2h_{\alpha_i}/(\alpha_i, \alpha_i)$ is a canonical basis for $\mathfrak{L}_i$. If we recall the form of the irreducible modules for such an algebra given in § 3.8 and use complete reducibility of finite-dimensional modules, we see that there exists a basis for $\mathfrak{M}$ such that $h'^R_{\alpha_i}$, hence also $h^R_{\alpha_i} = (1/2)(\alpha_i, \alpha_i)h'^R_{\alpha_i}$, has a diagonal matrix relative to this basis. This is equivalent to two statements about $h^R_{\alpha_i}$: the characteristic roots of $h^R_{\alpha_i}$ are in Φ and $h^R_{\alpha_i}$ is a semi-simple linear transformation. Since the $h^R_{\alpha_i}$ commute, a standard argument shows that there exists a basis $(u_1, u_2, \cdots, u_N)$ for $\mathfrak{M}$ such that every $h^R_{\alpha_i}$ has a diagonal matrix relative to this basis (cf. the proof of Theorem 3.10). Since the h_{α_i} form a basis for $\mathfrak{H}$ it follows that h^R has a diagonal matrix relative to (u_j), that is, we have: $u_jh = \Lambda_j(h)u_j$, $h \in \mathfrak{H}$, $j = 1, 2, \cdots, N$. Thus it is clear that $\mathfrak{H}^R$ is a split abelian Lie algebra of linear transformations and that the $\Lambda_j = \Lambda_j(h)$ are the weights of $\mathfrak{H}$. Moreover, h^R is the scalar $\Lambda(h)1$ in the weight space $\mathfrak{M}_\Lambda$ (whose basis is the set of u_j such that $\Lambda_j = \Lambda$). This proves the first statement. Now let α be a non-zero root and consider the subspace $\mathfrak{L}^{(\alpha)} = \mathfrak{H} + \Phi e_\alpha + \Phi e_{-\alpha}$. Since $\mathfrak{H}$ is an abelian subalgebra and $[e_\alpha h] = \alpha(h)e_\alpha$, $[e_{-\alpha}h] = -\alpha(h)e_{-\alpha}$, $\mathfrak{L}^{(\alpha)}$ is a subalgebra. Let $\mathfrak{H}_0$ be the subspace of $\mathfrak{H}$ defined by $\alpha(h) = 0$. Then $\mathfrak{H} = \mathfrak{H}_0 + \Phi h_\alpha$ and $\mathfrak{L}^{(\alpha)} = \mathfrak{H}_0 \oplus \mathfrak{K}$, where $\mathfrak{K}$ is the split three-dimensional simple Lie algebra with basis $(e_\alpha, e_{-\alpha}, h_\alpha)$. We have $[\mathfrak{H}_0\mathfrak{K}] = 0$, so $\mathfrak{H}_0$ is the center of $\mathfrak{L}^{(\alpha)}$. Now we have seen that h^R is semi-simple for every $h \in \mathfrak{H}$. Hence the main criterion for complete reducibility (Theorem 3.10) shows that $\mathfrak{M}$ is completely reducible as an $\mathfrak{L}^{(\alpha)}$-module. Let $\mathfrak{N}$ be an irreducible $\mathfrak{L}^{(\alpha)}$-submodule of $\mathfrak{M}$. Then $\mathfrak{N}$ contains a vector $y \neq 0$ such that $yh = M(h)y$ where M is a weight. If we replace y by one of the vectors in the sequence y, ye_α^R, $y(e_\alpha^R)^2, \cdots$ we may suppose also that $ye_\alpha = 0$. Let $E = e_\alpha'^R$, $F = e_{-\alpha}'^R$, $H = h_\alpha'^R$, where $(e_\alpha', e_{-\alpha}', h_\alpha')$ is the basis for $\mathfrak{K}$ given in (10). Then

$$yH = yh_\alpha' = \frac{2M(h_\alpha)}{(\alpha, \alpha)}y = \frac{2(M, \alpha)}{(\alpha, \alpha)}y$$

and $yE = 0$. The argument of § 3.8 shows that $2(M, \alpha)/(\alpha, \alpha)$ is a non-negative integer m (strictly speaking $m.1$) and that $(y, yF, \cdots, yF^m)$ is a basis for an irreducible $\mathfrak{K}$-module. Also we have $yF^{m+1} = 0$ and induction on i shows that $(yF^i)h = (M - i\alpha)(h)yF^i$. Hence $\sum \Phi yF^i$ is an $\mathfrak{L}^{(\alpha)}$-submodule of $\mathfrak{N}$ and since $\mathfrak{N}$ is irreducible as $\mathfrak{L}^{(\alpha)}$-module, $\mathfrak{N} = \sum \Phi yF^i$. We now modify the basis for $\mathfrak{N}$ by replacing yF^i by

$$y_i = y(e_{-\alpha}^R)^i = \left(\frac{(\alpha, \alpha)}{2}\right)^i yF^i .$$

Then we have $y_ih = (M - i\alpha)(h)y_i$, $y_ie_{-\alpha} = y_{i+1}$ for $i = 0, 1, \cdots, m-1$ and $y_me_{-\alpha} = 0$. Also the last formula of (3.36):

$$(yF^i)E = (-mi + i(i-1))yF^{i-1} , \qquad i = 1, \cdots, m ,$$

becomes

$$y_ie_\alpha = (-mi + i(i-1))\frac{(\alpha, \alpha)}{2}y_{i-1} ,$$

which is the same as the last equation in (12). Thus we have established (12). The argument shows also that any non-zero y in $\mathfrak{M}$ satisfying $yh = M(h)y$, $ye_\alpha = 0$ generates an irreducible $\mathfrak{L}^{(\alpha)}$-submodule of $\mathfrak{M}$. This proves our assertions about $\mathfrak{L}^{(\alpha)}$. Now let Λ be any weight and let Σ be the set of weights of the form $\Lambda + i\alpha$, α a fixed non-zero root, i an integer. The weight Λ is a weight for $\mathfrak{H}$ in one of the irreducible $\mathfrak{L}^{(\alpha)}$-modules $\mathfrak{N}$ into which $\mathfrak{M}$ can be decomposed and we may suppose that $\mathfrak{N}$ is generated by y such that $yh = M(h)y$, $ye_\alpha = 0$. Then $\Lambda = M - k\alpha$ where $2(M, \alpha)/(\alpha, \alpha) = m$ and $0 \leqq k \leqq m$. Let q be the largest integer such that $\Lambda + q\alpha$ is a weight. Then if $x \neq 0$ is chosen so that $xh = (\Lambda + q\alpha)(h)x$, we have $xe_\alpha = 0$, since $\Lambda + (q+1)\alpha$ is not a weight. Hence x generates an irreducible $\mathfrak{L}^{(\alpha)}$-submodule of dimensionality $s+1$ where

$$s = \frac{2(\Lambda + q\alpha, \alpha)}{(\alpha, \alpha)} = \frac{2(\Lambda, \alpha)}{(\alpha, \alpha)} + 2q .$$

On the other hand, $M = \Lambda + k\alpha$ is a weight, so $k \leqq q$, and we have $m = 2(M, \alpha)/(\alpha, \alpha) = 2(\Lambda + k\alpha, \alpha)/(\alpha, \alpha) = 2(\Lambda, \alpha)/(\alpha, \alpha) + 2k$. Then

$$(16) \qquad \begin{aligned} k - m &= -\frac{2(\Lambda, \alpha)}{(\alpha, \alpha)} - k \\ q - s &= -\frac{2(\Lambda, \alpha)}{(\alpha, \alpha)} - q , \end{aligned}$$

so $k - m \geqq q - s$. Now the weights in the module $\mathfrak{N}$ generated by y and those in the module generated by x are, respectively,

$$(17) \qquad \begin{array}{ll} \Lambda + k\alpha , & \Lambda + (k-1)\alpha, \cdots, \Lambda + (k-m)\alpha \\ \Lambda + q\alpha , & \Lambda + (q-1)\alpha, \cdots, \Lambda + (q-s)\alpha . \end{array}$$

Since $k \leqq q$ and $q - s \leqq k - m$, all those in the first progression are contained in the second and since Λ was arbitrary in Σ, it follows that Σ coincides with the second sequence. Since Λ is contained in this sequence we have $q - s \leqq 0$. If we set $r = -(q-s)$, then the last term becomes $\Lambda - r\alpha$ where $r \geqq 0$. Also $2(\Lambda, \alpha)/(\alpha, \alpha) = s - 2q = r - q$, which proves (13). We have $\Lambda' = \Lambda - [2(\Lambda, \alpha)/(\alpha, \alpha)]\alpha = \Lambda + (2q - s)\alpha$ and $-r \leqq 2q - s \leqq q$ by our inequalities. Hence $\Lambda' \in \Sigma$ and so Λ' is a weight. It remains to prove (15). We observe that the argument just used shows that Λ' also occurs in the first sequence in (17), that is, Λ' is a weight in every irreducible

$\mathfrak{L}^{(\alpha)}$-submodule $\mathfrak{N}$ which has Λ as weight. We have $\dim \mathfrak{N}_\Lambda = 1 = \dim \mathfrak{N}_{\Lambda'}$. Hence if $\mathfrak{M} = \Sigma \oplus \mathfrak{N}_j$ is a direct decomposition of $\mathfrak{M}$ into irreducible $\mathfrak{L}^{(\alpha)}$-submodules, then $\dim \mathfrak{M}_\Lambda$ is the number of $\mathfrak{N}_j$ which have Λ as weight for $\mathfrak{H}$. Hence $\dim \mathfrak{M}_\Lambda = \dim \mathfrak{M}_{\Lambda'}$.

This result applies in particular to the adjoint representation of $\mathfrak{L}$. Our result states that if β and α are any two roots, $\alpha \neq 0$, then the roots of the form $\beta + i\alpha$, i an integer, form an arithmetic progression with difference α. We call this the *α-string of roots containing β*. If the string is $\beta - r\alpha, \beta - (r-1)\alpha, \cdots, \beta + q\alpha$, then

$$\frac{2(\alpha, \beta)}{(\alpha, \alpha)} = (r - q)1 \,. \tag{18}$$

A number of consequences can now be drawn from Theorem 1. We continue the numbering of § 1.

IX. *If α, β and $\alpha + \beta$ are non-zero roots, then $[e_\alpha e_\beta] \neq 0$ for any $e_\alpha \neq 0$ in $\mathfrak{L}_\alpha$ and any $e_\beta \neq 0$ in $\mathfrak{L}_\beta$.*

Proof: The α-string containing β is $\beta - r\alpha, \cdots, \beta + q\alpha$ and $q \geqq 1$. None of these roots is 0 since no integral multiple of α is a root except $0, \pm\alpha$. Since the root spaces corresponding to non-zero roots are one-dimensional this holds for our string. Let x be a non-zero element of $\mathfrak{L}_{\beta+q\alpha}$. Then $[xh] = (\beta + q\alpha)(h)x$. If we choose $e_{-\alpha}$ so that $[e_\alpha e_{-\alpha}] = h_\alpha$, then the theorem shows that x, $x \operatorname{ad} e_{-\alpha}$, $x(\operatorname{ad} e_{-\alpha})^2, \cdots, x(\operatorname{ad} e_{-\alpha})^{r+q}$ are non-zero and these span the spaces $\mathfrak{L}_{\beta+q\alpha}, \mathfrak{L}_{\beta+(q-1)\alpha}, \cdots, \mathfrak{L}_{\beta-r\alpha}$. In particular, e_β is a non-zero multiple of $x(\operatorname{ad} e_{-\alpha})^q$. Formula (12) implies that

$$x(\operatorname{ad} e_{-\alpha})^q \operatorname{ad} e_\alpha = \frac{-q(r+1)}{2}(\alpha, \alpha)x(\operatorname{ad} e_{-\alpha})^{q-1} \,, \tag{19}$$

which is not zero since $r \geqq 0$ and $q > 0$. Hence $[e_\beta e_\alpha] \neq 0$.

We see also that $[[e_\beta e_\alpha]e_{-\alpha}] = -(q(r+1)/2)(\alpha, \alpha)e_\beta$. This formula is valid also if $\beta + \alpha$ is not a root since in this case $[e_\beta e_\alpha] = 0$ and $q = 0$. Also, if $\beta = -\alpha$, then the α-string containing β is α, 0, $-\alpha$, so $r = 0$, $q = 2$. The right-hand side of the formula is $-(\alpha, \alpha)e_{-\alpha}$ and the left-hand side is $[[e_{-\alpha}e_\alpha]e_{-\alpha}] = -[h_\alpha e_{-\alpha}] = [e_{-\alpha}h_\alpha] = -(\alpha, \alpha)e_{-\alpha}$, so again the formula holds. We have therefore proved

X. *Let α and β be non-zero roots and $e_\beta \in \mathfrak{L}_\beta$, $e_\alpha \in \mathfrak{L}_\alpha$, $e_{-\alpha} \in \mathfrak{L}_{-\alpha}$ satisfy $[e_\alpha e_{-\alpha}] = h_\alpha$. Suppose the α-string containing β is $\beta - r\alpha, \cdots, \beta, \cdots, \beta + q\alpha$. Then*

$$[[e_\beta e_\alpha]e_{-\alpha}] = \frac{-q(r+1)}{2}(\alpha, \alpha)e_\beta \,. \tag{20}$$

We remark also that if α and $-\alpha$ are interchanged, r and q are interchanged and we have $[e_{-\alpha}e_{\alpha}] = -h_{\alpha} = h_{-\alpha}$, $(-\alpha, -\alpha) = (\alpha, \alpha)$. Hence another form of (20) is

$$[[e_{\beta}e_{-\alpha}]e_{\alpha}] = \frac{-(q+1)r}{2}(\alpha, \alpha)e_{\beta} . \tag{20'}$$

XI. *No multiple of a non-zero root* α *is a root except* 0, α, $-\alpha$.

Proof: Let $\beta = k\alpha$ be a root. Then $2(\alpha, \beta)/(\alpha, \alpha) = 2k$ is an integer and we may assume this is odd and positive. Then it is immediate that the α-string containing β contains $\gamma = (1/2)\alpha$. This contradicts the fact that $\alpha = 2\gamma$ cannot be a root.

XII. *The* α*-string containing* β $(\alpha, \beta \neq 0)$ *contains at most four roots. Hence* $2(\alpha, \beta)/(\alpha, \alpha) = 0, \pm 1, \pm 2, \pm 3$.

Proof: We may assume that $\beta \neq \alpha, -\alpha$ since the α-string containing α consists of the three roots α, 0, $-\alpha$. Assume we have at least five roots. By re-labelling these we may suppose that $\beta - 2\alpha$, $\beta - \alpha$, β, $\beta + \alpha$, $\beta + 2\alpha$ are roots. Then $2\alpha = (\beta + 2\alpha) - \beta$ and $2(\beta + \alpha) = (\beta + 2\alpha) + \beta$ are not roots. Hence the β-string containing $\beta + 2\alpha$ has just the one term $\beta + 2\alpha$. Hence $(\beta + 2\alpha, \beta) = 0$. Similarly $\beta - 2\alpha - \beta$ and $\beta - 2\alpha + \beta$ are not roots, so that $(\beta - 2\alpha, \beta) = 0$. Adding we obtain $(\beta, \beta) = 0$, contradicting the fact that the non-zero roots are not isotropic. We now have $2(\alpha, \beta)/(\alpha, \alpha) = (r - q)$ while $r + q + 1 \leqq 4$. Hence $r \leqq 3$, $q \leqq 3$ and $2(\alpha, \beta)/(\alpha, \alpha) = 0, \pm 1, \pm 2, \pm 3$.

From now on we shall identify the prime field of Φ with the field Q of rational numbers. As before, let $\mathfrak{H}^*$ be the conjugate space of $\mathfrak{H}$ and now let $\mathfrak{H}_0^*$ denote the Q-space spanned by the roots. We have seen that the roots span the space $\mathfrak{H}^*$. We now prove

XIII. $\dim \mathfrak{H}_0^* = l = \dim \mathfrak{H}$.

Proof: It suffices to show that if $(\alpha_1, \alpha_2, \cdots, \alpha_l)$ is a basis for $\mathfrak{H}^*$ consisting of roots, then every root β is a linear combination of the α_i with rational coefficients. We have $\beta = \sum \lambda_i \alpha_i$, $\lambda_i \in \Phi$. Hence $(\beta, \alpha_j) = \sum \lambda_i(\alpha_i, \alpha_j)$, $j = 1, 2, \cdots, l$ and

$$\frac{2(\beta, \alpha_j)}{(\alpha_j, \alpha_j)} = \sum_{i=1}^{l} \frac{2(\alpha_i, \alpha_j)}{(\alpha_j, \alpha_j)} \lambda_i , \qquad j = 1, 2, \cdots, l . \tag{21}$$

This system of equations has integer coefficients $[2(\alpha_i, \alpha_j)/(\alpha_j, \alpha_j)]$, $[2(\beta, \alpha_j)/(\alpha_j, \alpha_j)]$ and the determinant

$$\det\left(\frac{2(\alpha_i, \alpha_j)}{(\alpha_j, \alpha_j)}\right) = \frac{2^l}{\prod_j (\alpha_j, \alpha_j)} \det((\alpha_i, \alpha_j)) \neq 0 , \tag{22}$$

since (ρ, σ) is non-degenerate and the α_i are a basis for $\mathfrak{H}^*$. Hence (21) has a unique solution which is rational. Thus the λ_i are rational numbers.

If Λ is a weight of $\mathfrak{H}$ for a representation of $\mathfrak{L}$, then we have seen that $2(\Lambda, \alpha)/(\alpha, \alpha)$ is an integer for every non-zero root α. The argument just given for roots shows that Λ is a rational linear combination of the roots α_j, $j = 1, 2, \cdots, l$. Thus the weights $\Lambda \in \mathfrak{H}_0^*$.

XIV. *(ρ, σ) is a rational number for $\rho, \sigma \in \mathfrak{H}_0^*$ and (ρ, σ) is a positive definite symmetric bilinear form on $\mathfrak{H}_0^*$.*

Proof: We recall the formula (3) for (h, k) $h, k \in \mathfrak{H}$. If we take into account the fact that the $\mathfrak{L}_\alpha$ are one-dimensional we can rewrite this as

$$(h, k) = \sum_{\alpha \text{ a root}} \alpha(h)\alpha(k) . \tag{23}$$

Let $\rho, \sigma \in \mathfrak{H}^*$. Then

$$\begin{aligned}(\rho, \sigma) = (h_\rho, h_\sigma) &= \sum_\alpha \alpha(h_\rho)\alpha(h_\sigma) \\ &= \sum_\alpha (h_\rho, h_\alpha)(h_\sigma, h_\alpha) .\end{aligned}$$

Hence

$$(\rho, \sigma) = \sum_\alpha (\rho, \alpha)(\alpha, \sigma) . \tag{24}$$

Now let β be a non-zero root. Then $(\beta, \beta) = \sum_\alpha (\beta, \alpha)^2$. Let the β-string containing α be

$$\alpha - r_{\alpha\beta}\beta, \alpha - (r_{\alpha\beta} - 1)\beta, \cdots, \alpha + q_{\alpha\beta}\beta .$$

Then, by (18), $2(\alpha, \beta)/(\beta, \beta) = r_{\alpha\beta} - q_{\alpha\beta}$ and $(\alpha, \beta) = [(r_{\alpha\beta} - q_{\alpha\beta})/2](\beta, \beta)$. Hence

$$(\beta, \beta) = \sum_\alpha \left(\frac{r_{\alpha\beta} - q_{\alpha\beta}}{2}\right)^2 (\beta, \beta)^2 .$$

Since $(\beta, \beta) \neq 0$, $\sum_\alpha (r_{\alpha\beta} - q_{\alpha\beta})^2 \neq 0$ and

$$(\beta, \beta) = \frac{4}{\sum_\alpha (r_{\alpha\beta} - q_{\alpha\beta})^2} \tag{25}$$

is a rational number and $(\alpha, \beta) = [(r_{\alpha\beta} - q_{\alpha\beta})/2](\beta, \beta)$ is rational for any roots α, β. If $\rho, \sigma \in \mathfrak{H}_0^*$, $\rho = \sum_1^l \mu_i \alpha_i$, $\sigma = \sum_1^l \nu_i \alpha_i$, $\mu_i, \nu_i \in Q$, α_i roots. Then $(\rho, \sigma) = \sum \mu_i \nu_j (\alpha_i, \alpha_j) \in Q$. Also $(\rho, \rho) = \sum (\rho, \alpha)^2 \geqq 0$ and $(\rho, \rho) = 0$ implies that every $(\rho, \alpha) = 0$. Then $\rho = 0$ since the roots α span $\mathfrak{H}^*$.

We recall that in a space with a non-degenerate bilinear form (ρ, σ), if α is a non-isotropic vector, then the mapping S_α:

$$\rho \to \rho - \frac{2(\rho, \alpha)}{(\alpha, \alpha)} \alpha \tag{26}$$

is a linear transformation which leaves fixed every vector in the hyperplane orthogonal to α and sends α into $-\alpha$. We call this the *reflection determined by* α. This belongs to the orthogonal group of the form (ρ, σ).

We have seen (Theorem 1) that if Λ is a weight of $\mathfrak{H}$ in a representation of $\mathfrak{L}$, then $\Lambda' = \Lambda S_\alpha = \Lambda - [2(\Lambda, \alpha)/(\alpha, \alpha)](\alpha)$ is also a weight. The reflections S_α, α a root, generate a group of linear transformations in $\mathfrak{H}_0^*$ called the *Weyl group* W of $\mathfrak{L}$ (relative to $\mathfrak{H}$). This group plays an important role in the representation theory for $\mathfrak{L}$. The result we have just noted is that the weights of a particular representation are a set of vectors which is invariant under the Weyl group. In particular, this holds for the roots. If two elements of W produce the same permutation of the roots they are identical since the roots span $\mathfrak{H}_0^*$. Since there are only a finite number of roots it follows that W is a finite group.

3. Simple systems of roots

We now introduce an ordering in the rational vector space $\mathfrak{H}_0^*$. For this purpose we choose a basis of roots $\alpha_1, \alpha_2, \cdots, \alpha_l$ and we call $\rho = \sum_1^l \lambda_i \alpha_i$, $\lambda_i \in Q$, *positive* if the first non-zero λ_i is positive. The set of positive vectors is closed under addition and under multiplication by positive rationals. If $\sigma, \rho \in \mathfrak{H}_0^*$ we write $\sigma > \rho$ if $\sigma - \rho > 0$. Then $\mathfrak{H}_0^*$ is totally ordered in this way and if $\sigma > \rho$ then $\sigma + \tau > \rho + \tau$ and $\lambda\sigma > \lambda\rho$ or $\lambda\sigma < \lambda\rho$ according as $\lambda > 0$ or $\lambda < 0$. We shall refer to the ordering of $\mathfrak{H}_0^*$ just defined as the *lexicographic ordering* determined by the ordered set of roots $(\alpha_1, \alpha_2, \cdots, \alpha_l)$ (which form a basis for $\mathfrak{H}_0^*$).

LEMMA 1. *Let* $\rho_1, \rho_2, \cdots, \rho_k \in \mathfrak{H}_0^*$. *Suppose the* $\rho_i > 0$ *and* $(\rho_i, \rho_j) \leqq 0$ *if* $i \neq j$. *Then the* ρ*'s are linearly independent over* Q.

Proof: Suppose $\rho_k = \sum_1^{k-1} \lambda_i \rho_i = \sum \lambda_q' \rho_q + \sum \lambda_s'' \rho_s$ where $1 \leqq q,\ s \leqq k-1,\ \lambda_q' > 0,\ \lambda_s'' \leqq 0$. Set $\sum \lambda_q' \rho_q = \sigma,\ \sum \lambda_s'' \rho_s = \tau$. Since $\rho_k > 0,\ \sigma \neq 0$. We have $(\sigma, \tau) = \sum \lambda_q' \lambda_s'' (\rho_q, \rho_s) \geqq 0$. Hence

$$(\rho_k, \sigma) = (\sigma, \sigma) + (\sigma, \tau) > 0 .$$

On the other hand, $(\rho_k, \sigma) = \sum \lambda_q' (\rho_k, \rho_q) \leqq 0$ which is a contradiction. Hence the ρ's are Q-independent.

DEFINITION 1. Relative to the ordering defined in $\mathfrak{H}_0^*$ we call a root α *simple* if $\alpha > 0$ and α cannot be written in the form $\beta + \gamma$ where β and γ are positive roots.

XV. *Let π be the collection of simple roots relative to a fixed lexicographic ordering of $\mathfrak{H}_0^*$. Then:*

(i) *If $\alpha, \beta \in \pi$, $\alpha \neq \beta$, then $\alpha - \beta$ is not a root.*

(ii) *If $\alpha, \beta \in \pi$ and $\alpha \neq \beta$, then $(\alpha, \beta) \leqq 0$.*

(iii) *The set π is a basis for $\mathfrak{H}_0^*$ over Q. If β is any positive root, then $\beta = \sum_{\alpha \in \pi} k_\alpha \alpha$ where the k_α are non-negative integers.*

(iv) *If β is a positive root and $\beta \notin \pi$, then there exists an $\alpha \in \pi$ such that $\beta - \alpha$ is a positive root.*

Proof: (i) If $\alpha, \beta \in \pi$ and $\alpha - \beta$ is a positive root, then $\alpha = \beta + (\alpha - \beta)$ contrary to the definition of π. If $\alpha - \beta$ is a negative root we again obtain a contradiction on writing $\beta = (\beta - \alpha) + \alpha$. (ii) Let $\beta - r\alpha,\ \beta - (r-1)\alpha, \cdots, \beta + q\alpha$ be the α-string containing β. Then $2(\alpha, \beta)/(\alpha, \alpha) = r - q$. Since $r = 0$ by (i) and $(\alpha, \alpha) > 0$, $(\alpha, \beta) \leqq 0$. (iii) The linear independence of the roots contained in π is clear from (ii) and the lemma. Let β be a positive root and suppose we already know that every root γ such that $\beta > \gamma > 0$ is of the form $\sum_{\alpha \in \pi} k_\alpha \alpha$, k_α a non-negative integer. We may suppose also that $\beta \notin \pi$, so that $\beta = \beta_1 + \beta_2$, $\beta_i > 0$. Then $\beta > \beta_i$ and $\beta_1 = \sum k_\alpha' \alpha$, $\beta_2 = \sum k_\alpha'' \alpha$, k_α', k_α'' non-negative integers; hence $\beta = \sum (k_\alpha' + k_\alpha'')\alpha$ which is the required form for β. If β is a negative root then $-\beta$ is a positive root. Hence $\beta = \sum k_\alpha \alpha$ where the k_α are integers $\leqq 0$. The first statement of (iii) follows from this and the linear independence of the elements of π. (iv) Let β be a positive root not in π. The lemma and (iii) imply that there is an $\alpha \in \pi$ such that $(\beta, \alpha) > 0$. Then $2(\beta, \alpha)/(\alpha, \alpha) = r - q > 0$ (r, q as before). Hence $r > 0$ and $\beta - \alpha$ is a root. If $\beta - \alpha < 0$, then $\alpha - \beta > 0$ and $\alpha = \beta + (\alpha - \beta)$ contrary to: $\alpha \in \pi$. Hence $\beta - \alpha > 0$ and $\beta = (\beta - \alpha) + \alpha$ where $\alpha \in \pi$.

We now write $\pi = (\alpha_1, \alpha_2, \cdots, \alpha_l)$ and we call this the *simple* *system of roots* for $\mathfrak{L}$ relative to $\mathfrak{H}$ and the given ordering in $\mathfrak{H}_0^*$. We have seen that every root $\beta = \sum k_i\alpha_i$ where the k_i are integers and either all $k_i \geqq 0$ or all $k_i \leqq 0$. This property is characteristic of simple systems. Thus let $\pi = (\alpha_1, \alpha_2, \cdots, \alpha_l)$ where $l = \dim \mathfrak{H}$ and the α_i are roots such that every root $\beta = \sum k_i\alpha_i$ where the k_i are integers of like sign (rationals would do too). Evidently our hypothesis implies that π is a basis for $\mathfrak{H}_0^*$. We introduce the lexicographic ordering of $\mathfrak{H}_0^*$ based on the α_i:

$$\sum \lambda_i\alpha_i > 0 \quad \text{if} \quad \lambda_1 = \cdots = \lambda_h = 0\,, \quad \lambda_{h+1} > 0\,, \quad h < l\,.$$

Then the positive roots $\beta = \sum k_i\alpha_i$ are those such that the $k_i \geqq 0$ and some $k_i > 0$. It is clear that no α_i is a sum of positive roots. Hence the α_i are simple. Since any simple system consists of l roots the set $\pi = (\alpha_1, \alpha_2, \cdots, \alpha_l)$ is the simple system defined by the ordering.

If $\pi = (\alpha_1, \alpha_2, \cdots, \alpha_l)$ is a simple system of roots, the matrix (A_{ij}), $A_{ij} = 2(\alpha_i, \alpha_j)/(\alpha_i, \alpha_i)$ is called a *Cartan matrix* for $\mathfrak{L}$ (relative to $\mathfrak{H}$). The diagonal entries of this matrix are $A_{ii} = 2$ and the off diagonal entries are $A_{ij} = 0, -1, -2$ or -3. (XII and XV (ii)). If $i \neq j$ the α_i and α_j are linearly independent so that if θ_{ij} is the angle between α_i and α_j then $0 \leqq \cos^2 \theta_{ij} < 1$. This gives $0 \leqq 4(\alpha_i, \alpha_j)^2/(\alpha_i, \alpha_i)(\alpha_j, \alpha_j) < 4$; hence $0 \leqq A_{ij}A_{ji} < 4$. This implies that either both A_{ij} and A_{ji} are 0 or one is -1 while the other is -1, -2, or -3. The determinant of the Cartan matrix (A_{ij}) is a non-zero multiple of that of $((\alpha_i, \alpha_j))$. Hence $\det (A_{ij}) \neq 0$.

We now choose $e_{\alpha_i} \in \mathfrak{L}_{\alpha_i}$, $e_{-\alpha_i} \in \mathfrak{L}_{-\alpha_i}$ so that $[e_{\alpha_i}e_{-\alpha_i}] = h_{\alpha_i}$ and we now write

$$(27) \qquad e_i = e_{\alpha_i}\,, \qquad f_i = 2e_{-\alpha_i}/(\alpha_i, \alpha_i)\,, \qquad h_i = 2h_{\alpha_i}/(\alpha_i, \alpha_i)\,.$$

These elements have the canonical multiplication table for a split three-dimensional simple Lie algebra: $[e_ih_i] = 2e_i$, $[f_ih_i] = -2f_i$, $[e_if_i] = h_i$. Also we have $[e_if_j] = 0$ if $i \neq j$ since $\alpha_i - \alpha_j$ is not a root, $[e_ih_j] = 2(\alpha_i, \alpha_j)/(\alpha_j, \alpha_j)e_i = A_{ji}e_i$ and $[f_ih_j] = -A_{ji}f_i$. The last relations include $[e_ih_i] = 2e_i$, $[f_ih_i] = -2f_i$. Thus we have the following relations for the e_i, f_i, h_i:

$$(28) \qquad \begin{aligned} &[h_ih_j] = 0 \\ &[e_if_j] = \delta_{ij}h_i \qquad (\delta_{ij} = 1 \text{ if } i = j\,, \quad \delta_{ij} = 0 \text{ if } i \neq j) \\ &[e_ih_j] = A_{ji}e_i \\ &[f_ih_j] = -A_{ji}f_i \end{aligned}$$

$i, j = 1, 2, \cdots, l$.

We wish to show that the $3l$ elements e_i, f_i, h_i (or the $2l$ elements e_i, f_i, since $h_i = [e_i f_i]$) generate $\mathfrak{L}$. Moreover, we shall show that we can obtain a basis for $\mathfrak{L}$ whose multiplication table is completely determined by the A_{ij}. This will prove that $\mathfrak{L}$ is determined by the Cartan matrix.

If $\beta = \sum k_i \alpha_i$ is a root then we define the *level* $|\beta| = \sum |k_i|$. The level is a positive integer and the positive roots of level one are just the $\alpha_i \in \pi$.

XVI. *The set of roots is determined by the simple system* π *and the Cartan matrix. In other words, the sequences* $(k_1, k_2, \cdots, k_l)$ *such that* $\sum k_i \alpha_i$ *are roots can be determined from the matrix* (A_{ij}).

Proof: It suffices to determine the positive roots. The positive roots of level one are just the $\alpha_i \in \pi$. Now suppose we already know the positive roots of level $\leqq n$, n a positive integer. We give a method for determining the positive roots of the next level. By XV these are of the form $\beta = \alpha + \alpha_j$, $\alpha > 0$ of level n, $\alpha_j \in \pi$. Hence the problem is to determine for a given $\alpha > 0$ of level n the $\alpha_j \in \pi$ such that $\alpha + \alpha_j$ is a root. If $\alpha = \alpha_j$, $\alpha + \alpha_j$ is not a root. Hence we may assume that $\alpha = \sum k_i \alpha_i$ and some $k_i > 0$ for $i \neq j$. Then the linear forms $\alpha - \alpha_j$, $\alpha - 2\alpha_j, \cdots$ which are roots are positive of level less than n. Hence one knows which of these are roots. Thus the number r such that the α_j-string containing α is $\alpha - r\alpha_j, \cdots, \alpha, \cdots, \alpha + q\alpha_j$ is known. We have

$$q = r - 2(\alpha, \alpha_j)/(\alpha_j, \alpha_j) = r - \sum_1^l k_i A_{ji} ;$$

hence q can be determined by the Cartan matrix. Since $\alpha + \alpha_j$ is a root if and only if $q > 0$ this gives a method of ascertaining whether or not $\alpha + \alpha_j$ is a root.

Example. Suppose the Cartan matrix is

$$\begin{pmatrix} 2 & -1 \\ -3 & 2 \end{pmatrix}, \tag{29}$$

that is,

$$2(\alpha_1, \alpha_2)/(\alpha_1, \alpha_1) = -1 , \qquad 2(\alpha_1, \alpha_2)/(\alpha_2, \alpha_2) = -3 .$$

Since $\alpha_1 - \alpha_2$ is not a root these relations imply that the α_1-string containing α_2 and the α_2-string containing α_1 are, respectively,

$$\begin{aligned} &\alpha_2 , \quad \alpha_2 + \alpha_1 \\ &\alpha_1 , \quad \alpha_1 + \alpha_2 , \quad \alpha_1 + 2\alpha_2 , \quad \alpha_1 + 3\alpha_2 . \end{aligned} \tag{30}$$

The only positive root of level two is $\alpha_1 + \alpha_2$. Since $\alpha_2 + 2\alpha_1$ is not a root the only positive root of level three is $\alpha_1 + 2\alpha_2$. Since $2\alpha_1 + 2\alpha_2$ is not a root the only positive root of level four is $\alpha_1 + 3\alpha_2$. We have $2(\alpha_1 + 3\alpha_2, \alpha_1)/(\alpha_1, \alpha_1) = 2 - 3 = -1$, which implies that $(\alpha_1 + 3\alpha_2) + \alpha_1 = 2\alpha_1 + 3\alpha_2$ is a root. Since $\alpha_1 + 4\alpha_2$ is not a root $2\alpha_1 + 3\alpha_2$ is the only positive root of level five. Since $(2\alpha_1 + 3\alpha_2) + \alpha_1 = 3(\alpha_1 + \alpha_2)$ and $(2\alpha_1 + 3\alpha_2) + \alpha_2 = 2(\alpha_1 + 2\alpha_2)$ are not roots there are no roots of level six or higher. Hence the roots are

$$\begin{aligned} &\pm\alpha_1\,,\quad \pm\alpha_2\,,\quad \pm(\alpha_1+\alpha_2)\,,\quad \pm(\alpha_1+2\alpha_2)\,,\\ &\pm(\alpha_1+3\alpha_2)\,,\quad \pm(2\alpha_1+3\alpha_2)\,. \end{aligned} \tag{31}$$

A simple induction on levels shows that any positive root β can be written as

$$\beta = \alpha_{i_1} + \alpha_{i_2} + \cdots + \alpha_{i_k}\,, \tag{32}$$

$\alpha_{i_j} \in \pi$ in such a way that every partial sum

$$\alpha_{i_1} + \cdots + \alpha_{i_m}\,, \qquad m \leqq k \tag{33}$$

is a root. The number k in (32) is the level of β. We shall now abbreviate $[\cdots[x_1x_2]\cdots x_r]$ as $[x_1x_2\cdots x_r]$, $x_i \in \mathfrak{L}$. Then the result IX implies that

$$[e_{i_1}e_{i_2}\cdots e_{i_k}] \neq 0\,; \tag{34}$$

hence this element spans $\mathfrak{L}_\beta$. It follows also that

$$[f_{i_1}f_{i_2}\cdots f_{i_k}] \neq 0 \tag{35}$$

and this element spans $\mathfrak{L}_{-\beta}$. Since the $\alpha_i \in \pi$ form a basis for $\mathfrak{H}^*$, the h_{α_i} and hence the h_i form a basis for $\mathfrak{H}$. Since $\mathfrak{L} = \mathfrak{H} + \sum(\mathfrak{L}_\beta + \mathfrak{L}_{-\beta})$ summed over the positive roots, this proves

XVII. *Let $\pi = (\alpha_1, \alpha_2, \cdots, \alpha_l)$ be a simple system of roots for $\mathfrak{L}$ relative to $\mathfrak{H}$ and let e_i, f_i, h_i be as in (27). Then the $3l$ elements e_i, f_i, h_i generate $\mathfrak{L}$. For each positive root β we can select a representation of $\beta = \alpha_{i_1} + \alpha_{i_2} + \cdots + \alpha_{i_k}$ so that $\alpha_{i_1} + \cdots + \alpha_{i_m}$ is a root for every $m \leqq k$. Then the elements*

$$h_i\,, \qquad [e_{i_1}\cdots e_{i_k}]\,, \qquad [f_{i_1}\cdots f_{i_k}] \tag{36}$$

determined by the positive roots β form a basis for $\mathfrak{L}$.

We shall be interested in the multiplication table of the basis (36). For this we require

XVIII. *Let β be a positive root and let the sequence $i_1, i_2, \cdots, i_k$ be determined by β as in* XVII. *Let $1', 2', \cdots, k'$ be a permutation of $1, 2, \cdots, k$. Then $[e_{i_{1'}} e_{i_{2'}} \cdots e_{i_{k'}}]$ is a rational multiple of $[e_{i_1} e_{i_2} \cdots e_{i_k}]$, the multiplier being determined by the A_{ij}. A similar statement holds for the f's.*

Proof: This is clear for $k = 1$, so we use induction and assume the result for positive roots of level $k - 1$. If $i_k = i_{k'} = j$, then we may assume that the sequence chosen for the root $\beta - \alpha_j$ is $i_1, i_2, \cdots, i_{k-1}$. Then the result for $k - 1$ implies that $[e_{i_{1'}} \cdots e_{i_{(k-1)'}}] = t[e_{i_1} \cdots e_{i_{k-1}}]$ where t is a rational number determined by the A_{ij}. Then $[e_{i_{1'}} \cdots e_{i_{k'}}] = [e_{i_{1'}} \cdots e_{i_{(k-1)'}} e_j] = t[e_{i_1} \cdots e_{i_{k-1}} e_j] = t[e_{i_1} \cdots e_{i_k}]$. Next suppose $i_k = j \neq i_{k'}$ and write

$$[e_{i_{1'}} \cdots e_{i_{k'}}] = [e_{i_{1'}} \cdots e_{i_{r'}} e_j \cdots e_{i_{k'}}]$$

where the displayed e_j is the last one occurring in the expression. If any of the partial sums $\alpha_{i_{1'}} + \cdots + \alpha_{i_{m'}}$ is not a root, then $[e_{i_{1'}} \cdots e_{i_{k'}}] = 0$. Since this fact can be ascertained from the knowledge of the Cartan matrix the result holds in this case. Now suppose every $\alpha_{i_{1'}} + \cdots + \alpha_{i_{m'}}$ is a root. Then $[e_{i_{1'}} \cdots e_{i_{k'}}] \neq 0$. Since $[f_j e_i] = 0$ if $i \neq j$, $[\operatorname{ad} f_j, \operatorname{ad} e_i] = 0$ and

$$[e_{i_{1'}} \cdots e_{i_{r'}} e_j \cdots e_{i_{k'}} f_j] = [e_{i_{1'}} \cdots e_{i_{r'}} e_j f_j \cdots e_{i_{k'}}] .$$

By (19), $[e_{i_{1'}} \cdots e_{i_{r'}} e_j f_j] = -q(r + 1)[e_{i_{1'}} \cdots e_{i_{r'}}]$ (since $\alpha_{i_{1'}} + \cdots + \alpha_{i_{r'}}$ is a root $\neq 0$) where the q and r are integers, $q > 0$, $r \geqq 0$ and q and r are determined by the α_j string containing $\beta \equiv \alpha_{i_{1'}} + \cdots + \alpha_{i_{r'}}$. This string is known from (A_{ij}). A similar argument shows that

$$[e_{i_{1'}} \cdots e_{i_{k'}} f_j e_j] = s[e_{i_{1'}} \cdots e_{i_{k'}}]$$

where s is a non-zero integer which can be determined from the A_{ij}. It follows that

$$\begin{aligned} s[e_{i_{1'}} \cdots e_{i_{k'}}] &= [e_{i_{1'}} \cdots e_{i_{k'}} f_j e_j] \\ &= -q(r + 1)[e_{i_{1'}} \cdots e_{i_{r'}} e_{i_{(r+2)'}} \cdots e_{i_{k'}} e_j] . \end{aligned}$$

Thus $[e_{i_{1'}} \cdots e_{i_{k'}}] = t[e_{i_{1'}} \cdots e_{i_{r'}} e_{i_{(r+2)'}} \cdots e_{i_{k'}} e_j]$ where t is a rational number which can be determined from the A_{ij}. This reduces the discussion to the first case. The f's can be treated similarly.

We can now prove the following basic theorem.

THEOREM 2. *Let $\pi = (\alpha_1, \alpha_2, \cdots, \alpha_l)$ be a simple system of roots for a split semi-simple Lie algebra $\mathfrak{L}$ relative to a splitting Cartan* *subalgebra* $\mathfrak{H}$. *Let the* e_i, f_i, h_i, $i = 1, 2, \cdots, l$, *be generators of* $\mathfrak{L}$ *defined in* (27) *and let the basis* h_i, $[e_{i_1} \cdots e_{i_k}]$, $[f_{i_1} \cdots f_{i_k}]$ *for* $\mathfrak{L}$ *be as specified in* XVII. *Then the multiplication table for this basis has rational coefficients which are determined by the Cartan matrix* (A_{ij}).

Proof: We have $[h_i h_j] = 0$ and $[e_{i_1} \cdots e_{i_k} h_j] = \sum_{m=1}^{k} A_{j i_m} [e_{i_1} \cdots e_{i_k}]$, by (28). Similarly, $[f_{i_1} \cdots f_{i_k} h_j] = -\sum A_{j i_m} [f_{i_1} \cdots f_{i_k}]$. It remains to consider products of e terms and f terms. Since $[e_{i_1} \cdots e_{i_k}] = [\cdots [e_{i_1} e_{i_2}] \cdots e_{i_k}]$, $\operatorname{ad} [e_{i_1} \cdots e_{i_k}] = [\cdots [\operatorname{ad} e_{i_1} \operatorname{ad} e_{i_2}] \cdots \operatorname{ad} e_{i_k}]$ and $[x, [e_{i_1} \cdots e_{i_k}]] = x[\cdots [\operatorname{ad} e_{i_1} \operatorname{ad} e_{i_2}] \cdots \operatorname{ad} e_{i_k}]$. This can be obtained by operating on x with a certain (non-commutative) polynomial in $\operatorname{ad} e_{i_1}, \cdots, \operatorname{ad} e_{i_k}$. It follows from this and a similar argument for the f's that it suffices to show that $[[e_{i_1} \cdots e_{i_k}] e_j]$, $[[f_{i_1} \cdots f_{i_k}] e_j]$, $[[e_{i_1} \cdots e_{i_k}] f_j]$, $[[f_{i_1} \cdots f_{i_k}] f_j]$ are rational combinations of our base elements where the coefficients can be determined by the A_{ij}. The argument is the same for the last two as for the first two so we consider the first two only. To evaluate $[[e_{i_1} \cdots e_{i_k}] e_j]$ we ascertain first whether or not $\beta + \alpha_j$, $\beta = \alpha_{i_1} + \cdots + \alpha_{i_k}$ is a root. If not, then the product is 0. On the other hand, if $\gamma = \beta + \alpha_j$ is a root then, by XVIII, $[[e_{i_1} \cdots e_{i_k}] e_j]$ is a rational multiple of the e-base element associated with γ, the multiplier determined by the Cartan matrix. Next consider $[[f_{i_1} \cdots f_{i_k}] e_j]$. If $k = 1$ the product is 0 unless $i_1 = j$, in which case $[f_{i_1} e_j] = -h_j$. If $k \geqq 2$ we shall show by induction that the product is a rational linear combination of f-base elements. Thus, if $k = 2$ the product is 0 unless $j = i_1$ or $j = i_2$. For $[[f_{i_1} f_j] e_j]$ we have $[[f_{i_1} f_j] e_j] = [f_{i_1} [f_j e_j]] + [[f_{i_1} e_j] f_j] = -[f_{i_1} h_j] = A_{j i_1} f_{i_1}$, since $i_1 \neq j$ and so $[f_{i_1} e_j] = 0$. The relation just derived implies also that $[[f_j f_{i_1}] e_j] = -A_{j i_1} f_{i_1}$. Now assume $k > 2$. If no $i_r = j$, then the product is 0. Otherwise, let i_{r+1} be the last index in $[f_{i_1} \cdots f_{i_k}]$ which equals j. Then

$$\begin{aligned} [[f_{i_1} \cdots f_{i_r} f_j \cdots f_{i_k}] e_j] &= [[f_{i_1} \cdots f_{i_r} f_j] e_j] f_{i_{r+2}} \cdots f_{i_k}] \\ &= -[\cdots [f_{i_1} \cdots f_{i_r}] h_j] \cdots f_{i_k}] + [[[f_{i_1} \cdots f_{i_r}] e_j] f_j \cdots f_{i_k}] . \end{aligned}$$

The first term is a rational multiple determined by the A_{ij} of an f-base element. The induction hypothesis establishes the same claim for the second element. This completes the proof.

SUMMARY. Before continuing our analysis, it will be well to summarize the results which we have obtained. For any split semi-simple Lie algebra $\mathfrak{L}$ with splitting Cartan subalgebra $\mathfrak{H}$ we have obtained a *canonical set of generators* e_i, f_i, h_i, $i = 1, 2, \cdots, l$, satisfying the *defining relations* (28). These were obtained by choosing a simple system π of roots. The characteristic property of π is that every root $\alpha = \sum_1^l k_i\alpha_i$, $\alpha_i \in \pi$, where the k_i are all either non-negative or non-positive integers. Such systems are obtained by introducing lexicographic orderings in the space $\mathfrak{H}_0^*$ of rational linear combinations of the roots and selecting the positive roots which are not of the form $\alpha + \beta$, $\alpha, \beta > 0$, in such an ordering. A canonical set of generators associated with π is $h_i = 2h_{\alpha_i}/(\alpha_i, \alpha_i)$, $e_i = e_{\alpha_i}$, $f_i = 2e_{-\alpha_i}/(\alpha_i, \alpha_i)$ where e_{α_i} is any non-zero element of $\mathfrak{L}_{\alpha_i}$ and $e_{-\alpha_i}$ is chosen in $\mathfrak{L}_{-\alpha_i}$ so that $[e_{\alpha_i}e_{-\alpha_i}] = h_{\alpha_i}$. We observe that h_i is uniquely determined by α_i while e_i can be replaced by any $\mu_i e_i$, $\mu_i \neq 0$, in Φ. Then f_i will be replaced by $\mu_i^{-1} f_i$. The elements e_i, f_i, h_i constitute a canonical basis for a split three-dimensional simple Lie algebra $\mathfrak{L}_i$. Moreover, it is easy to check that if (e_i', f_i', h_i') is any canonical basis for $\mathfrak{L}_i$ such that $h_i' \in \mathfrak{L}_i \cap \mathfrak{H}$, then $h_i' = h_i$, $e_i' = \mu_i e_i$, $f_i' = \mu_i^{-1} f_i$.

If e_i, f_i, h_i are canonical generators we obtain a *canonical basis* for $\mathfrak{L}$ in the following manner: the basis h_i for $\mathfrak{H}$ and for each non-zero root β a base element $e_\beta = [e_{i_1} \cdots e_{i_k}]$ or $[f_{i_1} \cdots f_{i_k}]$ according as $\beta > 0$ or $\beta < 0$ where $(i_1, \cdots, i_k)$ is a sequence such that $\alpha_{i_1} + \cdots + \alpha_{i_k} = \pm\beta$ and every partial sum $\alpha_{i_1} + \cdots + \alpha_{i_m}$ is a root.

The multiplication table for the canonical basis is rational and is determined by the Cartan matrix (A_{ij}), $A_{ij} = 2(\alpha_i, \alpha_j)/(\alpha_i, \alpha_i)$. The A_{ij} are integers, $A_{ii} = 2$ and if $i \neq j$, then either $A_{ij} = 0 = A_{ji}$ or one of the numbers A_{ij}, A_{ji} is -1 and the other is -1, -2 or -3. We observe also that the group of orthogonal linear transformations generated by the reflections

$$S_i \equiv S_{\alpha_i} : \quad \xi \to \xi - \frac{2(\xi, \alpha_i)}{(\alpha_i, \alpha_i)}\alpha_i$$

which is a subgroup of the Weyl group W is finite. (Later we shall see that this subgroup coincides with W.) The reflection S_i can be described by the Cartan matrix. Thus if we take the basis $(\alpha_1, \alpha_2, \cdots, \alpha_l)$ for $\mathfrak{H}_0^*$, then S_i is completely described by

$$\alpha_j S_i = \alpha_j - \frac{2(\alpha_i, \alpha_j)}{(\alpha_i, \alpha_i)}\alpha_i = \alpha_j - A_{ij}\alpha_i \,.$$

The conditions we have noted on the A_{ij} are redundant. However, sometimes one subset of these conditions will be used while at other times another will be used.

4. The isomorphism theorem. Simplicity

Theorem 2 of the last section makes the following isomorphism theorem almost obvious.

THEOREM 3. *Let $\mathfrak{L}$, $\mathfrak{L}'$ be split semi-simple Lie algebras over a field Φ of characteristic* 0 *with splitting Cartan subalgebras $\mathfrak{H}$, $\mathfrak{H}'$ of the same dimensionality l. Let $(\alpha_1, \alpha_2, \cdots, \alpha_l)$, $(\alpha_1', \alpha_2', \cdots, \alpha_l')$ be simple systems of roots for $\mathfrak{L}$ and $\mathfrak{L}'$ respectively. Suppose the Cartan matrices $(2(\alpha_i, \alpha_j)/(\alpha_i, \alpha_i))$, $(2(\alpha_i', \alpha_j')/(\alpha_i', \alpha_i'))$ are identical. Let $e_i, f_i, h_i, e_i', f_i', h_i'$, $i = 1, 2, \cdots, l$, be canonical generators for $\mathfrak{L}$ and $\mathfrak{L}'$ as in* (27). *Then there exists a unique isomorphism of $\mathfrak{L}$ onto $\mathfrak{L}'$ mapping e_i on e_i', f_i on f_i', h_i on h_i'.*

Proof: By XVI we know that $\sum k_i\alpha_i$ is a root for $\mathfrak{H}$ if and only if $\sum k_i\alpha_i'$ is a root for $\mathfrak{H}'$. If β is a positive root for $\mathfrak{H}$ we write $\beta = \alpha_{i_1} + \cdots + \alpha_{i_k}$ so that $\alpha_{i_1} + \cdots + \alpha_{i_m}$ is a root, $m \leqq k$. Then $\beta' = \alpha_{i_1}' + \cdots + \alpha_{i_k}'$ and every $\alpha_{i_1}' + \cdots + \alpha_{i_m}'$ is a root. We can choose as base elements in the canonical basis (36) for $\mathfrak{L}$ and $\mathfrak{L}'$ the elements $[e_{i_1} \cdots e_{i_k}]$, $[f_{i_1} \cdots f_{i_k}]$, $[e_{i_1}' \cdots e_{i_k}']$, $[f_{i_1}' \cdots f_{i_k}']$. Then Theorem 2 shows that the coefficients in the multiplication table for the basis for $\mathfrak{L}$ and $\mathfrak{L}'$ are identical. Hence the linear mapping which matches these base elements is the required isomorphism. The uniqueness is clear since the e_i, f_i, h_i are generators.

The result we have just proved is basic for the problem of determining the simple Lie algebras. It is useful also for the study of automorphisms (which we shall consider later) of a single Lie algebra. We note here that the result implies that there exists an automorphism of $\mathfrak{L}$ mapping $e_i \to f_i$, $f_i \to e_i$, $h_i \to -h_i$. This follows by observing that $\alpha_i' = -\alpha_i$, $i = 1, \cdots, l$, is a simple system and $e_i' = f_i$, $f_i' = e_i$, $h_i' = -h_i$ is a corresponding set of generators. We shall need this in § 7.

DEFINITION 2. A simple system of roots $\pi = (\alpha_1, \alpha_2, \cdots, \alpha_l)$ is called *indecomposable* if it is impossible to partition π into non-vacuous non-overlapping sets π', π'' such that $A_{ij} = 0$ for every $\alpha_i \in \pi'$, $\alpha_j \in \pi''$.

THEOREM 4. *$\mathfrak{L}$ is simple if and only if the associated simple* *system π of roots is indecomposable.*

Proof: Suppose first that $\pi = (\alpha_1, \cdots, \alpha_k) \cup (\alpha_{k+1}, \cdots, \alpha_l)$, $1 \leqq k < l$, so that $A_{ij} = 0$, $i \leqq k$, $j > k$. Choose canonical generators e_i, f_i, h_i and let $\mathfrak{L}_1$ denote the subalgebra generated by the e_j, f_j, h_j, $j \leqq k$. It is readily seen that $\mathfrak{L}_1 = \mathfrak{H}_1 + \sum' \mathfrak{L}_\gamma$ where $\mathfrak{H}_1$ is the subspace of $\mathfrak{H}$ spanned by the h_j and the summation is taken over the roots γ which are linearly dependent on the α_j. Hence $0 \subset \mathfrak{L}_1 \subset \mathfrak{L}$. If $r > k$, $j \leqq k$, then $A_{jr} = 0$ and since $\alpha_j - \alpha_r$ is not a root, this implies that $\alpha_j + \alpha_r$ is not a root. Hence $[e_j e_r] = 0$ as well as $[f_j, e_r] = 0$. Also $[h_j e_r] = 0$ and consequently e_r is in the normalizer of $\mathfrak{L}_1$. Similarly, f_r is in the normalizer of $\mathfrak{L}_1$. Since $\mathfrak{L}_1$ is contained in its own normalizer it follows that $\mathfrak{L}$ is the normalizer of $\mathfrak{L}_1$. Hence $\mathfrak{L}_1$ is an ideal and $\mathfrak{L}$ is not simple. Conversely, suppose $\mathfrak{L}$ is not simple. Then $\mathfrak{L} = \mathfrak{L}_1 \oplus \mathfrak{L}_2$ where the $\mathfrak{L}_i$ are non-zero ideals. Let α be a non-zero root and let $e_\alpha \in \mathfrak{L}_\alpha$. Then $e_\alpha = e_\alpha^{(1)} + e_\alpha^{(2)}$, $e_\alpha^{(i)} \in \mathfrak{L}_i$ and $[e_\alpha h] = \alpha(h) e_\alpha$ for $h \in \mathfrak{H}$ implies that $[e_\alpha^{(i)} h] = \alpha(h) e_\alpha^{(i)}$. Since $\mathfrak{L}_\alpha$ is one dimensional this implies that either $\mathfrak{L}_\alpha \subseteqq \mathfrak{L}_1$ or $\mathfrak{L}_\alpha \subseteqq \mathfrak{L}_2$. Since $[\mathfrak{L}_1 \mathfrak{L}_2] = 0$ and $[\mathfrak{L}_\alpha \mathfrak{L}_{-\alpha}] \neq 0$ we have either $\mathfrak{L}_\alpha + \mathfrak{L}_{-\alpha} \subseteqq \mathfrak{L}_1$ or $\mathfrak{L}_\alpha + \mathfrak{L}_{-\alpha} \subseteqq \mathfrak{L}_2$. In particular, we may order the canonical generators e_i, f_i so that $e_1, f_1, \cdots, e_k, f_k \in \mathfrak{L}_1$, $e_{k+1}, f_{k+1}, \cdots, e_l, f_l \in \mathfrak{L}_2$. Since the $\mathfrak{L}_i$ are non-zero ideals, $1 \leqq k < l$ and $0 = [e_j[e_r f_r]] = [e_j h_r] = A_{rj} e_j$ if $j \leqq k$ and $r > k$. Hence $A_{jr} = A_{rj} = 0$ and the simple system of roots is decomposable.

5. *The determination of the Cartan matrices*

The results of the last section reduce the problem of determining the simple Lie algebras to the following two problems: (1) determination of the Cartan matrices (A_{ij}) corresponding to indecomposable simple systems of roots and (2) determination of simple algebras associated with the Cartan matrices. We consider (1) here and (2) in the next section. We observe first that the indecomposability condition amounts to saying that it is impossible to order the indices (or the α_i) so that the matrix has the block form

$$\begin{pmatrix} B & 0 \\ 0 & C \end{pmatrix}$$

where B, C are not vacuous.

We now associate a diagram—the *Dynkin diagram*—with the Cartan matrix A_{ij}. We choose l points $\alpha_1, \alpha_2, \cdots, \alpha_l$ and we connect

α_i to α_j, $i \neq j$, by $A_{ij}A_{ji}$ lines. Also we attach to each point α_i the weight (α_i, α_i). If $A_{ij} = 0 = A_{ji}$, α_i and α_j are not connected and if $A_{ij} \neq 0$, $A_{ji} \neq 0$, then $A_{ji}/A_{ij} = (\alpha_i, \alpha_i)/(\alpha_j, \alpha_j)$. Hence A_{ji}/A_{ij} and $A_{ij}A_{ji}$ can be determined from the diagram. Since A_{ij} is non-positive this information determines A_{ij} and A_{ji}. Thus the matrix can be reconstructed from the diagram of points, lines and the weights. We consider two examples:

(37)

$$
\overset{3}{\underset{\alpha_1}{\circ}} \!\!\!\!\equiv\!\!\!\! \overset{1}{\underset{\alpha_2}{\circ}} \,, \qquad G_2
$$

$$
\overset{1}{\underset{\alpha_1}{\circ}} \!-\! \overset{1}{\underset{\alpha_2}{\circ}} \!-\! \overset{1}{\underset{\alpha_3}{\circ}} \, \cdot \, \cdot \, \cdot \, \overset{1}{\underset{\alpha_{l-1}}{\circ}} \!-\! \overset{1}{\underset{\alpha_l}{\circ}} \,, \qquad A_l \, .
$$

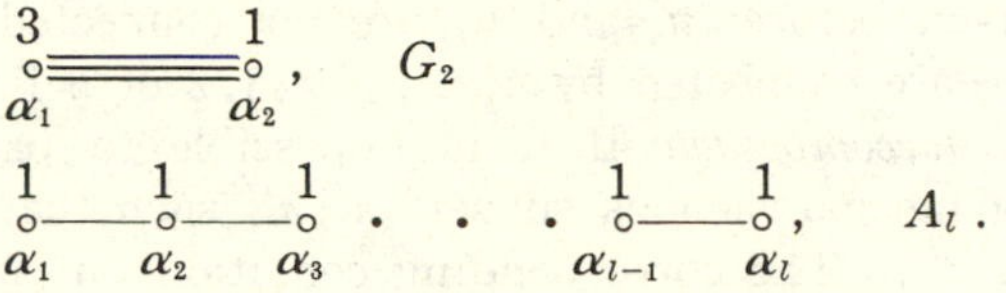

For G_2 we have $A_{21}/A_{12} = (\alpha_1, \alpha_1)/(\alpha_2, \alpha_2) = 3$, $A_{12}A_{21} = 3$ which implies that $A_{12} = -1$, $A_{21} = -3$. Hence the Cartan matrix is the matrix in (29). For A_l we have $A_{ii} = 2$, $A_{12} = A_{21} = A_{23} = A_{32} = \cdots = A_{l-1,l} = A_{l,l-1} = -1$ and all the other $A_{ij} = 0$. Hence the matrix is

(38)

$$
\begin{bmatrix}
2 & -1 & & & & & 0 \\
-1 & 2 & -1 & & & & \\
& -1 & 2 & & & & \\
& & & \cdot & & & \\
& & & & \cdot & & \\
& & & & & \cdot & -1 \\
0 & & & & & -1 & 2
\end{bmatrix} .
$$

In determining the Dynkin diagrams we at first drop the weights (α_i, α_i) on the points and consider only the collection of points and the lines joining these. We have l points $\alpha_1, \alpha_2, \cdots, \alpha_l$, and α_i and α_j, $i \neq j$, are not connected if $A_{ij}A_{ji} = 0$ and are connected by $A_{ij}A_{ji} = 1, 2$, or 3 lines if $A_{ij}A_{ji} \neq 0$. The α_i are linearly independent vectors in a Euclidian space E_0 over the rationals. This can be imbedded in the Euclidian space $E = E_{0R}$ over the field R of real numbers. If θ_{ij} denotes the angle between α_i and α_j then $A_{ij}A_{ji} = 4\cos^2\theta_{ij}$ and $\cos\theta_{ij} \leq 0$.

Any finite set $\alpha_1, \alpha_2, \cdots, \alpha_l$ of linearly independent vectors in a Euclidian space (over the reals) will be called an *allowable configuration* (a.c.) if $4\cos^2\theta_{ij} = 4(\alpha_i, \alpha_j)^2/(\alpha_i, \alpha_i)(\alpha_j, \alpha_j) = 0, 1, 2$ or 3 and $\cos\theta_{ij} \leq 0$ for every i, j, $i \neq j$. Thus $\cos\theta_{ij} = 0, -\frac{1}{2}, -\frac{1}{2}\sqrt{2}$ or $-\frac{1}{2}\sqrt{3}$ and accordingly $\theta_{ij} = 90°, 120°, 135°$, or $150°$. We may replace α_i by the unit vector u_i which is a positive multiple of α_i. Then the conditions are

$$(39)\qquad \begin{gathered}(u_i, u_i) = 1\,, \qquad 4(u_i, u_j)^2 = 0, 1, 2 \text{ or } 3\,, \qquad (u_i, u_j) \leqq 0\,, \\ i \neq j\,, \qquad i, j = 1, 2, \cdots, l\,.\end{gathered}$$

The Dynkin diagram (without weights) of an a.c. π is a collection of points u_i, $i = 1, \cdots, l$, and lines connecting these according to the rule given before: u_i and u_j are not connected if $(u_i, u_j) = 0$ and u_i and u_j are connected by $4(u_i, u_j)^2 = 1, 2$ or 3 lines otherwise. An a.c. is *indecomposable* if it is impossible to partition π into non-overlapping non-vacuous subsets π', π'' such that $(u_i, u_j) = 0$ if $u_i \in \pi'$, $u_j \in \pi''$. The corresponding condition on the Dynkin diagram is *connectedness*: If $u, v \in \pi$, then there exists a sequence $u_{i_1} = u$, $u_{i_2}, \cdots, u_{i_k} = v$ such that u_{i_j} and $u_{i_{j+1}}$ are connected in the diagram. If the Dynkin diagram is known, then all (u_i, u_j) will be known. We shall determine the Dynkin diagrams for all the connected a.c. after a few simple observations as follows.

1. *If S is a Dynkin diagram, the diagram obtained by suppressing a number of points and the lines incident with these is the Dynkin diagram of the* a.c. *obtained by dropping the vectors corresponding to the points.*

2. *If l is the number of vertices (points) of a Dynkin diagram, then the number of pairs of connected points $(u, v, (u, v) \neq 0)$ is less than l.*

Proof: Let $u = \sum_1^l u_i$. Then

$$0 < (u, u) = l + 2 \sum_{i<j} (u_i, u_j)\,.$$

If $(u_i, u_j) \neq 0$, then $2(u_i, u_j) \leqq -1$. Hence the inequality shows that the number of pairs u_i, u_j with $(u_i, u_j) \neq 0$ is less than l.

3. *A Dynkin diagram of an* a.c. *contains no cycles.* (A cycle is a sequence of points $u_1, \cdots, u_k$ such that u_i is connected to u_{i+1}, $i \leqq k - 1$ and u_k is connected to u_1.)

Proof: The subset forming a cycle is a diagram of an a.c. violating 2.

4. *The number of lines (counting multiplicities) issuing from a vertex does not exceed three.*

Proof: Let u be a vertex, $v_1, v_2, \cdots, v_k$ the vertices connected to u. No two v_i are connected since there are no cycles. Hence

$(v_i, v_j) = 0$, $i \neq j$. In the space spanned by u and the v_i we can choose a vector v_0 such that $(v_0, v_0) = 1$ and $v_0, v_1, \cdots, v_k$ are mutually orthogonal. Since u and the v_i, $i \geqq 1$, are linearly independent, u is not orthogonal to v_0 and so $(u, v_0) \neq 0$. Since $u = \sum_0^k (u, v_j)v_j$,

$$(u, u) = (u, v_0)^2 + (u, v_1)^2 + \cdots + (u, v_k)^2 = 1 .$$

Hence $\sum_1^k (u, v_i)^2 < 1$ and $\sum_1^k 4(u, v_i)^2 < 4$. Since $4(u, v_i)^2$ is the number of lines connecting u and v_i we have our result.

5. *The only connected* a.c. *containing a triple line is*

G_2 : 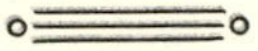 .

This is clear from 4.

6. *Let π be an* a.c. *and let $v_1, v_2, \cdots, v_k$ be vectors of π such that the corresponding points of the diagram form a simple chain in the sense that each one is connected to the next by a single line. Let π' be the collection of vectors of π which are not in the simple chain $v_1, \cdots, v_k$ together with the vector $v = \sum_1^k v_i$. Then π' is an* a.c.

Proof: We have $2(v_i, v_{i+1}) = -1$, $i = 1, \cdots, k-1$. Hence $(v, v) = k + 2\sum_{i<j}(v_i, v_j)$. Since there are no cycles $(v_i, v_j) = 0$ if $i < j$ unless $j = i + 1$. Hence $(v, v) = k - (k-1) = 1$ and v is a unit vector. Now let $u \in \pi$, $u \neq v_i$. Then u is connected with at most one of the v_i, say v_j, since there are no cycles. Then $(u, v) = (u, \sum_1^k v_i) = (u, v_j)$ and $4(u, v)^2 = 4(u, v_j)^2 = 0, 1, 2$ or 3 as required.

The diagram of π' is obtained from that of π by shrinking the simple chain to a point: Thus we replace all the vertices v_i by the single vertex v and we join this to any $u \in \pi$, $u \neq v_i$ by the total number of lines connecting u to any one of the v_j in the original diagram. Application of this to the following graphs

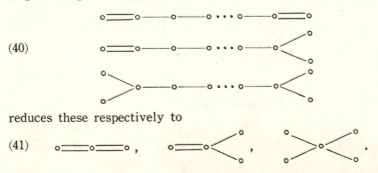

(40) 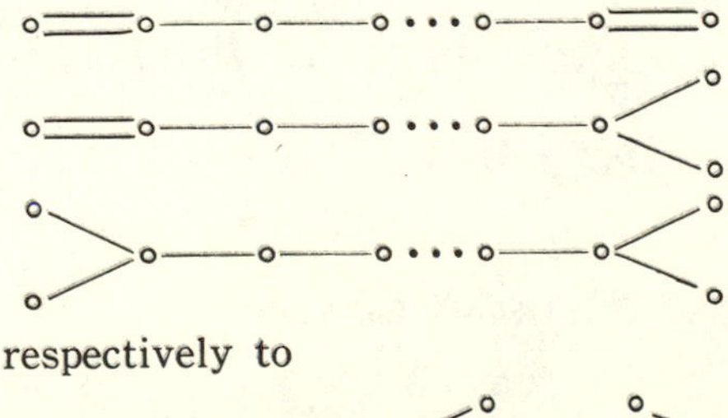

reduces these respectively to

(41)

Since the center vertex in each of these has four lines from this vertex, these cannot be diagrams of a.c., by 4. Hence we have

7. *No Dynkin diagram contains a subgraph of the form* (40).

8. *The only possible connected Dynkin diagrams have one of the following forms*

(42)

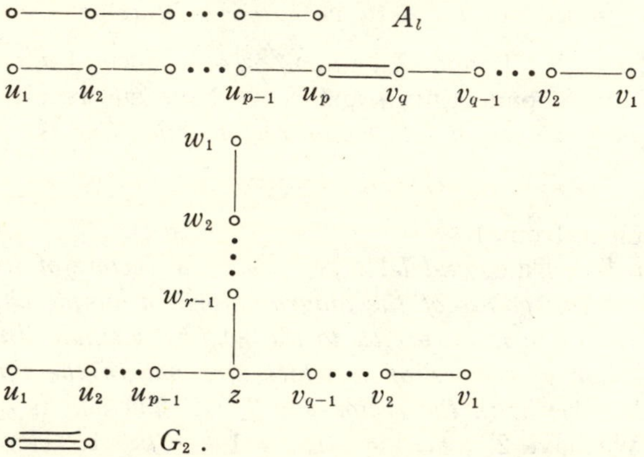

Proof: If a connected Dynkin diagram S contains a triple line then it must by G_2 by 5. If S contains a double line it contains only one such line and it contains no node, that is, graph of the form 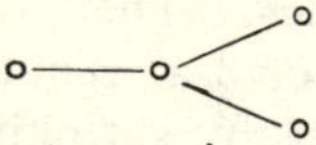 . This is clear from 7. Also it is clear that S cannot contain two nodes. This reduces the possibilities to those of (42).

We now investigate the possibilities for p, q, r in the second and third types in (42). For the second type set $u = \sum_1^p iu_i$, $v = \sum_1^q jv_j$. Since $2(u_i, u_{i+1}) = -1$ and $2(v_j, v_{j+1}) = -1$ we have

$$(u, u) = \sum_1^p i^2 - \sum_1^{p-1} i(i+1) = p^2 - p(p-1)/2 \tag{43}$$
$$= p(p+1)/2 ,$$

$$(v, v) = q(q+1)/2 . \tag{44}$$

Also $(u, v) = pq(u_p, v_q) = (pq/2)2(u_p, v_q)$ and

$$(u, v)^2 = p^2q^2/2 . \tag{45}$$

By Schwarz's inequality

$$\frac{p^2q^2}{2} < \frac{p(p+1)}{2}\,\frac{q(q+1)}{2} . \tag{46}$$

Since $pq > 0$ this gives $(p+1)(q+1) > 2pq$, which is equivalent to $(p-1)(q-1) < 2$. Hence the only possibilities for the positive integers p, q are

$$p = 1,\ q \text{ arbitrary}; \qquad q = 1,\ p \text{ arbitrary}; \qquad p = 2 = q\,. \tag{47}$$

The first two cases differ only in notation. Hence we have

9. *The only connected Dynkin diagrams of the second type in* (42) *are*

(48)

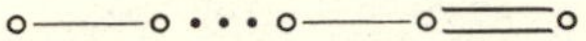

$B_l = C_l$

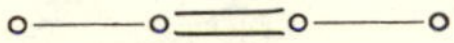

F_4 .

Finally we consider the third case in (42). Set $u = \sum_1^{p-1} iu_i$, $v = \sum_1^{q-1} jv_j$, $w = \sum_1^{r-1} kw_k$. The vectors u, v, w are mutually orthogonal and z is not in the space spanned by these vectors. Hence if $\theta_1, \theta_2, \theta_3$, are the angles between z and u, v and w then respectively, $\cos^2\theta_1 + \cos^2\theta_2 + \cos^2\theta_3 < 1$ (cf. the proof paragraph 4 above). Now

$$\begin{aligned} \cos^2\theta_1 &= (u, z)^2/(u, u) \\ &= \tfrac{1}{4}(p-1)^2/(p(p-1)/2) \qquad \text{(by (43))} \\ &= (p-1)/2p \\ &= \frac{1}{2}\left(1 - \frac{1}{p}\right). \end{aligned}$$

Similarly, $\cos^2\theta_2 = \frac{1}{2}(1 - 1/q)$, $\cos^2\theta_3 = \frac{1}{2}(1 - 1/r)$ so that we have

$$\frac{1}{2}\left(1 - \frac{1}{p} + 1 - \frac{1}{q} + 1 - \frac{1}{r}\right) < 1$$

or

$$\frac{1}{p} + \frac{1}{q} + \frac{1}{r} > 1\,. \tag{49}$$

We may suppose $p \geqq q \geqq r$ ($\geqq 2$). Then $p^{-1} \leqq q^{-1} \leqq r^{-1}$ and (49) implies that $3r^{-1} > 1$. Since $r \geqq 2$, this gives $r = 2$. Then (49) gives $p^{-1} + q^{-1} > \frac{1}{2}$. Hence $2q^{-1} > \frac{1}{2}$ and $q < 4$. Hence $2 \leqq q < 4$. If $q = 2$, then the condition is that $p^{-1} > 0$ which holds for all p. If $q = 3$, then the condition is $p^{-1} > 1/6$ and $p < 6$. Hence in this case $p = 3, 4, 5$. Thus the solutions for p, q, r are

$$\begin{aligned} &r = q = 2\,, \quad p \text{ arbitrary}; \\ &r = 2\,, \quad q = 3\,, \quad p = 3, 4, 5\,. \end{aligned} \tag{50}$$

This proves

10. *The only connected Dynkin diagrams of the third type in* (42) *are*

(51) 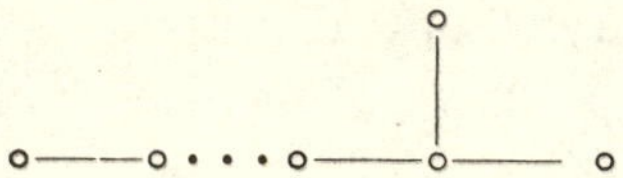 D_l

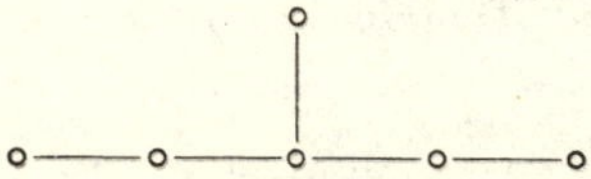 E_6

(52)

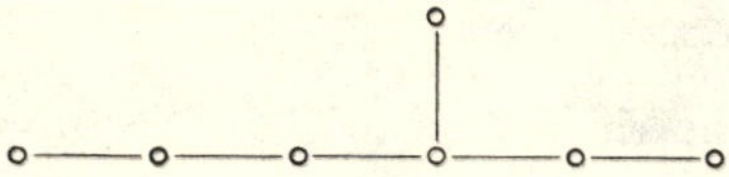 E_7

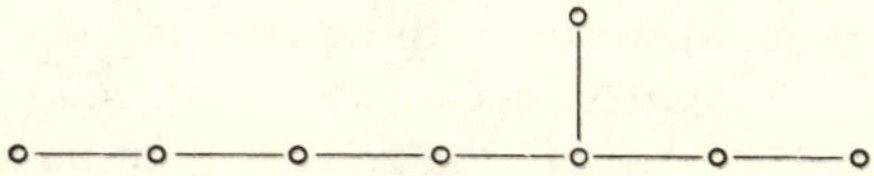 E_8 .

We have now completed the proof of the following

THEOREM 5. *The only connected Dynkin diagrams are* A_l, $l \geqq 1$, $B_l = C_l$, $l \geqq 2$, D_l, $l \geqq 4$ *and the five "exceptional" diagrams* G_2, F_4, E_6, E_7, E_8 *given in* (42), (48), (51), *and* (52).

We now re-introduce the weights on the diagrams. This will give all the possible Cartan matrices: We recall that in the Dynkin diagram obtained from the simple system $\pi = (\alpha_1, \alpha_2, \cdots, \alpha_l)$, $A_{ij}A_{ji}$, $i \neq j$, is the number of lines connecting α_i and α_j. If $A_{ij} \neq 0$, $A_{ji} \neq 0$, then $A_{ji}/A_{ij} = (\alpha_i, \alpha_i)/(\alpha_j, \alpha_j)$ and A_{ij} or $A_{ji} = -1$ while the other of these is -1, -2 or -3. Since nothing is changed in multiplying all the α_i by a fixed non-zero real number, we may take one of the α_i to be a unit vector. If the diagram has only single lines, then all the $(\alpha_i, \alpha_i) = 1$ since the diagram is connected. Hence the weighted diagrams for A_l, D_l, E_6, E_7, E_8 are

A_l : 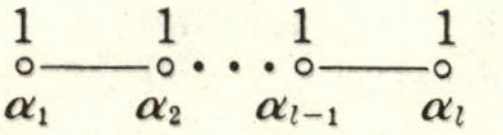

$l \geqq 1$,

D_l : 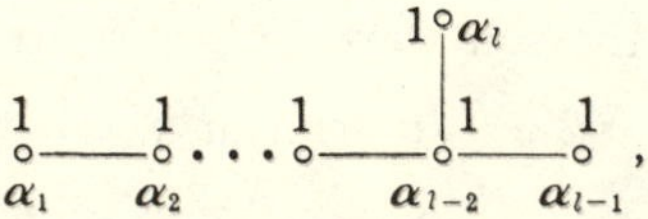

, $l \geqq 4$,

E_6:

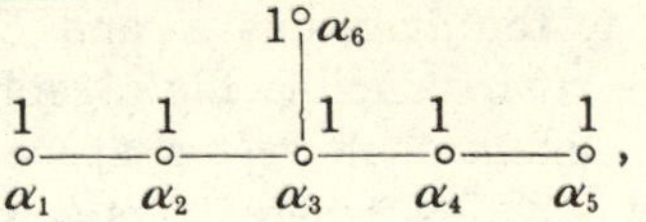

E_7:

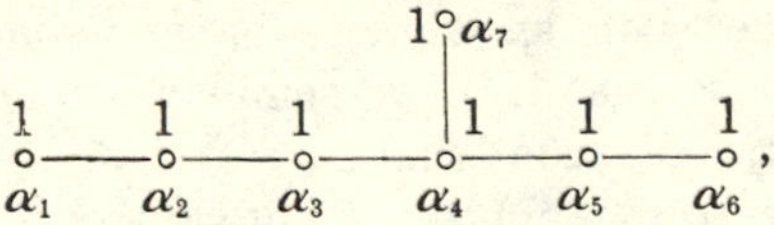

E_8:

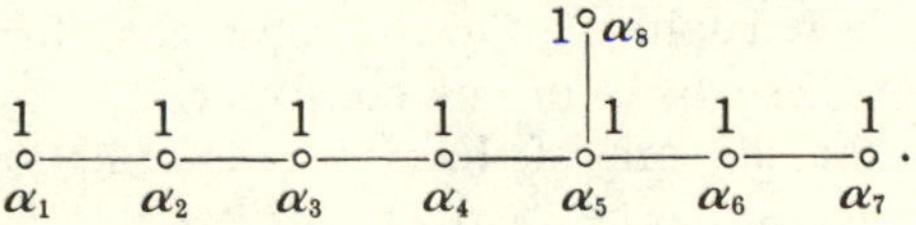

For G_2 we have chosen the notation so that

G_2:

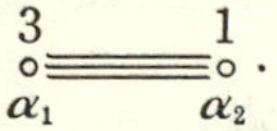

For F_4 we may take the weights as follows:

F_4:

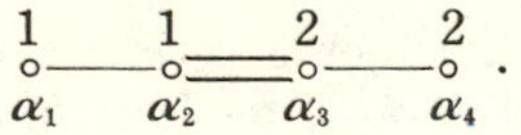

For B_l and C_l we take the following diagrams:

B_l:

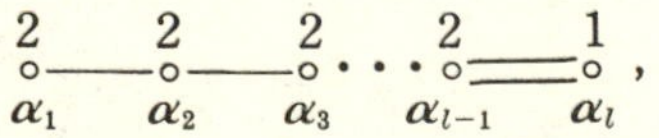

$l \geqq 2$

C_l:

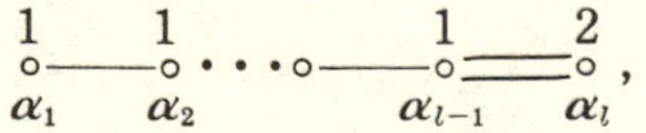

$l \geqq 3$.

These diagrams give the possible Cartan matrices.

6. Construction of the algebras

The time has now come to reveal the identity of the principal characters of our story—the split simple Lie algebras. With every connected Dynkin diagram of an a.c. which we determined in § 5 we obtain a corresponding Cartan matrix (A_{ij}) and there exists for this matrix a split simple Lie algebra with canonical generators $e_i, f_i\, h_i$, $i = 1, 2, \cdots, l$, such that $[e_ih_j] = A_{ji}e_i$. We shall give the simplest (linear) representation of the algebras corresponding to diagrams $A_l, B_l, C_l, D_l, G_2, F_4$ and E_6. Later (Chapter VII) another approach will be used to prove the existence of split simple Lie algebras corresponding to the diagrams E_7 and E_8.

We recall that if $\mathfrak{L}$ is an irreducible Lie algebra of linear transformations in a finite-dimensional vector space over a field of characteristic 0, then $\mathfrak{L} = \mathfrak{L}_1 \oplus \mathfrak{C}$ where $\mathfrak{L}_1$ is semi-simple and $\mathfrak{C}$ is the center. Hence such an algebra is semi-simple if and only if $\mathfrak{C} = 0$. We note also that if $\mathfrak{L}$ is semi-simple and $\mathfrak{L}$ contains an abelian subalgebra $\mathfrak{H}$ such that $\mathfrak{L} = \mathfrak{H} \oplus \Phi e_\alpha \oplus \Phi e_\beta \oplus \cdots$ where $\alpha, \beta, \cdots$ are non-zero mappings of $\mathfrak{H}$ into Φ such that $[e_\alpha h] = \alpha(h)e_\alpha$, $h \in \mathfrak{H}$, then $\mathfrak{H}$ is a splitting Cartan subalgebra for $\mathfrak{L}$ and $\mathfrak{L}$ is split. We shall use these facts in our constructions.

Let $\mathfrak{E}$ be the algebra of linear transformations in an $(l+1)$-dimensional vector space $\mathfrak{M}$ over Φ, $l \geqq 1$. It is well known (and easy to prove) that $\mathfrak{E}$ acts irreducibly on $\mathfrak{M}$. We have $\mathfrak{E}_L = \mathfrak{E}'_L \oplus \Phi 1$ where $\mathfrak{L} \equiv \mathfrak{E}'_L$ is the derived algebra and is the set of linear transformations of trace 0. Evidently, any $\mathfrak{E}'_L$-invariant subspace is $\mathfrak{E}_L$-invariant. Hence $\mathfrak{L} = \mathfrak{E}'_L$ is irreducible. Also the decomposition $\mathfrak{E}_L = \mathfrak{E}'_L \oplus \Phi 1$ shows that the center of $\mathfrak{E}'_L$ is 0. Hence $\mathfrak{E}'_L$ is semi-simple.

We now identify $\mathfrak{E}$ with the algebra Φ_{l+1} of $(l+1) \times (l+1)$ matrices with entries in Φ, $\mathfrak{L}$ with Φ'_{l+1}, the set of matrices of trace 0. We introduce the usual matrix basis (e_{ij}), $i, j = 1, \cdots, l+1$, in Φ_{l+1} so that

(53) $$e_{ij}e_{km} = \delta_{jk}e_{im}\,, \qquad \sum e_{ii} = 1\,.$$

A basis for $\mathfrak{L}$ is

(54) $$h_k = e_{kk} - e_{l+1,l+1}\,, \qquad k \leqq l\,; \qquad e_{ij}\,, \qquad i \neq j = 1, \cdots, l+1\,.$$

Set $h = \sum_1^l \omega_k h_k$. Then the set of h's is an abelian subalgebra $\mathfrak{H}$ of l dimensions and

(55) $$\begin{aligned} &[e_{rs}, h] = (\omega_s - \omega_r)e_{rs}\,, \\ &[e_{l+1,r}, h] = (\gamma + \omega_r)e_{l+1,r}\,, \qquad \gamma = \sum_1^l \omega_i \\ &[e_{r,l+1}, h] = -(\gamma + \omega_r)e_{r,l+1}\,, \qquad r \neq s = 1, \cdots, l\,. \end{aligned}$$

The $l^2 + l$ linear functions $h \to \omega_s - \omega_r$, $h \to \gamma + \omega_r$, $h \to -(\gamma + \omega_r)$ are distinct and are non-zero weights of $\text{ad}_{\mathfrak{L}}\mathfrak{H}$. We have $\mathfrak{L} = \mathfrak{H} + \sum \mathfrak{L}_\alpha$ where α runs over these weights. It follows that $\mathfrak{H}$ is a splitting Cartan subalgebra and the α are the non-zero roots. Set

(56) $$\begin{aligned} &\alpha_1 = \omega_1 - \omega_2\,, \qquad &&\alpha_2 = \omega_2 - \omega_3, \cdots\,, \\ &\alpha_{l-1} = \omega_{l-1} - \omega_l\,, \qquad &&\alpha_l = \gamma + \omega_l\,. \end{aligned}$$

Then

$$\begin{aligned}\alpha_k + \alpha_{k+1} + \cdots + \alpha_l &= (\omega_k - \omega_{k+1}) \\ &\quad + (\omega_{k+1} - \omega_{k+2}) + \cdots + (\gamma + \omega_l) \\ &= \gamma + \omega_k\,, \qquad k = 1, 2, \cdots, l-1\,,\end{aligned} \tag{57}$$

$$\begin{aligned}\alpha_i + \alpha_{i+1} + \cdots + \alpha_j &= (\omega_i - \omega_{i+1}) \\ &\quad + (\omega_{i+1} - \omega_{i+2}) + \cdots + (\omega_j - \omega_{j+1}) \\ &= \omega_i - \omega_{j+1}\,, \qquad 1 \leqq i \leqq j \leqq l-1\,.\end{aligned} \tag{58}$$

This shows that every root has the form $\sum k_i\alpha_i$ where the k_i are integers and $k_i \geqq 0$ for all i or $k_i \leqq 0$ for all i. Hence the α_i form a simple system of roots for $\mathfrak{L}$ relative to $\mathfrak{H}$. Equations (57) and (58) show that $\alpha_i + \alpha_{i+1}$ is a root, $1 \leqq i \leqq l-1$, $\alpha_i + 2\alpha_{i+1}$ is not a root, and $\alpha_i + \alpha_j$ is not a root if $j > i+1$. This means that the Cartan integers A_{ij} have the following values:

$$\begin{aligned}A_{i+1,i} &= -1 = A_{i,i+1}\,, \qquad 1 \leqq i < l \\ A_{ij} &= 0 \quad \text{if} \quad j > i+1 \quad \text{or} \quad j < i-1\,.\end{aligned} \tag{59}$$

Hence the Dynkin diagram is the connected simple chain

A_l :

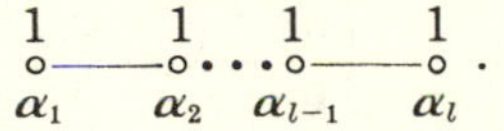

It follows that $\mathfrak{L}$ is simple of type A_l (with Dynkin diagram A_l). Hence we have

THEOREM 6. *Let $\mathfrak{L}$ be the Lie algebra of linear transformations of trace zero in an $(l+1)$-dimensional vector space over Φ. Then $\mathfrak{L}$ is a split simple Lie algebra of type A_l.*

Assume next that $\mathfrak{M}$ is n-dimensional over Φ and is equipped with a non-degenerate bilinear form (x, y) which is either symmetric or skew. Let $\mathfrak{L}$ be the Lie algebra of linear transformations A which are skew relative to (x, y), that is, $(xA, y) = -(x, yA)$ (cf. § 1.2). We require the following

LEMMA 2. *$\mathfrak{L}$ is irreducible if $n \geqq 3$.*

Proof: Let $\mathfrak{N}$ be a non-zero invariant subspace relative to $\mathfrak{L}$ and let z be a non-zero vector in $\mathfrak{N}$. Let u be any vector in the orthogonal complement $\Phi z^{\perp}$. Choose $v \notin \Phi z^{\perp}$ and consider the linear transformation A: $x \to (x, u)v - (v, x)u$. One checks that $A \in \mathfrak{L}$. Moreover, $zA = -(v, z)u \neq 0$ is in $\mathfrak{N}$. Hence $\mathfrak{N}$ contains $\Phi z^{\perp}$ for every $z \in \mathfrak{N}$. It follows that $\dim \mathfrak{N} \geqq n-1$ and $\dim \mathfrak{N} = n$ unless $\mathfrak{N} = \Phi z^{\perp}$ for every $z \in \mathfrak{N}$. Then $\mathfrak{N}$ is totally isotropic and $\dim \mathfrak{N} \leqq [n/2]$. (cf. the author's *Lectures in Abstract Algebra* II, p. 170). This implies that $n = 2$, contrary to our assumption.

We shall now distinguish the following three cases: B. (x, y) is symmetric and $n = 2l+1$, $l \geqq 1$. C. (x, y) is skew so necessarily n is even dimensional, say, $n = 2l$, $l > 1$. D. (x, y) is symmetric, $n = 2l$, $l > 1$. Moreover, we shall assume that in the symmetric cases (B and D) the bilinear form has maximal Witt index. This means that $\mathfrak{M}$ contains a totally isotropic subspace $\mathfrak{N}$ of l dimensions.

B. Let $(u_1, u_2, \cdots, u_n)$ be a basis for $\mathfrak{M}$ and let $(u_i, u_j) = \sigma_{ij}$, $s = (\sigma_{ij})$. A linear transformation $A \in \mathfrak{L}$ if and only if $(u_iA, u_j) = -(u_i, u_jA)$ for $i, j = 1, \cdots, n$. If $u_iA = \sum_k \alpha_{ik} u_k$, then these conditions are that $\sum_k \alpha_{ik}\sigma_{kj} = -\sum_k \sigma_{ik}\alpha_{jk}$, or in matrix form,

$$as = -sa', \qquad a = (\alpha_{ij}). \tag{60}$$

Since the form is of maximal Witt index the basis can be chosen so that

$$s = \begin{pmatrix} \sigma & 0 & 0 \\ 0 & 0 & \sigma\cdot 1_l \\ 0 & \sigma\cdot 1_l & 0 \end{pmatrix},$$

where 1_l denotes the identity matrix of l rows and columns and σ is a non-zero element in Φ (cf. the author [2], vol. II, p. 168). Since nothing is changed by replacing the form by a non-zero multiple of the form we may assume that

$$s = \begin{pmatrix} 1 & 0 & 0 \\ 0 & 0 & 1_l \\ 0 & 1_l & 0 \end{pmatrix}. \tag{61}$$

If we partition a in the same way as s:

$$a = \begin{pmatrix} \alpha_{11} & v_1 & v_2 \\ u_1 & a_{11} & a_{12} \\ u_2 & a_{21} & a_{22} \end{pmatrix},$$

where $a_{ij} \in \Phi_l$, u_1, u_2 are $l \times 1$ matrices and v_1, v_2 are $1 \times l$ matrices, then a simple computation shows that (60) holds if and only if

$$\begin{aligned} &u_1 = -v_2', \qquad u_2 = -v_1', \qquad a_{22} = -a_{11}', \\ &a_{12}' = -a_{12}, \qquad a_{21}' = -a_{21}, \qquad \alpha_{11} = 0. \end{aligned} \tag{62}$$

These conditions imply that the following set of elements is a basis for $\mathfrak{L}$, identified with the algebra of matrices satisfying (60)

$$
\begin{aligned}
h_i &= e_{i+1,\,i+1} - e_{i+l+1,\,i+l+1} & \\
e_{\omega_i-\omega_j} &= e_{j+1,\,i+1} - e_{i+l+1,\,j+l+1} & i \neq j \\
e_{\omega_i+\omega_j} &= e_{i+l+1,\,j+1} - e_{j+l+1,\,i+1} & i < j \\
e_{-\omega_i-\omega_j} &= e_{j+1,\,i+l+1} - e_{i+1,\,j+l+1} & i < j \\
e_{\omega_i} &= e_{1,\,i+1} - e_{i+l+1,\,1} & \\
e_{-\omega_i} &= e_{i+1,\,1} - e_{1,\,i+l+1} &
\end{aligned} \tag{63}
$$

where $i, j = 1, \cdots, l$. The set $\mathfrak{H} = \{\sum \omega_i h_i\}$ is an abelian subalgebra of $\mathfrak{L}$. The linear forms which are subscripts can be identified with the linear mappings $h \to \alpha(h)$, where $h = \sum \omega_i h_i$, and we have $[e_\alpha h] = \alpha e_\alpha$, $\alpha = \omega_i - \omega_j$, $\omega_i + \omega_j$, etc. It follows that $\mathfrak{H}$ is a splitting Cartan subalgebra and the α are the non-zero roots. $\mathfrak{L}$ acts irreducibly on $\mathfrak{M}$ and if $z = h_0 + \sum \rho_\alpha e_\alpha$ is a center element, then $[zh] = 0$ gives $\sum \rho_\alpha \alpha(h) e_\alpha = 0$. Since the e_α are linearly independent this implies that $\rho_\alpha \alpha(h) = 0$ for all h. Since $\alpha \neq 0$ we have $\rho_\alpha = 0$. Hence $z = h_0$. But then $[ze_\alpha] = 0$ gives $\alpha(h_0) = 0$ for all α. Since there are l linearly independent α (e.g. the ω_i, $i = 1, \cdots, l$) we have $h_0 = 0$. Hence the center is 0 and $\mathfrak{L}$ is semi-simple and split.

We now assume $l \geqq 2$ and we set

$$
\begin{aligned}
\alpha_1 &= \omega_1 - \omega_2\,, & \alpha_2 &= \omega_2 - \omega_3, \cdots, \\
\alpha_{l-1} &= \omega_{l-1} - \omega_l\,, & \alpha_l &= \omega_l\,.
\end{aligned} \tag{64}
$$

One checks that this is a simple system of roots with Dynkin diagram B_l:

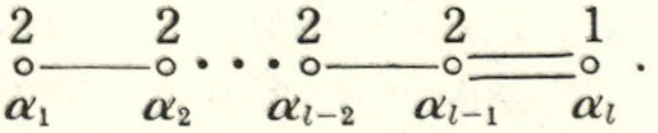

Hence $\mathfrak{L}$ is simple of type B_l. We therefore have the following

THEOREM 7. *Let $\mathfrak{L}$ be the Lie algebra of linear transformations in a $(2l+1)$-dimensional space, $l \geqq 2$, which are skew relative to a non-degenerate symmetric bilinear form of maximal Witt index. Then $\mathfrak{L}$ is a split simple Lie algebra of type B_l.*

C. Let $\mathfrak{M}$ be of dimension $2l$, $l \geqq 1$, (x, y) non-degenerate skew bilinear form in $\mathfrak{M}$. Let $\mathfrak{L}$ be the Lie algebra of linear transformations which are skew relative to (x, y). We can choose a basis

$(u_1, u_2, \cdots, u_{2l})$ so that the matrix $q = ((u_i, u_j))$ is

(65) $$q = \begin{pmatrix} 0 & 1_l \\ -1_l & 0 \end{pmatrix}.$$

As in B, one sees that $\mathfrak{L}$ can be identified with the Lie algebra of matrices $a \in \Phi_{2l}$ such that $aq = -qa'$. This implies that

(66) $$a = \begin{pmatrix} a_{11} & a_{12} \\ a_{21} & a_{22} \end{pmatrix}, \qquad a_{ij} \in \Phi_l \,,$$

where

(67) $$a_{22} = -a'_{11}\,, \qquad a'_{12} = a_{12}\,, \qquad a'_{21} = a_{21}\,.$$

Hence $\mathfrak{L}$ has the basis

(68) $$\begin{aligned} h_i &= e_{ii} - e_{l+i,\, l+i}\,, & \\ e_{\omega_j - \omega_i} &= e_{ij} - e_{j+l,\, i+l}\,, & i \neq j \\ e_{-\omega_i - \omega_j} &= e_{i,\, j+l} + e_{j,\, i+l}\,, & i < j \\ e_{\omega_i + \omega_j} &= e_{i+l,\, j} + e_{j+l,\, i}\,, & i < j \\ e_{-2\omega_i} &= e_{i,\, i+l}\,, \qquad e_{2\omega_i} = e_{i+l,\, i}\,, & \end{aligned}$$

where $i, j = 1, 2, \cdots, l$. As in B, one proves that $\mathfrak{H} = \{\sum \omega_i h_i\}$ is a Cartan subalgebra and the subscripts in (68) define the roots relative to $\mathfrak{H}$. Also $\mathfrak{L}$ has center 0 and so, by Lemma 2, $\mathfrak{L}$ is semi-simple. The roots

(69) $$\alpha_1 = \omega_1 - \omega_2, \cdots, \alpha_{l-1} = \omega_{l-1} - \omega_l\,, \qquad \alpha_l = 2\omega_l$$

form a simple system with Dynkin diagram C_l if $l \geqq 3$. This proves

THEOREM 8. *Let $\mathfrak{L}$ be the Lie algebra of linear transformations in a $2l$-dimensional space, $l \geqq 3$, which are skew relative to a non-degenerate skew bilinear form* (*the symplectic Lie algebra.*) *Then $\mathfrak{L}$ is a split simple Lie algebra of type C_l.*

D. $\mathfrak{M}$, $2l$-dimensional, $l \geqq 2$, (x, y) symmetric of maximal Witt index. Here we can choose the basis (u_i) so that $t = ((u_i, u_j))$ has the form

(70) $$t = \begin{pmatrix} 0 & 1_l \\ 1_l & 0 \end{pmatrix}$$

and $\mathfrak{L}$ can be identified with the Lie algebra of matrices a satisfying $at = -ta'$. These are the matrices

$$a = \begin{pmatrix} a_{11} & a_{12} \\ a_{21} & a_{22} \end{pmatrix}, \qquad a_{ij} \in \Phi_l\,, \tag{71}$$

such that

$$a_{22} = -a'_{11}\,, \qquad a'_{12} = -a_{12}\,, \qquad a'_{21} = -a_{21}\,. \tag{72}$$

Then $\mathfrak{L}$ has the basis

$$\begin{aligned} h_i &= e_{ii} - e_{l+i,l+i} \\ e_{\omega_i-\omega_j} &= e_{ji} - e_{i+l,j+l}\,, && i \neq j \\ e_{\omega_i+\omega_j} &= e_{i+l,j} - e_{j+l,i}\,, && i < j \\ e_{-\omega_i-\omega_j} &= e_{j,i+l} - e_{i,j+l}\,, && i < j \end{aligned} \tag{73}$$

where $i, j = 1, 2, \cdots, l$. $\mathfrak{H} = \{\sum \omega_i h_i\}$ is a splitting Cartan subalgebra and the subscripts in (73) define the roots. The center of $\mathfrak{L}$ is 0; hence $\mathfrak{L}$ is semi-simple. Set

$$\alpha_1 = \omega_1 - \omega_2, \cdots, \qquad \alpha_{l-1} = \omega_{l-1} - \omega_l\,, \qquad \alpha_l = \omega_{l-1} + \omega_l\,. \tag{74}$$

Then these α_i form a simple system of roots which has the Dynkin diagram D_l if $l \geqq 4$. We therefore have

THEOREM 9. *Let $\mathfrak{L}$ be the Lie algebra of linear transformations in a $2l$-dimensional space, $l \geqq 2$, which are skew symmetric relative to a non-degenerate symmetric bilinear form of maximal Witt index. Then if $l \geqq 4$, $\mathfrak{L}$ is a split simple Lie algebra of type D_l.*

The four classes of Lie algebras A, B, C and D are called the "great" classes of simple Lie algebras. These correspond to the linear groups which Weyl has called the classical groups in his book with this title. It is easy to see directly, or from the bases, that we have the following table of dimensionalities

(75)

type	*dimensionality*
A_l	$l(l+2)$
B_l	$l(2l+1)$
C_l	$l(2l+1)$
D_l	$l(2l-1)$.

The determination of the simple systems and the general isomorphism theorem (Theorem 3) and the criterion for simplicity (Theorem 4) yield a number of isomorphisms for the low dimensional orthogonal and symplectic Lie algebras. The verifications are left to the reader. These are

1. The orthogonal Lie algebra in 3-space (three-dimensional space

$\mathfrak{M}$) defined by a form of maximal Witt index and the symplectic Lie algebra in 2-space are split three-dimensional simple and so are isomorphic to the algebra of matrices of trace 0 in 2-space.

2. The orthogonal Lie algebra of a form of maximal Witt index in 4-space is a direct sum of two ideals isomorphic to split three-dimensional simple Lie algebras.

3. The symplectic Lie algebra in 4-space is isomorphic to the orthogonal Lie algebra in 5-space defined by a symmetric form of maximal Witt index.

4. The orthogonal Lie algebra in 6-space of a form of maximal Witt index is isomorphic to the Lie algebra of linear transformations of trace 0 in 4-space.

The remaining split simple Lie algebras: types G_2, F_4, E_6, E_7 and E_8 are called *exceptional*. We shall give irreducible representations for G_2, F_4, E_6 but we shall be content to state the results without proofs, even though some of these are not trivial. A complete discussion can be found in a forthcoming article by the author ([11]).

Our realizations of G_2 and F_4 will be as the derivation algebras of certain non-associative algebras, namely, an algebra of Cayley numbers $\mathfrak{C}$ and an exceptional simple Jordan algebra M_3^8.

Following Zorn, the definition of the *split Cayley algebra* or *vector-matrix* algebra $\mathfrak{C}$ is as follows. Let V be the three-dimensional vector algebra over Φ. Thus V has basis i, j, k over Φ and has bilinear scalar multiplication and skew symmetric vector multiplication $\times$ satisfying: i, j, k are orthogonal unit vectors and

$$
\begin{aligned}
&i \times i = j \times j = k \times k = 0\,, \qquad i \times j = k\,,\\
&j \times k = i\,, \qquad k \times i = j\,.
\end{aligned} \tag{76}
$$

Let $\mathfrak{C}$ be the set of 2×2 matrices of the form

$$
\begin{pmatrix} \alpha & a \\ b & \beta \end{pmatrix}, \qquad \alpha, \beta \in \Phi\,, \qquad a, b \in V\,. \tag{77}
$$

Addition and multiplication by elements of Φ are as usual, so that $\mathfrak{C}$ is eight-dimensional. We define an algebra product in $\mathfrak{C}$ by

$$
\begin{pmatrix} \alpha & a \\ b & \beta \end{pmatrix}\begin{pmatrix} \gamma & c \\ d & \delta \end{pmatrix} = \begin{pmatrix} \alpha\gamma - (a, d) & \alpha c + \delta a + b \times d \\ \gamma b + \beta d + a \times c & \beta\delta - (b, c) \end{pmatrix}. \tag{78}
$$

The split Cayley algebra is defined to be $\mathfrak{C}$ together with the vector space operations defined before and the multiplication of (78). $\mathfrak{C}$ is not associative but satisfies a weakening of the associative law called the *alternative law*:

$$x^2y = x(xy)\,, \qquad yx^2 = (yx)x\,. \tag{79}$$

Let $\mathfrak{L}$ be the Lie algebra of derivations in $\mathfrak{C}$. The unit matrix 1 is the identity in $\mathfrak{C}$ and since $1^2 = 1$, $1D = 0$ for every derivation D. The space $\mathfrak{C}_0$ of elements of trace 0 ($\alpha + \beta = 0$) coincides with the space spanned by the commutators $[xy] = xy - yx$, $x, y \in \mathfrak{C}$. Hence $\mathfrak{C}_0 D \subseteq \mathfrak{C}_0$ for a derivation. Thus $\mathfrak{C}_0$ is a seven-dimensional subspace of $\mathfrak{C}$ which is invariant under $\mathfrak{L}$. The representation in $\mathfrak{C}_0$ is faithful and irreducible.

If T is a linear transformation of trace 0 in V, and T^* is its adjoint relative to the scalar multiplication, then it can be verified that

$$\begin{pmatrix} \alpha & a \\ b & \beta \end{pmatrix} \to \begin{pmatrix} 0 & aT \\ -bT^* & 0 \end{pmatrix} \tag{80}$$

is a derivation in $\mathfrak{C}$. The set of these derivations is a subalgebra $\mathfrak{L}_0$ isomorphic to Φ'_{3L}. In any alternative algebra any mapping of the form $D_{a,b} = [a_L b_L] + [a_L b_R] + [a_R b_R]$, where a, b are in the algebra and a_L, a_R denote the left and the right multiplications ($x \to ax$, $x \to xa$) determined by a, is a derivation. In $\mathfrak{C}$ any derivation has the form $D_{e_1, a_{12}} + D_{e_2, b_{21}} + D_0$, where

$$\begin{aligned} e_1 &= \begin{pmatrix} 1 & 0 \\ 0 & 0 \end{pmatrix}, & e_2 &= \begin{pmatrix} 0 & 0 \\ 0 & 1 \end{pmatrix}, \\ a_{12} &= \begin{pmatrix} 0 & a \\ 0 & 0 \end{pmatrix}, & b_{21} &= \begin{pmatrix} 0 & 0 \\ b & 0 \end{pmatrix} \end{aligned} \tag{81}$$

and $D_0 \in \mathfrak{L}_0$. If $\mathfrak{H}$ is a Cartan subalgebra of $\mathfrak{L}_0$, $\mathfrak{H}$ is a Cartan subalgebra of $\mathfrak{L}$. If we identify $\mathfrak{L}_0$ with Φ'_{3L}, we can take $\mathfrak{H}$ to be the set of matrices of the form $\omega_1 h_1 + \omega_2 h_2$, $h_1 = e_{11} - e_{33}$, $h_2 = e_{22} - e_{33}$. Then $\mathfrak{H}$ is a splitting Cartan subalgebra of $\mathfrak{L}$ and the roots of $\mathfrak{H}$ in $\mathfrak{L}$ are: $\pm\omega_1$, $\pm\omega_2$, $\pm(\omega_1 - \omega_2)$, $\pm(\omega_1 + \omega_2)$, $\pm(2\omega_1 + \omega_2)$, $\pm(\omega_1 + 2\omega_2)$. The center of $\mathfrak{L}$ is 0 and since $\mathfrak{L}$ acts irreducibly in $\mathfrak{C}_0$, $\mathfrak{L}$ is semi-simple. The roots $\alpha_1 = \omega_1 - \omega_2$, $\alpha_2 = \omega_2$ form a simple system with Dynkin diagram G_2. Hence $\mathfrak{L}$ is split simple of type G_2.

The Cayley algebra $\mathfrak{C}$ has an anti-automorphism $x \to \bar{x}$ of period two such that $\bar{1} = 1$, $\bar{x} = -x$ if $x \in \mathfrak{C}_0$. Let M_3^8 denote the space of 3×3 hermitian Cayley matrices defined by this anti-automorphism. Thus M_3^8 is the set of matrices of the form

$$X = \begin{pmatrix} \xi_1 & x_3 & \bar{x}_2 \\ \bar{x}_3 & \xi_2 & x_1 \\ x_2 & \bar{x}_1 & \xi_3 \end{pmatrix}, \qquad x_i \in \mathfrak{C}, \qquad \xi_i \in \Phi . \tag{82}$$

If $X, Y \in M_3^8$, then

$$X \cdot Y = \tfrac{1}{2}(XY + YX) \in M_3^8 , \tag{83}$$

where the product XY is the usual matrix product. Then M_3^8 is a non-associative algebra relative to the usual vector space operations and the multiplication (83). Moreover, the multiplication in M_3^8 satisfies

$$X \cdot Y = Y \cdot X , \qquad (X^2 \cdot Y) \cdot X = X^2 \cdot (Y \cdot X) . \tag{84}$$

These identities are the defining properties of a class of algebras called *Jordan algebras*. A consequence of these identities is that if R_A denotes the mapping $X \to X \cdot A = A \cdot X$, then

$$D_{A,B} = [R_A R_B] \tag{85}$$

is a derivation in the algebra.

Let $\mathfrak{L}$ denote the Lie algebra of derivations in M_3^8 and let $\mathfrak{L}_0$ denote the subalgebra of those derivations which map the elements

$$\begin{aligned} e_1 &= \operatorname{diag}\{1, 0, 0\} , \qquad e_2 = \operatorname{diag}\{0, 1, 0\} , \\ e_3 &= \operatorname{diag}\{0, 0, 1\} \end{aligned} \tag{86}$$

into 0. Then $\mathfrak{L}_0$ is isomorphic to an orthogonal Lie algebra of a symmetric form of maximal Witt index in 8-space. Let $\mathfrak{H}$ be a Cartan subalgebra of $\mathfrak{L}_0$ corresponding to the Cartan subalgebra selected above for the orthogonal Lie algebra (type D_4). Then $\mathfrak{H}$ is a splitting Cartan subalgebra of $\mathfrak{L}$.

The derivation algebra maps the 26-dimensional space J_0 of matrices of M_3^8 of trace 0 into itself. The representation in J_0 is faithful and irreducible. The center of $\mathfrak{L}$ is 0, so $\mathfrak{L}$ is semi-simple. The roots of $\mathfrak{H}$ in $\mathfrak{L}$ are

$$\begin{aligned} &\pm \omega_i \pm \omega_j , && i < j = 1, 2, 3, 4 . \\ &\pm \omega_i , && \\ &\pm \Lambda_i , && \text{where } \Lambda_i = \tfrac{1}{2}(\omega_1 + \omega_2 + \omega_3 + \omega_4) - \omega_i , \\ &\pm M_i , && M_1 = \tfrac{1}{2}(\omega_1 + \omega_2 + \omega_3 + \omega_4) , \\ & && M_2 = \tfrac{1}{2}(\omega_1 + \omega_2 - \omega_3 - \omega_4) , \\ & && M_3 = \tfrac{1}{2}(\omega_1 - \omega_2 + \omega_3 - \omega_4) , \\ & && M_4 = \tfrac{1}{2}(\omega_1 - \omega_2 - \omega_3 + \omega_4) . \end{aligned} \tag{87}$$

The roots

$$\alpha_1 = \tfrac{1}{2}\omega_1 - \tfrac{1}{2}\omega_2 - \tfrac{1}{2}\omega_3 - \tfrac{1}{2}\omega_4 , \qquad \alpha_2 = \omega_4 ,$$
$$\alpha_3 = \omega_3 - \omega_4 , \qquad \alpha_4 = \omega_2 - \omega_3$$

form a simple system with the diagram $\overset{1}{\circ}\!-\!\!-\!\!-\overset{1}{\circ}\!=\!\!=\!\!=\overset{2}{\circ}\!-\!\!-\!\!-\overset{2}{\circ}$ of type F_4. Hence $\mathfrak{L}$ is split simple of type F_4.

We now use the notation $\mathfrak{D}$ for the derivation algebra of M_3^8 and we let $\mathfrak{L}$ denote the set of linear transformations in M_3^8 which have the form

$$L = R_A + D , \qquad \operatorname{tr} A = 0 , \qquad D \in \mathfrak{D} . \tag{88}$$

This is clearly a subspace of the space of linear transformations in M_3^8. Moreover, if $\operatorname{tr} B = 0$ and $E \in \mathfrak{D}$, then

$$[R_A R_B] \in \mathfrak{D} , \qquad [R_A E] = R_{AE} . \tag{89}$$

Since $[\mathfrak{D}, \mathfrak{D}] \subseteq \mathfrak{D}$ these relations imply that $\mathfrak{L}$ is a Lie algebra of linear transformations. Now $\mathfrak{L}$ acts irreducibly in M_3^8 and has 0 center. Hence $\mathfrak{L}$ is semi-simple. Let $\mathfrak{K}$ denote the Cartan subalgebra of $\mathfrak{D}$ defined above and let $\mathfrak{H} = \Phi R_{e_1-e_3} + \Phi R_{e_2-e_3} + \mathfrak{K}$. Then $\mathfrak{H}$ is a Cartan subalgebra of $\mathfrak{L}$ with roots

$$\begin{aligned} &\pm \omega_i \pm \omega_j , \qquad i < j = 1, 2, 3, 4 \\ &\pm (\omega_i \pm \tfrac{1}{2}(\omega_6 - \omega_5)) , \\ &\pm (\Lambda_i \pm \tfrac{1}{2}(\omega_7 - \omega_5)) , \\ &\pm (M_i \pm \tfrac{1}{2}(\omega_7 - \omega_6)) , \end{aligned} \tag{90}$$

where Λ_i and M_i are as in (87). The roots

$$\alpha_1 = -\,\omega_1 + \tfrac{1}{2}(\omega_6 - \omega_5) , \qquad \alpha_2 = \omega_1 - \omega_2 , \qquad \alpha_3 = \omega_2 - \omega_3 ,$$
$$\alpha_4 = \omega_3 + \omega_4 , \qquad \alpha_5 = -M_1 + \tfrac{1}{2}(\omega_7 - \omega_6) , \qquad \alpha_6 = \omega_3 - \omega_4$$

form a simple system of type E_6. Hence $\mathfrak{L}$ is simple of type E_6.

An enumeration of the roots gives the following table of dimensionalities:

(91)

type	*dimensionality*
G_2	14
F_4	52
E_6	78 .

We state also (without proof) that E_7 and E_8 exist and have dimensionalities 133 and 248 respectively.

Conclusion. Every split simple Lie algebra over an arbitrary field of characteristic 0 is isomorphic to one of the Lie algebras A_l, $l \geqq 1$, B_l, $l \geqq 2$, C_l, $l \geqq 3$, D_l, $l \geqq 4$ constructed above or to an exceptional Lie algebra G_2, F_4, E_6, E_7, or E_8. We shall see later (Chapter IX) that the algebras listed here are not isomorphic. Hence the results give a complete classification of split simple Lie algebras over any field of characteristic 0. In particular, we have a complete classification of simple Lie algebras over any algebraically closed field of characteristic 0.

7. Compact forms

One of the most fruitful and profound ideas in the theory of Lie groups is the method introduced by Weyl for transferring problems on representations of Lie groups and algebras to the case of compact Lie groups. Weyl gave this method the striking title of the "unitary trick" and he used it to give the first proof of complete reducibility of the representations of semi-simple Lie algebras over algebraically closed fields of characteristic 0. He used it also for studying the irreducible representations, determining the characters and dimensionalities of these representations. We shall consider this later on. Weyl's method permits the application of analysis—via the theory of compact groups—to Lie algebras. The essence of the method has been formalized by Chevalley and Eilenberg in the following form.

A property P of Lie algebras is called a *linear property* if (1) P holds for $\mathfrak{L}$ implies that P holds for $\mathfrak{L}_\Omega$, Ω any extension field of the base field of L and (2) P holds for $\mathfrak{L}_\Omega$ for *some* extension field Ω implies that P holds for $\mathfrak{L}$. A trivial example is the statement that $\dim \mathfrak{L} = n$. Cartan's criterion implies that the property of semi-simplicity is a linear property for Lie algebras of characteristic 0. Also it can be proved that the property that the finite-dimensional representations of such algebras are completely reducible is a linear property.

We now introduce a certain class of semi-simple Lie algebras over the field R of real numbers and we shall see that the validity of a linear property P for all of these algebras implies the validity of P for all semi-simple Lie algebras of characteristic zero. The class we require is given in the following

DEFINITION 3. A Lie algebra $\mathfrak{L}$ over the field R of real numbers is called *compact* if its Killing form is negative definite. (This implies semi-simplicity.)

The importance of compact Lie algebras is that the groups associated with these algebras in the Lie theory are compact groups. The following result is an immediate consequence of the classification—due to Cartan—of simple Lie algebras over the field of real numbers. The proof we shall give, which does not make use of this structure theory, is due to Weyl.

THEOREM 10. *Let $\mathfrak{L}$ be a semi-simple Lie algebra over the field C of complex numbers. Then there exists a compact Lie algebra $\mathfrak{L}_u$ over the field of real numbers such that the "complexification" $(\mathfrak{L}_u)_C \cong \mathfrak{L}$.*

Proof: Since C is algebraically closed our structure theory applies. Let e_i, f_i, h_i be canonical generators for $\mathfrak{L}$ and let the canonical basis be chosen as in §4. We have seen that there exists an automorphism σ of $\mathfrak{L}$ such that $e_i^\sigma = f_i$, $f_i^\sigma = e_i$. Then $e_i^{\sigma^2} = e_i$, $f_i^{\sigma^2} = f_i$ and since the e_i, f_i generate $\mathfrak{L}$, $\sigma^2 = 1$. If we recall the form of the canonical basis determined by the e_i, f_i, h_i it is clear that $\mathfrak{L}_\alpha^\sigma = \mathfrak{L}_{-\alpha}$ for every non-zero root α. Hence if we choose any $e_\alpha \neq 0$ in $\mathfrak{L}_\alpha$, then $e_\alpha^\sigma \neq 0$ is in $\mathfrak{L}_{-\alpha}$ and $(e_\alpha, e_\alpha^\sigma) = \eta_\alpha \neq 0$. If we replace e_α by $e'_\alpha = \lambda_\alpha e_\alpha$ we obtain $(e'_\alpha, e'^\sigma_\alpha) = \lambda_\alpha^2 \eta_\alpha$ and, since we are working in an algebraically closed field, we may choose λ_α so that $\lambda_\alpha^2 \eta_\alpha = -1$. Hence we may suppose that $(e_\alpha, e_\alpha^\sigma) = -1$ for every α. If we choose $e_{-\alpha} = e_\alpha^\sigma$ for the positive roots α (relative to some ordering in $\mathfrak{H}_0^*$), then we shall have this relation for all non-zero roots, since $\sigma^2 = 1$. We now choose as basis for $\mathfrak{L}$: $(h_1, h_2, \cdots, h_l, e_\alpha, e_{-\alpha}, \cdots)$ where $e_{-\alpha} = e_\alpha^\sigma$ and these elements satisfy

$$(e_\alpha, e_{-\alpha}) = -1 \tag{92}$$

for every non-zero root α. If α and β are non-zero roots and $\beta \neq \pm\alpha$, then

$$[e_\alpha e_\beta] = N_{\alpha\beta} e_{\alpha+\beta} \neq 0 \tag{93}$$

if $\alpha + \beta$ is a root and $[e_\alpha e_\beta] = 0$ otherwise. If we apply σ to this relation we obtain

$$[e_{-\alpha} e_{-\beta}] = N_{\alpha\beta} e_{-(\alpha+\beta)} \quad \text{or} \quad [e_{-\alpha} e_{-\beta}] = 0\,. \tag{94}$$

We know also that $(e_\alpha, e_{-\alpha}) = -1$ implies that $[e_\alpha e_{-\alpha}] = h_\alpha$ and this element is a rational linear combination of the h_i. Now if $\alpha + \beta$ is also a root, then

$$(95) \qquad [N_{\alpha\beta}e_{\alpha+\beta}, N_{\alpha\beta}e_{-\alpha-\beta}] = N_{\alpha\beta}^2 h_{\alpha+\beta} = N_{\alpha\beta}^2(h_\alpha + h_\beta) .$$

On the other hand,

$$\begin{aligned}[N_{\alpha\beta}e_{\alpha+\beta}, N_{\alpha\beta}e_{-\alpha-\beta}] &= [[e_\alpha e_\beta][e_{-\alpha}e_{-\beta}]] \\ &= [[[e_\alpha e_\beta]e_{-\alpha}]e_{-\beta}] - [[[e_\alpha e_\beta]e_{-\beta}]e_{-\alpha}] \\ &= -[[[e_\beta e_\alpha]e_{-\alpha}]e_{-\beta}] - [[[e_\alpha e_\beta]e_{-\beta}]e_{-\alpha}] \\ &= \frac{q(r+1)}{2}(\alpha, \alpha)[e_\beta e_{-\beta}] + \frac{q'(r'+1)}{2}(\beta, \beta)[e_\alpha e_{-\alpha}] ,\end{aligned}$$

by (20). This can be continued to

$$= \frac{q(r+1)}{2}(\alpha, \alpha)h_\beta + \frac{q'(r'+1)}{2}(\beta, \beta)h_\alpha .$$

Since α and β are linearly independent, h_α and h_β are linearly independent. Hence the last relation can be compared with (95) to see that $N_{\alpha\beta}^2 = [q(r+1)/2](\alpha, \alpha)$ is a positive rational number. It follows that all the $N_{\alpha\beta}$, α, β roots, are real numbers. Since $[e_\alpha h_i] = \alpha(h_i)e_\alpha$ and $\alpha(h_i)$ is rational it is now clear that the multiplication table for the basis we have chosen has real coefficients. Hence the set of real linear combinations of this basis is an algebra $\mathfrak{L}_1$ over the real field R such that $\mathfrak{L}_{1C} = \mathfrak{L}$. Since $e_\alpha^\sigma = e_{-\alpha}$, $h_i^\sigma = -h_i$, σ induces an automorphism of period two in $\mathfrak{L}_1$. We now modify this automorphism by intertwining it with the standard automorphism $\rho \to \bar{\rho}$ of the complex field. Thus if we denote the chosen basis by (u_i), we let τ be the mapping: $\sum \rho_i u_i \to \sum \bar{\rho}_i u_i^\sigma$. This is a *semi-automorphism* of $\mathfrak{L}$ in the sense that τ is a semi-linear transformation and $[xy]^\tau = [x^\tau y^\tau]$. In fact, we have $(x + y)^\tau = x^\tau + y^\tau$, $(\rho x)^\tau = \bar{\rho}x^\tau$, $\rho \in C$. The fact that τ satisfies $[xy]^\tau = [x^\tau y^\tau]$ is clear from the reality of the multiplication table of the u_i. It is clear also that τ is an automorphism of $\mathfrak{L}$ considered as an algebra over R; hence the set $\mathfrak{L}_u$ of fixed points under τ is an R-subalgebra of $\mathfrak{L}$. We shall show that $\mathfrak{L}_u$ is the required compact form of $\mathfrak{L}$. It is clear that $\tau^2 = 1$. Hence $\mathfrak{L}_u = \{x + x^\tau \mid x \in \mathfrak{L}\}$. It follows that every element of $\mathfrak{L}_u$ is a real linear combination of the elements $u_i + u_i^\tau$, $\sqrt{-1}(u_i - u_i^\tau)$. If we use the form of the u_i and the relations $h_i^\tau = h_i^\sigma = -h_i$, $e_\alpha^\tau = e_\alpha^\sigma = e_{-\alpha}$ we see that every element of $\mathfrak{L}_u$ is a real linear combination of the elements

$$\text{(96)} \qquad \begin{array}{ll} \sqrt{-1}h_j\,, & j = 1, \cdots, l\,, \\ e_\alpha + e_{-\alpha}\,, & \sqrt{-1}(e_\alpha - e_{-\alpha})\,, \end{array}$$

where α ranges over the positive roots. These elements form a basis for $\mathfrak{L}$ over C and this implies that $(\mathfrak{L}_u)_C = \mathfrak{L}$. Also we see that any element of $\mathfrak{L}_u$ has the form

$$x = \sum_1^l \xi_j \sqrt{-1} h_j + \sum_{\alpha>0} (\rho_\alpha e_\alpha + \bar{\rho}_\alpha e_{-\alpha})$$

where the ξ_j are real, ρ_α complex, $\bar{\rho}_\alpha$ the conjugate. Since $\mathfrak{L} = (\mathfrak{L}_u)_C$, the Killing form for $\mathfrak{L}_u$ is the restriction to $\mathfrak{L}_u$ of the Killing form in $\mathfrak{L}$. Hence we may use the orthogonality properties and the relations $(e_\alpha, e_{-\alpha}) = -1$ to calculate

$$\text{(97)} \qquad (x, x) = -(h, h) - 2 \sum_\alpha \rho_\alpha \bar{\rho}_\alpha$$

where $h = \sum \xi_i h_i$. Since $(h, h) > 0$ unless $h = 0$, it is clear that $(x, x) < 0$ unless $x = 0$. Hence the Killing form is negative definite.

Now let P be a linear property of Lie algebras which is valid for every compact Lie algebra over the field R of real numbers. Let $\mathfrak{L}$ be a semi-simple Lie algebra over a field Φ of characteristic 0 and let Ω be the algebraic closure of Φ. Then $\mathfrak{L}_\Omega$ is split and so this has the form $\mathfrak{L}_{0\Omega}$ where $\mathfrak{L}_0$ is a Lie algebra over the field Q of rational numbers. Then $\mathfrak{L}_{0C}$ is a semi-simple Lie algebra over the complex field C; hence, by Theorem 10, there exists a compact Lie algebra $\mathfrak{L}_u$ over the reals such that $\mathfrak{L}_{uC} = \mathfrak{L}_{0C}$. Since P is satisfied by $\mathfrak{L}_u$, it holds for $\mathfrak{L}_{uC} = \mathfrak{L}_{0C}$, hence for $\mathfrak{L}_0$ and for $\mathfrak{L}_{0\Omega} = \mathfrak{L}_\Omega$. It follows that P holds for $\mathfrak{L}$.

Exercises

In these exercises all base fields are of characteristic 0 and all algebras and modules are finite-dimensional. Unexplained notations are as in the text.

1. Show that $\mathfrak{H}$ is a Cartan subalgebra of a semi-simple Lie algebra $\mathfrak{L}$ if and only if: (1) $\mathfrak{H}$ is maximal commutative and (2) $\mathfrak{H}$ is reductive in $\mathfrak{L}$, that is, $\mathrm{ad}_{\mathfrak{L}}\mathfrak{H}$ is completely reducible.

2. Let $\mathfrak{H}$ be a Cartan subalgebra of the semi-simple Lie algebra $\mathfrak{L}$. Show that if R is any finite-dimensional representation of $\mathfrak{L}$, then h^R is semi-simple for every $h \in \mathfrak{H}$.

3. Let $\mathfrak{L}_1$ be a semi-simple subalgebra of the semi-simple Lie algebra $\mathfrak{L}$ and let $\mathfrak{H}_1$ be a Cartan subalgebra of $\mathfrak{L}_1$. Show that $\mathfrak{H}_1$ can be imbedded in a Cartan subalgebra of $\mathfrak{L}$.

4. Use the method of § 3 to obtain a canonical basis and multiplication table for A_2 and G_2.

5. Use the method of § 3 to obtain the roots for the split Lie algebra of type E_7 (assuming it exists) and show in this way that the dimensionality of this Lie algebra is 133.

6. Without using the determination of the connected Dynkin diagrams show that 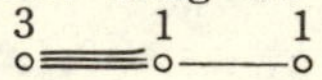

cannot be the Dynkin diagram of a Cartan matrix. (Hint: Show that the set of roots determined by this contains a positive root β such that 2β is a root.)

7. Show that the constants of multiplication of a canonical basis (§ 3) of a split semi-simple Lie algebra are rational numbers with denominators of the form $2^k 3^l$, k, l integers.

8. Prove that any split semi-simple Lie algebra can be generated by two elements.

9. Let $\mathfrak{L} = \mathfrak{G}'_L$, $\mathfrak{G}$, the algebra of linear transformations in an $(l+1)$-dimensional vector space and let (A, B) be the Killing form in $\mathfrak{L}$. Show that

$$(A, B) = 2(l+1) \operatorname{tr} AB .$$

(*Hint*: Use Exercise 3.9)

10. Express the Killing forms of the Lie algebras of types B_l, C_l and D_l in terms of the trace form $\operatorname{tr} AB$ where the representation is the one given in § 6.

11. Show that if $\mathfrak{L}$ is split, $\mathfrak{H}$ a splitting Cartan subalgebra, then $\mathfrak{H} + \sum_{\alpha>0}\mathfrak{L}_\alpha$ is a maximal solvable subalgebra and $\sum_{\alpha>0}\mathfrak{L}_\alpha$ is a maximal nilpotent subalgebra of $\mathfrak{L}$.

12. Determine the Weyl group for the split Lie algebra of type G_2 and the split Lie algebra of type A_l.

13. Let α_i be a simple root and let α be a positive root $\neq \alpha_i$. Show that every root of the form $\alpha + k\alpha_i$, k an integer, is positive. Show that $\sum'_k(\alpha + k\alpha_i, \alpha_i) = 0$ where the summation is taken over the k such that $\alpha + k\alpha_i$ is a root and α is any positive root. Hence prove that $\sum_{\alpha>0}(\alpha, \alpha_i) = (\alpha_i, \alpha_i)$, or $\sum_{\alpha>0}\alpha(h_i) = 2$.

CHAPTER V

Universal Enveloping Algebras

In this chapter we shall define the concept of the universal enveloping (associative) algebra $\mathfrak{U}$ of a Lie algebra $\mathfrak{L}$. The principal function of $\mathfrak{U}$ is to reduce the theory of representations of $\mathfrak{L}$ to that of representations of the associative algebra $\mathfrak{U}$. An important property of $\mathfrak{U}$ is that $\mathfrak{L}$ is isomorphic to a subalgebra of $\mathfrak{U}_L$. In this way one can obtain a faithful representation of every Lie algebra. For finite-dimensional Lie algebras we shall obtain in the next chapter a sharpening of this result, namely, that every such algebra has a faithful finite-dimensional representation. In this chapter we obtain the basic properties of $\mathfrak{U}$. Some of these will be used to prove an important formula due to Campbell and Hausdorff on the product of exponentials in an associative algebra. We shall give also the Cartan-Eilenberg definition of the cohomology groups of a Lie algebra.

The discussion here will not be confined to finite dimensional algebras nor to algebras of characteristic zero. In fact, a part of the chapter will be devoted to some notions which are peculiar to the characteristic $p \neq 0$ case. The main concept here is that of a restricted Lie algebra of characteristic p, which arises in considering subspaces of an associative algebra which are closed relative to the mapping $a \rightarrow a^p$ as well as the Lie product $[ab] = ab - ba$. Restricted Lie algebras have restricted representations, restricted derivations, etc., and we can define a "restricted" universal enveloping algebra called the u-algebra. We discuss these notions briefly and consider the theory of abelian restricted Lie algebras.

1. Definition and basic properties

The central notion of this chapter—the universal enveloping algebra of a Lie algebra—is a basic tool for the study of representations and more generally for the study of homomorphisms of a Lie algebra $\mathfrak{L}$ into a Lie algebra $\mathfrak{A}_L$ where $\mathfrak{A}$ is associative with an identity element. Any such homomorphism can be "extended" to a homomorphism of the (associative) universal enveloping algebra $\mathfrak{U}$ into $\mathfrak{A}$. Throughout this chapter we shall be concerned with Lie algebras and with associative algebras containing identities. We recall our conventions (of Chapter I) on terminology: "algebra" will mean associative algebra containing an identity element 1, "subalgebra" means subalgebra in the usual sense containing 1, and "homomorphism" for algebras means homomorphism in the usual sense mapping 1 into 1.

Definition 1. Let $\mathfrak{L}$ be a Lie algebra (arbitrary dimensionality and characteristic.) A pair $(\mathfrak{U}, i)$ where $\mathfrak{U}$ is an algebra and i is a homomorphism of $\mathfrak{L}$ into $\mathfrak{U}_L$ is called a *universal enveloping algebra* of $\mathfrak{L}$ if the following holds: If $\mathfrak{A}$ is any algebra and θ is a homomorphism of $\mathfrak{L}$ into $\mathfrak{A}_L$, then there exists a unique homomorphism θ' of $\mathfrak{U}$ into $\mathfrak{A}$ such that $\theta = i\theta'$. Diagramatically, we are given

$$\begin{array}{ccc} \mathfrak{U} = \mathfrak{U}_L & & \\ \uparrow i & & \\ \mathfrak{L} & \xrightarrow{\theta} & \mathfrak{A} = \mathfrak{A}_L \end{array}$$

where i and θ are homomorphisms of $\mathfrak{L}$ and we can complete this diagram to the commutative diagram

$$\begin{array}{ccc} \mathfrak{U} = \mathfrak{U}_L & & \\ \uparrow i & \searrow \theta' & \\ \mathfrak{L} & \xrightarrow{\theta} & \mathfrak{A} = \mathfrak{A}_L \end{array}$$

where θ' is a homomorphism of $\mathfrak{U}$.

A number of important properties of $(\mathfrak{U}, i)$ are built into this definition, or are easy consequences of basic facts on representations. We state these in the following

Theorem 1

1. *Let* $(\mathfrak{U}, i)$, $(\mathfrak{B}, j)$ *be universal enveloping algebras for* $\mathfrak{L}$. *Then there exists a unique isomorphism* j' *of* $\mathfrak{U}$ *onto* $\mathfrak{B}$ *such that* $j = ij'$.
2. $\mathfrak{U}$ *is generated by the image* $\mathfrak{L}i$.
3. *Let* $\mathfrak{L}_1, \mathfrak{L}_2$ *be Lie algebras with* $(\mathfrak{U}_1, i_1)$, $(\mathfrak{U}_2, i_2)$ *respective universal*

enveloping algebras and let α *be a homomorphism of* $\mathfrak{L}_1$ *into* $\mathfrak{L}_2$. *Then there exists a unique homomorphism* α' *of* $\mathfrak{U}_1$ *into* $\mathfrak{U}_2$ *such that* $\alpha i_2 = i_1\alpha'$, *that is, we have a commutative diagram*:

$$\begin{array}{ccc} \mathfrak{L}_1 & \xrightarrow{\alpha} & \mathfrak{L}_2 \\ {\scriptstyle i_1}\downarrow & & \downarrow{\scriptstyle i_2} \\ \mathfrak{U}_1 & \xrightarrow[\alpha']{} & \mathfrak{U}_2 \end{array}$$

4. *Let* $\mathfrak{B}$ *be an ideal in* $\mathfrak{L}$ *and let* $\mathfrak{K}$ *be the ideal in* $\mathfrak{U}$ *generated by* $\mathfrak{B}i$. *If* $l \in \mathfrak{L}$ *then* $j\colon l + \mathfrak{B} \to li + \mathfrak{K}$ *is a homomorphism of* $\mathfrak{L}/\mathfrak{B}$ *into* $\mathfrak{B}_L$ *where* $\mathfrak{B} = \mathfrak{U}/\mathfrak{K}$ *and* $(\mathfrak{B}, j)$ *is a universal enveloping algebra for* $\mathfrak{L}/\mathfrak{B}$.

5. $\mathfrak{U}$ *has a unique anti-automorphism* π *such that* $i\pi = -i$. *Moreover,* $\pi^2 = 1$.

6. *There is a unique homomorphism* δ' *called the diagonal mapping of* $\mathfrak{U}$ *into* $\mathfrak{U} \otimes \mathfrak{U}$ *such that* $ai\delta' = ai \otimes 1 + 1 \otimes ai$, $a \in \mathfrak{L}$.

7. *If* D *is a derivation in* $\mathfrak{L}$ *there exists a unique derivation* D' *in* $\mathfrak{U}$ *such that* $Di = iD'$.

Proof:

1. If we use the defining property of $(\mathfrak{U}, i)$ and the homomorphism $\theta = j$ of $\mathfrak{L}$ into $\mathfrak{B}_L$ we obtain a unique homomorphism j' of $\mathfrak{U}$ into $\mathfrak{B}$ such that $j = ij'$. Similarly, we have a homomorphism i' of $\mathfrak{B}$ into $\mathfrak{U}$ such that $i = ji'$. Hence $j = ji'j'$ and $i = ij'i'$. On the other hand, we have $j = j1_v$ where 1_v is the identity mapping in $\mathfrak{B}$. If we apply the uniqueness part of the defining property of $(\mathfrak{B}, j)$ to $\theta = j$ we see that $i'j' = 1_v$. Similarly, $j'i' = 1_u$ the identity in $\mathfrak{U}$. It follows that j' is an isomorphism of $\mathfrak{U}$ onto $\mathfrak{B}$.

2. Let $\mathfrak{B}$ be the subalgebra generated by $\mathfrak{L}i$. The mapping i can be considered as a mapping of $\mathfrak{L}$ into $\mathfrak{B}_L$. Hence there is a unique homomorphism i' of $\mathfrak{U}$ into $\mathfrak{B}$ such that $i = ii'$. Since $i = i1_u$ and i' can be considered as a mapping of $\mathfrak{U}$ into $\mathfrak{U}$, the uniqueness condition gives $i' = 1_u$. Hence $\mathfrak{U} = \mathfrak{U}1_u = \mathfrak{U}i' \subseteq \mathfrak{B}$. Hence $\mathfrak{B} = \mathfrak{U}$.

3. If α is a homomorphism of $\mathfrak{L}_1$ into $\mathfrak{L}_2$, αi_2 is a homomorphism of $\mathfrak{L}_1$ into $\mathfrak{U}_{2L}$. Hence there exists a unique homomorphism α' of $\mathfrak{U}_1$ into $\mathfrak{U}_2$ such that $i_1\alpha' = \alpha i_2$.

4. We note first that the mapping $l \to li + \mathfrak{K}$ of $\mathfrak{L}$ into $\mathfrak{B} = \mathfrak{U}/\mathfrak{K}$ is a homomorphism of $\mathfrak{L}$ into $\mathfrak{B}_L$. Since $\mathfrak{B}i \subseteq \mathfrak{K}$, $\mathfrak{B}$ is mapped into 0 by this homomorphism. Hence we have an induced homo-

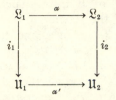

morphism $l + \mathfrak{B} \to li + \mathfrak{K}$ of $\mathfrak{L}/\mathfrak{B}$ into $\mathfrak{V}$. This is the mapping j. Now let θ be a homomorphism of $\mathfrak{L}/\mathfrak{B}$ into $\mathfrak{A}_L$, $\mathfrak{A}$ an algebra. Then η: $l \to (l + \mathfrak{B})\theta$ is a homomorphism of $\mathfrak{L}$ into $\mathfrak{A}_L$. Hence there exists a homomorphism η' of $\mathfrak{U}$ into $\mathfrak{A}$ such that $i\eta' = \eta$. If $b \in \mathfrak{B}$, $b\eta = 0$ so that bi is in the kernel of the homomorphism η'. It follows that $\mathfrak{K}$ is in the kernel of η' and consequently we have the induced homomorphism θ': $u + \mathfrak{K} \to u\eta'$ of $\mathfrak{V} = \mathfrak{U}/\mathfrak{K}$ into $\mathfrak{A}$. Now $(l + \mathfrak{B})\theta = l\eta = li\eta'$ and $(l + \mathfrak{B})j\theta' = (li + \mathfrak{K})\theta' = li\eta'$. Hence $\theta = j\theta'$ as required. It remains to prove that θ' is unique. This will follow by showing that $(\mathfrak{L}/\mathfrak{B})j$ generates $\mathfrak{V}$. Now, by 2., $\mathfrak{U}$ is generated by $\mathfrak{L}i$, which implies that $\mathfrak{V}$ is generated by the cosets $li + \mathfrak{K}$. Since $(l + \mathfrak{B})j = li + \mathfrak{K}$, it follows that $\mathfrak{V}$ is generated by the set of elements $(l + \mathfrak{B})j$, that is, by $(\mathfrak{L}/\mathfrak{B})j$. This completes the proof of 4.

5. If μ' is an anti-homomorphism of an algebra $\mathfrak{A}$ into an algebra $\mathfrak{B}$, then one verifies directly that $-\mu'$ is a homomorphism of $\mathfrak{A}_L$ into $\mathfrak{B}_L$. Let $\bar{\mathfrak{U}}$ be an algebra anti-isomorphic to $\mathfrak{U}$ under a mapping $u \to u\mu'$ of $\mathfrak{U}$. (Such algebras exist trivially.) The mapping $\theta = -i\mu'$ is a homomorphism of $\mathfrak{L}$ into $\bar{\mathfrak{U}}_L$. Hence there exists a homomorphism θ' of $\mathfrak{U}$ into $\bar{\mathfrak{U}}$ such that $\theta = -i\mu' = i\theta'$. Let $\pi = \theta'(\mu')^{-1}$. Then π is an anti-homomorphism of $\mathfrak{U}$ into itself such that $i\pi = -i$. Hence $i\pi^2 = i$ and since $\mathfrak{L}i$ generates $\mathfrak{U}$ and π^2 is a homomorphism, $\pi^2 = 1$. Hence π is an anti-automorphism in $\mathfrak{U}$. The uniqueness of π is also immediate from the fact that $\mathfrak{L}i$ generates $\mathfrak{U}$.

6. The reasoning used to define the product of representations and modules (§ 1.6) shows that $a \to ai \otimes 1 + 1 \otimes ai$, $a \in \mathfrak{L}$, is a homomorphism of $\mathfrak{L}$ into $(\mathfrak{U} \otimes \mathfrak{U})_L$. Hence there exists a unique homomorphism δ' of $\mathfrak{U}$ into $\mathfrak{U} \otimes \mathfrak{U}$ such that $ai\delta' = ai \otimes 1 + 1 \otimes ai$.

7. Let D be a derivation in $\mathfrak{L}$. We form the algebra $\mathfrak{U}_2$ of 2×2 matrices with entries in the universal enveloping algebra $\mathfrak{U}$ and we consider the mapping

$$\theta: \ a \to \begin{pmatrix} ai & aDi \\ 0 & ai \end{pmatrix} \tag{1}$$

of $\mathfrak{L}$ into $\mathfrak{U}_2$. This is a linear mapping and

$$\begin{aligned} &\begin{pmatrix} ai & aDi \\ 0 & ai \end{pmatrix}\begin{pmatrix} bi & bDi \\ 0 & bi \end{pmatrix} - \begin{pmatrix} bi & bDi \\ 0 & bi \end{pmatrix}\begin{pmatrix} ai & aDi \\ 0 & ai \end{pmatrix} \\ &\quad = \begin{pmatrix} [ai, bi] & [ai, bDi] + [aDi, bi] \\ 0 & [ai, bi] \end{pmatrix} = \begin{pmatrix} [ab]i & [a, b]Di \\ 0 & [ab]i \end{pmatrix}. \end{aligned} \tag{2}$$

Hence θ is a homomorphism of $\mathfrak{L}$ into $\mathfrak{U}_{2L}$. It follows that there is a homomorphism θ' of $\mathfrak{U}$ into $\mathfrak{U}_2$ such that $\theta = i\theta'$. Since

$$ai\theta' = \begin{pmatrix} ai & aDi \\ 0 & ai \end{pmatrix} \tag{3}$$

and the ai generate $\mathfrak{U}$, we have for any $x \in \mathfrak{U}$,

$$x\theta' = \begin{pmatrix} x & y \\ 0 & x \end{pmatrix}, \tag{4}$$

where y is uniquely determined by x. We write $y = xD'$ and a calculation like (2) shows that D' is a derivation in $\mathfrak{L}$. Then (3) shows that $aiD' = aDi$. Hence $iD' = Di$ as required. The uniqueness of D' follows from the fact that $\mathfrak{L}i$ generates $\mathfrak{U}$, and a derivation is determined by its effect on a set of generators.

We now give a construction of a universal enveloping algebra. Let $\mathfrak{T}$ denote the tensor algebra based on the vector space $\mathfrak{L}$. By definition,

$$\mathfrak{T} = \Phi 1 \oplus \mathfrak{L}_1 \oplus \mathfrak{L}_2 \oplus \cdots \oplus \mathfrak{L}_i \oplus \cdots \tag{5}$$

where $\mathfrak{L}_1 = \mathfrak{L}$ and $\mathfrak{L}_i = \mathfrak{L} \otimes \mathfrak{L} \cdots \otimes \mathfrak{L}$, i times. The vector space operations in $\mathfrak{T}$ are as usual and multiplication in $\mathfrak{T}$ is indicated by $\otimes$ and is characterized by

$$\begin{aligned} &(x_1 \otimes \cdots \otimes x_i) \otimes (y_1 \otimes \cdots \otimes y_j) \\ &\quad = x_1 \otimes \cdots \otimes x_i \otimes y_1 \otimes \cdots \otimes y_j . \end{aligned} \tag{6}$$

Let $\mathfrak{R}$ be the ideal in $\mathfrak{T}$ which is generated by all the elements of the form

$$[ab] - a \otimes b + b \otimes a , \qquad a, b \in \mathfrak{L} , \tag{7}$$

and let $\mathfrak{U} = \mathfrak{T}/\mathfrak{R}$. Let i denote the restriction to $\mathfrak{L} = \mathfrak{L}_1$ of the canonical homomorphism of $\mathfrak{T}$ onto $\mathfrak{U}$. We have

$$\begin{aligned} &[ab]i - ai \otimes bi + bi \otimes ai \\ &\quad = ([ab] - a \otimes b + b \otimes a) + \mathfrak{R} \\ &\quad = \mathfrak{R} = 0 \quad (\text{in } \mathfrak{U}) . \end{aligned}$$

Hence i is a homomorphism of $\mathfrak{L}$ into $\mathfrak{U}_L$. We shall now prove

THEOREM 2. *$(\mathfrak{U}, i)$ is a universal enveloping algebra for $\mathfrak{L}$.*

Proof: We recall first the basic property of the tensor algebra, that any linear mapping θ of $\mathfrak{L}$ into an algebra $\mathfrak{A}$ can be extended to a homomorphism of $\mathfrak{T}$ into $\mathfrak{A}$. Thus let $\{u_j \mid j \in J\}$ be a basis for $\mathfrak{L}$, then it is well-known and easy to prove that the distinct "monomials" $u_{j_1} \otimes u_{j_2} \otimes \cdots \otimes u_{j_n}$ of degree n form a basis for $\mathfrak{L}_n$. Here $u_{j_1} \otimes u_{j_2} \otimes \cdots \otimes u_{j_n} = u_{k_1} \otimes u_{k_2} \otimes \cdots \otimes u_{k_n}$ if and only if $j_r = k_r$, $r = 1, 2, \cdots, n$. The element 1 and the different monomials of degrees $1, 2, \cdots$ form a basis for $\mathfrak{T}$. It is easy to check that the linear mapping θ'' of $\mathfrak{T}$ into $\mathfrak{A}$ such that $1\theta'' = 1$, $(u_{j_1} \otimes u_{j_2} \otimes \cdots \otimes u_{j_n})\theta'' = (u_{j_1}\theta)(u_{j_2}\theta) \cdots (u_{j_n}\theta)$ is a homomorphism of $\mathfrak{T}$ into $\mathfrak{A}$ such that $a\theta'' = a\theta$ for $a \in \mathfrak{L}$ $(= \mathfrak{L}_1 \subseteq \mathfrak{T})$. Now let θ be a homomorphism of $\mathfrak{L}$ into $\mathfrak{A}_L$ and let θ'' be its extension to a homomorphism of $\mathfrak{T}$ into $\mathfrak{A}$. If $a, b \in \mathfrak{L}$,

$$
\begin{aligned}
&[ab]\theta'' - (a\theta'')(b\theta'') + (b\theta'')(a\theta'') \\
&= [ab]\theta - (a\theta)(b\theta) + (b\theta)(a\theta) \\
&= [a\theta, b\theta] - (a\theta)(b\theta) + (b\theta)(a\theta) \\
&= 0 .
\end{aligned}
$$

Hence the generators (7) of $\mathfrak{R}$ belong to the kernel of θ''. We therefore have an induced homomorphism θ' of $\mathfrak{U}$ into $\mathfrak{A}$ such that $ai\theta' = (a + \mathfrak{R})\theta' = a\theta'' = a\theta$. Thus $\theta = i\theta'$ as required. The tensor algebra $\mathfrak{T}$ is generated by $\mathfrak{L}$ and this implies that $\mathfrak{U}$ is generated by $\mathfrak{L}i$. Since two homomorphisms which coincide on generators are necessarily identical there is only one homomorphism θ' such that $i\theta' = \theta$.

2. *The Poincaré-Birkhoff-Witt theorem*

We have noted that if $\{u_j \mid j \in J\}$, where J is a set, is a basis for $\mathfrak{L}$, then the monomials $u_{j_1} \otimes u_{j_2} \otimes \cdots \otimes u_{j_n}$ of degree n form a basis for $\mathfrak{L}_n$, $n \geqq 1$. We suppose now that the set J of indices is ordered and we proceed to use this ordering to introduce a partial order in the set of monomials of any given degree $n \geqq 1$. We define the *index* of a monomial $u_{j_1} \otimes u_{j_2} \otimes \cdots \otimes u_{j_n}$ as follows. For i, k, $i < k$, set

$$
\eta_{ik} = \begin{cases} 0 & \text{if} \quad j_i \leqq j_k \\ 1 & \text{if} \quad j_i > j_k \end{cases}
$$

and define the index

$$
\text{ind}\,(u_{j_1} \otimes u_{j_2} \otimes \cdots \otimes u_{j_n}) = \sum_{i<k} \eta_{ik} . \tag{8}
$$

Note that ind $= 0$ if and only if $j_1 \leqq j_2 \leqq \cdots \leqq j_n$. Monomials having this property will be called *standard*. We now suppose $j_k > j_{k+1}$ and we wish to compare

$$\text{ind}\,(u_{j_1} \otimes u_{j_2} \otimes \cdots \otimes u_{j_n}) \qquad \text{and}$$
$$\text{ind}\,(u_{j_1} \otimes \cdots \otimes u_{j_{k+1}} \otimes u_{j_k} \otimes \cdots \otimes u_{j_n})\,,$$

where the second monomial is obtained by interchanging u_{j_k}, $u_{j_{k+1}}$. Let η'_{ik} denote the η's for the second monomial. Then we have $\eta'_{ij} = \eta_{ij}$ if $i, j \neq k,\ k+1$; $\eta'_{ik} = \eta_{i,k+1}$, $\eta'_{i,k+1} = \eta_{i,k}$ if $i < k$; $\eta'_{k,j} = \eta_{k+1,j}$, $\eta'_{k+1,j} = \eta_{k,j}$ if $j > k+1$ and $\eta'_{k,k+1} = 0$, $\eta_{k,k+1} = 1$, Hence

$$\text{ind}\,(u_{j_1} \otimes \cdots \otimes u_{j_n}) = 1$$
$$+ \text{ind}\,(u_{j_1} \otimes \cdots \otimes u_{j_{k+1}} \otimes u_{j_k} \otimes \cdots \otimes u_{j_n})\,.$$

We apply these remarks to the study of the algebra $\mathfrak{U} = \mathfrak{T}/\mathfrak{K}$ for which we prove first the following

LEMMA 1. *Every element of* $\mathfrak{T}$ *is congruent mod* $\mathfrak{K}$ *to a* Φ*-linear combination of* 1 *and standard monomials.*

Proof: It suffices to prove the statement for monomials. We order these by degree and for a given degree by the index. To prove the assertion for a monomial $u_{j_1} \otimes u_{j_2} \otimes \cdots \otimes u_{j_n}$ it suffices to assume it for monomials of lower degree and for those of the same degree n which are of lower index than the given monomial. Assume the monomial is not standard and suppose $j_k > j_{k+1}$. We have

$$u_{j_1} \otimes \cdots \otimes u_{j_n} = u_{j_1} \otimes \cdots \otimes u_{j_{k+1}} \otimes u_{j_k} \otimes \cdots \otimes u_{j_n}$$
$$+ u_{j_1} \otimes \cdots \otimes (u_{j_k} \otimes u_{j_{k+1}} - u_{j_{k+1}} \otimes u_{j_k}) \otimes \cdots \otimes u_{j_n}\,.$$

Since $u_{j_k} \otimes u_{j_{k+1}} - u_{j_{k+1}} \otimes u_{j_k} \equiv [u_{j_k} u_{j_{k+1}}] \pmod{\mathfrak{K}}$,

$$u_{j_1} \otimes \cdots \otimes u_{j_n} \equiv u_{j_1} \otimes \cdots \otimes u_{j_{k+1}} \otimes u_{j_k} \otimes \cdots \otimes u_{j_n}$$
$$+ u_{j_1} \otimes \cdots \otimes [u_{j_k} u_{j_{k+1}}] \otimes \cdots \otimes u_{j_n} \pmod{\mathfrak{K}}\,.$$

The first term on the right-hand side is of lower index than the given monomial while the second is a linear combination of monomials of lower degree. The result follows from the induction hypothesis.

We wish to show that the cosets of 1 and the standard monomials are linearly independent and so form a basis for $\mathfrak{U}$. For this purpose we introduce the vector space $\mathfrak{P}_n$ with the basis $u_{i_1} u_{i_2} \cdots u_{i_n}$, $i_1 \leqq i_2 \leqq \cdots \leqq i_n$, $i_j \in J$, and the vector space $\mathfrak{P} = \Phi 1 \oplus \mathfrak{P}_1 \oplus \mathfrak{P}_2 \oplus \cdots$. The required independence will follow easily from the following

LEMMA 2. *There exists a linear mapping σ of $\mathfrak{T}$ into $\mathfrak{P}$ such that*

$$1\sigma = 1\,, \qquad (u_{i_1} \otimes u_{i_2} \otimes \cdots \otimes u_{i_n})\sigma = u_{i_1}u_{i_2} \cdots u_{i_n}\,, \quad \text{if} \quad i_1 \leqq i_2 \leqq \cdots \leqq i_n \tag{9}$$

$$(u_{j_1} \otimes \cdots \otimes u_{j_n} - u_{j_1} \otimes \cdots \otimes u_{j_{k+1}} \otimes u_{j_k} \otimes \cdots \otimes u_{j_n})\sigma = (u_{j_1} \otimes \cdots \otimes [u_{j_k}u_{j_{k+1}}] \otimes \cdots \otimes u_{j_n})\sigma\,. \tag{10}$$

Proof: Set $1\sigma = 1$ and let $\mathfrak{L}_{n,j}$ be the subspace of $\mathfrak{L}_n$ spanned by the monomials of degree n and index $\leqq j$. Suppose a linear mapping σ has already been defined for $\Phi 1 \oplus \mathfrak{L}_1 \oplus \cdots \oplus \mathfrak{L}_{n-1}$ satisfying (9) and (10) for the monomials in this space. We extend σ linearly to $\Phi 1 \oplus \mathfrak{L}_1 \oplus \cdots \oplus \mathfrak{L}_{n-1} \oplus \mathfrak{L}_{n,0}$ by requiring that $(u_{i_1} \otimes u_{i_2} \otimes \cdots \otimes u_{i_n})\sigma = u_{i_1}u_{i_2} \cdots u_{i_n}$ for the standard monomials of degree n. Next assume σ has already been defined for $\Phi 1 \oplus \mathfrak{L}_1 \oplus \cdots \oplus \mathfrak{L}_{n-1} \oplus \mathfrak{L}_{n,i-1}$, satisfying (9) and (10) for the monomials belonging to this space and let $u_{j_1} \otimes \cdots \otimes u_{j_n}$ be of index $i \geqq 1$. Suppose $j_k > j_{k+1}$. Then we set

$$(u_{j_1} \otimes \cdots \otimes u_{j_n})\sigma = (u_{j_1} \otimes \cdots \otimes u_{j_{k+1}} \otimes u_{j_k} \otimes \cdots \otimes u_{j_n})\sigma + (u_{j_1} \otimes \cdots \otimes [u_{j_k}u_{j_{k+1}}] \otimes \cdots \otimes u_{j_n})\sigma\,. \tag{11}$$

This makes sense since the two terms on the right are in $\Phi 1 \oplus \mathfrak{L}_1 \oplus \cdots \oplus \mathfrak{L}_{n-1} \oplus \mathfrak{L}_{n,i-1}$. We show first that (11) is independent of the choice of the pair $(j_k j_{k+1})$, $j_k > j_{k+1}$. Let $(j_l j_{l+1})$ be a second pair with $j_l > j_{l+1}$. There are essentially two cases: I. $l > k+1$, II. $l = k+1$.

I. Set $u_{j_k} = u$, $u_{j_{k+1}} = v$, $u_{j_l} = w$, $u_{j_{l+1}} = x$. Then the induction hypothesis permits us to write for the right hand side of (11)

$$\begin{aligned}
&(\cdots v \otimes u \otimes \cdots \otimes x \otimes w \otimes \cdots)\sigma \\
&\quad + (\cdots \otimes v \otimes u \otimes \cdots \otimes [wx] \otimes \cdots)\sigma \\
&\quad + (\cdots [uv] \otimes \cdots \otimes x \otimes w \otimes \cdots)\sigma \\
&\quad + (\cdots \otimes [uv] \otimes \cdots \otimes [wx] \otimes \cdots)\sigma\,.
\end{aligned}$$

If we start with $(j_l j_{l+1})$ we obtain

$$\begin{aligned}
&(\cdots \otimes u \otimes v \otimes \cdots \otimes x \otimes w \otimes \cdots)\sigma \\
&\quad + (\cdots \otimes u \otimes v \otimes \cdots \otimes [wx] \otimes \cdots)\sigma \\
&= (\cdots \otimes v \otimes u \otimes \cdots \otimes x \otimes w \otimes \cdots)\sigma \\
&\quad + (\cdots \otimes [uv] \otimes \cdots \otimes x \otimes w \otimes \cdots)\sigma \\
&\quad + (\cdots \otimes v \otimes u \otimes \cdots \otimes [wx] \otimes \cdots)\sigma \\
&\quad + (\cdots \otimes [uv] \otimes \cdots \otimes [wx] \otimes \cdots)\sigma\,.
\end{aligned}$$

This is the same as the value obtained before.

II. Set $u_{j_k} = u$, $u_{j_{k+1}} = v = u_{j_l}$, $u_{j_{l+1}} = w$. If we use the induction hypothesis we can change the right hand side of (11) to

$$(12)\qquad \begin{aligned} &(\cdots w \otimes v \otimes u \cdots)\sigma + (\cdots [vw] \otimes u \cdots)\sigma \\ &\quad + (\cdots v \otimes [uw] \cdots)\sigma + (\cdots [uv] \otimes w \cdots)\sigma \,. \end{aligned}$$

Similarly, if we start with

$$(\cdots u \otimes w \otimes v \cdots)\sigma + (\cdots u \otimes [vw] \cdots)\sigma \,,$$

we can wind up with

$$(13)\qquad \begin{aligned} &(\cdots w \otimes v \otimes u \cdots)\sigma + (\cdots w \otimes [uv] \cdots)\sigma \\ &\quad + (\cdots [uw] \otimes v \cdots)\sigma + (\cdots u \otimes [vw] \cdots)\sigma \,. \end{aligned}$$

Hence we have to show that σ annihilates the following element of $\Phi 1 \oplus \mathfrak{L}_1 \oplus \cdots \oplus \mathfrak{L}_{n-1}$:

$$(14)\qquad \begin{aligned} &(\cdots [vw] \otimes u \cdots) - (\cdots u \otimes [vw] \cdots) \\ &\quad + (\cdots v \otimes [uw] \cdots) - (\cdots [uw] \otimes v \cdots) \\ &\quad + (\cdots [uv] \otimes w \cdots) - (\cdots w \otimes [uv] \cdots) \,. \end{aligned}$$

Now, it follows easily from the properties of σ in $\Phi 1 \oplus \cdots \oplus \mathfrak{L}_{n-1}$ that if $(\cdots a \otimes b \cdots) \in \mathfrak{L}_{n-1}$, where $a, b \in \mathfrak{L}_1$, then

$$(15)\qquad (\cdots a \otimes b \cdots)\sigma - (\cdots b \otimes a \cdots)\sigma - (\cdots [ab] \cdots)\sigma = 0 \,.$$

Hence σ applied to (14) gives

$$(16)\qquad (\cdots [[vw]u] \cdots)\sigma + (\cdots [v[uw]] \cdots)\sigma + (\cdots [[uv]w] \cdots)\sigma \,.$$

Since $[[vw]u] + [v[uw]] + [[uv]w] = [[vw]u] + [[wu]v] + [[uv]w] = 0$, (16) has the value 0. Hence in this case, too, the right hand side of (11) is uniquely determined. We now apply (11) to define σ for the monomials of degree n and index i. The linear extension of this mapping to the space $\mathfrak{L}_{n,i}$ gives a mapping on $\Phi 1 \oplus \cdots \oplus \mathfrak{L}_{n-1} \oplus \mathfrak{L}_{n,i}$ satisfying our conditions. This completes the proof of the lemma.

We can now prove the following

THEOREM 3 (*Poincaré-Birkhoff-Witt*). *The cosets of* 1 *and the standard monomials form a basis for* $\mathfrak{U} = \mathfrak{T}/\mathfrak{R}$.

Proof: Lemma 1 shows that every coset is a linear combination of $1 + \mathfrak{R}$ and the cosets of the standard monomials. Lemma 2 gives a linear mapping σ of $\mathfrak{T}$ into $\mathfrak{P}$ satisfying (9) and (10). It is easy to see that every element of the ideal $\mathfrak{R}$ is a linear combination of elements of the form

$$u_{j_1} \otimes \cdots \otimes u_{j_n} - u_{j_1} \otimes \cdots \otimes u_{j_{k+1}} \otimes u_{j_k} \otimes \cdots \otimes u_{j_n}$$
$$- u_{j_1} \otimes \cdots \otimes [u_{j_k} u_{j_{k+1}}] \otimes \cdots \otimes u_{j_n} .$$

Since σ maps these elements into 0, $\mathfrak{K}\sigma = 0$ and so σ induces a linear mapping of $\mathfrak{U} = \mathfrak{T}/\mathfrak{K}$ into $\mathfrak{P}$. Since (9) holds, the induced mapping sends the cosets of 1 and the standard monomial $u_{i_1} \otimes \cdots \otimes u_{i_n}$ into 1 and $u_{i_1}u_{i_2} \cdots u_{i_n}$ respectively. Since these images are linearly independent in $\mathfrak{P}$, we have the linear independence in $\mathfrak{U}$ of the cosets of 1 and the standard monomials. This completes the proof.

COROLLARY 1. *The mapping i of $\mathfrak{L}$ into $\mathfrak{U}$ is* $1:1$ *and* $\Phi 1 \cap \mathfrak{L}i = 0$.

Proof: If (u_j) is a basis for $\mathfrak{L}$ over Φ, then $1 = 1 + \mathfrak{K}$ and the cosets $u_j i = u_j + \mathfrak{K}$ are linearly independent. This implies both statements.

We shall now simplify our notations in the following way: We write the product in $\mathfrak{U}$ in the usual way for associative algebras: xy. We write 1 for the identity in $\mathfrak{U}$ and we identify $\mathfrak{L}$ with its image $\mathfrak{L}i$ in $\mathfrak{U}$. This is a subalgebra of $\mathfrak{U}_L$ since the identity mapping is an isomorphism of $\mathfrak{L}$ into $\mathfrak{U}_L$. Also $\mathfrak{L}$ generates $\mathfrak{U}$ and the Poincaré-Birkhoff-Witt theorem states that if $\{u_j \mid j \in J\}$, J ordered, is a basis for $\mathfrak{L}$, then the elements

(17) $$1, u_{i_1}u_{i_2} \cdots u_{i_r}, \qquad i_1 \leqq i_2 \leqq \cdots$$

form a basis for $\mathfrak{U}$. In particular, if $\mathfrak{L}$ has the finite basis $u_1, u_2, \cdots, u_n$ then the elements

(18) $$u_1^{k_1}u_2^{k_2} \cdots u_n^{k_n}, \qquad k_i = 0, 1, 2, \cdots$$

$(u_i^0 = 1)$ form a basis for $\mathfrak{U}$. The defining property of $\mathfrak{U}$ can be re-stated in the following way: If θ is a homomorphism of $\mathfrak{L}$ into $\mathfrak{A}_L$, $\mathfrak{A}$ an algebra, then θ can be extended to a unique homomorphism θ (formerly θ') of $\mathfrak{U}$ into $\mathfrak{A}$. In particular, a representation R of $\mathfrak{L}$ can be extended to a unique representation R of $\mathfrak{U}$. This implies that any module $\mathfrak{M}$ for $\mathfrak{L}$ can be considered in one and only one way as a right $\mathfrak{U}$-module in which xl, $x \in \mathfrak{M}$, $l \in \mathfrak{L}$, is as defined for $\mathfrak{M}$ as $\mathfrak{L}$-module. Conversely, the restriction to $\mathfrak{L}$ of a representation of $\mathfrak{U}$ is a representation of $\mathfrak{L}$ and any right $\mathfrak{U}$-module defines a right $\mathfrak{L}$-module on restricting the multiplication to $\mathfrak{L}$. In the sequel we shall pass freely from $\mathfrak{L}$-modules to $\mathfrak{U}$-modules and conversely, without comment.

The Poincaré-Birkhoff-Witt theorem (hereafter referred to as the P-B-W theorem) gives a characterization of the universal enveloping algebra in the following sense: Let $\mathfrak{L}$ be a subalgebra of $\mathfrak{A}_L$, $\mathfrak{A}$ an algebra having the property that if $\{u_j \,|\, j \in J\}$ is a certain ordered basis for $\mathfrak{L}$, then the elements 1 and the standard monomials $u_{i_1} u_{i_2} \cdots u_{i_r}$, $i_1 \leqq i_2 \leqq \cdots \leqq i_r$, form a basis for $\mathfrak{A}$. Then $\mathfrak{A}$ (and the identity mapping) is a universal enveloping algebra for $\mathfrak{L}$. Thus we have a homomorphism of the universal enveloping algebra $\mathfrak{U}$ into $\mathfrak{A}$ which is the identity on $\mathfrak{L}$. The condition shows that this is $1:1$ and surjective. Hence $\mathfrak{A}$ can be taken as a universal enveloping algebra. Now suppose that $\mathfrak{B}$ is a subalgebra of $\mathfrak{L}$ and let $\mathfrak{U}$ be the universal enveloping algebra of $\mathfrak{L}$. We may choose an ordered basis $\{u_j \,|\, j \in J\}$ for $\mathfrak{L}$ so that $J = K \cup L$, $K \cap L = \varnothing$, $k < l$ if $k \in K$, $l \in L$ and $\{u_k \,|\, k \in K\}$ is an ordered basis for $\mathfrak{B}$. Let $\mathfrak{V}$ be the subalgebra of $\mathfrak{U}$ generated by $\mathfrak{B}$ (or by the u_k.) Then it is clear that 1 and the standard monomials $u_{k_1} u_{k_2} \cdots u_{k_s}$, $k_1 \leqq k_2 \leqq \cdots \leqq k_s$, form a basis for $\mathfrak{V}$. Hence $\mathfrak{V}$ can be taken to be the universal enveloping algebra of $\mathfrak{B}$.

Next assume $\mathfrak{B}$ is an ideal in $\mathfrak{L}$ and let the notations be as before. Let $\mathfrak{K}$ be the ideal in $\mathfrak{U}$ generated by $\mathfrak{B}$. By Theorem 1, part 4., we know that $\mathfrak{B} = \mathfrak{U}/\mathfrak{K}$ and the mapping $a + \mathfrak{B} \to a + \mathfrak{K}$ define a universal enveloping algebra for $\mathfrak{L}/\mathfrak{B}$. By Corollary 1, $a + \mathfrak{B} \to a + \mathfrak{K}$ is $1:1$ so we may identify $\mathfrak{L}/\mathfrak{B}$ with the subalgebra $(\mathfrak{L} + \mathfrak{K})/\mathfrak{K}$ of $\mathfrak{B}_L$. This subalgebra is the set of cosets $a + \mathfrak{K}$ and it has the basis $\{u_l + \mathfrak{K} \,|\, l \in L\}$. Hence by the P-B-W theorem, the cosets $1 + \mathfrak{K}$ and $u_{l_1} u_{l_2} \cdots u_{l_t} + \mathfrak{K}$, $l_1 \leqq l_2 \leqq \cdots \leqq l_t$, form a basis for $\mathfrak{B}$. Hence if $\mathfrak{D}$ is the subspace spanned by the elements 1, and the standard monomials $u_{l_1} u_{l_2} \cdots u_{l_t}$, then $\mathfrak{D} \cap \mathfrak{K} = 0$. We note next that any standard monomial of the form

$$(19) \qquad u_{k_1} u_{k_2} \cdots u_{k_s} u_{l_1} \cdots u_{l_t} ,$$

where $s \geqq 1$ and $t \geqq 0$, is in $\mathfrak{K}$ and these monomials together with 1 and $u_{l_1} \cdots u_{l_t}$, $l_1 \leqq l_2 \leqq \cdots \leqq l_t$, constitute all the elements of the standard basis for $\mathfrak{U}$. It follows that the elements (19) form a basis for $\mathfrak{K}$.

The main results on subalgebras and ideals we have noted can be stated as follows:

COROLLARY 2.   *Let be $\mathfrak{L}$ a Lie algebra, $\mathfrak{U}$ its universal enveloping al-*

gebra. If $\mathfrak{B}$ is a subalgebra of $\mathfrak{L}$, then the subalgebra of $\mathfrak{U}$ generated by $\mathfrak{B}$ can be taken to be the universal enveloping algebra of $\mathfrak{B}$. If $\mathfrak{B}$ is an ideal in $\mathfrak{L}$, then $\mathfrak{L}/\mathfrak{B}$ can be identified with $(\mathfrak{L}+\mathfrak{K})/\mathfrak{K}$ where $\mathfrak{K}$ is the ideal in $\mathfrak{U}$ generated by $\mathfrak{B}$ and $\mathfrak{B} = \mathfrak{U}/\mathfrak{K}$ is the universal enveloping algebra of $\mathfrak{L}/\mathfrak{B}$. Moreover, the set of standard monomials (19) *forms a basis for $\mathfrak{K}$, if (u_j) is an ordered basis for $\mathfrak{L}$ such that (u_k) is a basis for $\mathfrak{B}$.*

If we take $\mathfrak{B} = \mathfrak{L}$ in this corollary we see that the ideal $\mathfrak{U}_0$ generated by $\mathfrak{L}$ in $\mathfrak{U}$ has the basis consisting of the standard monomials $u_{i_1}u_{i_2}\cdots u_{i_r}$, $i_1 \leqq i_2 \leqq \cdots \leqq i_r$. Since these elements together with 1 form a basis for $\mathfrak{U}$ we have the following:

COROLLARY 3. *Let $\mathfrak{U}_0$ be the ideal in $\mathfrak{U}$ generated by $\mathfrak{L}$. Then $\mathfrak{U} = \Phi 1 \oplus \mathfrak{U}_0$.*

Since the imbedding of $\mathfrak{L}$ in $\mathfrak{U}$ is 1 : 1 and since any algebra has a faithful representation, we have:

COROLLARY 4. *Any Lie algebra has a faithful representation by linear transformations.*

The main properties of $\mathfrak{U}$ given in Theorem 1 which remain to be re-stated are:

2′. $\mathfrak{L}$ generates $\mathfrak{U}$.

3′. Any homomorphism of a Lie algebra $\mathfrak{L}_1$ into a Lie algebra $\mathfrak{L}_2$ can be extended to a unique homomorphism of the universal enveloping algebra $\mathfrak{U}_1$ of $\mathfrak{L}_1$ into the universal enveloping algebra $\mathfrak{U}_2$ of $\mathfrak{L}_2$.

5′. There exists a unique anti-automorphism π of $\mathfrak{U}$ satisfying $a\pi = -a$ for $a \in \mathfrak{L}$.

6′. There exists a homomophism δ (the diagonal mapping) of $\mathfrak{U}$ into $\mathfrak{U} \otimes \mathfrak{U}$ such that $a\delta = a \otimes 1 + 1 \otimes a$, $a \in \mathfrak{L}$.

7′. Any derivation D of $\mathfrak{L}$ has a unique extension to a derivation D of $\mathfrak{U}$.

We prove next:

COROLLARY 5. *The diagonal mapping δ of $\mathfrak{U}$ into $\mathfrak{U} \otimes \mathfrak{U}$ is* 1 : 1.

Proof: The decomposition $\mathfrak{U} = \Phi 1 \oplus \mathfrak{U}_0$ gives a decomposition $\mathfrak{U} \otimes \mathfrak{U} = \Phi(1 \otimes 1) \oplus (1 \otimes \mathfrak{U}_0) \oplus (\mathfrak{U}_0 \otimes 1) \oplus (\mathfrak{U}_0 \otimes \mathfrak{U}_0)$ where $1 \otimes \mathfrak{U}_0$ is the subspace of elements $1 \otimes b$, $b \in \mathfrak{U}_0$, $\mathfrak{U}_0 \otimes 1$ is the subspace of elements $b \otimes 1$, $b \in \mathfrak{U}_0$ and $\mathfrak{U}_0 \otimes \mathfrak{U}_0$ is the subspace of elements $\sum b \otimes b'$, $b, b' \in \mathfrak{U}_0$. We have $1\delta = 1 \otimes 1$, $a\delta = a \otimes 1 + 1 \otimes a$, $a \in \mathfrak{L}$. Since $\mathfrak{U}_0 \otimes \mathfrak{U}_0$ is an ideal in $\mathfrak{U} \otimes \mathfrak{U}$, it is easy to prove by induction on r that

$$(u_{i_1}u_{i_2}\cdots u_{i_r})\delta \equiv u_{i_1}\cdots u_{i_r}\otimes 1 + 1\otimes u_{i_1}\cdots u_{i_r} \qquad (\text{mod } \mathfrak{U}_0\otimes\mathfrak{U}_0)$$

where (u_i) is a basis for $\mathfrak{L}$. Is follows that the set of images under δ of the standard basis 1, $u_{i_1}\cdots u_{i_r}$, $i_1 \leqq i_2 \leqq \cdots \leqq i_r$, is linearly independent. Hence δ is 1 : 1.

Examples. (1) Let $\mathfrak{L}$ have the basis u, v with $[uv] = u$. Then $\mathfrak{U}$ has the basis $v^j u^k$ and we have the commutation relation

$$uv - vu = u\,. \tag{20}$$

Hence

$$uv = (v+1)u\,. \tag{21}$$

The elements of $\mathfrak{U}$ are the polynomials

$$v_0 + v_1u + v_2u^2 + \cdots + v_mu^m\,,$$

where the $v_i \in \Phi[v]$. Multiplication for such polynomials is defined in the usual way except that we have

$$u^kf(v) = f(v+k)u^k\,, \qquad k = 0, 1, 2, \cdots\,, \tag{22}$$

which is a consequence of (21). This is a type of ring of difference polynomials.

(2) Let $\mathfrak{L}$ be abelian with basis (u_j). Then $\mathfrak{U}$ is commutative and has the basis 1 and the monomials $u_{i_1}u_{i_2}\cdots u_{i_n}$, $i_1 \leqq i_2 \leqq \cdots \leqq i_n$. This means that the u_j are algebraically independent and $\mathfrak{U}$ is the polynomial algebra in these u_j.

3. *Filtration and graded algebra*

An algebra $\mathfrak{A}$ is said to be *graded* if $\mathfrak{A} = \sum_{i=0}^{\infty}\oplus\mathfrak{A}_i$ where $\mathfrak{A}_i$ is a subspace and $\mathfrak{A}_i\mathfrak{A}_j \subseteq \mathfrak{A}_{i+j}$. An example is the tensor algebra $\mathfrak{T} = \Phi 1 \oplus \mathfrak{L}_1 \oplus \mathfrak{L}_2 \oplus \cdots$ where we take $\mathfrak{T}_0 = \Phi 1$, $\mathfrak{T}_i = \mathfrak{L}_i$ if $i > 0$. Another example is the algebra of polynomials in algebraically independent elements in which $\mathfrak{A}_i$ is the space of homogeneous elements of degree i. If $\mathfrak{A}$ is any graded algebra, then the elements of $\mathfrak{A}_i$ are called *homogeneous of degree i* and every $a \in \mathfrak{A}$ has a unique representation $a = \sum_0^{\infty} a_i$, $a_i \in \mathfrak{A}_i$, and $a_i = 0$ for all but a finite number of i. The a_i are called the *homogeneous parts* of a. A left (right) ideal $\mathfrak{B}$ of $\mathfrak{A}$ is called *homogeneous* if $\mathfrak{B} = \sum_{i=0}^{\infty}(\mathfrak{B}\cap\mathfrak{A}_i)$. This is equivalent to saying that $b \in \mathfrak{B}$ if and only if its homogeneous parts $b_i \in \mathfrak{B}$.

An algebra $\mathfrak{A}$ is said to be *filtered* if for each non-negative integer i there is defined a subspace $\mathfrak{A}^{(i)}$ such that (1) $\mathfrak{A}^{(i)} \subseteqq \mathfrak{A}^{(j)}$ if $i \leqq j$; (2) $\cup \mathfrak{A}^{(i)} = \mathfrak{A}$; (3) $\mathfrak{A}^{(i)}\mathfrak{A}^{(j)} \subseteqq \mathfrak{A}^{(i+j)}$. If $\mathfrak{A}$ is graded and we set $\mathfrak{A}^{(i)} = \sum_{j \leqq i}\mathfrak{A}_j$ then this defines a filtration in $\mathfrak{A}$ and $\mathfrak{A}$ becomes a filtered algebra. Another standard way of obtaining a filtered algebra is the following. Let $\mathfrak{A}$ be any algebra and let $\mathfrak{M}$ be a subspace of $\mathfrak{A}$ which generates $\mathfrak{A}$ as an algebra. This implies that we can write $\mathfrak{A} = \Phi 1 + \mathfrak{M} + \mathfrak{M}^2 + \cdots$ where $\mathfrak{M}^i$ is the subspace spanned by all products of i elements taken out of $\mathfrak{M}$. Set $\mathfrak{A}^{(i)} = \Phi 1 + \mathfrak{M} + \cdots + \mathfrak{M}^i$. Then it is clear that these $\mathfrak{A}^{(i)}$ define a filtration in $\mathfrak{A}$. This applies in particular to the universal enveloping algebra $\mathfrak{U}$ of a Lie algebra and the space $\mathfrak{M} = \mathfrak{L}$ of generators of $\mathfrak{U}$.

An important notion associated with a filtered algebra $\mathfrak{A}$ is the *associated graded algebra* $\bar{\mathfrak{A}} = G(\mathfrak{A})$. One obtains this algebra by forming the vector space

$$G(\mathfrak{A}) = \bar{\mathfrak{A}} = \sum_{i=0}^{\infty} \oplus \bar{\mathfrak{A}}_i\,, \qquad \bar{\mathfrak{A}}_i = \mathfrak{A}^{(i)}/\mathfrak{A}^{(i-1)}\,, \tag{23}$$

where we take $\mathfrak{A}^{(-1)} = 0$. A multiplication in $\bar{\mathfrak{A}}$ is defined component-wise by

$$(a_i + \mathfrak{A}^{(i-1)})(a_j + \mathfrak{A}^{(j-1)}) = a_i a_j + \mathfrak{A}^{(i+j-1)}\,, \tag{24}$$

if $a_i \in \mathfrak{A}^{(i)}$, $a_j \in \mathfrak{A}^{(j)}$. If $a_i \equiv b_i \pmod{\mathfrak{A}^{(i-1)}}$ and $a_j \equiv b_j \pmod{\mathfrak{A}^{(j-1)}}$ then $a_i a_j \equiv b_i b_j \pmod{\mathfrak{A}^{(i+j-1)}}$. Hence (24) gives a single-valued product for an element of $\bar{\mathfrak{A}}_i$ and an element of $\bar{\mathfrak{A}}_j$ with result in $\bar{\mathfrak{A}}_{i+j}$. This is extended by addition to $\bar{\mathfrak{A}}$. It is easy to see that this gives a graded algebra $\bar{\mathfrak{A}}$ with $\bar{\mathfrak{A}}_i$ as the space of homogeneous elements of degree i.

Let $b \in \mathfrak{A}$ and suppose $b \in \mathfrak{A}^{(n)}$, $\notin \mathfrak{A}^{(n-1)}$. The element $\bar{b} = b + \mathfrak{A}^{(n-1)}$ is a homogeneous element of degree n in $\bar{\mathfrak{A}}$ and is called the *leading term* of b. If $b = 0$ we take the leading term to be 0. If b and c have the same leading term of degree n, then $b - c \in \mathfrak{A}^{(n-1)}$. Any homogeneous element of $\bar{\mathfrak{A}}$ is a leading term. If $\bar{b}$ is a leading term of b and $\bar{c}$ is a leading term of c, then either $\bar{b}\bar{c} = 0$ or $\bar{b}\bar{c}$ is the leading term of bc. These remarks imply that if $\mathfrak{B}$ is a left ideal in $\mathfrak{A}$, then the set of sums of the leading terms of the elements of $\mathfrak{B}$ is a homogeneous left ideal $\bar{\mathfrak{B}}$ in $\bar{\mathfrak{A}}$. We can now prove the following

THEOREM 4. *Let $\mathfrak{A}$ be a filtered algebra with $\bar{\mathfrak{A}}$ as the associated* *graded algebra. If* $\bar{\mathfrak{A}}$ *has no zero divisors* $\neq 0$, *then* $\mathfrak{A}$ *has no zero divisors* $\neq 0$. *If* $\bar{\mathfrak{A}}$ *is left (right) Noetherian, then* $\mathfrak{A}$ *is left (right) Noetherian (definition below).*

Proof: Let $b, c \in \mathfrak{A}, b \neq 0, c \neq 0$. Suppose $\bar{b}$ is the leading term of b, $\bar{c}$ that of c. Then, by definition, $\bar{b} \neq 0$, $\bar{c} \neq 0$ and so $\bar{b}\bar{c} \neq 0$. Then $\bar{b}\bar{c}$ is the leading term of bc and this is $\neq 0$. Hence $bc \neq 0$. The statement that a ring is left Noetherian means that the ascending chain condition holds for left ideals of the ring. As is well known, this is equivalent to the property that every left ideal is finitely generated. Suppose $\bar{\mathfrak{A}}$ has this property. Let $\mathfrak{B}$ be a left ideal in $\mathfrak{A}$ and let $\bar{\mathfrak{B}}$ be the associated homogeneous left ideal in $\bar{\mathfrak{A}}$ consisting of the sums of the leading terms of the elements $b \in \mathfrak{B}$. By hypothesis, $\bar{\mathfrak{B}}$ has generators $(\bar{b}_1, \bar{b}_2, \cdots, \bar{b}_r)$ and we may assume that $\bar{b}_i$ is the leading term of $b_i \in \mathfrak{B}$. We assert that the b_i generate $\mathfrak{B}$, that is, every $b \in \mathfrak{B}$ has the form $\sum c_i b_i$, $c_i \in \mathfrak{A}$. We assume that the leading term $\bar{b}$ of b is of degree n and we may suppose that the result holds for elements with leading terms of degree less than n. Now $\bar{b} = \sum \bar{c}_i \bar{b}_i$, $\bar{c}_i \in \bar{\mathfrak{A}}$. By dropping some of the terms $\bar{c}_i\bar{b}_i$ (equating homogeneous parts) we obtain $\bar{b} = \sum \bar{c}_j \bar{b}_j$ where $\bar{c}_j$ is homogeneous and $\bar{c}_j\bar{b}_j$ is homogeneous of degree n. Then $\bar{c}_j$ is the leading term of an element c_j and the leading term of $\sum c_j b_j$ is $\bar{b}$. Hence $d = b - \sum c_j b_j \in \mathfrak{A}^{(n-1)}$ so that d is in the left ideal generated by the b_j. Hence b is in the left ideal generated by the b_j.

We prove next the following ring theoretic result which we shall apply to the universal enveloping algebra of a finite-dimensional Lie algebra.

THEOREM 5 (*Goldie-Ore*). *Let* $\mathfrak{A}$ *be a ring (associative with* 1) *without zero-divisor* $\neq 0$ *satisfying the ascending chain condition for left ideals. Then* $\mathfrak{A}$ *has a left quotient division ring (definition below).*

Proof: Goldie's part of this result is that any two non-zero elements $a, b \in \mathfrak{A}$ have a non-zero common left multiple $m = b'a = a'b$. This is equivalent to saying that the intersection of the principal left ideals $\mathfrak{A}a \cap \mathfrak{A}b \neq 0$. To prove this we consider (following Lesieur and Croisot) the sequence of ideals $\mathfrak{A}a$, $\mathfrak{A}a + \mathfrak{A}ab$, $\mathfrak{A}a + \mathfrak{A}ab + \mathfrak{A}ab^2, \cdots$. By the ascending chain condition there exists a k such that $ab^{k+1} \in \mathfrak{A}a + \mathfrak{A}ab + \cdots + \mathfrak{A}ab^k$. Then $ab^{k+1} = x_0 a + x_1 ab + \cdots x_k ab^k$, $x_i \in \mathfrak{A}$. Since $\mathfrak{A}$ has no zero divisors $ab^{k+1} \neq 0$ so not every $x_i = 0$.

If x_h is the first one of these which is not zero we have $ab^{k+1} = x_h ab^h + x_{h+1}ab^{h+1} + \cdots + x_k ab^k$, $x_h \neq 0$. Cancellation of b^h gives

$$0 \neq x_h a = ab^{k+1-h} - x_{h+1}ab - \cdots - x_k ab^{k-h} \in \mathfrak{A}a \cap \mathfrak{A}b .$$

Hence $\mathfrak{A}$ has the (left) common multiple property. Now Ore has shown that any ring without zero divisors $\neq 0$ has a left quotient division ring if and only if every pair of non-zero elements a, b have a non-zero common left multiple $m = a'b = b'a$. We recall that $\mathfrak{D}$ is called a *left quotient division* ring for $\mathfrak{A}$ if: (1) $\mathfrak{D}$ is a division ring; (2) $\mathfrak{A}$ is a subring of $\mathfrak{D}$; (3) every element of $\mathfrak{D}$ has the form $a^{-1}b$, $a, b \in \mathfrak{A}$ (cf. Jacobson [1], p. 118.)

We now apply our results to the universal enveloping algebra $\mathfrak{U}$ of a Lie algebra $\mathfrak{L}$. As before, we employ the filtration of $\mathfrak{U}$ defined by

$$\mathfrak{U}^{(0)} = \Phi 1 , \qquad \mathfrak{U}^{(i)} = \Phi 1 + \mathfrak{L} + \mathfrak{L}^2 + \cdots + \mathfrak{L}^i , \qquad i \geqq 1 . \tag{25}$$

Let $\bar{\mathfrak{U}} = G(\mathfrak{U})$ be the associated graded algebra. It is easy to see from the definition of $G(\mathfrak{U})$ that, since $\mathfrak{L}$ generates $\mathfrak{U}$, $\bar{\mathfrak{L}} = \mathfrak{U}^{(1)}/\mathfrak{U}^{(0)} = (\Phi 1 + \mathfrak{L})/\Phi 1$ generates $\bar{\mathfrak{U}}$. It follows that if $\{u_j \mid j \in J\}$, J ordered, is a basis for $\mathfrak{L}$, then the cosets $\bar{u}_j = u_j + \Phi 1$ in $\bar{\mathfrak{U}}_1$ generate $\bar{\mathfrak{U}}$. We have $\bar{u}_j\bar{u}_k = u_ju_k + \mathfrak{U}^{(1)}$ and $\bar{u}_k\bar{u}_j = u_ku_j + \mathfrak{U}^{(1)}$ and $u_ju_k - u_ku_j = [u_ju_k] \in \mathfrak{U}^{(1)}$. Hence $\bar{u}_k\bar{u}_j = \bar{u}_j\bar{u}_k$. Thus the generators commute and consequently $\bar{\mathfrak{U}}$ is a commutative algebra. It follows that every element of $\bar{\mathfrak{U}}$ is a linear combination of the elements $\bar{1}(= 1)$, $\bar{u}_{i_1}\bar{u}_{i_2} \cdots \bar{u}_{i_n}$, $i_1 \leqq i_2 \leqq \cdots \leqq i_n$. It is easily seen from the definition and the Poincaré-Birkhoff-Witt theorem that the different "standard" monomials indicated here form a basis for $\bar{\mathfrak{U}}$. This means that the $\bar{u}_j$ are algebraically independent and $\bar{\mathfrak{U}}$ is the ordinary algebra of polynomials in these elements. The general results we have derived and the properties of polynomial rings now give the following theorem on $\mathfrak{U}$.

THEOREM 6

1. *The universal enveloping algebra* $\mathfrak{U}$ *of any Lie algebra* $\mathfrak{L}$ *has no zero divisors* $\neq 0$.

2. *If* $\mathfrak{L}$ *is finite-dimensional, then* $\mathfrak{U}$ *satisfies the ascending chain condition for left or right ideals and* $\mathfrak{U}$ *has a left or right quotient division ring.*

Proof:

1. Since $\bar{\mathfrak{U}}$ is a polynomial ring in algebraically independent elements over a field, $\bar{\mathfrak{U}}$ has no zero divisors $\neq 0$. Consequently $\mathfrak{U}$

has no zero-divisors $\neq 0$, by Theorem 4.

2. If $\mathfrak{L}$ has the finite basis $u_1, u_2, \cdots, u_n$, then $\bar{\mathfrak{U}}$ is the polynomial algebra in $\bar{u}_1, \bar{u}_2, \cdots, \bar{u}_n$. This satisfies the ascending chain condition on ideals by the Hilbert basis theorem. Hence $\mathfrak{U}$ is left and right (Noetherian) by Theorem 4. Hence $\mathfrak{U}$ has a left and right quotient division ring by the Goldie—Ore theorem.

4. Free Lie algebras

The notion of a *free algebra* (*free Lie algebra*) generated by a set $X = \{x_j \mid j \in J\}$ can be formulated in a manner similar to that of the definition of a universal enveloping algebra of a Lie algebra. We define this to consist of a pair $(\mathfrak{F}, i)$ $((\mathfrak{F}\mathfrak{L}, i))$ consisting of an algebra $\mathfrak{F}$ (Lie algebra $\mathfrak{F}\mathfrak{L}$) and a mapping i of X into $\mathfrak{F}\,(\mathfrak{F}\mathfrak{L})$ such that if θ is any mapping of X into an algebra $\mathfrak{A}$ (Lie algebra $\mathfrak{B}$), then there exists a unique homomorphism θ' of $\mathfrak{F}\,(\mathfrak{F}\mathfrak{L})$ into $\mathfrak{A}\,(\mathfrak{B})$ such that $\theta = i\theta'$. It is easy to construct a free algebra generated by any set X. For this purpose one forms a vector space $\mathfrak{M}$ with basis X and one forms the tensor algebra $\mathfrak{F}(=\mathfrak{T}) = \Phi 1 \oplus \mathfrak{M} \oplus (\mathfrak{M} \otimes \mathfrak{M}) \oplus \cdots$ based on $\mathfrak{M}$. The mapping i is taken to be the injection of X into $\mathfrak{F}$. Now let θ be a mapping of X into an algebra $\mathfrak{A}$. Since X is a basis for $\mathfrak{M}$, θ can be extended to a unique linear mapping of $\mathfrak{M}$ into $\mathfrak{A}$ and this can be extended to a unique homomorphism θ of $\mathfrak{F}$ into $\mathfrak{A}$. Hence $\mathfrak{F}$ and the injection mapping of X into $\mathfrak{F}$ is a free algebra generated by X.

It is somewhat awkward to give a direct construction for a free Lie algebra generated by X. Instead, one obtains the desired Lie algebra by using the free algebra $\mathfrak{F}$ generated by X. Let $\mathfrak{F}\mathfrak{L}$ denote the subalgebra of the Lie algebra $\mathfrak{F}_L$ generated by the subset X. Let θ be a mapping of X into a Lie algebra $\mathfrak{B}$ and let $\mathfrak{U}$ be the universal enveloping algebra of $\mathfrak{B}$, which (by the Poincaré-Birkhoff-Witt theorem) we suppose contains $\mathfrak{B}$. Then θ can be considered as a mapping of X into $\mathfrak{U}$, so this can be extended to a homomorphism θ of $\mathfrak{F}$ into $\mathfrak{U}$. Moreover, θ is a homomorphism of $\mathfrak{F}_L$ into $\mathfrak{U}_L$ and since θ maps X into a subset of $\mathfrak{B}$ $(\subseteq \mathfrak{U})$, the restriction of θ to the subalgebra $\mathfrak{F}\mathfrak{L}$ of $\mathfrak{F}_L$ generated by X is a homomorphism of $\mathfrak{F}\mathfrak{L}$ into $\mathfrak{B}$. We have therefore shown that θ can be extended to a homomorphism of $\mathfrak{F}\mathfrak{L}$ into $\mathfrak{B}$. Since X generates $\mathfrak{F}\mathfrak{L}$, θ is unique. Hence $\mathfrak{F}\mathfrak{L}$ and the injection mapping of X into $\mathfrak{F}\mathfrak{L}$ is a free Lie algebra generated by X.

We note next that $\mathfrak{F}$ (and the injection mapping) is the universal enveloping algebra of $\mathfrak{F}\mathfrak{L}$. Thus let θ be a homomorphism of $\mathfrak{F}\mathfrak{L}$ into a Lie algebra $\mathfrak{A}_L$, $\mathfrak{A}$ an algebra. Then there exists a homomorphism θ of $\mathfrak{F}$ into $\mathfrak{A}$ which coincides with the restriction of θ to X. Then θ is a homomorphism of $\mathfrak{F}_L$ into $\mathfrak{A}_L$ and so the restriction θ' of θ to $\mathfrak{F}\mathfrak{L}$ is a homomorphism of $\mathfrak{F}\mathfrak{L}$ into $\mathfrak{A}_L$. Since $x\theta' = x\theta$ for $x \in X$ and X generates $\mathfrak{F}\mathfrak{L}$, it is clear that θ' coincides with the given homomorphism θ of $\mathfrak{F}\mathfrak{L}$ into $\mathfrak{A}_L$. Thus we have extended θ to a homomorphism of $\mathfrak{F}$ into $\mathfrak{A}$. Since $\mathfrak{F}\mathfrak{L}$ generates $\mathfrak{F}$ it is clear that the extension is unique. Hence $\mathfrak{F}$ is the universal enveloping algebra of $\mathfrak{F}\mathfrak{L}$.

The two results which we have established can be stated in the following

THEOREM 7 (Witt). *Let X be an arbitrary set and let $\mathfrak{F}$ denote the free algebra (freely) generated by X. Let $\mathfrak{F}\mathfrak{L}$ denote the subalgebra of $\mathfrak{F}_L$ generated by the elements of X. Then $\mathfrak{F}\mathfrak{L}$ is a free Lie algebra generated by X and $\mathfrak{F}$ is the universal enveloping algebra of $\mathfrak{F}\mathfrak{L}$.*

For the sake of simplicity we shall now restrict our attention to the case of a finite set $X = \{x_1, x_2, \cdots, x_r\}$. Then

$$\mathfrak{M} = \Phi x_1 \oplus \Phi x_2 \oplus \cdots \oplus \Phi x_r$$

and $\mathfrak{F} = \Phi 1 \oplus \mathfrak{M} \oplus (\mathfrak{M} \otimes \mathfrak{M}) \oplus \cdots$, and we write $\mathfrak{F} = \Phi\{x_1, \cdots, x_r\}$. The algebra $\mathfrak{F}$ is graded with $\mathfrak{M}_m = \mathfrak{M} \otimes \mathfrak{M} \otimes \cdots \otimes \mathfrak{M}$ (m-times) as the space of homogeneous elements of degree m. A basis for this space is the set of monomials of the form $x_{i_1}x_{i_2}\cdots x_{i_m}$, $i_j = 1, 2, \cdots, r$; hence $\dim \mathfrak{M}_m = r^m$.

An element $a \in \mathfrak{F}$ is called a *Lie element* if $a \in \mathfrak{F}\mathfrak{L}$. We proceed to obtain two important criteria that an element of $\mathfrak{F}$ be a Lie element. We observe first that it is enough to treat the case in which the given element a is homogeneous. Thus consider the collection of linear combinations of the Lie elements of the form

$$[\cdots[[x_{i_1}x_{i_2}]x_{i_3}]\cdots x_{i_m}], \tag{26}$$

$i_j = 1, 2, \cdots, r$, $m = 1, 2, \cdots$. The Jacobi identity shows that this subspace is a subalgebra of $\mathfrak{F}_L$. Since it contains the x_i it coincides with $\mathfrak{F}\mathfrak{L}$. We therefore see that every element of $\mathfrak{F}\mathfrak{L}$ is a sum of homogeneous Lie elements. Hence an element is a Lie element if and only if its homogeneous parts are Lie elements. We see also that if a is a Lie element which is homogeneous of degree m then

a is a linear combination of elements of the form (26).

Let $\mathfrak{F}'$ denote the ideal $\mathfrak{M} \oplus (\mathfrak{M} \otimes \mathfrak{M}) \oplus \cdots$ in $\mathfrak{F}$. An element of $\mathfrak{F}$ is in $\mathfrak{F}'$ if and only if it is a linear combination of monomials $x_{i_1} \cdots x_{i_m}$, $m \geqq 1$. Since the different monomials of this type form a basis for $\mathfrak{F}'$ we have a linear mapping σ of $\mathfrak{F}'$ into $\mathfrak{F}\mathfrak{L}$ such that

$$x_i\sigma = x_i\,, \qquad (x_{i_1} \cdots x_{i_m})\sigma = [\cdots [x_{i_1}x_{i_2}] \cdots x_{i_m}]\,, \qquad m > 1\,. \tag{27}$$

We consider also the adjoint representation of $\mathfrak{F}\mathfrak{L}$. Since $\mathfrak{F}$ is the universal enveloping algebra of $\mathfrak{F}\mathfrak{L}$, this representation can be extended to a homomorphism θ of $\mathfrak{F}$ into the algebra $\mathfrak{E}(\mathfrak{F}\mathfrak{L})$ of linear transformations in the space $\mathfrak{F}\mathfrak{L}$. We have

$$\begin{aligned}(x_{i_1} \cdots x_{i_m}x_{j_1} \cdots x_{j_t})\sigma &= [\cdots [x_{i_1}x_{i_2}] \cdots x_{i_m}]x_{j_1}] \cdots x_{j_t}] \\ &= (x_{i_1} \cdots x_{i_m})\sigma \operatorname{ad} x_{j_1} \operatorname{ad} x_{j_2} \cdots \operatorname{ad} x_{j_t} \\ &= (x_{i_1} \cdots x_{i_m})\sigma(x_{j_1} \cdots x_{j_t})\theta\,.\end{aligned}$$

This implies that

$$(uv)\sigma = (u\sigma)(v\theta)\,, \qquad u \in \mathfrak{F}'\,, \qquad v \in \mathfrak{F}\,. \tag{28}$$

If $a, b \in \mathfrak{F}\mathfrak{L}$, then

$$\begin{aligned}[ab]\sigma &= (ab)\sigma - (ba)\sigma = (a\sigma)(b\theta) - (b\sigma)(a\theta) \\ &= (a\sigma) \operatorname{ad} b - (b\sigma) \operatorname{ad} a \\ &= [a\sigma, b] + [a, b\sigma]\,.\end{aligned} \tag{29}$$

Thus the restriction of σ to $\mathfrak{F}\mathfrak{L} \subseteqq \mathfrak{F}'$ is a derivation in $\mathfrak{F}\mathfrak{L}$. We can use this result to obtain the following criterion.

THEOREM 8 (*Dynkin-Specht-Wever*). *If Φ is of characteristic* 0, *then a homogeneous element a of degree $m > 0$ is a Lie element if and only if $a\sigma = ma$ where σ is the linear mapping of $\mathfrak{F}'$ into $\mathfrak{F}\mathfrak{L}$ defined by* (27).

Proof: The condition is evidently sufficient (even for characteristic p, provided $p \nmid m$). Now let a be a Lie element which is homogeneous of degree m. We use induction on m. We have seen that a is a linear combination of terms of the form (26). Hence it suffices to prove that $a\sigma = ma$ for a as in (26). We have

$$\begin{aligned}[\cdots [x_{i_1}x_{i_2}] \cdots x_{i_m}]\sigma &= [[\cdots [x_{i_1}x_{i_2}] \cdots x_{i_{m-1}}]\sigma, x_{i_m}] \\ &\quad + [[\cdots [x_{i_1}x_{i_2}] \cdots x_{i_{m-1}}], x_{i_m}\sigma] \\ &= (m-1)[\cdots [x_{i_1}x_{i_2}] \cdots x_{i_m}] + [\cdots [x_{i_1}x_{i_2}] \cdots x_{i_m}] \\ &= m[\cdots [x_{i_1}x_{i_2}] \cdots x_{i_m}]\,,\end{aligned}$$

which is the required result.

Our next criterion for Lie elements does not require the reduction to homogeneous elements. This is

THEOREM 9 (*Friedrichs*). *Let* $\mathfrak{F} = \Phi\{x_1, \cdots, x_r\}$ *be the free algebra generated by the* x_i *over a field of characteristic* 0. *Let* δ *be the diagonal mapping of* $\mathfrak{F}$, *that is, the homomorphism of* $\mathfrak{F}$ *into* $\mathfrak{F} \otimes \mathfrak{F}$ *such that* $x_i\delta = x_i \otimes 1 + 1 \otimes x_i$. *Then* $a \in \mathfrak{F}$ *is a Lie element if and only if* $a\delta = a \otimes 1 + 1 \otimes a$.

Proof: We have $[a \otimes 1 + 1 \otimes a, b \otimes 1 + 1 \otimes b] = [ab] \otimes 1 + 1 \otimes [ab]$ which implies that the set of elements a satisfying $a\delta = a \otimes 1 + 1 \otimes a$ is a subalgebra of $\mathfrak{F}_L$. This includes the x_i; hence it contains $\mathfrak{F}\mathfrak{L}$. Let $y_1, y_2, \cdots$ be a basis for $\mathfrak{F}\mathfrak{L}$. Then since $\mathfrak{F}$ is the universal enveloping algebra of $\mathfrak{F}\mathfrak{L}$ the elements $y_1^{k_1} y_2^{k_2} \cdots y_m^{k_m}$, m arbitrary, $k_i \geqq 0$ $(y_i^0 = 1)$ form a basis for $\mathfrak{F}$. Hence the products

$$(y_1^{k_1} y_2^{k_2} \cdots y_m^{k_m}) \otimes (y_1^{l_1} y_2^{l_2} \cdots y_n^{l_n})$$

form a basis for $\mathfrak{F} \otimes \mathfrak{F}$. We have

$$\begin{aligned}(y_1^{k_1} y_2^{k_2} \cdots y_m^{k_m})\delta &= (y_1 \otimes 1 + 1 \otimes y_1)^{k_1}(y_2 \otimes 1 + 1 \otimes y_2)^{k_2} \\ &\qquad \cdots (y_m \otimes 1 + 1 \otimes y_m)^{k_m} \\ &= y_1^{k_1} y_2^{k_2} \cdots y_m^{k_m} \otimes 1 + k_1 y_1^{k_1-1} y_2^{k_2} \cdots y_m^{k_m} \otimes y_1 \\ &\quad + k_2 y_1^{k_1} y_2^{k_2-1} \cdots y_m^{k_m} \otimes y_2 + \cdots + k_m y_1^{k_1} \cdots y_m^{k_m-1} \otimes y_m + * \end{aligned} \tag{30}$$

where * represents a linear combination of base elements of the form $y_1^{j_1} y_2^{j_2} \cdots y_s^{j_s} \otimes y_1^{l_1} y_2^{l_2} \cdots y_t^{l_t}$ with $\sum l_i > 1$. The second through the $(m + 1)$-st term do not occur in the expressions of this type for any other base element $y_1^{j_1} y_2^{j_2} \cdots y_s^{j_s}$. It follows that in order that $a\delta$ shall be a linear combination of the base elements of the form $y_1^{k_1} \cdots y_m^{k_m} \otimes 1$ and $1 \otimes y_1^{j_1} \cdots y_s^{j_s}$ it is necessary that in the expression for a in terms of the chosen basis only base elements $y_1^{k_1} \cdots y_m^{k_m}$ with one $k_i = 1$ and all the other $k_j = 0$ occur with nonzero coefficients. This means that a is a linear combination of the y_i; hence $a \in \mathfrak{F}\mathfrak{L}$. Hence $a\delta = a \otimes 1 + 1 \otimes a$ if and only if $a \in \mathfrak{F}\mathfrak{L}$.

5. *The Campbell-Hausdorff formula*

We shall now use our criteria for Lie elements to derive a formula due to Campbell and to Hausdorff for the product of exponentials in an algebra. For this purpose we need to extend the free algebra $\mathfrak{F} = \Phi\{x_1, \cdots, x_r\}$ to the algebra $\bar{\mathfrak{F}}$ of formal power series in the x_i. More generally, let $\mathfrak{A}$ be any graded algebra with $\mathfrak{A}_i$ as the subspace of homogeneous elements of degree i, $i = 0, 1, 2, \cdots$. Let $\bar{\mathfrak{A}}$ be the complete direct sum of the spaces $\mathfrak{A}_i$. Thus the elements of $\bar{\mathfrak{A}}$ are the expressions $\sum_0^\infty a_i = a_0 + a_1 + \cdots$ such that $\sum a_i = \sum b_i$ if and only if $a_i = b_i$, $i = 0, 1, 2, \cdots$. Addition and scalar multiplication are defined component-wise. We introduce a multiplication in $\bar{\mathfrak{A}}$ by $(\sum_0^\infty a_i)(\sum_0^\infty b_j) = \sum_0^\infty c_k$ where

$$c_k = a_0 b_k + a_1 b_{k-1} + \cdots + a_k b_0 \in \mathfrak{A}_k .$$

It is easy to check that $\bar{\mathfrak{A}}$ is an algebra. Moreover, the subset of $\bar{\mathfrak{A}}$ of elements $\sum a_i$ such that $a_i = 0$ if $i > m$, $m = 0, 1, 2, \cdots$ is a subalgebra which may be identified with $\mathfrak{A}$. The subset $\bar{\mathfrak{A}}^{(i)}$ of elements of the form $a_i + a_{i+1} + \cdots$ is an ideal in $\bar{\mathfrak{A}}$ and $\cap \bar{\mathfrak{A}}^{(i)} = 0$. We define a valuation in $\bar{\mathfrak{A}}$ by setting $|0| = 0$, $|a| = 2^{-i}$ if $a \neq 0$ and $a \in \bar{\mathfrak{A}}^{(i)}$, $\notin \bar{\mathfrak{A}}^{(i+1)}$. Then we have the following properties:

(i) $\quad |a| \geqq 0 , \quad |a| = 0$ if and only if $a = 0$.

(ii) $\quad |ab| \leqq |a|\,|b|$.

(iii) $\quad |a + b| \leqq \max(|a|, |b|)$.

This valuation makes $\bar{\mathfrak{A}}$ a topological algebra. Convergence of series $x_1 + x_2 + \cdots$, $x_i \in \bar{\mathfrak{A}}$, is defined in the usual way. The non-archimedean property (iii) of the valuation implies the very simple criterion that $x_1 + x_2 + \cdots$ converges if and only if $|x_i| \to 0$. It is clear that the subalgebra $\mathfrak{A}$ is dense in $\bar{\mathfrak{A}}$.

If the characteristic of the base field is 0 and $z \in \bar{\mathfrak{A}}^{(1)}$, then

$$\exp z = 1 + z + \frac{z^2}{2!} + \frac{z^3}{3!} + \cdots \tag{31}$$

and

$$\log(1 + z) = z - \frac{z^2}{2} + \frac{z^3}{3} - \cdots \tag{32}$$

are well-defined elements of $\bar{\mathfrak{A}}$ (that is, the series indicated converge). A direct calculation shows that

$$\exp(\log(1 + z)) = 1 + z , \quad \log(\exp z) = z . \tag{33}$$

Moreover, if $z_1, z_2 \in \bar{\mathfrak{A}}^{(1)}$ and $[z_1 z_2] = 0$, then

$$\exp z_1 \exp z_2 = \exp(z_1 + z_2) \tag{34}$$

and

(35) $$\log(1+z_1)(1+z_2) = \log(1+z_1) + \log(1+z_2)$$

We consider all of this, in particular, for $\mathfrak{A} = \mathfrak{F} = \Phi\{x_1, \cdots, x_r\}$. The resulting algebra $\bar{\mathfrak{F}}$ is called the algebra of *formal power series* in the x_i. We shall also apply the construction to the algebra $\mathfrak{F} \otimes \mathfrak{F}$. Since $\mathfrak{F} = \sum_{i=0}^{\infty} \oplus \mathfrak{F}_i$, where $\mathfrak{F}_i$ is the subspace of homogeneous elements of degree i, $\mathfrak{F} \otimes \mathfrak{F} = \sum \oplus (\mathfrak{F}_i \otimes \mathfrak{F}_j)$. We have $(\mathfrak{F}_i \otimes \mathfrak{F}_j)(\mathfrak{F}_{i'} \otimes \mathfrak{F}_{j'}) \subseteq \mathfrak{F}_{i+i'} \otimes \mathfrak{F}_{j+j'}$. Hence, if we set $(\mathfrak{F} \otimes \mathfrak{F})_k = \mathfrak{F}_k \otimes \mathfrak{F}_0 + \mathfrak{F}_{k-1} \otimes \mathfrak{F}_1 + \cdots + \mathfrak{F}_0 \otimes \mathfrak{F}_k$, then $(\mathfrak{F} \otimes \mathfrak{F})_k (\mathfrak{F} \otimes \mathfrak{F})_l \subseteq (\mathfrak{F} \otimes \mathfrak{F})_{k+l}$ and $\mathfrak{F} \otimes \mathfrak{F} = \sum \oplus (\mathfrak{F} \otimes \mathfrak{F})_k$. It follows that $\mathfrak{F} \otimes \mathfrak{F}$ is graded with $(\mathfrak{F} \otimes \mathfrak{F})_k$ as subspace of homogeneous elements of degree k. We can therefore construct the algebra $\overline{\mathfrak{F} \otimes \mathfrak{F}}$.

If $a = \sum_0^{\infty} a_i$, $a_i \in \mathfrak{F}_i$, then $\sum (a_i \otimes 1)$ and $\sum (1 \otimes a_i)$ are elements of $\overline{\mathfrak{F} \otimes \mathfrak{F}}$ which we denote as $a \otimes 1$ and $1 \otimes a$, respectively. The mappings $a \to a \otimes 1$, $a \to 1 \otimes a$ are isomorphisms and homeomorphisms of $\bar{\mathfrak{F}}$ into $\overline{\mathfrak{F} \otimes \mathfrak{F}}$. We have $[a \otimes 1, 1 \otimes b] = 0$ for any $a, b \in \bar{\mathfrak{F}}$. Also, in the characteristic 0 case we have $\exp(a \otimes 1) = \exp a \otimes 1$, $\exp(1 \otimes a) = 1 \otimes \exp a$ and similar formulas for the log function.

Let $\overline{\mathfrak{F}\mathfrak{L}}$ denote the subset of $\bar{\mathfrak{F}}$ of elements of the form $b_1 + b_2 + \cdots$ where b_i is a Lie element in $\mathfrak{F}_i$. It is clear that $\overline{\mathfrak{F}\mathfrak{L}}$ is a subalgebra of $\bar{\mathfrak{F}}_L$. The diagonal isomorphism δ of $\mathfrak{F}$ into $\mathfrak{F} \otimes \mathfrak{F}$ has an extension to an isomorphism of $\bar{\mathfrak{F}}$ into $\overline{\mathfrak{F} \otimes \mathfrak{F}}$. We note first that if $a_i \in \mathfrak{F}_i$ then $a_i\delta \in (\mathfrak{F} \otimes \mathfrak{F})_i$. This is immediate by induction on i, using the formulas $\mathfrak{F}_i = \sum_{j=1}^{r} x_j \mathfrak{F}_{i-1}$, $x_j\delta = x_j \otimes 1 + 1 \otimes x_j$. Hence if $a = \sum_0^{\infty} a_i$, $a_i \in \mathfrak{F}_i$, then $\sum a_i\delta$ is a well-defined element of $\bar{\mathfrak{F}} \otimes \bar{\mathfrak{F}}$. We denote this as $a\delta$ and it is clear that δ: $a \to a\delta$ is an isomorphism and homeomorphism of $\bar{\mathfrak{F}}$ into $\bar{\mathfrak{F}} \otimes \bar{\mathfrak{F}}$. It is clear also from Friedrichs' theorem applied to the a_i that in the characteristic 0 case $a = \sum a_i$, $a_i \in \mathfrak{F}_i$, is in $\overline{\mathfrak{F}\mathfrak{L}}$ if and only if $a\delta = a \otimes 1 + 1 \otimes a$.

We assume now that $r = 2$ and we denote the generators as x, y and write $\mathfrak{F} = \Phi\{x, y\}$. Assume also that the characteristic is 0. Consider the element $\exp x \exp y$ of $\bar{\mathfrak{F}}$. We can write this as $1 + z$, where $z = z_1 + z_2 + \cdots$, $z_i \in \mathfrak{F}_i$, and we can write

(36) $$\exp x \exp y = 1 + z = \exp w\ ,$$

where $w = \log(1 + z)$. We shall prove

Proposition 1. *The element* $w = \log(\exp x \exp y)$ *is a Lie element, that is,* $w \in \overline{\mathfrak{F}\mathfrak{L}}$.

Proof: Consider $(\exp x \exp y)\delta$. We have

$$\begin{aligned}(\exp x \exp y)\delta &= (\exp x\delta)(\exp y\delta)\\ &= \exp(x \otimes 1 + 1 \otimes x)\exp(y \otimes 1 + 1 \otimes y)\\ &= \exp(x \otimes 1)\exp(1 \otimes x)\exp(y \otimes 1)\exp(1 \otimes y)\\ &= (\exp x \otimes 1)(1 \otimes \exp x)(\exp y \otimes 1)(1 \otimes \exp y)\\ &= (\exp x \otimes 1)(\exp y \otimes 1)(1 \otimes \exp x)(1 \otimes \exp y)\\ &= (\exp x \exp y \otimes 1)(1 \otimes \exp x \exp y)\,.\end{aligned}$$

Hence $(1 + z)\delta = ((1 + z) \otimes 1)(1 \otimes (1 + z))$ and so

$$\begin{aligned}(\log(1 + z))\delta &= \log(1 + z)\delta\\ &= \log((1 + z) \otimes 1)(1 \otimes (1 + z))\\ &= \log((1 + z) \otimes 1) + \log(1 \otimes (1 + z))\\ &= \log(1 + z) \otimes 1 + 1 \otimes \log(1 + z)\,.\end{aligned}$$

This shows that $\log(1 + z)$ satisfies Friedrichs' condition for a Lie element. Hence $\log(1 + z) \in \overline{\mathfrak{F}\mathfrak{L}}$.

Now that we know that $w = \log(1 + z)$ is a Lie element, we can obtain an explicit formula for this element by using the Specht-Wever theorem. We have

$$z = \exp x \exp y - 1 = \sum_{p,q} \frac{x^p}{p!}\frac{x^q}{q!}\,, \qquad p + q > 0$$

and

$$z^m = \sum_{p_i,q_i} \frac{x^{p_1}}{p_1!}\frac{y^{q_1}}{q_1!}\frac{x^{p_2}}{p_2!}\frac{y^{q_2}}{q_2!}\cdots\frac{x^{p_m}}{p_m!}\frac{y^{q_m}}{q_m!}\,, \quad p_i + q_i > 0\,. \tag{37}$$

Hence

$$\log(1 + z) = \sum_m \sum_{p_i,q_i} \frac{(-1)^{m-1}}{m}\frac{x^{p_1}}{p_1!}\frac{y^{q_1}}{q_1!}\cdots\frac{x^{p_m}}{p_m!}\frac{y^{q_m}}{q_m!}\,, \qquad p_i + q_i > 0\,. \tag{38}$$

Since this is a Lie element, if we apply the operator σ to the terms of degree k we obtain k times the homogeneous part of degree k in this element. It follows that we have the following expression of $\log(1 + z)$ as a Lie element.

$$\log(1 + z) = \sum_m \sum_{p_i,q_i} \frac{(-1)^{m-1}}{m(\sum p_i + q_i)} \frac{[\cdots[\overbrace{[xx]\cdots x}^{p_1}]\overbrace{y]\cdots y]}^{q_1}\cdots \overbrace{y]\cdots y]}^{q_m}}{p_1!\,q_1!\cdots p_m!\,q_m!}\,. \tag{39}$$

It is easy to calculate the first few terms of this series and obtain

$$(40)\qquad \log(1+z) = x + y + \frac{1}{2}[xy] + \frac{1}{12}[[xy]y] - \frac{1}{12}[[xy]x] + \cdots .$$

6. Cohomology of Lie algebras. The standard complex

In this section we shall give the Cartan-Eilenberg definition of the cohomology groups of a Lie algebra and we shall show that this is equivalent to the definition given in § 3.10. To obtain the Cartan-Eilenberg definition one begins with the field Φ which one regards as a trivial module for $\mathfrak{L}$, that is, one sets $\xi l = 0$, $\xi \in \Phi$, $l \in \mathfrak{L}$. This module and all $\mathfrak{L}$-modules can be considered as right modules for the universal enveloping algebra $\mathfrak{U}$ of $\mathfrak{L}$, since every representation of $\mathfrak{L}$ has a unique extension to a representation of $\mathfrak{U}$. The term $\mathfrak{U}$-module will mean "right $\mathfrak{U}$-module" throughout our discussion. We recall that a module is called *free* if it is a direct sum of submodules isomorphic to the module $\mathfrak{U}$. One seeks a sequence of free $\mathfrak{U}$-modules $X_0, X_1, X_2, \cdots$ with $\mathfrak{U}$-homomorphism ε of X_0 into Φ and $\mathfrak{U}$-homomorphisms d_{i-1} of X_i into X_{i-1} such that the sequence

$$(41)\qquad 0 \longleftarrow \Phi \underset{\varepsilon}{\longleftarrow} X_0 \underset{d_0}{\longleftarrow} X_1 \underset{d_1}{\longleftarrow} X_2 \longleftarrow \cdots$$

is exact, that is, ε is surjective, the kernel, Ker ε = image, Im d_0, and for every $i \geqq 1$, Ker d_{i-1} = Im d_i. A sequence (41) is called a *free resolution* of the module Φ. Now let $\mathfrak{M}$ be an arbitrary $\mathfrak{L}$-, hence $\mathfrak{U}$-, module. Let Hom $(X_i, \mathfrak{M})$, $i \geqq 0$, denote the set of $\mathfrak{U}$-homomorphisms of X_i into $\mathfrak{M}$. The usual definitions of addition and scalar multiplication can be used to make Hom $(X_i, \mathfrak{M})$ a vector space over Φ. If $\eta_i \in$ Hom $(X_i, \mathfrak{M})$, then $d_i\eta_i \in$ Hom $(X_{i+1}, \mathfrak{M})$. We therefore obtain a mapping δ_i: $\eta_i \rightarrow d_i\eta_i$ of Hom $(X_i, \mathfrak{M})$ into Hom $(X_{i+1}, \mathfrak{M})$, $i \geqq 0$. This mapping is linear. Moreover, the exactness of (41) implies that $d_{i+1}d_i = 0$, $i \geqq 0$, and this implies that $\delta_i\delta_{i+1} = 0$. Hence Im $\delta_i \subseteqq$ Ker δ_{i+1}. One defines the i-th *cohomology group of* $\mathfrak{L}$ *relative to the module* $\mathfrak{M}$ to be the factor space Ker δ_i/Im δ_{i-1}, $i \geqq 1$, and Ker δ_0 for $i = 0$. The Cartan-Eilenberg theory shows that these groups are independent of the particular resolution (41) for Φ.

We shall not go into the details of this theory here but shall be content to give the construction of what appears to be the most useful resolution (41). The space $X \equiv \sum_0^\infty X_i$ we shall obtain is called the *standard complex* for $\mathfrak{L}$. It will turn out that the standard complex has an algebra structure as well as a vector space structure. This has the consequence that in suitable circumstances one can define a cohomology algebra in place of the cohomology groups.

We now suppose we have a representation R of a Lie algebra $\mathfrak{L}$ by derivations in an algebra $\mathfrak{A}$. Let $\mathfrak{U}$ be the universal enveloping algebra of $\mathfrak{L}$. Then we shall define a new algebra $(\mathfrak{A}, \mathfrak{U}, R)$, the *algebra of differential operators* of the representation R of $\mathfrak{L}$. The space of $(\mathfrak{A}, \mathfrak{U}, R)$ will be $\mathfrak{A} \otimes \mathfrak{U}$. Since the mapping of elements $v \in \mathfrak{U}$ into their left multiplications $v_L(u \rightarrow vu)$ is an anti-homomorphism, the mapping $l \rightarrow -l_L$ is a representation of $\mathfrak{L}$ acting in $\mathfrak{U}$. We now form the tensor product of this representation and the given representation R in $\mathfrak{A}$. The resulting representation maps l into the linear transformation sending $a \otimes u$ into $al^R \otimes u - a \otimes lu$. We can extend this representation to a representation of $\mathfrak{U}$, which we shall now write as πL where π is the anti-homomorphism of $\mathfrak{U}$ such that $l\pi = -l$, $l \in \mathfrak{L}$, and L is an anti-homomorphism of $\mathfrak{U}$ into the algebra of linear transformations of $\mathfrak{A} \otimes \mathfrak{U}$. We write the image of $v \in \mathfrak{U}$ under L as L_v. Then we have $(a \otimes u)(l(\pi L)) = al^R \otimes u - a \otimes lu$; hence,

$$(a \otimes u)L_l = a \otimes lu - al^R \otimes u \, . \tag{42}$$

Next if $b \in \mathfrak{A}$, we define L_b to be the linear mapping in $\mathfrak{A} \otimes \mathfrak{U}$ such that

$$(a \otimes u)L_b = ba \otimes u \, . \tag{43}$$

The mapping $b \rightarrow L_b$ is an anti-homomorphism of $\mathfrak{A}$ into the algebra of linear transformations in $\mathfrak{A} \otimes \mathfrak{U}$. It follows from (42) and (43) that

$$\begin{aligned} (a \otimes u)L_bL_l &= ba \otimes lu - (ba)l^R \otimes u \\ &= ba \otimes lu - bl^Ra \otimes u - b(al^R) \otimes u \\ (a \otimes u)L_lL_b &= ba \otimes lu - b(al^R) \otimes u \\ (a \otimes u)L_{bl^R} &= (bl^R)a \otimes u \, . \end{aligned}$$

Hence we have

$$[L_bL_l] = -L_{bl^R} \, , \tag{44}$$

which shows that the set of linear transformations L_b is an ideal in the Lie algebra of linear transformations of the form $L_b + L_l$.

It follows that the enveloping algebra $\mathfrak{X}$ of this set of transformations is the set of mappings of the form $\sum L_{u_i}L_{a_i}$ (cf. § 2.2).

We have the canonical mapping of $\mathfrak{A} \otimes \mathfrak{U}$ into $\mathfrak{X}$ sending $\sum a_i \otimes u_i$ into $\sum L_{u_i}L_{a_i}$. If $l \in \mathfrak{L}$, $1l^R = 0$, since l^R is a derivation. Hence $(1 \otimes v)L_l = 1 \otimes lv$ and, consequently, $(1 \otimes v)L_u = 1 \otimes uv$, $u, v \in \mathfrak{U}$. This implies that $(1 \otimes 1)L_uL_a = a \otimes u$ so that, if $\sum L_{u_i}L_{a_i} = 0$, then $\sum a_i \otimes u_i = 0$. Thus the mapping $\sum a_i \otimes u_i \to \sum L_{u_i}L_{a_i}$ is a vector space isomorphism. Since the set of mappings of the form $\sum L_{u_i}L_{a_i}$ is an algebra we can use this mapping to convert $\mathfrak{A} \otimes \mathfrak{U}$ into an algebra by specifying that our mapping is an algebra anti-isomorphism. The resulting algebra whose space is $\mathfrak{A} \otimes \mathfrak{U}$ is the algebra $(\mathfrak{A}, \mathfrak{U}, R)$ which we wished to define.

It is immediate from our definitions that the subset of $(\mathfrak{A}, \mathfrak{U}, R)$ of elements of the form $a \otimes 1$ is a subalgebra isomorphic to $\mathfrak{A}$. We identify this with $\mathfrak{A}$ and write a for $a \otimes 1$. Similarly, the set of elements of the form $1 \otimes u$, $u \in \mathfrak{U}$, is a subalgebra isomorphic to $\mathfrak{U}$. We identify this with $\mathfrak{U}$ and write u for $1 \otimes u$. The mapping of $(\mathfrak{A}, \mathfrak{U}, R) = \mathfrak{A} \otimes \mathfrak{U}$ into $\mathfrak{X}$ sends $a = a \otimes 1$ into $L_1L_a = L_a$ and sends $u = 1 \otimes u$ into $L_uL_1 = L_u$. The formula (44) gives the following basic commutation formula in $(\mathfrak{A}, \mathfrak{U}, R)$:

$$(45) \qquad [bl] \equiv bl - lb = bl^R , \qquad b \in \mathfrak{A} , \qquad l \in \mathfrak{L} .$$

Since every element of $\mathfrak{X}$ has the form $\sum L_{u_i}L_{a_i}$, every element of $(\mathfrak{A}, \mathfrak{U}, R)$ has the form $\sum a_iu_i$, $a_i \in \mathfrak{A}$, $u_i \in \mathfrak{U}$. Also $\sum a_iu_i = 0$ if and only if $\sum a_i \otimes u_i = 0$ in $\mathfrak{A} \otimes \mathfrak{U}$.

We establish next a "universal" property of the algebra $(\mathfrak{A}, \mathfrak{U}, R)$ in the following

PROPOSITION 2. *Let $\mathfrak{D}$ be an algebra and let θ_1 be a homomorphism of $\mathfrak{A}$ into $\mathfrak{D}$, θ_2 a homomorphism of $\mathfrak{U}$ into $\mathfrak{D}$ such that*

$$(46) \quad [b\theta_1, l\theta_2] \equiv (b\theta_1)(l\theta_2) - (l\theta_2)(b\theta_1) = (bl^R)\theta_1 , \qquad b \in \mathfrak{A} , \qquad l \in \mathfrak{L} ,$$

holds. Then there exists a unique homomorphism θ of $(\mathfrak{A}, \mathfrak{U}, R)$ into $\mathfrak{D}$ such that $b\theta = b\theta_1$, $u\theta = u\theta_2$.

Proof: We form $(\mathfrak{A}, \mathfrak{U}, R) \oplus \mathfrak{D}$ which we consider as an algebra of pairs (x, d), $x \in (\mathfrak{A}, \mathfrak{U}, R)$, $d \in \mathfrak{D}$, with component addition and multiplication. We have the homomorphisms π_1, π_2 of this algebra onto $(\mathfrak{A}, \mathfrak{U}, R)$ and $\mathfrak{D}$ respectively defined by $(x, d)\pi_1 = x$, $(x, d)\pi_2 = d$. Let $\mathfrak{Y}$ be the subalgebra of the direct sum generated by the elements $(a, a\theta_1)$ and $(u, u\theta_2)$, $a \in \mathfrak{A}$, $u \in \mathfrak{U}$. The mapping π_1 induces a homomorphism of $\mathfrak{Y}$ onto $(\mathfrak{A}, \mathfrak{U}, R)$ and the mapping π_2 induces a homomorphism of $\mathfrak{Y}$ onto the subalgebra of $\mathfrak{D}$ generated by the elements $a\theta_1$ and $u\theta_2$, $a \in \mathfrak{A}$, $u \in \mathfrak{U}$. By (45) and (46), we have the following relation in $\mathfrak{Y}$

$$(a, a\theta_1)(l, l\theta_2) - (l, l\theta_2)(a, a\theta_1) = (al^R, (al^R)\theta_1) . \tag{47}$$

This permits us to carry out a collecting process such as indicated in the discussion of the construction of the algebra $(\mathfrak{A}, \mathfrak{U}, R)$ to write the elements of $\mathfrak{Y}$ in the form $\sum (a_i, a_i\theta_1)(u_i, u_i\theta_2)$, $a_i \in \mathfrak{A}$, $u_i \in \mathfrak{U}$. The property of $\mathfrak{A} \otimes \mathfrak{U}$ $(=(\mathfrak{A}, \mathfrak{U}, R))$ implies that we have a vector space homomorphism of $(\mathfrak{A}, \mathfrak{U}, R)$ onto $\mathfrak{Y}$ sending $\sum a_i u_i$ into $\sum (a_i, a_i\theta_1)(u_i, u_i\theta_2)$. The existence of this mapping implies that π_1 is an isomorphism of $\mathfrak{Y}$ onto $(\mathfrak{A}, \mathfrak{U}, R)$. Then $\theta = \pi_1^{-1}\pi_2$ is a homomorphism of $(\mathfrak{A}, \mathfrak{U}, R)$ into $\mathfrak{D}$ such that $a\theta = a\pi_1^{-1}\pi_2 = (a, a\theta_1)\pi_2 = a\theta_1$ and $u\theta = u\pi_1^{-1}\pi_2 = (u, u\theta_2)\pi_2 = u\theta_2$. The uniqueness of θ follows from the fact that $\mathfrak{A}$ and $\mathfrak{U}$ generate $(\mathfrak{A}, \mathfrak{U}, R)$.

Remark. If $\mathfrak{M}$ is a set of generators for $\mathfrak{A}$ then the conclusion of Proposition 2 will hold provided that (46) holds for all $b \in \mathfrak{M}$. This follows from the fact that the set of b satisfying this for all $l \in \mathfrak{L}$ is a subalgebra of $\mathfrak{A}$—as can be verified directly.

We consider next an extension of the notion of a derivation:

Definition 2. Let $\mathfrak{A}$ and $\mathfrak{B}$ be algebras, s_1 and s_2 homomorphisms of $\mathfrak{A}$ into $\mathfrak{B}$. Then a linear mapping d of $\mathfrak{A}$ into $\mathfrak{B}$ is called an (s_1, s_2)-*derivation* if

$$(ab)d = (as_1)(bd) + (ad)(bs_2) . \tag{48}$$

It is easy to check that the conditions on d are just those which insure that the mapping

$$a \to \begin{pmatrix} as_1 & ad \\ 0 & as_2 \end{pmatrix} \tag{49}$$

is a homomorphism of $\mathfrak{A}$ into the two-rowed matrix algebra $\mathfrak{B}_2$. (A special case of this remark was used in the proof of 7 of Th. 1.) If $\mathfrak{A}$ is a subalgebra of $\mathfrak{B}$ and s_1 is the inclusion mapping, then we call d an s_2-*derivation* and if s_2 is also the inclusion mapping, then we have a *derivation of* $\mathfrak{A}$ *into* $\mathfrak{B}$. One can verify directly, or use (49) to see, that if d is an (s_1, s_2)-derivation, then the kernel of d is a subalgebra. It follows that $d = 0$ if $\mathfrak{M}d = 0$ for a set of generators $\mathfrak{M}$ of $\mathfrak{A}$. Also if d_1 and d_2 are (s_1, s_2)-derivations then

$d_1 - d_2$ is an (s_1, s_2)-derivation. Hence it is clear that $d_1 = d_2$ if $xd_1 = xd_2$ for every x in a set of generators.

We consider again the algebra $(\mathfrak{A}, \mathfrak{U}, R)$ and we can prove the following result on derivations for this algebra.

PROPOSITION 3. *Let* $\mathfrak{D}$ *be an algebra,* s_1, s_2 *homomorphisms of* $(\mathfrak{A}, \mathfrak{U}, R)$ *into* $\mathfrak{D}$. *Let* d_1 *be an* (s_1, s_2)*-derivation of* $\mathfrak{A}$ *into* $\mathfrak{D}$ *and* d_2 *an* (s_1, s_2)*-derivation of* $\mathfrak{U}$ *into* $\mathfrak{D}$ *(*s_1, s_2 *are the restrictions to* $\mathfrak{A}, \mathfrak{U}$ *respectively of* s_1, s_2 *on* $(\mathfrak{A}, \mathfrak{U}, R)$.*) Suppose that for* $b \in \mathfrak{A}$, $l \in \mathfrak{L}$:

$$(bs_1)(ld_2) + (bd_1)(ls_2) - (ls_1)(bd_1) - (ld_2)(bs_2) = (bl^R)d_1 . \tag{50}$$

Then there exists a unique (s_1, s_2)*-derivation* d *of* $(\mathfrak{A}, \mathfrak{U}, R)$ *into* $\mathfrak{D}$ *such that* $bd = bd_1$, $ud = ud_2$.

Proof: Consider the mappings θ_1 and θ_2 of $\mathfrak{A}$ and $\mathfrak{U}$, respectively, such that

$$b\theta_1 = \begin{pmatrix} bs_1 & bd_1 \\ 0 & bs_2 \end{pmatrix}, \qquad u\theta_2 = \begin{pmatrix} us_1 & ud_2 \\ 0 & us_2 \end{pmatrix}.$$

Since d_i is an (s_1, s_2)-derivation, θ_1 and θ_2 are homomorphisms into the matrix algebra $\mathfrak{D}_2$. A direct calculation shows that the condition (46) is a consequence of (50). Hence Proposition 2 implies that there exists a homomorphism θ of $(\mathfrak{A}, \mathfrak{U}, R)$ into $\mathfrak{D}_2$ such that $b\theta = b\theta_1$, $u\theta = u\theta_2$, $a \in \mathfrak{A}$, $u \in \mathfrak{U}$. It is clear that θ has the form

$$\theta\colon\ x \to \begin{pmatrix} xs_1 & y \\ 0 & xs_2 \end{pmatrix}.$$

Set $y = xd$; then since θ is a homomorphism, d is an (s_1, s_2)-derivation of $(\mathfrak{A}, \mathfrak{U}, R)$ into $\mathfrak{D}$. We have $bd = bd_1$, $ud = ud_2$, as required. The uniqueness is clear.

The proof of Proposition 3 and the remark following Proposition 2 show that it is enough to suppose that (50) holds for all b in a set of generators $\mathfrak{M}$ of $\mathfrak{A}$. We shall need this sharper form of Proposition 3 later on.

Let $\mathfrak{M}$ be a vector space over Φ and let $E(\mathfrak{M})$ be the *exterior algebra* (or *Grassmann algebra*) over $\mathfrak{M}$. We recall that E is the difference algebra of the tensor algebra $\mathfrak{T} = \Phi 1 \oplus \mathfrak{M} \oplus (\mathfrak{M} \otimes \mathfrak{M}) \oplus \cdots$ with respect to the ideal generated by the the elements $x \otimes x$. It follows immediately that if s is a linear mapping of $\mathfrak{M}$ into an algebra $\mathfrak{B}$ and $(xs)^2 = 0$ for every $x \in \mathfrak{M}$, then s can be extended to a unique homomorphism of E into $\mathfrak{B}$. The canonical mapping of $\mathfrak{T}$ onto E is an isomorphism on $\Phi 1 \oplus \mathfrak{M}$. One identifies $\Phi 1 + \mathfrak{M}$ with its image. It is clear that $\mathfrak{M}$ generates E. We shall denote the multiplication in E simply as ab (rather than the more complicated $a \wedge b$ which is customary in differential geometry.) If $x, y \in \mathfrak{M}$, then $xy + yx = (x + y)^2 - x^2 - y^2 = 0$. The algebra E is graded with $\mathfrak{M}^m$ as the space E_m of homogeneous elements of degree m. If $\{u_j \mid j \in J\}$ is an ordered basis for $\mathfrak{M}$, then the set of monomials of the form $u_{j_1}u_{j_2} \cdots u_{j_m}$, $j_1 < j_2 < \cdots < j_m$ is a basis for E_m. In particular, if $\dim \mathfrak{M} = n$, then $E_m = 0$ if $m > n$ and $\dim E_m = \binom{n}{m}$ if $m \leqq n$. Hence $\dim E = 2^n$. The exterior algebra has an automorphism η such that $x\eta = -x$ for all $x \in \mathfrak{M}$. If $x_m \in E_m$ then $x_m\eta = (-1)^m x_m$.

We are interested in derivations and η-derivations of $E(\mathfrak{M})$ into algebras $\mathfrak{B}$ containing E as a subalgebra. The η-derivations will be called *anti-derivations* of E into $\mathfrak{B}$. We shall need the following criterion.

Proposition 4. *Let $\mathfrak{B}$ be an algebra which contains the exterior algebra $E = E(\mathfrak{M})$ as a subalgebra. Let d be a linear mapping of $\mathfrak{M}$ into $\mathfrak{B}$. Then d can be extended to a derivation (anti-derivation) of E into $\mathfrak{B}$ if and only if $x(xd) + (xd)x = 0$. $(x(xd) - (xd)x = 0)$ for all $x \in \mathfrak{M}$.*

Proof: The conditions are necessary since $x^2 = 0$ in E implies $x(xd) + (xd)x = 0$ or $x(xd) - (xd)x = 0$ according as d is a derivation or an anti-derivation. Conversely, suppose the conditions hold. In the first case we consider the mapping

$$\theta\colon\ x \to \begin{pmatrix} x & xd \\ 0 & x \end{pmatrix}$$

and in the second, the mapping

$$\theta\colon\ x \to \begin{pmatrix} x & xd \\ 0 & -x \end{pmatrix}$$

of $\mathfrak{M}$ into the matrix algebra $\mathfrak{B}_2$. In both cases one checks that $(x\theta)^2 = 0$. Hence θ can be extended to a homomorphism of E. The rest of the argument is like that of Proposition 3.

It is clear that when the condition is satisfied the extension is unique.

We apply this first to the following situation. Let $\mathfrak{L}$ be a Lie algebra and $\mathfrak{M}$ a module for $\mathfrak{L}$. Let $E = E(\mathfrak{M})$ be the exterior algebra defined by $\mathfrak{M}$. If $x \in \mathfrak{M}$, $l \in \mathfrak{L}$, $xl \in \mathfrak{M}$ so that $x(xl) + (xl)x = 0$ in E. Hence there exists a unique derivation d_l in E such that $xd_l = xl$, $x \in \mathfrak{M}$. We have $xd_{l_1+l_2} = xd_{l_1} + xd_{l_2}$, $xd_{\alpha l} = \alpha xd_l$, $xd_{[l_1l_2]} = x[d_{l_1}d_{l_2}]$ for $x \in \mathfrak{M}$, $\alpha \in \Phi$, $l, l_1, l_2, \in \mathfrak{L}$, since $\mathfrak{M}$ is an $\mathfrak{L}$-module. Since all the mappings considered here are derivations and $\mathfrak{M}$ generates E, we have $d_{l_1+l_2} = d_{l_1} + d_{l_2}$, $d_{\alpha l} = \alpha d_l$, $d_{[l_1l_2]} = [d_{l_1}d_{l_2}]$ in E. Consequently, $l \to d_l$ is a representation of $\mathfrak{L}$ by derivations in $E(\mathfrak{M})$.

We consider the special case of this in which $\mathfrak{M} = \mathfrak{L}$ and the representation is the adjoint representation. We form the algebra $X = (E(\mathfrak{L}), \mathfrak{U}, A)$ where A denotes the extension of the adjoint representation to a representation by derivations in $E(\mathfrak{L})$. There is a slight difficulty here in our notations since we now have two copies of $\mathfrak{L}$, one contained in $\mathfrak{U}$, the other in E. We shall therefore denote the elements of the copy in $\mathfrak{U}$ by $\bar{l}$, $l \in \mathfrak{L}$ and we write $\overline{\mathfrak{L}} = \{\bar{l}\}$. We denote the corresponding elements in E by l and we denote this copy as $\mathfrak{L}$. In order to avoid ambiguity we shall denote the Lie composition in $\mathfrak{L}$—which does not coincide with the Lie product in E_L—by $[l_1 \circ l_2]$. Then we have the Lie isomorphism $l \to \bar{l}$ of $\mathfrak{L}$ onto $\overline{\mathfrak{L}}$; hence we have $[\overline{l_1 \circ l_2}] = [\bar{l}_1\bar{l}_2] = \bar{l}_1\bar{l}_2 - \bar{l}_2\bar{l}_1$. Since $\overline{\mathfrak{L}}$ and $\mathfrak{L}$ generate $\mathfrak{U}$ and E, respectively, $\mathfrak{L} \cup \overline{\mathfrak{L}}$ generates X. The following relations connecting these generators are decisive:

$$
(51) \qquad \begin{aligned} &[l_1, \bar{l}_2] = [\overline{l_1 \circ l_2}]\,, \qquad l^2 = 0 \\ &[l_1, \bar{l}_2] = [l_1 \circ l_2]\,, \end{aligned}
$$

$l_i \in \mathfrak{L}$. The last of these is a consequence of (45).

The last relation in (51) implies that if $x_i \in E_i$, then $[x_i, \bar{l}] \in E_i$. It follows easily from this that $\overline{\mathfrak{L}}^{(j)} E_i = E_i \overline{\mathfrak{L}}^{(j)}$ for $\overline{\mathfrak{L}}^{(j)} = \Phi 1 + \overline{\mathfrak{L}} + \cdots + \overline{\mathfrak{L}}^j$ in $\mathfrak{U}$. Hence $\mathfrak{U}E_i = E_i\mathfrak{U}$. If we set $X_i = E_i\mathfrak{U}$, then we have $X_iX_j = E_i\mathfrak{U}E_j\mathfrak{U} \subseteq E_{i+j}\mathfrak{U} = X_{i+j}$. Also since $X = E\mathfrak{U} \cong E \otimes \mathfrak{U}$ it follows from the properties of tensor products that $X = \sum \oplus X_i = \sum \oplus E_i\mathfrak{U}$. This shows that X is a graded algebra with $X_i = E_i\mathfrak{U}$ as the space of homogeneous elements of degree i.

PROPOSITION 5. *There exists a unique automorphism η of $X = (E(\mathfrak{L}), \mathfrak{U}, A)$ and a unique η-derivation d of X into X such that*

$$
(52) \qquad \begin{aligned} &l\eta = -l\,, \qquad \bar{l}\eta = \bar{l} \\ &ld = \bar{l}\,, \qquad \bar{l}d = 0\,, \qquad l \in \mathfrak{L}\,. \end{aligned}
$$

Moreover, $\eta d + d\eta = 0$ and $d^2 = 0$.

Proof: Evidently we have a homomorphism η_1 of E into X such that $l\eta_1 = -l$, $l \in \mathfrak{L}$ and a homomorphism η_2 of $\mathfrak{U}$ into X such that $\bar{l}\eta = \bar{l}$. The condition (46) for $b = l_1$, $l = \bar{l}_2$, $\theta_1 = \eta_1$, $\theta_2 = \eta_2$ reads $[l_1\eta_1, \bar{l}_2\eta_2] = [l_1 \circ l_2]\eta_1$ which is equivalent to $[-l_1, \bar{l}_2] = -[l_1 \circ l_2]$. This holds by (51). Hence Proposition 2 can be applied to prove the existence and uniqueness of η. It is evident that the mapping $d_2 = 0$ is an η-derivation of $\mathfrak{U}$ into X. Since $l\bar{l} - \bar{l}l = [l\bar{l}] = [l \circ l] = 0$ the condition of Proposition 4 is satisfied and so there exists an anti-derivation d_1 of $E(\mathfrak{L})$ into X such that $ld_1 = \bar{l}$, $l \in \mathfrak{L}$. We now check (50) for $b = l_1$, $l = \bar{l}_2$, $s_1 = 1$, $s_2 = \eta$, $d_1 = d_1$, $d_2 = d_2$. The left-hand side is

$$l_1 0 + \bar{l}_1\bar{l}_2 - \bar{l}_2\bar{l}_1 - 0(-l_1) = [\bar{l}_1\bar{l}_2]\ .$$

The right-hand side is $[l_1\bar{l}_2]d_1 = [l_1 \circ l_2]d_1 = \overline{[l_1 \circ l_2]} = [\bar{l}_1\bar{l}_2]$. Hence (50) holds for these elements, so by the remark following Proposition 3, there exists a unique η-derivation d in X satisfying the conditions (52). We have $X_0 d = 0$ and $X_i d \subseteq X_{i-1}$ where $X_i = E_i\mathfrak{U}$, $i \geqq 1$. This can be established by induction since $X_i = X_{i-1}X_1$. Also it is immediate that $x_i\eta = (-1)^i x_i$ if $x_i \in X_i$. Hence $x_i\eta d = (-1)^i x_i d$ and $x_i d\eta = (-1)^{i-1} x_i d$. Hence $\eta d + d\eta = 0$. If $x, y \in X$, $(xy)d^2 = (x(yd) + (xd)(y\eta))d = x(yd^2) + (xd)(yd\eta) + (xd)(y\eta d) + (xd^2)(y\eta^2) = x(yd^2) + (xd^2)(y\eta^2)$. Since $\eta^2 = 1$ this shows that d^2 is a derivation. On the other hand, (52) implies that $ld^2 = 0 = \bar{l}d^2$. Hence $d^2 = 0$. This concludes the proof.

Let d_{i-1}, $i = 1, 2, \cdots$, denote the restriction of d to X_i. Then d_{i-1} maps X_i into X_{i-1} and $d_i d_{i-1} = 0$. Since $\mathfrak{U}d = 0$, d and the d_i commute with the right multiplications by elements of $\mathfrak{U}$. Hence if we consider X_i in this way as a $\mathfrak{U}$-module, then d_i is a $\mathfrak{U}$-homomorphism. We have $X_i = E_i\mathfrak{U} \cong E_i \otimes \mathfrak{U}$ and $X = \sum \oplus X_i$, by the properties of tensor products. Hence any Φ-basis for E_i is a set of free generators for X_i relative to $\mathfrak{U}$. Thus every X_i is a free $\mathfrak{U}$-module. We have $X_0 = \mathfrak{U}$. If $\mathfrak{U}'$ denotes the ideal in $\mathfrak{U}$ generated by the $\bar{l}$, $l \in \mathfrak{L}$, then $\mathfrak{U}/\mathfrak{U}' \cong \Phi$. If $\{u_j \mid j \in J\}$ is an ordered basis for $\mathfrak{L}$, then we know that the elements 1, $\bar{u}_{i_1}\bar{u}_{i_2}\cdots\bar{u}_{i_r}$, $i_1 \leqq i_2 \leqq \cdots \leqq i_r$, form a basis for $\mathfrak{U}$. If $\bar{a} = \alpha 1 + \sum \alpha_{i_1\cdots i_r}\bar{u}_{i_1}\bar{u}_{i_2}\cdots\bar{u}_{i_r}$, then we define a mapping ε of $\mathfrak{U}$ into Φ by $\bar{a}\varepsilon = \alpha$. This mapping is a $\mathfrak{U}$-homomorphism of $X_0 = \mathfrak{U}$ into Φ considered as $\mathfrak{U}$-module where $\alpha\mathfrak{U}' = 0$, $\alpha 1 = \alpha$, $\alpha \in \Phi$. Evidently ε is surjective.

Let $b \in X_1$. Then $b = \sum u_j\bar{v}_j$, $\{u_j\}$ the basis for $\mathfrak{L}$, $\bar{v}_j \in \mathfrak{U}$. Then $bd_0 = bd = \sum \bar{u}_j\bar{v}_j \in \mathfrak{U}'$ and $bd_0\varepsilon = 0$. Conversely, let $\bar{a} \in X_0$ and assume $\bar{a}\varepsilon = 0$. Then $\bar{a} = \sum \alpha_{i_1 \cdots i_r} \bar{u}_{i_1} \cdots \bar{u}_{i_r} \in \mathfrak{U}'$ and $\bar{a} = bd_0$ for $b = \sum u_{i_1}(\sum \alpha_{i_1} \cdots {}_{i_r} \bar{u}_{i_2} \cdots \bar{u}_{i_r})$. We have therefore shown that $0 \longleftarrow \Phi \underset{\varepsilon}{\longleftarrow} X_0 \underset{d_0}{\longleftarrow} X_1$ is exact. It remains to show that $\operatorname{Ker} d_{i-1} = \operatorname{Im} d_i$, $i \geqq 1$. For this purpose we consider first the case in which $\mathfrak{L}$ is abelian; hence $\mathfrak{U}$ is the polynomial algebra in the $\bar{u}_j$ which commute and are algebraically independent. Also in the abelian case $[m\bar{l}] = [m \circ l] = 0$, $m, l \in \mathfrak{L}$. Hence the elements of $\mathfrak{U}$ commute with those of $E(\mathfrak{L})$. Consequently, X is the tensor product of E and $\mathfrak{U}$ in the sense of algebras.

Now let $\mathfrak{L} = \mathfrak{M} \oplus \mathfrak{N}$ where $\mathfrak{M}, \mathfrak{N}$ are subspaces, hence subalgebras of the abelian Lie algebra $\mathfrak{L}$. Let $F = E(\mathfrak{M})$, $\mathfrak{V} = \mathfrak{U}(\mathfrak{M})$, $Y = X(\mathfrak{M})$, $G = E(\mathfrak{N})$, $\mathfrak{W} = \mathfrak{U}(\mathfrak{N})$, $Z = X(\mathfrak{N})$. It is easy to see that F can be identified with the subalgebra of E generated by $\mathfrak{M}$, $\mathfrak{V}$ with the subalgebra of $\mathfrak{U}$ generated by $\bar{\mathfrak{M}} = \{\bar{m} \in \mathfrak{M}\}$ and Y with the subalgebra of X generated by $\mathfrak{M} + \bar{\mathfrak{M}}$. Similar statements hold for $G, \mathfrak{W}, Z$. These results follow easily by looking at bases. Similarly, we leave it to the reader to check that $X = YZ$ where Y and Z are the subalgebras we have indicated and that we have a vector space isomorphism of $Y \otimes Z$ onto X sending $y \otimes z$ into yz, $y \in Y$, $z \in Z$. This result implies that if μ is a linear mapping in Y and ν is a linear mapping in Z then we have a unique linear mapping λ in X such that $(yz)\lambda = (y\mu)(z\nu)$.

We now regard the augmentation ε of X_0 into Φ as a mapping of X_0 into the subalgebra $\Phi 1$ of X and we extend this to a linear mapping in X such that $X_i\varepsilon = 0$, $i \geqq 1$. This is an algebra homomorphism of X onto the subalgebra Φ. We now prove

PROPOSITION 6. *There exists a linear mapping D in $X = X(\mathfrak{L})$, $\mathfrak{L}$ abelian, such that $Dd + dD = 1 - \varepsilon$.*

Proof: Suppose first that $\dim \mathfrak{L} = 1$ and (u) is a basis for $\mathfrak{L}$. Then X has the basis $1, \bar{u}, \bar{u}^2, \bar{u}^3, \cdots; u, u\bar{u}, u\bar{u}^2, \cdots$. We have $\bar{u}^i d = 0$, $u\bar{u}^i d = \bar{u}^{i-1}$, $i \geqq 0$; $1\varepsilon = 1$, $\bar{u}^{i-1}\varepsilon = 0$, $u\bar{u}^i\varepsilon = 0$. Hence if we define D to be the linear mapping such that $1D = 0$, $\bar{u}^{i+1}D = u\bar{u}^i$, $u\bar{u}^i D = 0$, $i \geqq 0$, then one checks that $dD + Dd = 1 - \varepsilon$ as required. We note also that $\eta D + D\eta = 0$ if η is the automorphism previously defined in X ($\bar{u}^i\eta = \bar{u}^i$, $u\bar{u}^i\eta = -u\bar{u}^i$). Moreover, we have the relations $\varepsilon d = 0 = d\varepsilon$ and $\varepsilon\eta = \varepsilon = \eta\varepsilon$. Now suppose $\mathfrak{L} = \mathfrak{M} \oplus \mathfrak{N}$ where $\mathfrak{N}$ is one-dimensional. Then $X = YZ$, $Y = X(\mathfrak{M})$, $Z = X(\mathfrak{N})$ where Y and Z are the algebras defined before. It is clear that Y and Z are invariant under d, ε and η, and that their restrictions are just the corresponding mappings in $X(\mathfrak{M})$ and $X(\mathfrak{N})$. Let D_2 be the mapping just defined in Z and let D_1 be any mapping in Y such that $dD_1 + D_1 d = 1 - \varepsilon$ in Y. There exists a unique linear mapping D in X such that

$$(53) \qquad (yz)D = y(zD_2) + (yD_1)(z\varepsilon) , \qquad y \in Y , \qquad z \in Z .$$

Then we have

$$\begin{aligned}(yz)Dd &= y(zD_2d) + (yd)(zD_2\eta) + (yD_1)(z\varepsilon d) + (yD_1d)(z\varepsilon\eta)\\ &= y(zD_2d) + (yd)(zD_2\eta) + (yD_1d)(z\varepsilon)\\ (yz)dD &= (y(zd) + (yd)(z\eta))D\\ &= y(zdD_2) + (yD_1)(zd\varepsilon) + (yd)(z\eta D_2) + (ydD_1)(z\eta\varepsilon)\\ &= y(zdD_2) + (yd)(z\eta D_2) + (ydD_1)(z\varepsilon) .\end{aligned}$$

Hence,

$$\begin{aligned}(yz)(Dd + dD) &= y(z(1-\varepsilon)) + (y(1-\varepsilon))(z\varepsilon)\\ &= yz - y(z\varepsilon) + y(z\varepsilon) - (y\varepsilon)(z\varepsilon)\\ &= yz(1-\varepsilon) .\end{aligned}$$

The inductive step we have just established implies the result in the finite-dimensional case by ordinary induction and in the infinite-dimensional case by transfinite induction or by Zorn's lemma.

This proposition implies the exactness of $X_0 \xleftarrow[d_0]{} X_1 \xleftarrow[d_1]{} X_2 \longleftarrow \cdots$. Thus let $x \in \sum_{i \geqq 1} X_i$ satisfy $xd = 0$. Then $x = x(1-\varepsilon) = x(dD + Dd) = (xD)d \in \operatorname{Im} d$. This implies that $\operatorname{Ker} d_{i-1} = \operatorname{Im} d_i$, $i \geqq 1$.

We consider now the general case of an arbitrary Lie algebra $\mathfrak{L}$ and we introduce a filtration in X using the space of generators $\mathfrak{N} = \mathfrak{L} + \bar{\mathfrak{L}}$. Thus we define $X^{(j)} = \Phi 1 + \mathfrak{N} + \mathfrak{N}^2 + \cdots + \mathfrak{N}^j = \sum_{h+k \leqq j} E_h \mathfrak{L}^k$. Since $(E_h\mathfrak{L}^k)d \subseteq E_{h-1}\mathfrak{L}^{k+1}$, $X^{(j)}$ is a subcomplex. Also it is clear that $X^{(j)}$ is a sum of homogeneous subspaces relative to the grading in X. We can form the difference complex $X^{(j)}/X^{(j-1)}$ which has a grading induced by that of X. If $\{u_j \mid j \in J\}$ is an ordered basis for $\mathfrak{L}$ then the cosets relative to $X^{(j-1)}$ of the base elements $u_{j_1} \cdots u_{j_h}\bar{u}_{i_1} \cdots \bar{u}_{i_k}$, $j_1 < j_2 < \cdots < j_h$, $i_1 \leqq i_2 \leqq \cdots \leqq i_k$, $h + k = j$ form a basis for $X^{(j)}/X^{(j-1)}$. We identify these cosets with the base elements. Then we can say that d in $X^{(j)}/X^{(j-1)}$ is determined by the rule

$$(54) \qquad \begin{aligned}&(u_{j_1} \cdots u_{j_h}\bar{u}_{i_1} \cdots \bar{u}_{i_k})d\\ &\quad = \sum (-1)^{h-q} u_{j_1} \cdots \hat{u}_{j_q} \cdots u_{j_h}\bar{u}_{i_1} \cdots \bar{u}_{j_q} \cdots \bar{u}_{i_k}\end{aligned}$$

where the $\hat{}$ indicates that u_{j_q} is omitted, and the order of the subscripts in the product of the $\bar{u}$'s is non-decreasing. This differentiation does not make use of the structure of $\mathfrak{L}$. Thus it is the same as that which one obtains in the abelian Lie algebra. In the latter case the space spanned by the monomials on the left-hand side of (54) satisfying $h + k = j$ fixed form a subcomplex of X and X is a direct sum of these subcomplexes for $j = 0, 1, \cdots$. It follows that if x is in one of these subcomplexes and is in X_{i-1}, $i \geqq 1$, and $xd = 0$, then $x = yd$, $y \in X_i$ and where y is the given complex containing x. The result we wished to prove on the exactness of (41) will now follow by induction on the index j in $X^{(j)}$ by the following

LEMMA 3. *Let Y be a subcomplex of a graded complex $X = \sum_{i=0}^{\infty} \oplus X_i$ such that $Y = \sum_{i=0}^{\infty}(Y \cap X_i)$. Set $Z = X/Y = \sum_{i=0}^{\infty} \oplus Z_i$, $Z_i \cong X_i/(Y \cap X_i)$. Suppose $\text{Ker}\, d_{i-1} = \text{Im}\, d_i$, $i \geqq 1$, holds in Y and in Z. Then this holds also in X.*

Proof: Let $x \in \sum_{i\geqq 1} X_i$ satisfy $xd = 0$. Then $(x + Y)d = 0$. Hence there exists $x' + Y$ such that $(x' + Y)d = x + Y$. Thus $x'd = x + y$, $y \in Y$. Then $yd = x'd^2 - xd = 0$ and so there exists a $y' \in Y$ such that $y = y'd$. Hence $x = x'd - y = x'd - y'd = (x' - y')d$.

We have now completed the proof of the following

THEOREM 10. *Let $X = (E(\mathfrak{L}), \mathfrak{U}, A)$ be the algebra of differential operators determined by the exterior algebra of $E(\mathfrak{L})$ and the extension of the adjoint representation of $\mathfrak{L}$ to E. Let d be the antiderivation in X defined in Proposition 5 and let $X_i = E_i\mathfrak{U}$. Then $\Phi \xleftarrow[\varepsilon]{} X_0 \xleftarrow[d_0]{} X_1 \longleftarrow \cdots$ is a free resolution of the module Φ.*

It remains to show that the cohomology groups defined by this resolution coincide with those of § 3.10. Let there be a $\mathfrak{U}$-homomorphism φ of X_i into a (right) $\mathfrak{U}$-module $\mathfrak{M}$. Let $(l_1, l_2, \cdots, l_i)$ be an ordered set of i elements of $\mathfrak{L}$ and define a mapping f of $\mathfrak{L} \times \cdots \times \mathfrak{L}$ (i times) into $\mathfrak{M}$ by $f(l_1, \cdots, l_i) = (l_1 l_2 \cdots l_i)\varphi$ where $l_1 l_2 \cdots l_i \in E_i \subseteqq X_i$. It is clear that this f is multilinear and alternate and it is easy to see that if f is any multilinear alternate mapping of $\mathfrak{L} \times \cdots \times \mathfrak{L}$ into $\mathfrak{M}$ then there exists a linear mapping φ of $E_i(\mathfrak{L})$ into $\mathfrak{M}$ such that $(l_1 l_2 \cdots l_i)\varphi = f(l_1, \cdots, l_i)$. This φ has a unique extension to a $\mathfrak{U}$-homomorphism of X_i into $\mathfrak{M}$. We therefore have a $1:1$ linear isomorphism of the space of $\mathfrak{U}$-homomorphisms of X_i into $\mathfrak{M}$ and the space of multilinear alternate mappings of the i-fold product set $\mathfrak{L} \times \cdots \times \mathfrak{L}$ into $\mathfrak{M}$. If f is a mapping of the latter type we define $f\delta$ by

$$(f\delta)(l_1, \cdots, l_{i+1}) = ((l_1 \cdots l_{i+1})d)\varphi \,. \tag{55}$$

We have

$$(l_1 \cdots l_{i+1})d = \sum_{q=1}^{i+1} (-1)^{i+1-q} l_1 \cdots \bar{l}_q \cdots l_{i+1} \,,$$

and, by (45),

$$\begin{aligned}\bar{l}_q l_{q+1} \cdots l_{i+1} &= l_{q+1} \cdots l_{i+1}\bar{l}_q - \sum_{r>q} l_{q+1} \cdots [l_r l_q] \cdots l_{i+1} \\ &= l_{q+1} \cdots l_{i+1}\bar{l}_q + \sum_{r>q} l_{q+1} \cdots [l_q l_r] \cdots l_{i+1} \\ &= l_{q+1} \cdots l_{i+1}\bar{l}_q + \sum_{r>q} (-1)^{i+1-r} l_{q+1} \cdots \hat{l}_r \cdots l_{i+1}[l_q l_r] \,.\end{aligned}$$

Hence

$$\begin{aligned}(l_1 \cdots l_{i+1})d = &\sum_{q=1}^{i+1} (-1)^{i+1-q} l_1 \cdots \hat{l}_q \cdots l_{i+1}\bar{l}_q \\ &+ \sum_{q<r=1}^{i+1} (-1)^{r+q} l_1 \cdots \hat{l}_q \cdots \hat{l}_r \cdots l_{i+1}[l_q l_r] \,.\end{aligned}$$

It follows that

$$\begin{aligned}(f\delta)(l_1, \cdots l_{i+1}) &= \sum_{q=1}^{i+1} (-1)^{i+1-q} f(l_1, \cdots, \hat{l}_q, \cdots l_{i+1})\, l_q \\ &= \sum_{q<r=1}^{i+1} (-1)^{r+q} f(l_1, \cdots, \hat{l}_q, \cdots, \hat{l}_r, \cdots, l_{i+1}, [l_q l_r]) \,,\end{aligned} \tag{56}$$

which are the same definitions as given in § 3.10.

7. Restricted Lie algebras of characteristic p

In many connections in which Lie algebras arise naturally one encounters in the characteristic $p \neq 0$ case structures which are somewhat richer than ordinary Lie algebras. For example, let $\mathfrak{A}$ be an arbitrary non-associative algebra and let $\mathfrak{D}(\mathfrak{A})$ be the set of derivations of $\mathfrak{A}$. Then we know that $\mathfrak{D}(\mathfrak{A})$ is a subalgebra of the Lie algebra of linear transformations in $\mathfrak{A}$. We note also that one has the Leibniz formula

$$(ab)D^k = \sum_{i=0}^{k} \binom{k}{i} (aD^i)(bD^{k-i}) \tag{57}$$

for any derivation D. This can be established by induction on k. Now assume the base field of $\mathfrak{A}$, hence of $\mathfrak{D}(\mathfrak{A})$ is of characteristic p and take $k = p$ in (57). Then the binomial coefficients $\binom{p}{i} = 0$ if $1 \leqq i \leqq p-1$. Hence (57) reduces to

(58) $$(ab)D^p = (aD^p)b + a(bD^p) ,$$

which implies that D^p is a derivation. Thus $\mathfrak{D}(\mathfrak{A})$ is closed under the mapping $D \to D^p$ as well as the Lie algebra compositions. Similarly, let $\mathfrak{A}$ be an associative algebra and $a \to \bar{a}$ an anti-automorphism in $\mathfrak{A}$. Let $\mathfrak{L}$ be the subset of skew elements relative to $a \to \bar{a}$. Then we know that $\mathfrak{L}$ is a subalgebra of $\mathfrak{A}_L$. Moreover, if the characteristic is p then $\bar{a} = -a$ implies that $\overline{a^p} = \bar{a}^p = (-a)^p = -a^p$. Hence $a^p \in \mathfrak{L}$, so again we have a Lie algebra closed under the p-mapping. The systems $\mathfrak{D}(\mathfrak{A})$ and $\mathfrak{L}$ just considered are examples of restricted Lie algebras which we shall define abstractly.

For this purpose we need some relations connecting a^p with the compositions in a Lie algebra $\mathfrak{A}_L$, $\mathfrak{A}$ associative of characteristic p. We recall first the following two identities in $\phi[\lambda, \mu]$, λ, μ algebraically independent indeterminates, Φ of characteristic p:

(59) $$\begin{aligned} (\lambda - \mu)^p &= \lambda^p - \mu^p \\ (\lambda - \mu)^{p-1} &= \sum_{i=0}^{p-1} \lambda^i \mu^{p-1-i} . \end{aligned}$$

The first of these is well-known and the second is a consequence of the first and the identity $\lambda^p - \mu^p = (\lambda - \mu)(\sum_{i=0}^{p-1} \lambda^i \mu^{p-1-i})$. These relations imply corresponding relations for commuting elements a, b in any associative algebra. In particular we may take $a = b_R$, $b = b_L$ the right and left multiplications determined by an element $b \in \mathfrak{A}$. These give

$$\begin{aligned} (b_R - b_L)^p &= b_R^p - b_L^p = (b^p)_R - (b^p)_L \\ (b_R - b_L)^{p-1} &= \sum_{i=0}^{p-1} b_R^i b_L^{p-1-i} = \sum_{i=0}^{p-1} (b^i)_R (b^{p-1-i})_L , \end{aligned}$$

or,

(60) $$[\cdots \overbrace{[ab]b] \cdots b}^{p}] = [ab^p] ,$$

(61) $$[\cdots \overbrace{[ab]b] \cdots b}^{p-1}] = \sum_{i=0}^{p-1} b^{p-1-i} a b^i .$$

Also it is clear that

$$(\alpha a)^p = \alpha^p a^p \, , \tag{62}$$

and we shall use (61) to prove that

$$(a+b)^p = a^p + b^p + \sum_{i=1}^{p-1} s_i(a, b) \, , \tag{63}$$

where $is_i(a, b)$ is the coefficient of λ^{i-1} in

$$a(\mathrm{ad}\,(\lambda a + b))^{p-1} \, , \tag{64}$$

λ an indeterminate. To prove this we introduce the polynomial ring $\mathfrak{A}[\lambda]$ and we write

$$(\lambda a + b)^p = \lambda^p a^p + b^p + \sum_{i=1}^{p-1} s_i(a, b)\lambda^i \, , \tag{65}$$

where $s_i(a, b)$ is a polynomial in a, b of total degree p. If we differentiate (65) with respect to λ we obtain

$$\sum_{i=0}^{p-1} (\lambda a + b)^i a(\lambda a + b)^{p-i-1} = \sum_{i=1}^{p-1} is_i(a, b)\lambda^{i-1} \, .$$

By (61), this gives

$$a(\mathrm{ad}\,(\lambda a + b))^{p-1} = \sum_{i=1}^{p-1} is_i(a, b)\lambda^{i-1} \, . \tag{66}$$

Thus we see that $is_i(a, b)$ is the coefficient of λ^{i-1} in $a(\mathrm{ad}\,(\lambda a + b))^{p-1}$. On the other hand, substitution of $\lambda = 1$ in (65) gives the relation (63). It is clear that $s_i(a, b)$ is obtained by applying addition and commutation to a, b and so is in the Lie subalgebra of $\mathfrak{A}_L$ generated by a, b. For example,

$$s_1(a, b) = [ab] \quad \text{if} \quad p = 2 \, ;$$

$$s_1(a, b) = [[ab]b] \, , \quad 2s_2(a, b) = [[ab]a] \quad \text{if} \quad p = 3 \, ,$$

$$s_1(a, b) = [[[[ab]b]b]b] \, ,$$

$$2s_2(a, b) = [[[[ab]a]b]b] + [[[[ab]b]a]b] + [[[[ab]b]b]a] \, ,$$

$$3s_3(a, b) = [[[[ab]a]a]b] + [[[[ab]a]b]a] + [[[[ab]b]a]a] \, ,$$

$$4s_4(a, b) = [[[[ab]a]a]a] \quad \text{if} \quad p = 5 \, .$$

These considerations lead to the following

Definition 4. A *restricted Lie algebra* $\mathfrak{L}$ of characteristic $p \neq 0$ is a Lie algebra of characteristic p in which there is defined a mapping $a \to a^{[p]}$ such that

(i) $$(\alpha a)^{[p]} = \alpha^p a^{[p]}$$

(ii) $$(a + b)^{[p]} = a^{[p]} + b^{[p]} + \sum_{i=1}^{p-1} s_i(a, b) ,$$

where $is_i(a, b)$ is the coefficient of λ^{i-1} in $a(\operatorname{ad}(\lambda a + b))^{p-1}$ and

(iii) $$[ab^{[p]}] = a(\operatorname{ad} b)^p .$$

If $\mathfrak{A}$ is an associative algebra of characteristic p, then the foregoing discussion shows that $\mathfrak{A}$ defines a restricted Lie algebra in which the vector space compositions are as in $\mathfrak{A}$, $[ab] = ab - ba$ and $a^{[p]} = a^p$. We use the notation $\mathfrak{A}_L$ for this restricted Lie algebra (as well as for the ordinary Lie algebra). A *homomorphism* S of a restricted Lie algebra into a second restricted Lie algebra is, by definition, a mapping satisfying $(a + b)^S = a^S + b^S$, $(\alpha a)^S = \alpha a^S$, $[ab]^S = [a^S b^S]$, $(a^{[p]})^S = (a^S)^{[p]}$. Ideals and subalgebras are defined in the obvious way. A *representation* of $\mathfrak{L}$ is a homomorphism of $\mathfrak{L}$ into the restricted Lie algebra $\mathfrak{E}_L$, $\mathfrak{E}$, the algebra of linear transformations of a vector space $\mathfrak{M}$ over Φ. If R is a representation acting in the space $\mathfrak{M}$, then $\mathfrak{M}$ is an $\mathfrak{L}$-module relative to $xa \equiv xa^R$, $x \in \mathfrak{M}$, $a \in \mathfrak{L}$. The module product xa satisfies the usual conditions in a Lie module and the additional condition that $xa^{[p]} = (\cdots (xa)a) \cdots a$, p a's.

We consider now the following two basic questions: (1) Does every restricted Lie algebra have a 1 : 1 representation? (2) What are the conditions that an ordinary Lie algebra be restricted relative to a suitable definition of $a^{[p]}$? In connection with (2) it is clear that a necessary condition is that for every $a \in \mathfrak{L}$ the derivation $(\operatorname{ad} a)^p$ is inner; for, in a restricted Lie algebra $(\operatorname{ad} a)^p = \operatorname{ad} a^{[p]}$. We shall show that this condition is sufficient. In fact, we shall see that it will be enough to have $(\operatorname{ad} u_i)^p$ inner for every u_i in a basis $\{u_i\}$ of $\mathfrak{L}$. We note also that if $\mathfrak{L}$ is restricted relative to two p-mappings $a \to a^{[p]_1}$ and $a \to a^{[p]_2}$, then $f(a) = a^{[p]_1} - a^{[p]_2}$ is in the center of $\mathfrak{L}$ since $[b, f(a)] = [ba^{[p]_1}] - [ba^{[p]_2}] = b(\operatorname{ad} a)^p - b(\operatorname{ad} a)^p = 0$. It is clear from (ii) and (i) that

(67) $$f(a + b) = f(a) + f(b) , \qquad f(\alpha a) = \alpha^p f(a) .$$

A mapping of one vector space of characteristic $p \neq 0$ into a second one having these properties is called a *p-semi-linear mapping*. Conversely, if $\mathfrak{L}$ is restricted relative to $a \to a^{[p]_1}$ and $a \to f(a)$ is a

p-semi-linear mapping of $\mathfrak{L}$ into the center $\mathfrak{C}$ of $\mathfrak{L}$, then $\mathfrak{L}$ is also restricted relative to $a \to a^{[p]_2} \equiv a^{[p]_1} + f(a)$. The kernel of a p-semi-linear mapping is a subspace. Hence if $f(u_i) = 0$ for u_i in a basis, then $f = 0$. It follows that if two p-mappings $a \to a^{[p]_1}$ and $a \to a^{[p]_2}$ making $\mathfrak{L}$ restricted coincide on a basis, then they are identical.

Suppose now that $\mathfrak{L}$ is a Lie algebra with the basis $\{u_i \mid i \in I\}$ where I is an ordered set and let $\mathfrak{U}$ be the universal enveloping algebra of $\mathfrak{L}$. The Poincaré-Birkhoff-Witt theorem states that the "standard monomials" $u_{i_1}^{k_1} u_{i_2}^{k_2} \cdots u_{i_r}^{k_r}$, $i_1 < i_2 < \cdots < i_r$, $k_j \geqq 0$ form a basis for $\mathfrak{U}$. We have the filtration of $\mathfrak{U}$ defined by $\mathfrak{U}^{(k)} = \Phi 1 + \mathfrak{L} + \mathfrak{L}^2 + \cdots + \mathfrak{L}^k$. It is easy to see (induction on k and the usual "straightening" argument) that the monomials $u_{i_1}^{k_1} u_{i_2}^{k_2} \cdots u_{i_r}^{k_r}$ such that $k_1 + k_2 + \cdots + k_r \leqq k$ form a basis for $\mathfrak{U}^{(k)}$. We assume now that for each base element u_i there exists a positive integer n_i and an element z_i in the center $\mathfrak{C}$ of $\mathfrak{U}$ such that

$$v_i = u_i^{n_i} - z_i \tag{68}$$

is in $\mathfrak{U}^{(n_i - 1)}$. Then we have the following

Lemma 4. *The elements of the form*

$$z_{i_1}^{h_1} z_{i_2}^{h_2} \cdots z_{i_r}^{h_r} u_{i_1}^{\lambda_1} u_{i_2}^{\lambda_2} \cdots u_{i_r}^{\lambda_r} \tag{69}$$

such that $i_1 < i_2 < \cdots < i_r$, $h_j \geqq 0$, $0 \leqq \lambda_j < n_{i_j}$, $r = 1, 2, 3, \cdots$ *form a basis for* $\mathfrak{U}$.

Proof: We show first that every element $u_{i_1}^{k_1} \cdots u_{i_r}^{k_r}$, $k_j \geqq 0$ is a linear combination of the elements (69). If $k = k_1 + k_2 + \cdots + k_r$, then we employ an induction on k. If every $l_j < n_{i_j}$, then the result is clear. Hence we may assume $k_j \geqq n_{i_j}$ for some j. Then we may replace $u_{i_j}^{n_{i_j}}$ by $v_{i_j} + z_{i_j}$ and obtain

$$u_{i_1}^{k_1} \cdots u_{i_r}^{k_r} = z_{i_j} u_{i_1}^{k_1} \cdots u_{i_j}^{k_j - n_{i_j}} \cdots u_{i_r}^{k_r} + u_{i_1}^{k_1} \cdots u_{i_j}^{k_j - n_{i_j}} v_{i_j} \cdots u_{i_r}^{k_r}.$$

The elements $u_{i_1}^{k_1} \cdots u_{i_j}^{k_j - n_{i_j}} \cdots u_{i_r}^{k_r}$ and the second term on the right are in $\mathfrak{U}^{(k-1)}$. Hence these elements are linear combinations of the elements (69). Since the set of elements in (69) is closed under multiplication by any z_i, the result asserted is clear. We show next that the elements (69) are linearly independent. We replace z_{i_j} by $z_{i_j} = u_{i_j}^{n_{i_j}} - v_{i_j}$ in $z_{i_1}^{h_1} \cdots z_{i_r}^{h_r} u_{i_1}^{\lambda_1} \cdots u_{i_r}^{\lambda_r}$. This gives

$$
\begin{aligned}
z_{i_1}^{h_1} \cdots z_{i_r}^{h_r} u_{i_1}^{\lambda_1} \cdots u_{i_r}^{\lambda_r} &= z_{i_1}^{h_1} u_{i_1}^{\lambda_1} z_{i_2}^{h_2} u_{i_2}^{\lambda_2} \cdots z_{i_r}^{h_r} u_{i_r}^{\lambda_r} \\
&= (u_{i_1}^{n_{i_1}} - v_{i_1})^{h_1} u_{i_1}^{\lambda_1} \cdots (u_{i_r}^{n_{i_r}} - v_{i_r})^{h_r} u_{i_r}^{\lambda_r} \qquad (70) \\
&\equiv u_{i_1}^{h_1 n_{i_1} + \lambda_1} u_{i_2}^{h_2 n_{i_2} + \lambda_2} \cdots u_{i_r}^{h_r n_{i_r} + \lambda_r} \qquad (\text{mod } \mathfrak{U}^{(k-1)})
\end{aligned}
$$

if $k = \sum_{j=1}^{r}(h_j n_{i_j} + \lambda_j)$. Hence the element $z_{i_1}^{h_1} \cdots z_{i_r}^{h_r} u_{i_1}^{\lambda_1} \cdots u_{i_r}^{\lambda_r} = u_{i_1}^{h_1 n_{i_1} + \lambda_1} \cdots u_{i_r}^{h_r n_{i_r} + \lambda_r} + *$, where $*$ is a linear combination of standard monomials belonging to $\mathfrak{U}^{(k-1)}$. It follows that if we have a non-trivial linear relation connecting the elements in (69), then we have such a relation for terms for which the "degree" $k = \sum_1^r(h_j n_{i_j} + \lambda_j)$ is fixed. This gives a relation with the same coefficients in the corresponding standard monomials $u_{i_1}^{h_1 n_{i_1} + \lambda_1} \cdots u_{i_r}^{h_r n_{i_r} + \lambda_r}$. Since the standard monomials are linearly independent we must have $h_j n_{i_j} + \lambda_j = k_j$, $j = 1, \cdots, r$, for the elements of (69) which appear in the relation with non-zero coefficients. Since $\lambda_j < n_{i_j}$ this implies that h_j, λ_j are determined by the equation $h_j n_{i_j} + \lambda_j = k_j$. Hence there is just one term in the relation. This is impossible in view of (70). Hence the elements of (69) are linearly independent and so form a basis for $\mathfrak{U}$.

We can now prove the following

THEOREM 11. *Let $\mathfrak{L}$ be a Lie algebra of characteristic $p \neq 0$ with ordered basis $\{u_i\}$ such that for every u_i, $(\text{ad}\, u_i)^p$ is an inner derivation. For each u_i let $u_i^{[p]}$ be an element of $\mathfrak{L}$ such that $(\text{ad}\, u_i)^p = \text{ad}\, u_i^{[p]}$. Then there exists a unique mapping $a \to a^{[p]}$ of $\mathfrak{L}$ into $\mathfrak{L}$ such that $u_i^{[p]}$ is as given and $\mathfrak{L}$ is a restricted Lie algebra relative to the mapping $a \to a^{[p]}$.*

Proof: Let $\mathfrak{U}$ be the universal enveloping algebra of $\mathfrak{L}$. Since $(\text{ad}\, u_i)^p = \text{ad}\, u_i^{[p]}$, $z_i = u_i^p - u_i^{[p]}$ commutes with every $l \in \mathfrak{L}$. Hence z_i is in the center of $\mathfrak{U}$ and $u_i^p = z_i + u_i^{[p]}$ where $u_i^{[p]} \in \mathfrak{L}$. We can therefore conclude from the lemma that the elements

$$
z_{i_1}^{h_1} z_{i_2}^{h_2} \cdots z_{i_r}^{h_r} u_{i_1}^{\lambda_1} \cdots u_{i_r}^{\lambda_r} \qquad (71)
$$

such that $i_1 < i_2 < \cdots < i_r$, $h_j \geqq 0$, $0 \leqq \lambda_j \leqq p - 1$, form a basis for $\mathfrak{U}$. Let $\mathfrak{B}$ denote the ideal in $\mathfrak{U}$ generated by the z_i. Then it is clear that the subset of our basis consisting of the elements (71) with some $h_j > 0$ is a basis for $\mathfrak{B}$. Hence the cosets of the elements $u_{i_1}^{\lambda_1} \cdots u_{i_r}^{\lambda_r}$, $0 \leqq \lambda_j \leqq p - 1$, form a basis for the algebra $\bar{\mathfrak{U}} = \mathfrak{U}/\mathfrak{B}$. Since the canonical homomorphism $x \to \bar{x} = x + \mathfrak{B}$ is a homomorphism of $\mathfrak{U}_L$ into $\bar{\mathfrak{U}}_L$, the restriction to $\mathfrak{L}$ is a homomorphism of

$\mathfrak{L}$ onto $\bar{\mathfrak{L}} = (\mathfrak{L} + \mathfrak{B})/\mathfrak{B}$. Since the $\bar{u}_i = u_i + \mathfrak{B}$ are linearly independent, $l \to \bar{l} = l + \mathfrak{B}$ is an isomorphism of $\mathfrak{L}$ onto $\bar{\mathfrak{L}}$. We note next that $\bar{\mathfrak{L}}$ is a subalgebra of $\bar{\mathfrak{U}}_L$ considered as a restricted Lie algebra. Thus we have $\bar{u}_i^p = (u_i + \mathfrak{B})^p = u_i^p + \mathfrak{B} = u_i^{[p]} + \mathfrak{B} \in \bar{\mathfrak{L}}$. It follows from (63) and (62) that $(\sum \alpha_i \bar{u}_i)^p \in \bar{\mathfrak{L}}$ which proves the assertion. The isomorphism of $\mathfrak{L}$ and $\bar{\mathfrak{L}}$ permits us to consider $\mathfrak{L}$ as a restricted Lie algebra by defining $l^{[p]_1}$ by $\overline{l^{[p]_1}} = \bar{l}^p$. Then we have $u_i^{[p]_1} = u_i^p + \mathfrak{B} = u_i^{[p]} + \mathfrak{B}$ so that $u_i^{[p]_1} = u_i^{[p]}$ as required. We now write $l^{[p]} = l^{[p]_1}$ and the result is proved.

We recall the result of § 3.6 that if $\mathfrak{L}$ is a finite-dimensional Lie algebra with a non-degenerate Killing form, then every derivation of $\mathfrak{L}$ is inner. It is clear also that the center of $\mathfrak{L}$ is 0. Hence if the characteristic is p, then for every $a \in \mathfrak{L}$ there exists a unique element $a^{[p]}$ such that the derivation $(\operatorname{ad} a)^p = \operatorname{ad} a^{[p]}$. It follows that we can introduce a p-operator in $\mathfrak{L}$ in one and only one way so that $\mathfrak{L}$ is a restricted Lie algebra. We therefore have the following

Corollary. *If $\mathfrak{L}$ is a finite-dimensional Lie algebra of characteristic $p \neq 0$ with non-degenerate Killing form, then there is a unique p-mapping in $\mathfrak{L}$ which makes $\mathfrak{L}$ restricted.*

We suppose next that $\mathfrak{L}$ is an arbitrary restricted Lie algebra and we prove the following

Theorem 12. *Let $\mathfrak{L}$ be a restricted Lie algebra of characteristic $p \neq 0$, $\mathfrak{U}$ the universal enveloping algebra, $\mathfrak{B}$ the ideal in $\mathfrak{U}$ generated by the elements $a^p - a^{[p]}$, $a \in \mathfrak{L}$, $\bar{\mathfrak{U}} = \mathfrak{U}/\mathfrak{B}$. Then the mapping $a \to \bar{a} = a + \mathfrak{B}$ is an isomorphism of $\mathfrak{L}$ into the restricted Lie algebra $\bar{\mathfrak{U}}_L$. If S is a homomorphism of $\mathfrak{L}$ into the restricted Lie algebra $\mathfrak{A}_L$, $\mathfrak{A}$ an algebra (associative with 1), then there exists a unique homomorphism of $\bar{\mathfrak{U}}$ into $\mathfrak{A}$ such that $\bar{a} \to a^S$. If $\mathfrak{L}$ is of finite dimensionality n, then $\dim \bar{\mathfrak{U}} = p^n$.*

Proof: Since $\overline{a^{[p]}} = \bar{a}^p = a^p + \mathfrak{B}$, $a \to \bar{a}$ is a homomorphism of the restricted Lie algebra $\mathfrak{L}$ into the restricted Lie algebra $\bar{\mathfrak{U}}_L$. If $\{u_i\}$ is a basis for $\mathfrak{L}$ over Φ, then it is clear from the rules for p-powers and the operation $a \to a^{[p]}$ that $a^p - a^{[p]}$ is a linear combination of the elements $u_i^p - u_i^{[p]}$. Hence $\mathfrak{B}$ is also the ideal generated by the elements $u_i^p - u_i^{[p]}$. The proof of the preceding theorem shows that the cosets $\bar{u}_i = u_i + \mathfrak{B}$ are linearly independent; hence $a \to \bar{a}$ is an isomorphism of $\mathfrak{L}$ into $\bar{\mathfrak{U}}_L$. Now let S be a homomorphism of $\mathfrak{L}$

into a restricted Lie algebra $\mathfrak{A}_L$, $\mathfrak{A}$ an algebra. Then we have a homomorphism of $\mathfrak{U}$ into $\mathfrak{A}$ sending $a \in \mathfrak{L}$ into a^S. Under this mapping $a^p \rightarrow (a^S)^p = (a^{[p]})^S$. Thus $a^p - a^{[p]}$ is in the kernel and so $\mathfrak{B}$ is in the kernel. Consequently, we have an induced homomorphism of $\bar{\mathfrak{U}} = \mathfrak{U}/\mathfrak{B}$ into $\mathfrak{A}$ such that $\bar{a} \rightarrow a^S$. If $\mathfrak{L}$ has the finite basis $u_1, u_2, \cdots, u_n$ then we have seen that the cosets of the elements $u_1^{\lambda_1} u_2^{\lambda_2} \cdots u_n^{\lambda_n}$, $0 \leqq \lambda_i \leqq p-1$ form a basis for $\bar{\mathfrak{U}}$. This proves the last statement of the theorem.

The algebra $\bar{\mathfrak{U}} = \mathfrak{U}/\mathfrak{B}$ of the theorem will be called the *u-algebra* of the restricted Lie algebra $\mathfrak{L}$. It plays the same role for $\mathfrak{L}$ as restricted Lie algebra as is played by $\mathfrak{U}$ for $\mathfrak{L}$ considered as an ordinary Lie algebra. In particular, a representation of $\mathfrak{L}$ as restricted Lie algebra defines a representation of $\bar{\mathfrak{U}}$ and conversely. Since $\bar{\mathfrak{U}}$ has a faithful representation it follows that every restricted Lie algebra has a faithful representation. Moreover, if $\mathfrak{L}$ is finite-dimensional, then $\bar{\mathfrak{U}}$ is finite-dimensional and so has a faithful representation acting in a finite-dimensional space. Consequently this holds also for $\mathfrak{L}$.

8. Abelian restricted Lie algebras

An abelian restricted Lie algebra is a vector space $\mathfrak{L}$ with a mapping $a \rightarrow a^p$ (we use this notation for the p-operator from now on) in $\mathfrak{L}$ such that

$$(a + b)^p = a^p + b^p , \qquad (\alpha a)^p = \alpha^p a^p . \tag{72}$$

The theory of these algebras is a special case of the theory of semi-linear transformations. This is equivalent to a theory of modules over certain types of non-commutative polynomial domains (cf. Jacobson, *Theory of Rings*, Chapter 3). In the present instance the polynomial ring is the set of polynomials $\alpha_0 + t\alpha_1 + \cdots + t^m\alpha_m$, $\alpha_i \in \Phi$, t an indeterminate such that $\alpha t = t\alpha^p$. If Φ is perfect it can be shown that the ring has no zero divisors and every left ideal and right ideal in the ring is a principal ideal. The study of this ring and its modules is a natural tool for studying abelian restricted Lie algebras. However, we shall not undertake this here. Instead we shall derive one or two basic results on the algebras without using the polynomial rings.

THEOREM 13. *Let $\mathfrak{L}$ be a finite-dimensional abelian restricted Lie algebra over an algebraically closed field of characteristic p. Assume* *that the p-mapping in* $\mathfrak{L}$ *is* 1 : 1. *Then* $\mathfrak{L}$ *has a basis* $(a_1, a_2, \cdots, a_n)$ *such that* $a_i^p = a_i$.

Proof: Let $a \neq 0$ be in $\mathfrak{L}$ and let m be the smallest positive integer such that $a^{p^m} = \alpha_1 a + \alpha_2 a^p + \cdots + \alpha_m a^{p^{m-1}}$. Then $a, a^p, \cdots, a^{p^{m-1}}$ are linearly independent and every a^{p^k} is a linear combination of these elements. If $\alpha_1 = 0$, set $\beta_i = \alpha_i^{1/p}$, $i = 2, \cdots, m$. Then $(a^{p^{m-1}} - \beta_2 a - \cdots - \beta_{m-1} a^{p^{m-2}})^p = 0$ which implies that $a^{p^{m-1}} = \beta_2 \alpha + \cdots + \beta_{m-1} a^{p^{m-2}}$ contrary to the choice of m. Hence $\alpha_1 \neq 0$. This implies that a is a linear combination of $a^p, a^{p^2}, \cdots$. We shall now show that there exists a $b = \beta_1 a + \beta_2 a^p + \cdots + \beta_m a^{p^{m-1}} \neq 0$ such that $b^p = b$. This will be the case if the β_i satisfy the following system of equations:

$$(73) \qquad \begin{aligned} \beta_1 &= \beta_m^p \alpha_1 \\ \beta_i &= \beta_m^p \alpha_i + \beta_{i-1}^p, \qquad i = 2, \cdots, m, \end{aligned}$$

and not all the β_i are 0. Successive substitution gives

$$(74) \qquad \beta_m = \beta_m^{p^m} \alpha_1^{p^{m-1}} + \beta_m^{p^{m-1}} \alpha_2^{p^{m-2}} + \cdots + \beta_m^p \alpha_m .$$

Since $\alpha_1 \neq 0$ and Φ is algebraically closed this has a non-zero solution for β_m. Then the remaining β_i can be determined from (73) for $i \leqq m - 1$ and the equation for $i = m$ will hold by (74). Now suppose we have already determined $a_1, a_2, \cdots, a_r$ which are linearly independent and satisfy $a_i^p = a_i$. Then $\mathfrak{L}_1 = \sum \Phi a_i$ is a subalgebra of $\mathfrak{L}$. Suppose a is an element of $\mathfrak{L}$ such that $a^p \in \mathfrak{L}_1$. Since a is a linear combination of $a^p, a^{p^2}, \cdots$ it follows that $a \in \mathfrak{L}_1$. Thus we have shown that $\mathfrak{L}/\mathfrak{L}_1$ is a restricted Lie algebra satisfying the hypotheses of the theorem. Hence if $\mathfrak{L} \neq \mathfrak{L}_1$, then we can find a $b \notin \mathfrak{L}_1$ such that $b^p \equiv b \pmod{\mathfrak{L}_1}$. Thus $b^p - b = \sum_1^r \gamma_i a_i$. We can determine δ_i so that $\delta_i^p - \delta_i + \gamma_i = 0$, $i = 1, 2, \cdots, r$. Then $a_{r+1} = b + \sum \delta_i a_i$ satisfies $a_{r+1} \notin \mathfrak{L}_1$, $a_{r+1}^p = a_{r+1}$. Hence the result follows by induction.

THEOREM 14. *Let* $\mathfrak{L}$ *be a commutative restricted Lie algebra of finite-dimensionality over an algebraically closed field. Suppose the p-mapping* $a \to a^p$ *is* 1 : 1 *and let* $\mathfrak{M}$ *be a finite-dimensional module for* $\mathfrak{L}$. *Then* $\mathfrak{M}$ *is completely reducible into one-dimensional submodules. If* $(a_1, a_2, \cdots, a_n)$ *is a basis such that* $a_i^p = a_i$ *and* $a = \sum \lambda_i a_i$, *then every weight in* $\mathfrak{M}$ *has the form* $\Lambda(a) = \sum m_i \lambda_i$, m_i *in the prime field.*

Proof: Let the basis $(a_1, a_2, \cdots, a_n)$ be as indicated and let

$a_i \rightarrow A_i$ in the given finite-dimensional representation. Then $A_i^p = A_i$ so the minimum polynomial of A_i is a factor of $\lambda^p - \lambda = \prod_{m=0}^{p-1}(\lambda - m)$. Thus the minimum polynomial of A_i has distinct roots and these are in the prime field. It follows that there exists a basis $(x_1, x_2, \cdots, x_n)$ for $\mathfrak{M}$ such that $x_j A_i = m_{ij} x_j$, m_{ij} in the prime field. Since the A_i commute we can find a basis which has this property simultaneously for the A_i, $i = 1, 2, \cdots, n$. Then $x_j(\sum \lambda_i A_i) = (\sum m_{ij}\omega_i)x_j$ so that $\mathfrak{M}$ is a direct sum of the irreducible invariant subspaces Φx_j and the weights are $\Lambda_j = \sum_i m_{ij}\lambda_i$.

Remark. It is easy to extend the first part of Theorem 14 to arbitrary base fields of characteristic p, that is, complete reducibility holds if the p-mapping is $1:1$. On the other hand, it has been shown by Hochschild ([4]) that if all the modules for a restricted Lie algebra (everything finite-dimensional) are completely reducible, then $\mathfrak{L}$ is abelian with non-singular p-mapping.

Exercises

1. Let $\mathfrak{L}$ be a Lie algebra over a field of characteristic zero, $\mathfrak{U}$ the universal enveloping algebra. Show that every element of $\mathfrak{U}$ is a linear combination of powers of elements of $\mathfrak{L}$.

2 (Witt). Let $\mathfrak{FL}$ be the free Lie algebra with r (free) generators $x_1, x_2, \cdots, x_r$ over a field of characteristic 0, $\mathfrak{F}$ the universal enveloping algebra of $\mathfrak{FL}$. Let $(\mathfrak{FL})_n = \mathfrak{FL} \cap \mathfrak{F}_n$, $\mathfrak{F}_n$ the space of homogeneous elements of degree n in $\mathfrak{F}$. Show that

$$\dim(\mathfrak{FL})_n = \frac{1}{n}\sum_{d/n} \mu(d) r^{n/d}$$

where μ is the Möbius function.

3. Let $\mathfrak{L}$ be finite-dimensional of characteristic 0, $\mathfrak{N}$ the nil radical of $\mathfrak{L}$. Show that the collection of linear transformations $\exp(\operatorname{ad} z)$, $z \in \mathfrak{N}$, is a group under multiplication.

4. Show that if Z is a nilpotent linear transformation in a finite dimensional vector space over a field of characteristic 0, then $\exp Z$ is unimodul ar ($\det \exp Z = 1$). Show that if Z is skew relative to a non-degenerate symmetric or skew bilinear form, then $\exp Z$ is orthogonal relative to this form.

5. If $\mathfrak{A}$ is an algebra there exists a unique automorphism τ of period two in $\mathfrak{A} \otimes \mathfrak{A}$ such that $(a \otimes b)\tau = b \otimes a$. Show that if $\mathfrak{U}$ is the universal enveloping algebra of a Lie algebra, then $\mathfrak{U}\delta$ is contained in the subalgebra of τ-fixed elements of $\mathfrak{U} \otimes \mathfrak{U}$.

6. Let $\mathfrak{U}$ be the universal enveloping algebra of a Lie algebra and let $\mathfrak{U}^*$ be the conjugate space of $\mathfrak{U}$. If $\varphi, \psi \in \mathfrak{U}^*$, then there exists a unique linear function $\varphi \otimes \psi$ on $\mathfrak{U} \otimes \mathfrak{U}$ such that $(\varphi \otimes \psi)(\sum u_i \otimes v_i) = \sum \varphi(u_i)\psi(v_i)$. Define

$\varphi\psi \in \mathfrak{U}^*$ by $(\varphi\psi)(u) = (\varphi \otimes \psi)(u\delta)$, δ the diagonal mapping of $\mathfrak{U}$ into $\mathfrak{U} \otimes \mathfrak{U}$. This makes $\mathfrak{U}^*$ an algebra. Show that this algebra is commutative and associative.

7. Prove the following analogue of Friedrichs' theorem in the characteristic $p \neq 0$ case: an element a of $\mathfrak{F}$ satisfies $a\delta = a \otimes 1 + 1 \otimes a$ if and only if a is in the restricted Lie algebra generated by the x_i.

8. Let $\mathfrak{L}$ be a finite-dimensional Lie algebra, $\mathfrak{L}_1$ and $\mathfrak{L}_2$ ideals in $\mathfrak{L}$ which are contragredient modules for $\mathfrak{L}$ relative to the adjoint representation. Let $(u_1, \cdots, u_n)$, $(u^1, \cdots, u^n)$ be dual bases for $\mathfrak{L}_1$ and $\mathfrak{L}_2$ as in the definition of a Casimir element of a representation (§ 3.7). Show that $\gamma = \sum u_i u^i$ is in the center of the universal enveloping algebra $\mathfrak{U}$ of $\mathfrak{L}$.

9. Let the notations be as in 8. Let R be a finite-dimensional representation for $\mathfrak{L}$ (hence for $\mathfrak{U}$) and let $f(u) = \operatorname{tr} u^R$, $u \in \mathfrak{U}$. Show that the element

$$\sum_{i_1, \cdots, i_r = 1}^{n} f(u_{i_1} u_{i_2} \cdots u_{i_r}) u^{i_1} u^{i_2} \cdots u^{i_r}$$

is in the center of $\mathfrak{U}$. More generally, show that if θ is a permutation of $1, 2, \cdots, r$, then

$$\sum_{i_1 \cdots i_r = 1}^{n} f(u_{i_1\theta} u_{i_2\theta} \cdots u_{i_r\theta}) u^{i_1} u^{i_2} \cdots u^{i_r}$$

is in the center of $\mathfrak{U}$.

10. Determine the center of the universal enveloping algebra of the three-dimensional split simple algebra over a field of characteristic zero.

11. Let Φ be a field, $\Phi[t_i] \equiv \Phi[t_1, t_2, \cdots, t_r]$ be the algebra of polynomials in indeterminates t_i, $\Phi < t_i > = \Phi < t_1, t_2, \cdots, t_r > = \overline{\Phi[t_i]}$ the closure of $\Phi[t_i]$ regarded as a graded algebra in the usual way (an element is homogeneous of degree k if it is a homogeneous polynomial of degree k in the usual sense). $\Phi < t_i >$ is called the *algebra of formal power series* in the t_i and its quotient field P is the *field of formal power* series in the t_i. $\Phi < t_i >$ has a valuation as defined in § 5. Thus if $a = a_k + a_{k+1} + a_{k+2} + \cdots$, a_j homogeneous of degree j, $a_k \neq 0$ then $|a| = 2^{-k}$. This valuation has a unique extension to a valuation in P satisfying $|ab| = |a||b|$. Let P_n be the algebra of $n \times n$ matrices with entries in P. Define $|(a_{ij})| = \max |a_{ij}|$. Show that this defines a valuation in P_n and show that if $D_1, D_2, \cdots, D_r$ are matrices with entries in Φ, then there exists a unique continuous homomorphism of the algebra $\mathfrak{F} = \Phi\{x_1, \cdots, x_r\}$ of § 5 into P_n mapping $x_i \to t_i D_i$, $i = 1, 2, \cdots, r$.

12. Let $\mathfrak{L}$ be a finite dimensional Lie algebra over a field Φ of characteristic 0 and let P be defined as in 11. Let $D_1, D_2, \cdots, D_r$ be derivations in $\mathfrak{L}$ over Φ and denote their extensions to $\mathfrak{L}_P$ over P by $D_1, D_2, \cdots, D_r$ again. Show that $G = \exp t_1 D_1 \exp t_2 D_2 \cdots \exp t_r D_r$ is a well defined automorphism in $\mathfrak{L}_P$ and that $G = \exp D$ where D is in the subalgebra generated by the D_i in the Lie algebra of derivations of $\mathfrak{L}_P$.

13. Let $\mathfrak{L}$ be a finite-dimensional restricted Lie algebra in which every element is nilpotent: $a^{p^k} = 0$ for some $k > 0$. Show that the u-algebra of $\mathfrak{L}$ has the form $\Phi 1 + \mathfrak{R}$ where $\mathfrak{R}$ is the radical.

14. A polynomial of the form $\lambda^{p^m} + \alpha_1\lambda^{p^{m-1}} + \cdots + \alpha_m\lambda$ is called a *p-polynomial* and it is called *regular* if $\alpha_m \neq 0$. Let $\mathfrak{L}$ be a restricted Lie algebra (possibly infinite-dimensional) with the property that for every $a \in \mathfrak{L}$ there exists a regular p-polynomial $\mu_a(\lambda)$ such that $\mu_a(a) = 0$. Show that if c is an element of $\mathfrak{L}$ such that all the roots of $\mu_c(\lambda)$ are in the base field Φ, then c is in the center $\mathfrak{C}$ of $\mathfrak{L}$. Hence show that $\mathfrak{L}$ is abelian if Φ is algebraically closed. Show that any finite dimensional nonabelian restricted Lie algebra over an algebraically closed field contains an element $z \neq 0$ such that $z^p = 0$.

15. Let $\mathfrak{L}$ be restricted with the property that $a^p = \alpha a$, α fixed and $\neq 0$. Prove that $\mathfrak{L}$ is abelian.

16. Use 14 and 15 to prove that if $a^{p^2} = a$ in a restricted Lie algebra, then $\mathfrak{L}$ is abelian. Conjecture: If $a^{p^{n(a)}} = a$, $n(a) > 0$, then $\mathfrak{L}$ is abelian.

17. Prove that if $\mathfrak{L}$ is restricted of characteristic three and $a^3 = 0$ for all a, then any finitely generated subalgebra of $\mathfrak{L}$ is finite-dimensional.

Conjecture (probably false but probably true under additional hypotheses): If $\mathfrak{L}$ is finitely generated and every element of $\mathfrak{L}$ (restricted of characteristic p) is algebraic in the sense that there exists a non-zero p-polynomial $\mu_a(\lambda)$ such that $\mu_a(a) = 0$, then $\mathfrak{L}$ is finite-dimensional.

18. Call a derivation D of a restricted Lie algebra *restricted* if $a^pD = (aD)(\text{ad } a)^{p-1}$. Note that every inner derivation is restricted. Show that a derivation is restricted if and only if it can be extended to a derivation of the u-algebra. Show that if $\mathfrak{L}$ has center 0, then every derivation is restricted.

19. Let $\mathfrak{L}$ be an abelian finite-dimensional restricted Lie algebra over a perfect field. Show that $\mathfrak{L} = \mathfrak{L}_0 \oplus \mathfrak{L}_1$ where $\mathfrak{L}_0$ is the space of nilpotent elements ($a^{p^k} = 0$) and $\mathfrak{L}_1 = \cap \mathfrak{L}^{p^i}$.

20. $\mathfrak{L}$ as in 19, base field infinite and perfect. Assume $\mathfrak{L}_0 = 0$. Show that $\mathfrak{L}$ is cyclic in the sense that there exists an element b such that $\mathfrak{L} = \sum \Phi b^{p^i}$.

21. Let $\mathfrak{A}$ be the group algebra over a field of characteristic p of the cyclic group of order p. Then $\mathfrak{A}$ has the basis $(1, x, x^2, \cdots, x^{p-1})$ with $x^p = 1$. Show that the derivation algebra $\mathfrak{D}$ of $\mathfrak{A}$ has a basis D_i, $i = 0, 1, \cdots, p-1$, where $xD_i = x^{i+1}$. Verify that

$$[D_iD_j] = (i-j)D_{i+j}$$
$$D_0^p = D_0\,, \qquad D_i^p = 0\,, \qquad i > 0$$

Prove that $\mathfrak{D}$ is simple. $\mathfrak{D}$ is called the *Witt algebra*.

22. Generalize 21 by considering the derivation algebra of the group algebra $\mathfrak{A}$ of the direct product of r cyclic groups of order p, Φ of characteristic p. These derivation algebras are simple too.

23. Prove the following identity for Lie algebras of characteristic $p \neq 0$:

$$b \sum_{i=0}^{p-2} (\text{ad } a)^{p-2-i}(\text{ad } c)(\text{ad } a)^i$$

$$= c \sum_{i=0}^{p-2} (\text{ad } a)^{p-2-i}(\text{ad } b)(\text{ad } a)^i \, .$$

24. Let $\mathfrak{L}$ be a nilpotent Lie algebra of linear transformations in a finite-dimensional vector space over a field of characteristic $p \neq 0$ such that $\mathfrak{L}^p = 0$. Show that if $A, B \in \mathfrak{L}$ then $(A + B)^p = A^p + B^p$. Use this to prove that if the base field is algebraically closed, then the weight functions are linear.

25. Let $\mathfrak{T}_0$ be the Lie algebra of $n \times n$ triangular matrices of trace 0 over a field of characteristic $p \neq 0$, $p \times n$. Prove $\mathfrak{T}_0$ complete. (*Hint*: Show that every derivation is restricted if $\mathfrak{T}_0$ is considered as a restricted Lie algebra and study the effect of a derivation on a diagonal matrix with distinct diagonal entries. (Φ may be assumed infinite.))

CHAPTER VI

The Theorem of Ado-Iwasawa

In this chapter we shall prove that every finite-dimensional Lie algebra has a faithful finite-dimensional representation. We shall treat the two cases: characteristic 0 and characteristic p separately. The result in the first case is known as Ado's theorem. For this we shall give a proof which is essentially a simplification of one due to Harish-Chandra. For the characteristic p case the result is due to Iwasawa. The proof we shall give is simpler than his and leads to several other results on representations in the characteristic p case.

1. Preliminary results

If R is a homomorphism of a Lie algebra $\mathfrak{L}$ into $\mathfrak{A}_L$, where $\mathfrak{A}$ is a finite-dimensional algebra (associative with 1), then we know that R has a unique extension to a homomorphism R of $\mathfrak{U}$ into $\mathfrak{A}$. If $\mathfrak{X}$ is the kernel, $\mathfrak{U}/\mathfrak{X} \cong \mathfrak{U}^R \subseteq \mathfrak{A}$ so $\mathfrak{U}/\mathfrak{X}$ is finite-dimensional. In general, if $\mathfrak{N}$ is a subspace of a vector space $\mathfrak{M}$, then the dimension of $\mathfrak{M}/\mathfrak{N}$ will be called the *co-dimension* of $\mathfrak{N}$ *in* $\mathfrak{M}$. Thus R determines an ideal $\mathfrak{X}$ in $\mathfrak{U}$ of finite co-dimension. The homomorphism R is an isomorphism of $\mathfrak{L}$ if and only if $\mathfrak{X} \cap \mathfrak{L} = 0$. Conversely, let $\mathfrak{X}$ be an ideal in $\mathfrak{U}$ such that $\mathfrak{X} \cap \mathfrak{L} = 0$ and $\mathfrak{X}$ has finite co-dimension in $\mathfrak{U}$. Then the restriction to $\mathfrak{L}$ of the canonical homomorphism of $\mathfrak{U}$ into $\overline{\mathfrak{U}} = \mathfrak{U}/\mathfrak{X}$ is an isomorphism of $\mathfrak{L}$ into the finite-dimensional algebra $\overline{\mathfrak{U}}_L$. Since any finite-dimensional algebra has a faithful finite-dimensional representation it is clear that a Lie algebra $\mathfrak{L}$ will have a faithful finite-dimensional representation if and only if the universal enveloping algebra $\mathfrak{U}$ of $\mathfrak{L}$ contains an ideal $\mathfrak{X}$ of finite co-dimension satisfying $\mathfrak{X} \cap \mathfrak{L} = 0$.

We recall that an element a of an algebra $\mathfrak{A}$ is called *algebraic* if there exists a non-zero polynomial $\varphi(\lambda)$ such that $\varphi(a) = 0$. This is equivalent to the assumption that the subalgebra $\mathfrak{A}$ generated by a is finite-dimensional. Consequently, every element of a finite-dimensional algebra is algebraic. If $\mathfrak{X}$ is an ideal in $\mathfrak{A}$ we shall say that $a \in \mathfrak{A}$ is *algebraic modulo* $\mathfrak{X}$ if there exists a non-zero polynomial $\varphi(\lambda)$ such that $\varphi(a) \in \mathfrak{X}$. This is equivalent to saying that the coset $\bar{a} = a + \mathfrak{X}$ is algebraic in $\bar{\mathfrak{A}} = \mathfrak{A}/\mathfrak{X}$. It follows that if $\mathfrak{X}$ is of finite co-dimension, then every element of $\mathfrak{A}$ is algebraic modulo $\mathfrak{X}$. We now state the following criterion for the universal enveloping algebra $\mathfrak{U}$ of a Lie algebra.

Lemma 1. *Let $\mathfrak{L}$ be a finite-dimensional Lie algebra,* $(u_1, u_2, \cdots, u_n)$ *a basis for $\mathfrak{L}$, $\mathfrak{U}$ the universal enveloping algebra of $\mathfrak{L}$, $\mathfrak{X}$ an ideal in $\mathfrak{U}$. Then $\mathfrak{X}$ is of finite co-dimension in $\mathfrak{U}$ if and only if every u_i is algebraic modulo $\mathfrak{X}$.*

Proof: The necessity of the condition has been established above. Now let $\varphi_i(\lambda)$ be a non-zero polynomial such that $\varphi_i(u_i) \in \mathfrak{X}$ and let $n_i = \deg \varphi_i(\lambda)$. Then every u_i^k is congruent modulo $\mathfrak{X}$ to a linear combination of the elements $1, u_i, u_i^2, \cdots, u_i^{n_i-1}$. The set of standard monomials $u_1^{k_1}u_2^{k_2}\cdots u_n^{k_n}$, $k_i \geqq 0$, is a basis for $\mathfrak{U}$ and the remark just made implies that these monomials are congruent modulo $\mathfrak{X}$ to a linear combination of the monomials $u_1^{\lambda_1}u_2^{\lambda_2}\cdots u_n^{\lambda_n}$ with $0 \leqq \lambda_i < n_i$. Since this is a finite set, $\mathfrak{U}/\mathfrak{X}$ is finite-dimensional.

Lemma 2. *Same assumptions as Lemma* 1, *$\mathfrak{X}$ and $\mathfrak{Y}$ ideals in $\mathfrak{U}$. If $\mathfrak{X}$ and $\mathfrak{Y}$ are of finite co-dimension, then $\mathfrak{X}\mathfrak{Y}$ is of finite co-dimension in $\mathfrak{U}$.*

Proof: Let $\varphi_i(\lambda)$, $\psi_i(\lambda)$ be non-zero polynomials such that $\varphi_i(u_i) \in \mathfrak{X}$, $\psi_i(u_i) \in \mathfrak{Y}$. Then $\varphi_i(\lambda)\psi_i(\lambda)$ has the property that $\varphi_i(u_i)\psi_i(u_i) \in \mathfrak{X}\mathfrak{Y}$ and the result follows from Lemma 1.

Lemma 3. *Let $\mathfrak{A}$ be an algebra, B a set of generators for $\mathfrak{A}$ and D a derivation in $\mathfrak{A}$. Suppose that for every $u \in B$ there exists a positive integer $n(u)$ such that $uD^{n(u)} = 0$. Then for every $a \in \mathfrak{A}$ there exists a positive integer $n(a)$ such that $aD^{n(a)} = 0$. If $\mathfrak{A}$ is finite-dimensional then D is nilpotent.*

Proof: Let $\mathfrak{B}$ be the subset of elements b such that $bD^{n(b)} = 0$ for some integer $n(b) \geqq 0$. If $b_1, b_2 \in \mathfrak{B}$ and $b_1D^{n_1} = 0 = b_2D^{n_2}$ then $(b_1 + b_2)D^n = 0$ for $n = \max(n_1, n_2)$. $(\alpha b_1)D^{n_1} = 0$, $\alpha \in \Phi$ and

$$(b_1b_2)D^N = \sum \binom{N}{i} (b_1D^i)(b_2D^{N-i}) = 0$$

if $N = n_1 + n_2 - 1$. Hence $\mathfrak{B}$ is a subalgebra and since $\mathfrak{B} \supseteqq B$,

$\mathfrak{B} = \mathfrak{A}$. This proves the first statement. The second is an immediate consequence.

2. The characteristic zero case

The key lemma for our proof of Ado's theorem is the following

LEMMA 4. *Let $\mathfrak{S}$ be a finite-dimensional solvable Lie algebra over a field of characteristic zero, $\mathfrak{N}$ the nil radical of $\mathfrak{S}$, $\mathfrak{U}$ the universal enveloping algebra of $\mathfrak{S}$. Suppose $\mathfrak{X}$ is an ideal of $\mathfrak{U}$ of finite co-dimension such that every element of $\mathfrak{N}$ is nilpotent modulo $\mathfrak{X}$. Then there exists an ideal $\mathfrak{Z}$ in $\mathfrak{U}$ such that*: (1) $\mathfrak{Z} \subseteq \mathfrak{X}$, (2) *$\mathfrak{Z}$ is of finite co-dimension*, (3) *$\mathfrak{Z}D \subseteq \mathfrak{Z}$ for every derivation D of $\mathfrak{S}$ (extended to $\mathfrak{U}$)*, (4) *every element of $\mathfrak{N}$ is nilpotent modulo $\mathfrak{Z}$.*

Proof: Let $\mathfrak{Y}$ be the ideal in $\mathfrak{U}$ generated by $\mathfrak{X}$ and $\mathfrak{N}$. Then $\mathfrak{Y} \supseteq \mathfrak{X}$ and $\mathfrak{Y}/\mathfrak{X}$ is the ideal in $\mathfrak{U}/\mathfrak{X}$ generated by $(\mathfrak{N} + \mathfrak{X})/\mathfrak{X}$. Since $(\mathfrak{N} + \mathfrak{X})/\mathfrak{X}$ is an ideal in the Lie algebra $(\mathfrak{S} + \mathfrak{X})/\mathfrak{X}$ and the elements of this ideal are nilpotent in the finite-dimensional enveloping associative algebra $\mathfrak{U}/\mathfrak{X}$ of $(\mathfrak{S} + \mathfrak{X})/\mathfrak{X}$ it follows from Theorem 2.2 that $(\mathfrak{N} + \mathfrak{X})/\mathfrak{X}$ is contained in the radical of $\mathfrak{U}/\mathfrak{X}$. Hence $\mathfrak{Y}/\mathfrak{X}$ is in the radical. This implies that there exists an integer r such that $\mathfrak{Z} \equiv \mathfrak{Y}^r \subseteq \mathfrak{X}$. If $l \in \mathfrak{S}$ and D is a derivation in $\mathfrak{S}$ then we know that $lD \in \mathfrak{N}$ (Theorem 3.7). It follows that $\mathfrak{U}D \subseteq \mathfrak{Y}$. Hence $\mathfrak{Y}D \subseteq \mathfrak{Y}$, which implies that $\mathfrak{Z}D = \mathfrak{Y}^rD \subseteq \mathfrak{Y}^r = \mathfrak{Z}$. Hence (3) holds for $\mathfrak{Z}$ and we have already noted that (1) holds. Since $\mathfrak{Y}$ contains $\mathfrak{X}$, $\mathfrak{Y}$ is of finite co-dimension and $\mathfrak{Z} = \mathfrak{Y}^r$ is of finite co-dimension in $\mathfrak{U}$. This proves (2). If $z \in \mathfrak{N}$, $z^n \in \mathfrak{X}$ for some positive integer n. Hence $z^n \in \mathfrak{Y}$ and $z^{nr} \in \mathfrak{Y}^r = \mathfrak{Z}$. This proves (4).

THEOREM 1. *Let $\mathfrak{L} = \mathfrak{S} \oplus \mathfrak{L}_1$ where $\mathfrak{S}$ is a solvable ideal and $\mathfrak{L}_1$ is a subalgebra of the finite dimensional Lie algebra $\mathfrak{L}$ of characteristic 0. Suppose we have a finite dimensional representation S of $\mathfrak{S}$ such that z^S is nilpotent for every z in the nil radical $\mathfrak{N}$ of $\mathfrak{S}$. Then there exists a finite-dimensional representation R of $\mathfrak{L}$ such that*: (1) *if $x^R = 0$ for x in $\mathfrak{S}$ then $x^S = 0$*, (2) *y^R is nilpotent for every y of the form $y = z + u$ where $z \in \mathfrak{N}$ and $u \in \mathfrak{L}_1$ is such that $\mathrm{ad}_{\mathfrak{S}}u$ is nilpotent.*

Proof: S defines a homomorphism of the universal enveloping algebra $\mathfrak{U}$ of $\mathfrak{S}$ whose kernel $\mathfrak{X}$ is of finite co-dimension. Also if $z \in \mathfrak{N}$ then $(z^S)^n = 0$ so $z^n \in \mathfrak{X}$ and $\mathfrak{X}$ satisfies the hypothesis of

Lemma 4. Let $\mathfrak{Z}$ be the ideal in the conclusion of this lemma. We shall define the required representation R in $\mathfrak{U}/\mathfrak{Z}$. We first define a representation R' of $\mathfrak{L} = \mathfrak{S} \oplus \mathfrak{L}_1$ acting in the space $\mathfrak{U}$. If $s \in \mathfrak{S}$ we set $s^{R'} = s_R$, the right multiplication in $\mathfrak{U}$ determined by s. If $l \in \mathfrak{L}_1$ we define $l^{R'}$ to be the derivation in $\mathfrak{U}$ which extends the derivation $s \rightarrow [sl]$ of $\mathfrak{S}$. The R''s defined on $\mathfrak{S}$ and on $\mathfrak{L}_1$ define a unique linear transformation R' on $\mathfrak{L}$ which is a representation of $\mathfrak{S}$ and $\mathfrak{L}_1$ separately. To prove that R' is a representation for $\mathfrak{L}$ it suffices to show that $[sl]^{R'} = [s^{R'}, l^{R'}]$, $s \in \mathfrak{S}$, $l \in \mathfrak{L}_1$. Now $[sl] \in \mathfrak{S}$ so $[sl]^{R'} = [sl]_R$. On the other hand, if D is a derivation in $\mathfrak{U}$ and $a \in \mathfrak{U}$, then the derivation condition gives $[a_R D] = (aD)_R$. Hence we have $[sl]^{R'} = [sl]_R = (sl^{R'})_R = [s_R, l^{R'}] = [s^{R'}, l^{R'}]$, as required. Since $\mathfrak{Z}$ is an ideal in $\mathfrak{U}$ such that $\mathfrak{Z}D \subseteq \mathfrak{Z}$ for any derivation D of $\mathfrak{S}$, $\mathfrak{Z}$ is a subspace of $\mathfrak{U}$ which is invariant relative to the representation R' of $\mathfrak{L}$ acting in $\mathfrak{U}$. Hence we have an induced representation R in the finite-dimensional factor space $\mathfrak{U}/\mathfrak{Z}$. Let $x \in \mathfrak{S}$ satisfy $x^R = 0$. This means that x^R maps $\mathfrak{U}$ into $\mathfrak{Z}$. Hence $x \in \mathfrak{Z}$, $x \in \mathfrak{X}$ and so $x^S = 0$. Let $z \in \mathfrak{N}$. Then, by Lemma 4, z is nilpotent modulo $\mathfrak{Z}$. Hence z^R is nilpotent. Since $\mathfrak{N}$ is an ideal in $\mathfrak{L}$ (Theorem 3.7) it follows that z^R is in the radical $\mathfrak{R}$ of the algebra of linear transformations generated by $\mathfrak{L}^R$. Now let $y = z + u$ where $z \in \mathfrak{N}$, $u \in \mathfrak{L}_1$ and $\mathrm{ad}_{\mathfrak{S}} u$ is nilpotent. Since $z^R \in \mathfrak{R}$, in order to prove that y^R is nilpotent, it suffices to prove that u^R is nilpotent. By definition $u^{R'}$ is the derivation in $\mathfrak{U}$ which coincides with $\mathrm{ad}_{\mathfrak{S}} u$ on $\mathfrak{S}$ and $\mathrm{ad}_{\mathfrak{S}} u$ is nilpotent. Since $\mathfrak{S}$ generates $\mathfrak{U}$ it follows from Lemma 3 that for every $a \in \mathfrak{U}$ there is an integer $n(a)$ such that $a(u^{R'})^{n(a)} = 0$. Hence for every $\bar{a} \in \mathfrak{U}/\mathfrak{Z}$ we have $n(\bar{a})$ such that $\bar{a}(u^R)^{n(\bar{a})} = 0$. Since $\mathfrak{U}/\mathfrak{Z}$ is finite-dimensional this implies that u^R is nilpotent. Thus R satisfies the conditions (1) and (2).

We can now prove

Ado's theorem. Every finite-dimensional Lie algebra $\mathfrak{L}$ of characteristic zero has a faithful finite-dimensional representation.

Proof: We recall that the kernel of the adjoint representation A is the center $\mathfrak{C}$ of $\mathfrak{L}$. It will therefore suffice to prove the existence of a finite-dimensional representation R of $\mathfrak{L}$ which is faithful on the center $\mathfrak{C}$. For then we can form the direct sum representation of R and A. The kernel of this is the intersection of the kernels of R and of A. Hence this representation is faithful as well as finite-dimensional. We proceed to construct R. Let $\mathfrak{S}$

be the radical, $\mathfrak{N}$ the nil radical. Let $\mathfrak{N}_1 = \mathfrak{C} \subset \mathfrak{N}_2 \subset \cdots \subset \mathfrak{N}_h = \mathfrak{N}$ where each $\mathfrak{N}_i$ is an ideal in the next and $\dim \mathfrak{N}_{i+1} = \dim \mathfrak{N}_i + 1$. Such a sequence exists since $\mathfrak{N}$ is solvable and contains $\mathfrak{C}$. If $\dim \mathfrak{C} = c$ then in a $c+1$-dimensional vector space there exists a nilpotent linear transformation z such that $z^c \neq 0$. Then $\mathfrak{C}$ is isomorphic to the Lie algebra with basis $(z, z^2, \cdots, z^c)$, so $\mathfrak{C}$ has a faithful representation by nilpotent linear transformations in a finite-dimensional space. Since each $\mathfrak{N}_i$ is nilpotent and $\mathfrak{N}_{i+1} = \mathfrak{N}_i \oplus \Phi u_{i+1}$ where Φu_{i+1} is a subalgebra, the preceding theorem can be applied successively to obtain a finite-dimensional representation T of $\mathfrak{N}$ by nilpotent linear transformations such that T is faithful on $\mathfrak{C}$. Next we obtain a sequence of subspaces, $\mathfrak{S}_1 = \mathfrak{N} \subset \mathfrak{S}_2 \subset \cdots \subset \mathfrak{S}_k = \mathfrak{S}$ such that $\mathfrak{S}_{i+1}$ is an ideal in $\mathfrak{S}_i$ and $\dim \mathfrak{S}_{i+1} = \dim \mathfrak{S}_i + 1$. Then $\mathfrak{S}_{i+1} = \mathfrak{S}_i \oplus \Phi v_{i+1}$. Also $\mathfrak{N}$ is the nil radical of every $\mathfrak{S}_i$ (Theorem 3.7). Hence the theorem can be applied again beginning with T to obtain a representation S of $\mathfrak{S}$ which is finite-dimensional, faithful on $\mathfrak{C}$ and represents the elements of $\mathfrak{N}$ by nilpotent linear transformations. Finally we write $\mathfrak{L} = \mathfrak{S} \oplus \mathfrak{L}_1$, $\mathfrak{L}_1$ a subalgebra (Levi's theorem). Then we can apply Theorem 1 again to obtain the required representation R of $\mathfrak{L}$.

Remark: The R constructed has the property that z^R is nilpotent for every $z \in \mathfrak{N}$. The same holds for the adjoint representation. Hence the direct sum has the property too. We therefore have a faithful finite-dimensional representation such that the transformations corresponding to the elements of $\mathfrak{N}$ are nilpotent—and hence are in the radical of the enveloping associative algebra.

3. The characteristic p case

We recall that if Φ is of characteristic p then a polynomial of the form $\alpha_0\lambda^{p^m} + \alpha_1\lambda^{p^{m-1}} + \cdots + \alpha_m\lambda$, $\alpha_i \in \Phi$ is called a p-polynomial. If $\mu(\lambda)$ is a polynomial of degree m, then we can write

$$\lambda^{p^i} = \mu(\lambda)q_i(\lambda) + r_i(\lambda)\,, \qquad i = 0, 1, 2, \cdots, m \tag{1}$$

where the $r_i(\lambda)$ are of degree $< m$. Since the space of polynomials of degree $< m$ is m-dimensional there exist α_i, $i = 0, \cdots, m$, not all 0 such that $\sum \alpha_i r_i(\lambda) = 0$. Then (1) implies that $\sum \alpha_i \lambda^{p^i} = \mu(\lambda)(\sum \alpha_i q_i(\lambda))$. We have therefore proved that every polynomial is a factor of a suitable non-zero p-polynomial.

Now let $\mathfrak{L}$ be a finite-dimensional Lie algebra over Φ, $\mathfrak{U}$ the universal enveloping algebra. Let $a \in \mathfrak{L}$ and let $\mu(\lambda)$ be a non-zero polynomial such that $\mu(\text{ad}\, a) = 0$. Such a polynomial exists since the algebra of transformations in $\mathfrak{L}$ is finite-dimensional. Let $m(\lambda) = \lambda^{p^m} + \alpha_1\lambda^{p^{m-1}} + \cdots + \alpha_m\lambda$ be a p-polynomial divisible by $\mu(\lambda)$. Then we have

(2) $$(\text{ad}\, a)^{p^m} + \alpha_1(\text{ad}\, a)^{p^{m-1}} + \cdots + \alpha_m(\text{ad}\, a) = 0\,.$$

In other words, for every $b \in \mathfrak{L}$ we have

(2′) $$[\cdots[b\overbrace{a]\cdots a}^{p^m}] + \alpha_1[\cdots[b\overbrace{a]\cdots a}^{p^{m-1}}] + \cdots + \alpha_m[ba] = 0\,.$$

On the other hand, we know that $[\cdots[b\overbrace{a]\cdots a}^{p}] = [ba^p]$. Iteration of this gives

(3) $$[\cdots[b\overbrace{a]\cdots a}^{p^k}] = [ba^{p^k}]$$

Hence (2′) implies

(4) $$[b, a^{p^m} + \alpha_1 a^{p^{m-1}} + \cdots + \alpha_m a] = 0\,,$$

$b \in \mathfrak{L}$, which implies that the element

(5) $$z \equiv a^{p^m} + \alpha_1 a^{p^{m-1}} + \cdots + \alpha_m a$$

is in the center $\mathfrak{C}$ of $\mathfrak{U}$. We have therefore proved the following

Lemma 5. *Let $\mathfrak{L}$ be a finite-dimensional Lie algebra over a field of characteristic $p \neq 0$ and let $\mathfrak{U}$ be the universal enveloping algebra. Then for every $a \in \mathfrak{L}$ there exists a polynomial $m_a(\lambda)$ such that $m_a(a)$ is in the center $\mathfrak{C}$ of $\mathfrak{U}$.*

The result just proved and Lemma 5.4 are the main steps in our proof of

Iwasawa's theorem. Every finite-dimensional Lie algebra of characteristic $p \neq 0$ has a faithful finite-dimensional representation.

Proof: Let $(u_1, u_2, \cdots, u_n)$ be a basis for $\mathfrak{L}$ and let $m_i(\lambda)$ be a p-polynomial such that $m_i(u_i) = z_i \in \mathfrak{C}$, the center of the universal enveloping algebra. If $\deg m_i(\lambda) = p^{m_i}$ then $z_i = u_i^{p^{m_i}} + v_i$ where $v_i \in \mathfrak{U}_{p^{m_i}-1}$. Hence, by Lemma 5.4, the elements $z_1^{h_1}z_2^{h_2}\cdots z_n^{h_n}u_1^{\lambda_1}\cdots u_n^{\lambda_n}$, $h_i \geqq 0$, $0 \leqq \lambda_i < p^{m_i}$ form a basis for $\mathfrak{U}$. Let $\mathfrak{B}$ be the ideal in $\mathfrak{U}$ generated by the z_i. As in the proof of Theorem 5.11, the cosets of the elements $u_1^{\lambda_1}\cdots u_n^{\lambda_n}$, $0 \leqq \lambda_i < p^{m_i}$ form a basis for $\mathfrak{U}/\mathfrak{B}$. Hence this algebra is finite-dimensional and the canonical mapping $a \to \bar{a} = a + \mathfrak{B}$, $a \in \mathfrak{L}$, is an isomorphism of $\mathfrak{L}$ into $\overline{\mathfrak{U}}_L$, $\overline{\mathfrak{U}} = \mathfrak{U}/\mathfrak{B}$. It follows that there exists a faithful finite-dimensional representation of $\mathfrak{L}$.

We shall show next that in the characteristic $p \neq 0$ case there is no connection between structure of Lie algebras and complete reducibility of modules. In the following theorem we shall need a result proved in Chapter II (Theorem 2.10) that an algebra $\mathfrak{A}$ of linear transformations in a finite-dimensional vector space which has a non-zero radical cannot be completely reducible. We shall need also a result which is somewhat more difficult to prove, namely, that if z is an element of a finite-dimensional algebra and z does not belong to the radical of the algebra, then there exists an irreducible representation R of the algebra such that $z^R \neq 0$ (See, for example, Jacobson [3], Theorem 3.1 and Definition 1.1.)

THEOREM 2. *Every finite-dimensional Lie algebra over a field of characteristic $p \neq 0$ has a* $1:1$ *finite-dimensional representation which is not completely reducible and a* $1:1$ *finite-dimensional completely reducible representation.*

Proof: Let the u_i and $z_i = m_i(u_i)$ be as in the proof of Iwasawa's theorem. Let $\mathfrak{B}_1$ be the ideal in $\mathfrak{U}$ generated by $(z_1^2, z_2, \cdots z_n)$. Then the argument shows that $z_1 + \mathfrak{B}_1 \neq 0$ in $\mathfrak{U}/\mathfrak{B}_1$ but $(z_1 + \mathfrak{B}_1)^2 = 0$. Hence $z_1 + \mathfrak{B}_1$ is a non-zero center nilpotent element in the finite-dimensional algebra $\mathfrak{U}/\mathfrak{B}_1$. The ideal generated by such an element is nilpotent. Hence $\mathfrak{U}/\mathfrak{B}_1$ is not semi-simple. Hence any $1:1$ representation of this algebra is not completely reducible. Since $(\mathfrak{L} + \mathfrak{B}_1)/\mathfrak{B}_1$ generates $\mathfrak{U}/\mathfrak{B}_1$ this representation provides a representation for $\mathfrak{L}$ which is not completely reducible. The argument used before shows that the canonical mapping of $\mathfrak{L}$ into $\mathfrak{U}/\mathfrak{B}_1$ is an isomorphism. Hence the representation we have indicated is $1:1$ for $\mathfrak{L}$ and this proves our first assertion. Next let a be any non-zero element of $\mathfrak{L}$ and take $u_1 = a$ in the basis $(u_1, u_2, \cdots, u_n)$ for $\mathfrak{L}$. Let $\alpha \neq 0$ be in Φ. Then $m_1(\lambda) - \alpha$ is not divisible by λ and this is the minimum polynomial of $a + \mathfrak{B}_2$ in $\mathfrak{U}/\mathfrak{B}_2$ where $\mathfrak{B}_2$ is the ideal generated by $m_1(u_1) - \alpha$, $m_2(u_2), \cdots, m_n(u_n)$. Thus $a + \mathfrak{B}_2$ is not nilpotent and so it does not belong to the radical. It follows that there exists a finite-dimensional irreducible representation of $\mathfrak{U}/\mathfrak{B}_2$ such that $a + \mathfrak{B}_2$ is not represented by 0. This gives a finite-dimensional irreducible representation R_a of $\mathfrak{L}$ such that $a^{R_a} \neq 0$. Let $\mathfrak{R}_a$ denote the kernel of R_a (in $\mathfrak{L}$). Then $\bigcap_{a \in \mathfrak{L}} \mathfrak{R}_a = 0$. Since $\mathfrak{L}$ is finite-dimensional we can find a finite number $a_1, a_2, \cdots, a_m$ of the a's in $\mathfrak{L}$ such that $\cap_j \mathfrak{K}_{a_j} = 0$. We now form the module $\mathfrak{M}$ which is a direct sum of the m irreducible modules $\mathfrak{M}_j$ corresponding to the representations R_{a_j}. Then evidently $\mathfrak{M}$ is completely reducible and the kernel of the associated representation is $\cap \mathfrak{K}_{a_j} = 0$. Hence this gives a faithful finite-dimensional completely reducible representation for $\mathfrak{L}$.

Exercises

1. Show that any finite-dimensional Lie algebra of characteristic p has indecomposable modules of arbitrarily high finite-dimensionalities.

Exercises 2–4 are designed to prove the following theorem: Let $\mathfrak{A}$ be an algebra over an algebraically closed field of characteristic 0, $\mathfrak{L}$ a finite-dimensional simple subalgebra of $\mathfrak{A}_L$ which contains a non-zero algebraic element. Then the subalgebra of $\mathfrak{A}$ generated by $\mathfrak{L}$ is finite dimensional. We may as well assume that this subalgebra is $\mathfrak{A}$ itself and it suffices to show that $\mathfrak{L}$ has a basis consisting of algebraic elements.

2. Show that $\mathfrak{L}$ contains a non-zero nilpotent element e. (*Hint*: Use Exercise 3.11.)

3. Show that $\mathfrak{L}$ contains a non-zero algebraic element h which is contained in some Cartan subalgebra $\mathfrak{H}$ of $\mathfrak{L}$. (*Hint*: Use Theorem 3.17, and Exercise 3.13.)

4. If e_α have the usual significance relative to $\mathfrak{H}$ show that there exists a root $\alpha \neq 0$ such that $h_\alpha, e_\alpha, e_{-\alpha}$ are algebraic. Then show that this holds for every root α and hence that $\mathfrak{L}$ has a basis of algebraic elements. Use this to prove the theorem stated.

5. Extend the theorem stated above to $\mathfrak{L}$ semi-simple under the stronger hypothesis that $\mathfrak{L}$ contains a set of algebraic elements such that the ideal in $\mathfrak{L}$ generated by this set is all of $\mathfrak{L}$.

6. Extend the result in 5 to the case in which the base field is any field of characteristic 0.

7. (Harish-Chandra). Let $\mathfrak{L}$ be a finite-dimensional Lie algebra over a field of characteristic 0 and let R be a faithful finite-dimensional representation of $\mathfrak{L}$ by linear transformations of trace 0 in $\mathfrak{M}$. Let R_i, $i = 1, 2, \cdots$ denote the representation in $\mathfrak{M} \otimes \mathfrak{M} \otimes \cdots \otimes \mathfrak{M}$, i times and let $\mathfrak{X}_i$ denote the kernel in $\mathfrak{U}$ of R_i. Prove that $\cap_1^\infty \mathfrak{X}_i = 0$.

8. Show that every finite-dimensional Lie algebra has a faithful finite-dimensional representation by linear transformations of trace 0.

CHAPTER VII

Classification of Irreducible Modules

The principal objective of this chapter is the classification of the finite-dimensional irreducible modules for a finite-dimensional split semi-simple Lie algebra $\mathfrak{L}$ over a field of characteristic 0. The main result—due to Cartan—gives a $1:1$ correspondence between the modules of the type specified and the "dominant integral" linear functions on a splitting Cartan subalgebra $\mathfrak{H}$ of $\mathfrak{L}$. The existence of a finite-dimensional irreducible module corresponding to any dominant integral function was established by Cartan by separate case investigations of the simple Lie algebras and so it depended on the classification of these algebras. A more elegant method for handling this question was devised by Chevalley and by Harish-Chandra (independently). This does not require case considerations. Moreover, it yields a uniform proof of the existence of a split semi-simple Lie algebra corresponding to every Cartan matrix or Dynkin diagram and another proof of the uniqueness (in the sense of isomorphism) of this algebra.

Harish-Chandra's proof of these results is quite complicated.* The version we shall give is a comparatively simple one which is based on an explicit definition of a certain infinite-dimensional Lie algebra $\tilde{\mathfrak{L}}$ defined by an integral matrix (A_{ij}) satisfying certain conditions which are satisfied by the Cartan matrices, and the study of certain cyclic modules, "e-extreme modules" for $\tilde{\mathfrak{L}}$. The principal tools which are needed in our discussion are the Poincaré-Birkhoff-Witt theorem and the representation theory for split three-dimensional simple Lie algebras.

1. *Definition of certain Lie algebras*

Let (A_{ij}), $i, j = 1, 2, \cdots, l$, be a matrix of integers A_{ij} having the following properties (which are known to hold for the Cartan matrix of any finite-dimensional split semi-simple Lie algebra over

* Chevalley's proof has not been published.

a field of characteristic 0)

(α) $A_{ii} = 2$, $A_{ij} \leqq 0$ if $i \neq j$, $A_{ij} = 0$ implies $A_{ji} = 0$.

(β) $\det (A_{ij}) \neq 0$.

(γ) If $(\alpha_1, \alpha_2, \cdots, \alpha_l)$ is a basis for an l-dimensional vector space $\mathfrak{H}_0^*$ over the rationals, then the group W generated by the l linear transformations S_{α_i} defined by

$$(1) \qquad \alpha_j S_{\alpha_i} = \alpha_j - A_{ij}\alpha_i \,, \qquad j = 1, \cdots, l \,,$$

is a finite group.

Let Φ be an arbitrary field of characteristic 0. We shall define a Lie algebra $\tilde{\mathfrak{L}}$ over Φ which is determined by the matrix (A_{ij}). We begin with the free Lie algebra $\mathfrak{FL}$ (§ 5.4) generated by the free generators e_i, f_i, h_i, $i = 1, 2, \cdots, l$, and let $\mathfrak{K}$ be the (Lie) ideal in $\mathfrak{FL}$ generated by the elements

$$(2) \qquad \begin{aligned} &[h_i h_j] \\ &[e_i f_j] - \delta_{ij} h_i \\ &[e_i h_j] - A_{ji} e_i \\ &[f_i h_j] + A_{ji} f_i \,. \end{aligned}$$

Let $\tilde{\mathfrak{L}} = \mathfrak{FL}/\mathfrak{K}$. Let $\mathfrak{H}$ be the subspace of $\mathfrak{FL}$ spanned by the h_i and let α_i be the linear function on $\mathfrak{H}$ such that

$$(3) \qquad \alpha_i(h_j) = A_{ji} \,, \qquad j = 1, \cdots, l \,.$$

The condition (β) implies that the l α_i form a basis for the conjugate space $\mathfrak{H}^*$ of $\mathfrak{H}$.

Since $\mathfrak{FL}$ is freely generated by the e_i, f_i, h_i, $i = 1, 2, \cdots, l$, any mapping $e_i \to E_i$, $f_i \to F_i$, $h_i \to H_i$ of the generators into linear transformations of a vector space defines a (unique) representation of $\mathfrak{FL}$. In other words, if $\mathfrak{X}$ is any vector space with basis $\{u_j\}$, then $\mathfrak{X}$ can be made into an $\mathfrak{FL}$-module by defining the module products $u_j e_i$, $u_j f_i$, $u_j h_i$ in a completely arbitrary manner as elements of $\mathfrak{X}$. We now let $\mathfrak{X}$ be the free (associative) algebra generated by l free generators $x_1, x_2, \cdots, x_l$. Then $\mathfrak{X}$ has the basis $1, x_{i_1} \cdots x_{i_r}$, $i_j = 1, 2, \cdots, l$, $r = 1, 2, \cdots$. Let $\Lambda \equiv \Lambda(h)$ be a linear function on $\mathfrak{H}$. Then the foregoing remark implies that we can turn $\mathfrak{X}$ into an $\mathfrak{FL}$-module by defining

$$
\begin{aligned}
&1h = \Lambda 1\,,\\
&(x_{i_1}\cdots x_{i_r})h = (\Lambda - \alpha_{i_1} - \cdots - \alpha_{i_r})x_{i_1}\cdots x_{i_r}\,;\\
(4)\qquad &1f_i = x_i, \quad x_{i_1}\cdots x_{i_r}f_i = x_{i_1}\cdots x_{i_r}x_i\,;\\
&1e_i = 0\,,\\
&x_{i_1}\cdots x_{i_r}e_i = (x_{i_1}\cdots x_{i_{r-1}}e_i)x_{i_r}\\
&\qquad - \delta_{i_r i}(\Lambda - \alpha_{i_1} - \cdots - \alpha_{i_{r-1}})(h_i)x_{i_1}\cdots x_{i_{r-1}}\,.
\end{aligned}
$$

In these equations and in those which will appear subsequently we abbreviate $\Lambda(h)$ by Λ, etc., but write in full $\Lambda(h_i)$, etc. Let $\mathfrak{K}'$ be the kernel of the representation of $\mathfrak{F}\mathfrak{L}$ in $\mathfrak{X}$. We proceed to show that $\mathfrak{K}' \supseteq \mathfrak{K}$, the ideal defining $\tilde{\mathfrak{L}}$. This will imply that $\mathfrak{X}$ can be considered as defining a representation, hence a module, for $\tilde{\mathfrak{L}}$. The linear transformation in $\mathfrak{X}$ corresponding to h has a diagonal matrix relative to the chosen basis. Hence any two of these transformations commute and so $[h_ih_j] \in \mathfrak{K}'$. We note next that the linear transformation corresponding to f_i is the right multiplication x_{i_R} in $\mathfrak{X}$ determined by x_i. The last equation and fourth equation in (4) imply that

$$
\begin{aligned}
&(x_{i_1}\cdots x_{i_{r-1}})f_{i_r}e_i - (x_{i_1}\cdots x_{i_{r-1}})e_if_{i_r}\\
&\quad = -\delta_{i_r i}(\Lambda - \alpha_{i_1} - \cdots - \alpha_{i_{r-1}})(h_i)x_{i_1}\cdots x_{i_{r-1}}\\
&\quad = -\delta_{i_r i}(x_{i_1}\cdots x_{i_{r-1}})h_i\,.
\end{aligned}
$$

This implies that $[e_if_j] - \delta_{ij}h_i$ is in the kernel $\mathfrak{K}'$. We have

$$
\begin{aligned}
1[f_ih] &= 1(f_ih - hf_i) = x_ih - \Lambda x_i = (\Lambda - \alpha_i)x_i - \Lambda x_i\\
&= -\alpha_ix_i = -1(\alpha_if_i)\,,\\
x_{i_1}\cdots x_{i_r}[f_ih] &= x_{i_1}\cdots x_{i_r}(f_ih - hf_i)\\
&= x_{i_1}\cdots x_{i_r}x_ih - (\Lambda - \alpha_{i_1} - \cdots - \alpha_{i_r})x_{i_1}\cdots x_{i_r}x_i\\
&= (\Lambda - \alpha_{i_1} - \cdots - \alpha_{i_r} - \alpha_i)x_{i_1}\cdots x_{i_r}x_i\\
&\qquad - (\Lambda - \alpha_{i_1} - \cdots - \alpha_{i_r})x_{i_1}\cdots x_{i_r}x_i\\
&= -\alpha_ix_{i_1}\cdots x_{i_r}x_i = -x_{i_1}\cdots x_{i_r}(\alpha_if_i)\,.
\end{aligned}
$$

Hence $[f_ih] + \alpha_if_i \in \mathfrak{K}'$. This implies that if x is an element of $\mathfrak{X}$ such that $xh = M(h)x \equiv Mx$, then $(xf_i)h = (M - \alpha_i)(xf_i)$, or $(xx_i)h = (M - \alpha_i)(xx_i)$. We now assert that $(x_{i_1}\cdots x_{i_r}e_i)h = (\Lambda - \alpha_{i_1} - \cdots - \alpha_{i_r} + \alpha_i)x_{i_1}\cdots x_{i_r}e_i$. This is clear for $r = 0$ if we adopt the convention that the corresponding base element is 1. We now assume the result for $r - 1$. Then

$$
\begin{aligned}
(x_{i_1}\cdots x_{i_r}e_i)h &= ((x_{i_1}\cdots x_{i_{r-1}}e_i)x_{i_r})h \\
&\quad - \delta_{i_r i}(\Lambda - \alpha_{i_1} - \cdots - \alpha_{i_{r-1}})(h_i)x_{i_1}\cdots x_{i_{r-1}}h \\
&= (\Lambda - \alpha_{i_1} - \cdots - \alpha_{i_{r-1}} + \alpha_i - \alpha_{i_r})((x_{i_1}\cdots x_{i_{r-1}}e_i)x_{i_r}) \\
&\quad - \delta_{i_r i}(\Lambda - \alpha_{i_1} - \cdots - \alpha_{i_{r-1}})(h_i)(\Lambda - \alpha_{i_1} - \cdots - \alpha_{i_{r-1}})x_{i_1}\cdots x_{i_{r-1}} \\
&= (\Lambda - \alpha_{i_1} - \cdots - \alpha_{i_{r-1}} + \alpha_i - \alpha_{i_r})(x_{i_1}\cdots x_{i_r}e_i)\,.
\end{aligned}
$$

This proves our assertion and it implies that

$$
\begin{aligned}
x_{i_1}\cdots x_{i_r}[e_i h] &= x_{i_1}\cdots x_{i_r}(e_i h - h e_i) \\
&= (\Lambda - \alpha_{i_1} - \cdots - \alpha_{i_r} + \alpha_i)x_{i_1}\cdots x_{i_r}e_i \\
&\quad - (\Lambda - \alpha_{i_1} - \cdots - \alpha_{i_r})x_{i_1}\cdots x_{i_r}e_i \\
&= \alpha_i x_{i_1}\cdots x_{i_r}e_i\,.
\end{aligned}
$$

Hence $[e_i h] - \alpha_i e_i \in \mathfrak{K}'$. We have therefore proved that all the generators of $\mathfrak{K}$ are contained in $\mathfrak{K}'$. Consequently $\mathfrak{K} \subseteq \mathfrak{K}'$ and so $\mathfrak{X}$ can be regarded as a module for $\tilde{\mathfrak{L}} = \mathfrak{FL}/\mathfrak{K}$. We can now prove

THEOREM 1. *Let $\mathfrak{FL}$ denote the free Lie algebra generated by $3l$ elements e_i, f_i, h_i, $i = 1, 2, \cdots, l$, let $\mathfrak{K}$ be the ideal in $\mathfrak{FL}$ generated by the elements* (2), *and let $\tilde{\mathfrak{L}} = \mathfrak{FL}/\mathfrak{K}$. Then*: (i) *The canonical homomorphism of $\mathfrak{FL}$ onto $\tilde{\mathfrak{L}}$ maps $\sum_1^l \Phi e_i + \sum_1^l \Phi f_i + \sum_1^l \Phi h_i$ isomorphically into $\tilde{\mathfrak{L}}$, so we can identify the corresponding subspaces.* (ii) *The subspace $\mathfrak{L}_i \equiv \Phi e_i + \Phi f_i + \Phi h_i$ of $\tilde{\mathfrak{L}}$ is a subalgebra which is a split three-dimensional simple algebra.* (iii) *The subalgebra $\tilde{\mathfrak{L}}_-$ of $\tilde{\mathfrak{L}}$ generated by the f_i is the free Lie algebra generated by these elements and a similar statement holds for the subalgebra $\tilde{\mathfrak{L}}_+$ generated by the e_i. $\mathfrak{H} \equiv \sum_1^l \Phi h_i$ is an abelian subalgebra and we have the vector space decomposition*

$$
\tilde{\mathfrak{L}} = \mathfrak{H} \oplus \tilde{\mathfrak{L}}_- \oplus \tilde{\mathfrak{L}}_+\,. \tag{5}
$$

(iv) *If $\tilde{\mathfrak{U}}$ is the universal enveloping algebra of $\tilde{\mathfrak{L}}$, then $\tilde{\mathfrak{U}} = \mathfrak{B}\tilde{\mathfrak{U}}_+\tilde{\mathfrak{U}}_-$ where $\mathfrak{B}$ is the subalgebra generated by $\mathfrak{H}$, $\tilde{\mathfrak{U}}_+$ the subalgebra generated by $\tilde{\mathfrak{L}}_+$ and $\tilde{\mathfrak{U}}_-$ the subalgebra generated by $\tilde{\mathfrak{L}}_-$.*

Proof: For the moment let $\tilde{x} = x + \mathfrak{K}$, $x \in \mathfrak{FL}$. Consider the representation R of $\tilde{\mathfrak{L}}$ determined in $\mathfrak{X}$ by the linear function $\Lambda = 0$ on $\mathfrak{H}$. If $h \in \mathfrak{H}$ and $\tilde{h} = 0$, then $\tilde{h}^R = h^R = 0$, which implies that $(\alpha_{i_1} + \cdots + \alpha_{i_r})(h) = 0$ for all choices of the α_{i_j}. Since $\alpha_1, \alpha_2, \cdots, \alpha_l$ form a basis for the conjugate space this implies that $h = 0$. Hence $h \rightarrow \tilde{h}$ is $1:1$. We have $[\tilde{e}_i\tilde{h}_i] = 2\tilde{e}_i$, $[\tilde{f}_i\tilde{h}_i] = -2\tilde{f}_i$, $[\tilde{e}_i, \tilde{f}_i] = \tilde{h}_i$, hence $\Phi\tilde{e}_i + \Phi\tilde{f}_i + \Phi\tilde{h}_i$ is a subalgebra of $\tilde{\mathfrak{L}}$ which is a homomorphic image of the split three-dimensional simple Lie algebra. Because of the simplicity of the latter the image is either 0 or is isomorphic to the three-dimensional split simple algebra. Since $\tilde{h}_i \neq 0$ by our first result, we have an isomorphism. This proves (ii). We have $[\tilde{h}\tilde{h}'] = 0$ for any $h, h' \in \sum \Phi h_i$ and $[\tilde{e}_i\tilde{h}] = \alpha_i(h)\tilde{e}_i$, $[\tilde{f}_i\tilde{h}] = -\alpha_i(h)\tilde{f}_i$. Since the linear functions $0, \pm\alpha_i$ are all different the usual weight argument implies that a relation of the form $\sum_1^l \xi_i\tilde{e}_i + \sum_1^l \eta_i\tilde{f}_i + \tilde{h}' = 0$, $h' \in \sum \Phi h_i$, implies $\xi_i = 0$, $\eta_i = 0$ for all i and $\tilde{h}' = 0$. Then $h' = 0$ by our first result. Thus we see that $x \to \tilde{x}$ is an isomorphism on $\sum \Phi e_i + \sum \Phi f_i + \sum \Phi h_i$. We make the identification of this space with its image and from now on we write e_i, f_i, h_i etc. for $\tilde{e}_i, \tilde{f}_i, \tilde{h}_i$ etc. We write $\mathfrak{H} = \sum_1^l \Phi h_i$ and we have seen that this is an l-dimensional abelian subalgebra of $\tilde{\mathfrak{L}}$. We have noted before that in the representation R of $\tilde{\mathfrak{L}}$ acting in $\mathfrak{X}$, $f_i^R = x_{iR}$, the right multiplication in $\mathfrak{X}$ determined by x_i. Let $\tilde{\mathfrak{U}}_-$ denote the universal enveloping algebra of $\tilde{\mathfrak{L}}_-$, the subalgebra of $\tilde{\mathfrak{L}}$ generated by the f_i. Then we have a homomorphism of $\tilde{\mathfrak{U}}_-$ into the algebra of linear transformations in $\mathfrak{X}$ mapping f_i into x_{iR}. If we combine this with the inverse of the isomorphism $a \to a_R$ of $\mathfrak{X}$ (the regular representation) we obtain a homomorphism of $\tilde{\mathfrak{U}}_-$ into $\mathfrak{X}$ sending f_i into x_i. On the other hand, since $\mathfrak{X}$ is freely generated by the x_i we have a homomorphism of $\mathfrak{X}$ into $\tilde{\mathfrak{U}}$ mapping x_i into f_i. It follows that both our homomorphisms are surjective isomorphisms. Since the free Lie algebra is obtained by taking the Lie algebra generated by the generators of a free associative algebra it is now clear that $\tilde{\mathfrak{L}}_-$ is the free Lie algebra generated by the f_i and $\tilde{\mathfrak{U}}_-$ is the free associative algebra generated by the f_i. Also the Poincaré-Birkhoff-Witt theorem permits the identification of $\tilde{\mathfrak{U}}_-$ with the subalgebra of $\tilde{\mathfrak{U}}$ generated by $\tilde{\mathfrak{L}}_-$, hence by the f_i. The basic property of free generators implies that we have an automorphism of $\mathfrak{F}\mathfrak{L}$ sending $e_i \to f_i$, $f_i \to e_i$, $h_i \to -h_i$. This maps the generators (2) of $\mathfrak{K}$ into $\mathfrak{K}$ and so it induces an automorphism in $\tilde{\mathfrak{L}} = \mathfrak{F}\mathfrak{L}/\mathfrak{K}$ which maps f_i into e_i. Since the subalgebra generated by the f_i is free it follows that the subalgebra $\tilde{\mathfrak{L}}_+$ generated by the e_i is free and its universal enveloping algebra $\tilde{\mathfrak{U}}_+$ is the free associative algebra generated by the e_i. This algebra can be identified with the subalgebra of $\tilde{\mathfrak{U}}$ generated by the e_i. It remains to prove (5); for once this is done then the relation $\tilde{\mathfrak{U}} = \mathfrak{B}\tilde{\mathfrak{U}}_+\tilde{\mathfrak{U}}_-$ follows from the Poincaré-Birkhoff-Witt theorem by choosing an ordered basis for $\tilde{\mathfrak{L}}$ to consist of an ordered basis for

$\mathfrak{H}$ followed by one for $\tilde{\mathfrak{L}}_+$ followed by one for $\tilde{\mathfrak{L}}_-$. To prove (5) we show first that $\tilde{\mathfrak{L}}_1 \equiv \mathfrak{H} + \tilde{\mathfrak{L}}_+ + \tilde{\mathfrak{L}}_-$ is a subalgebra of $\tilde{\mathfrak{L}}$. The argument is similar to one we have used before (§ 4.3). We observe first that every element of $\tilde{\mathfrak{L}}_+$ is a linear combination of the elements $[e_{i_1}e_{i_2}\cdots e_{i_r}] \equiv [\cdots[e_{i_1}e_{i_2}]\cdots e_{i_r}]$ and every element of $\tilde{\mathfrak{L}}_-$ is a linear combination of the elements $[f_{i_1}\cdots f_{i_r}]$. The Jacobi identity and induction on r implies that

$$(6)\qquad \begin{aligned} [[e_{i_1}\cdots e_{i_r}]h] &= (\alpha_{i_1} + \alpha_{i_2} + \cdots + \alpha_{i_r})[e_{i_1}\cdots e_{i_r}] \\ [[f_{i_1}\cdots f_{i_r}]h] &= -(\alpha_{i_1} + \alpha_{i_2} + \cdots + \alpha_{i_r})[f_{i_1}\cdots f_{i_r}]\,. \end{aligned}$$

Hence $[\tilde{\mathfrak{L}}_1\mathfrak{H}] \subseteq \tilde{\mathfrak{L}}_1$. We have $[e_if_j] \in \mathfrak{H}$ and by induction on $r \geqq 2$ we can show that $[[e_{i_1}\cdots e_{i_r}]f_j] \in \tilde{\mathfrak{L}}_+$. It follows that $\tilde{\mathfrak{L}}_1 \mathrm{ad} f_j \subseteq \tilde{\mathfrak{L}}_1$. Iteration of this and the Jacobi identity implies that $[\tilde{\mathfrak{L}}_1\tilde{\mathfrak{L}}_-] \subseteq \tilde{\mathfrak{L}}_1$. Similarly, $[\tilde{\mathfrak{L}}_1\tilde{\mathfrak{L}}_+] \subseteq \tilde{\mathfrak{L}}_1$. These and $[\tilde{\mathfrak{L}}_1\mathfrak{H}] \subseteq \tilde{\mathfrak{L}}_1$ imply that $\tilde{\mathfrak{L}}_1$ is a subalgebra. Since $e_i, f_i, h_i \in \tilde{\mathfrak{L}}_1$ it follows that $\tilde{\mathfrak{L}}_1 = \tilde{\mathfrak{L}}$, that is, $\tilde{\mathfrak{L}} = \tilde{\mathfrak{L}}_+ + \mathfrak{H} + \tilde{\mathfrak{L}}_-$. Equations (6) imply that $\tilde{\mathfrak{L}}$ is a direct sum of root spaces relative to $\mathfrak{H}$ and the non-zero roots are the functions $\pm(\alpha_{i_1} + \cdots + \alpha_{i_r})$. It is clear that $\tilde{\mathfrak{L}}_+$ is the sum of the root spaces corresponding to the roots $\alpha_{i_1} + \cdots + \alpha_{i_r}$ and $\tilde{\mathfrak{L}}_-$ is the sum of those corresponding to the roots $-(\alpha_{i_1} + \cdots + \alpha_{i_r})$. It follows that $\tilde{\mathfrak{L}} = \tilde{\mathfrak{L}}_+ \oplus \mathfrak{H} \oplus \tilde{\mathfrak{L}}_-$. This completes the proof.

2. *On certain cyclic modules for* $\tilde{\mathfrak{L}}$

A Lie module $\mathfrak{M}$ is *cyclic* with *generator* x if $\mathfrak{M}$ is the smallest submodule of $\mathfrak{M}$ containing x. If $\mathfrak{U}$ is the universal enveloping algebra of the Lie algebra, then $x\mathfrak{U} = \{xu \mid u \in \mathfrak{U}\}$ is the smallest submodule containing x. Hence $\mathfrak{M}$ is cyclic with x as generator if and only if $\mathfrak{M} = x\mathfrak{U}$. The module $\mathfrak{X}$ for $\tilde{\mathfrak{L}} = \mathfrak{F}\mathfrak{L}/\mathfrak{R}$ which we constructed in § 1 is cyclic with 1 as generator since $1f_{i_1}\cdots f_{i_r} = x_{i_1}\cdots x_{i_r}$ and these elements and 1 form a basis for $\mathfrak{X}$.

We shall call a module $\mathfrak{M}$ for $\tilde{\mathfrak{L}}$ *e-extreme* if it is cyclic and the generator x can be chosen so that $xh = \Lambda(h)x$ and $xe_i = 0$, $i = 1, 2, \cdots, l$. It is apparent from (4) that $\mathfrak{X}$ is e-extreme with 1 as generator of the required type. Thus we see that for every linear function $\Lambda(h)$ on $\mathfrak{H}$ there exists an e-extreme module for which the generator x satisfies $xh = \Lambda(h)x$, $xe_i = 0$. We shall now consider the theory of e-extreme modules for $\tilde{\mathfrak{L}}$. A similar theory can be developed for f-extreme modules which are defined to be cyclic with generator y such that $yh = \Lambda(h)y$, $yf_i = 0$, $i = 1, 2, \cdots, l$. We shall stick to the e-extreme modules but shall make use of the corresponding results for f-extreme modules when needed.

Let $\mathfrak{M}$ be e-extreme with generator x satisfying $xh = \Lambda(h)x$, $xe_i = 0$. We know that the universal enveloping algebra $\tilde{\mathfrak{U}}$ of $\tilde{\mathfrak{L}}$ can be factored as $\tilde{\mathfrak{U}} = \mathfrak{B}\tilde{\mathfrak{U}}_+\tilde{\mathfrak{U}}_-$ where $\mathfrak{B}, \tilde{\mathfrak{U}}_+$ and $\tilde{\mathfrak{U}}_-$ are the subalgebras generated by $\mathfrak{H}$, $\tilde{\mathfrak{L}}_+$ and $\tilde{\mathfrak{L}}_-$ respectively. Then $\mathfrak{M} = x\tilde{\mathfrak{U}} = x\mathfrak{B}\tilde{\mathfrak{U}}_+\tilde{\mathfrak{U}}_-$. Since $xh = \Lambda(h)x$, $x\mathfrak{B} = \Phi x$ and since $xe_i = 0$, $x\tilde{\mathfrak{U}}_+ = \Phi x$. Hence $\mathfrak{M} = x\mathfrak{B}\tilde{\mathfrak{U}}_+\tilde{\mathfrak{U}}_- = x\tilde{\mathfrak{U}}_-$. Moreover, $\tilde{\mathfrak{U}}_-$ is generated by the elements f_i. Hence every element of $\mathfrak{M}$ is a linear combination of the elements

$$xf_{i_1}\cdots f_{i_r}\,, \tag{7}$$

where we now adopt the convention that $f_{i_1}\cdots f_{i_r} = 1$ if $r = 0$. We have $[f_{i_1}\cdots f_{i_r}, h] = -(\alpha_{i_1} + \cdots + \alpha_{i_r})f_{i_1}\cdots f_{i_r}$ (induction on r) and $xh = \Lambda x$ and these relations imply that $(xf_{i_1}\cdots f_{i_r})h = x[f_{i_1}\cdots f_{i_r}, h] + (xh)f_{i_1}\cdots f_{i_r} = (\Lambda - \alpha_{i_1} - \cdots - \alpha_{i_r})xf_{i_1}\cdots f_{i_r}$. Hence

$$(xf_{i_1}\cdots f_{i_r})h = (\Lambda - \alpha_{i_1} - \cdots - \alpha_{i_r})(xf_{i_1}\cdots f_{i_r})\,. \tag{8}$$

Thus $\mathfrak{M}$ is a direct sum of weight spaces relative to $\mathfrak{H}$ and the weights are of the form

$$\Lambda - (\alpha_{i_1} + \alpha_{i_2} + \cdots + \alpha_{i_r}) = \Lambda - \sum_1^l k_i\alpha_i\,, \tag{9}$$

where the k_i are non-negative integers. Also it is clear that the restriction to a weight space of the linear transformation corresponding to any h is a scalar multiplication by a field element. It is clear also that the weight space $\mathfrak{M}_\Lambda$ corresponding to Λ has x as basis and so is one-dimensional. The weight space $\mathfrak{M}_M$ corresponding to the weight $M = \Lambda - \sum k_i\alpha_i$ is spanned by the vectors (7) such that $\alpha_{i_1} + \cdots + \alpha_{i_r} = \sum k_i\alpha_i$. Clearly there are only a finite number of sequences $(i_1, i_2, \cdots, i_r)$ such that $\alpha_{i_1} + \cdots + \alpha_{i_r} = \sum k_i\alpha_i$ where the k_i are fixed non-negative integers. Hence $\mathfrak{M}_M$ is finite-dimensional.

The weight Λ can be characterized as the only weight of $\mathfrak{H}$ in $\mathfrak{M}$ such that every weight has the form $\Lambda - \sum k_i\alpha_i$, k_i non-negative integers. We shall call Λ *the highest weight* of $\mathfrak{H}$ in $\mathfrak{M}$ or of $\mathfrak{M}$. If $\mathfrak{N}$ is isomorphic to $\mathfrak{M}$, then $\mathfrak{N}$ is also e-extreme and has the highest weight Λ. It follows that two e-extreme $\tilde{\mathfrak{L}}$-modules having distinct highest weights cannot be isomorphic.

Let $\mathfrak{N}$ be a submodule of $\mathfrak{M} = \sum \oplus \mathfrak{M}_M$ where $\mathfrak{M}_M$ is the weight module corresponding to M. If $y \in \mathfrak{N}$, $y \in \mathfrak{M}_{M_1} + \cdots + \mathfrak{M}_{M_k}$ where

$\{M_1, \cdots, M_k\}$ is a finite subset of the set of weights. Hence $y \in \mathfrak{N} \cap (\mathfrak{M}_{M_1} + \cdots + \mathfrak{M}_{M_k})$ and this is a submodule of the finite-dimensional $\mathfrak{H}$-module $\sum_{j=1}^{k} \mathfrak{M}_{M_j}$. Such an $\mathfrak{H}$-module is split and is a direct sum of weight modules whose weights are in the set $\{M_j\}$ (cf. § 2.4). This means that $\mathfrak{N} \cap (\sum \mathfrak{M}_{M_j}) = \sum (\mathfrak{N} \cap \mathfrak{M}_{M_j})$. Since y is any element of $\mathfrak{N}$ we have also that $\mathfrak{N} = \sum \oplus \mathfrak{N}_M$ where $\mathfrak{N}_M = \mathfrak{N} \cap \mathfrak{M}_M$. If $\mathfrak{N}_M \neq 0$, M is a weight for $\mathfrak{H}$ in $\mathfrak{N}$. At any rate, it is clear that $\mathfrak{N}$ is a direct sum of weight modules and the weights of $\mathfrak{H}$ in $\mathfrak{N}$ are among the weights of $\mathfrak{H}$ in $\mathfrak{M}$.

Now let $\mathfrak{M}'$ be the subspace of $\mathfrak{M}$ spanned by the $\mathfrak{M}^M$ with $M \neq \Lambda$ and assume that the submodule $\mathfrak{N} \neq \mathfrak{M}$. In this case, we must have $\mathfrak{N}_\Lambda = 0$ since, otherwise, $\mathfrak{N}_\Lambda = \mathfrak{M}_\Lambda$ which is one-dimensional. Then $x \in \mathfrak{N}$ and $\mathfrak{N} = x\tilde{\mathfrak{U}} = \mathfrak{M}$ contrary to assumption. Thus we see that any proper submodule $\mathfrak{N} = \sum_{M \neq \Lambda} \mathfrak{N}_M \subseteq \mathfrak{M}' \subset \mathfrak{M}$. It follows from this that the sum $\mathfrak{P}$ of all the proper submodules of $\mathfrak{M}$ is contained in $\mathfrak{M}' \subset \mathfrak{M}$ and so this is a proper submodule. This proves the existence of a maximal (proper) submodule $\mathfrak{P}$ of $\mathfrak{M}$. Moreover, $\mathfrak{P}$ is unique.

We consider again the module $\mathfrak{X}$ constructed in § 1 which we shall now show is a "universal" e-extreme module with highest weight Λ in the sense that every module $\mathfrak{M}$ of this type is a homomorphic image of $\mathfrak{X}$. For this purpose we define θ to be the linear mapping of $\mathfrak{X}$ onto $\mathfrak{M}$ such that $(x_{i_1} \cdots x_{i_r})\theta = xf_{i_1} \cdots f_{i_r}$ ($x_{i_1} \cdots x_{i_r} = 1$ if $r = 0$). Then

$$(x_{i_1} \cdots x_{i_r} f_i)\theta = (x_{i_1} \cdots x_{i_r} x_i)\theta = xf_{i_1} \cdots f_{i_r} f_i = ((x_{i_1} \cdots x_{i_r})\theta) f_i \,, \tag{10}$$

$$\begin{aligned}(x_{i_1} \cdots x_{i_r} h)\theta &= (\Lambda - \alpha_{i_1} - \cdots - \alpha_{i_r})(x_{i_1} \cdots x_{i_r})\theta \\ &= (\Lambda - \alpha_{i_1} - \cdots - \alpha_{i_r}) x f_{i_1} \cdots f_{i_r} \\ &= (xf_{i_1} \cdots f_{i_r})h = ((x_{i_1} \cdots x_{i_r})\theta)h \,,\end{aligned} \tag{11}$$

$$\begin{aligned}&(x_{i_1} \cdots x_{i_r} e_i)\theta \\ &= ((x_{i_1} \cdots x_{i_{r-1}} e_i) x_{i_r})\theta - \delta_{i_r i}(\Lambda - \alpha_{i_1} - \cdots - \alpha_{i_{r-1}})(h_i)(x_{i_1} \cdots x_{i_{r-1}})\theta \\ &= ((x_{i_1} \cdots x_{i_{r-1}} e_i) f_{i_r})\theta - \delta_{i_r i}(\Lambda - \alpha_{i_1} - \cdots - \alpha_{i_{r-1}})(h_i)(x_{i_1} \cdots x_{i_{r-1}})\theta \\ &= ((x_{i_1} \cdots x_{i_{r-1}} e_i)\theta) f_{i_r} - \delta_{i_r i}(\Lambda - \alpha_{i_1} - \cdots - \alpha_{i_{r-1}})(h_i)(x_{i_1} \cdots x_{i_{r-1}})\theta \,.\end{aligned}$$

If we use induction on r we can use this to establish the formula

$$(x_{i_1} \cdots x_{i_r} e_i)\theta = (x_{i_1} \cdots x_{i_r} \theta) e_i \,. \tag{12}$$

Since the elements e_i, f_i, h_i generate $\tilde{\mathfrak{L}}$, equations (10), (11) and (12) imply that θ is a module homomorphism of $\mathfrak{X}$ onto $\mathfrak{M}$.

The result just established shows that any e-extreme $\tilde{\mathfrak{L}}$-module with highest weight Λ is isomorphic to a module of the form $\mathfrak{X}/\mathfrak{N}$, $\mathfrak{N}$ a submodule of the module $\mathfrak{X}$. If $\mathfrak{M}$ is irreducible we have $\mathfrak{M} \cong \mathfrak{X}/\mathfrak{P}$ where $\mathfrak{P}$ is a maximal submodule of $\mathfrak{X}$. We have seen that there is only one such submodule. Hence it is clear that any two irreducible e-extreme modules with the same highest weight are isomorphic. The existence of an irreducible e-extreme module with highest weight Λ is clear also, for, the module $\mathfrak{M} = \mathfrak{X}/\mathfrak{P}$ satisfies these requirements.

We summarize our main results in the following

THEOREM 2. *Let the notations be as in Theorem* 1 *and let* $\Lambda(h)$ *be a linear function on* $\mathfrak{H}$. *Then there exists an irreducible e-extreme* $\tilde{\mathfrak{L}}$*-module with highest weight* Λ. *The weights for such a module are of the form* $\Lambda - \sum k_i\alpha_i$, k_i *a non-negative integer. The weight space corresponding to* Λ *is one-dimensional and all the weight spaces are finite-dimensional. Every* $h \in \mathfrak{H}$ *acts as a scalar multiplication in every weight space. Two irreducible e-extreme* $\tilde{\mathfrak{L}}$*-modules are isomorphic if and only if they have the same highest weight.*

3. Finite-dimensional irreducible modules

We shall call a linear function Λ on $\mathfrak{H}$ *integral* if $\Lambda(h_i)$ is an integer for every $i = 1, 2, \cdots, l$, and we shall call an integral linear function *dominant* if $\Lambda(h_i) \geqq 0$ for all i. In this section we establish a $1:1$ correspondence between these linear functions and the isomorphism classes of finite-dimensional irreducible modules for $\tilde{\mathfrak{L}}$. In view of the correspondence between the isomorphism classes of irreducible e-extreme $\tilde{\mathfrak{L}}$-modules and the highest weights which we established in Theorem 2 it suffices to prove two things: (1) Every finite-dimensional irreducible module is e-extreme with highest weight dominant integral, (2) Any irreducible e-extreme module with highest weight a dominant integral linear function is finite-dimensional. We prove first

THEOREM 3. *Let* $\tilde{\mathfrak{L}}$ *be as before and let* $\mathfrak{M}$ *be a finite-dimensional irreducible* $\tilde{\mathfrak{L}}$*-module. Then* $\mathfrak{M}$ *is e-extreme and its highest weight is a dominant integral linear function on* $\mathfrak{H}$.

Proof: $\mathfrak{M}$ is a finite-dimensional module for the subalgebra $\mathfrak{L}_i =$

$\Phi e_i + \Phi f_i + \Phi h_i$ which is a split three-dimensional simple Lie algebra. Hence $\mathfrak{M}$ is completely reducible as $\mathfrak{L}_i$-module. The form of the irreducible modules for $\mathfrak{L}_i$ (§ 3.8) shows that there exists a basis $(x_1^{(i)}, x_2^{(i)}, \cdots, x_N^{(i)})$ for $\mathfrak{M}$ such that $x_k^{(i)} h_i = m_{ik} x_k^{(i)}$ where the m_{ik} are integers. Since $[h_i h_j] = 0$ the linear transformations associated with the different h_i commute; hence, we can find a basis $(x_1, x_2, \cdots, x_N)$ such that $x_k h_i = m_{ik} x_k$, $i = 1, \cdots, l$, $k = 1, \cdots, N$. If Λ_k denotes the linear function on $\mathfrak{H}$ such that $\Lambda_k(h_i) = m_{ik}$, then Λ_k is integral and

$$x_k h = \Lambda_k x_k \,, \qquad k = 1, 2, \cdots, N \,. \tag{13}$$

Then Λ_k are the weights of $\mathfrak{H}$ in $\mathfrak{M}$. Let $\mathfrak{H}_0^*$ be the rational vector space spanned by the linear functions α_i. It is easy to see that a linear function $\alpha \in \mathfrak{H}_0^*$ if and only if the values $\alpha(h_i)$ are rational for $i = 1, 2, \cdots, l$ (cf. the proof of XIII in § 4.2). Hence the weights $\Lambda_k \in \mathfrak{H}_0^*$ and so we may pick out among these weights the highest weight Λ in the ordering of $\mathfrak{H}_0^*$ which is specified by saying that $\sum_1^l \lambda_i \alpha_i > 0$ if the first $\lambda_i \neq 0$ is positive. In this case it is clear that $\Lambda + \alpha_i$ is not a weight for any α_i. Let x be a non-zero vector such that $xh = \Lambda x$. Then $(xe_i)h = (\Lambda + \alpha_i)(xe_i)$ so $xe_i = 0$ by the maximality of Λ. Since $\mathfrak{M}$ is irreducible and $x\tilde{\mathfrak{U}}$ is a submodule of $\mathfrak{M}$ it is clear that $\mathfrak{M} = x\tilde{\mathfrak{U}}$. Hence $\mathfrak{M}$ is e-extreme with Λ as its highest weight also in the sense of the last section. The results of § 2 show that every weight is of the form $\Lambda - \sum k_i \alpha_i$, k_i a non-negative integer. On the other hand, the proof of the representation theorem, Theorem 4.1, (applied to $\mathfrak{H} + \Phi e_i + \Phi f_i$) shows that if M is a weight for $\mathfrak{H}$ in $\mathfrak{M}$, then $M - M(h_i)\alpha_i$ is also a weight. Hence for each i, $\Lambda - \Lambda(h_i)\alpha_i$ is a weight and so has the form $\Lambda - \sum k_i \alpha_i$. It follows that $\Lambda(h_i) = k_i \geqq 0$. Thus we see that Λ is a dominant integral function. This completes the proof.

Next let Λ be any dominant integral linear function on $\mathfrak{H}$ and let $\mathfrak{M}$ be the irreducible module furnished by Theorem 2 with maximum weight Λ. The weights of $\mathfrak{H}$ in $\mathfrak{M}$ have the form $\Lambda - \sum k_i \alpha_i$, k_i integral and non-negative. Hence these are integral and so they can be ordered by the ordering in $\mathfrak{H}_0^*$. We shall prove that $\mathfrak{M}$ is finite-dimensional. The proof will be based on several lemmas, as follows.

LEMMA 1. *Let $\theta_{ij} = f_j(\mathrm{ad}\, f_i)^{-A_{ij}+1}$, $i \neq j = 1, 2, \cdots, l$. Then $[\theta_{ij} e_k] = 0$, $k = 1, 2, \cdots, l$, and $\mathfrak{M}\theta_{ij} = 0$ for any e-extreme irreducible $\tilde{\mathfrak{L}}$-module $\mathfrak{M}$.*

Proof: If $k \neq i$, $[f_ie_k] = 0$; hence $[\mathrm{ad}\, f_i\, \mathrm{ad}\, e_k] = 0$. Then

$$\begin{aligned}[\theta_{ij}e_k] &= f_j(\mathrm{ad}\, f_i)^{-A_{ij}+1}\,\mathrm{ad}\, e_k = f_j\,\mathrm{ad}\, e_k(\mathrm{ad}\, f_i)^{-A_{ij}+1} \\ &= [f_je_k](\mathrm{ad}\, f_i)^{-A_{ij}+1} = -\,\delta_{jk}h_j(\mathrm{ad}\, f_i)^{-A_{ij}+1}\end{aligned}$$

If $k \neq j$ this is 0. If $k = j$ we obtain $[\theta_{ij}e_k] = -\,A_{ji}f_i(\mathrm{ad}\, f_i)^{-A_{ij}}$. If $A_{ij} = 0$, $A_{ji} = 0$ by the hypothesis (α) of § 1, so the result is 0 in this case. Otherwise, $-A_{ij} > 0$ and $f_i(\mathrm{ad}\, f_i)^{-A_{ij}} = 0$. Next let $k = i$. Then $[\theta_{ij}e_k] = f_j(\mathrm{ad}\, f_i)^{-A_{ij}+1}\,\mathrm{ad}\, e_i$. We recall the commutation formula: $a^kx = xa^k - \binom{k}{1}x'a^{k-1} + \binom{k}{2}x''a^{k-2} - \cdots$, where $x' = [xa]$, $x'' = [x'a], \cdots$. (eq. 2.6). If we make use of this and the table: $[e_if_i] = h_i$, $[[e_if_i]f_i] = 2f_i$, $[[[e_if_i]f_i]f_i] = 0$, we obtain

$$\begin{aligned}f_j(\mathrm{ad}\, f_i)^{-A_{ij}+1}\,\mathrm{ad}\, e_i &= f_j\,\mathrm{ad}\, e_i(\mathrm{ad}\, f_i)^{-A_{ij}+1} \\ &\quad -(-A_{ij}+1)f_j\,\mathrm{ad}\, h_i(\mathrm{ad}\, f_i)^{-A_{ij}}\,, \qquad \text{if } A_{ij} = 0 \\ &= f_j(\mathrm{ad}\, e_i)(\mathrm{ad}\, f_i)^{-A_{ij}+1} - (-A_{ij}+1)f_j\,\mathrm{ad}\, h_i(\mathrm{ad}\, f_i)^{-A_{ij}} \\ &\quad + \tfrac{1}{2}(-A_{ij}+1)(-A_{ij})f_j(\mathrm{ad}\, 2f_i)(\mathrm{ad}\, f_i)^{-A_{ij}-1}\,, \qquad \text{if } -A_{ij} > 0\,.\end{aligned}$$

The first term in both of these formulas can be dropped since $f_j\,\mathrm{ad}\, e_i = [f_je_i] = 0$. If $A_{ij} = 0$ we have $f_j\,\mathrm{ad}\, h_i = [f_jh_i] = 0$. If $-A_{ij} > 0$ we use the second formula and $f_j\,\mathrm{ad}\, h_i = [f_jh_i] = -\,A_{ij}f_j$ to obtain

$$\begin{aligned}&-(-A_{ij}+1)(-A_{ij})f_j(\mathrm{ad}\, f_i)^{-A_{ij}} \\ &\quad + (-A_{ij}+1)(-A_{ij})f_j(\mathrm{ad}\, f_i)^{-A_{ij}} = 0\,.\end{aligned}$$

This completes the proof of $[\theta_{ij}e_k] = 0$. Now let $\mathfrak{M}$ be an irreducible e-extreme $\tilde{\mathfrak{L}}$-module and, as before, let $\mathfrak{M}'$ be the subspace spanned by the weight spaces corresponding to the weights other than the highest weight Λ. Consider the submodule $\mathfrak{M}\theta_{ij}\tilde{\mathfrak{U}} = \mathfrak{M}\theta_{ij}\mathfrak{B}\tilde{\mathfrak{U}}_+\tilde{\mathfrak{U}}_-$ where $\mathfrak{B}, \tilde{\mathfrak{U}}_+$ and $\tilde{\mathfrak{U}}_-$ are as in Theorem 1. If we use the fact that $\mathrm{ad}\, h$ is a derivation and that $[f_ih] = -\,\alpha_if_i$ we see that

$$\begin{aligned}\theta_{ij}h &= h\theta_{ij} + [\theta_{ij}h] \\ &= h\theta_{ij} + (-\alpha_j + (A_{ij} - 1)\alpha_i)\theta_{ij}\,,\end{aligned}$$

which implies that $\mathfrak{M}\theta_{ij}\mathfrak{B} = \mathfrak{M}\theta_{ij}$. Also $\theta_{ij}e_k = e_k\theta_{ij}$ and this implies that $\mathfrak{M}\theta_{ij}\tilde{\mathfrak{U}}_+ = \mathfrak{M}\theta_{ij}$. Hence $\mathfrak{M}\theta_{ij}\tilde{\mathfrak{U}} = \mathfrak{M}\theta_{ij}\tilde{\mathfrak{U}}_-$. If x is a canonical generator of $\mathfrak{M}$ such that $xh = \Lambda x$, $xe_i = 0$, $i = 1, 2, \cdots, l$, then every element of $\mathfrak{M}$ is a linear combination of the elements of the form $xf_{i_1}\cdots f_{i_r}$ and every element of $\mathfrak{M}'$ is a linear combination of these elements for which $r \geqq 1$. It follows from the definition of θ_{ij} that $\mathfrak{M}\theta_{ij} \subseteqq \mathfrak{M}'$. Also it is clear that $\mathfrak{M}'\tilde{\mathfrak{U}}_- \subseteqq \mathfrak{M}'$. Hence

$\mathfrak{M}\theta_{ij}\tilde{\mathfrak{U}} \subseteq \mathfrak{M}'$ and so $\mathfrak{M}\theta_{ij}\tilde{\mathfrak{U}}$ is a proper submodule of $\mathfrak{M}$. Since $\mathfrak{M}$ is irreducible we must have $\mathfrak{M}\theta_{ij} = 0$ and the proof is complete.

Lemma 2. *Let $\mathfrak{M}$ be an irreducible e-extreme module for $\tilde{\mathfrak{L}}$ whose highest weight Λ is a dominant integral linear function. Then for any $y \in \mathfrak{M}$ there exist positive integers r_i, s_i such that $ye_i^{r_i} = 0 = yf_i^{s_i}$, $i = 1, 2, \cdots, l$.*

Proof: Let x be a generator of $\mathfrak{M}$ such that $xh = \Lambda x$, $xe_i = 0$, $i = 1, 2, \cdots, l$. Then every element of $\mathfrak{M}$ is a linear combination of elements of the form $xf_{i_1}f_{i_2}\cdots f_{i_r}$. It suffices to prove the result for every $y = xf_{i_1}f_{i_2}\cdots f_{i_r}$. We have $yh = My$, $M = \Lambda - \sum k_i\alpha_i$, k_i an integer $\geqq 0$. Then $(ye_i^{k_i+1})h = (M + (k_i + 1)\alpha_i)ye_i^{k_i+1}$. Since $M + (k_i + 1)\alpha_i$ is not a weight, $ye_i^{k_i+1} = 0$ which proves the assertion for e_i. (This argument is valid for arbitrary e-extreme $\tilde{\mathfrak{L}}$-modules.) Let $\Lambda(h_i) = m_i \geqq 0$. We show next that $xf_i^{m_i+1} = 0$. We have $xh_i = m_ix$, $xe_i = 0$. Hence if we apply the theory of e-extreme modules to the algebra $\mathfrak{L}_i = \Phi e_i + \Phi f_i + \Phi h_i$ (in place of $\tilde{\mathfrak{L}}$) we see that the $\mathfrak{L}_i$-submodule $\mathfrak{M}_i$ generated by x is the space spanned by $x, xf_i, xf_i^2, \cdots$. Suppose this has a proper submodule $\mathfrak{N}_i \neq 0$. Since $(xf_i^k)h_i = (m_i - 2k)xf_i^k$, the spaces Φxf_i^k are the weight spaces relative to Φh_i. Hence $\mathfrak{N}_i$ is spanned by certain of the subspaces Φxf_i^k and $k \geqq 1$ since $\mathfrak{N}_i \subset \mathfrak{M}_i$. Moreover, if $xf_i^k \in \mathfrak{N}_i$ then $xf_i^q \in \mathfrak{N}_i$ for all $q \geqq k$. It follows that $\mathfrak{N}_i = \sum_{q\geqq k\geqq 1}\Phi xf_i^q$ where k is the least positive integer such that $xf_i^k \in \mathfrak{N}_i$. Evidently $\mathfrak{N}_i \subseteq \mathfrak{M}'$. We now observe that $\mathfrak{N}_ih \subseteq \mathfrak{N}_i$ since $(xf_i^q)h = (\Lambda - q\alpha_i)(xf_i^q)$ and $\mathfrak{N}_ie_k \subseteq \mathfrak{N}_i$ since this is clear for $k = i$ and it holds for $k \neq i$ since $xf_i^qe_k = xe_kf_i^q = 0$. It follows now that $\mathfrak{N}_i\mathfrak{B} = \mathfrak{N}_i$ and $\mathfrak{N}_i\tilde{\mathfrak{U}}_+ = \mathfrak{N}_i$. Then $\mathfrak{N}_i\tilde{\mathfrak{U}} = \mathfrak{N}_i\mathfrak{B}\tilde{\mathfrak{U}}_+\tilde{\mathfrak{U}}_- = \mathfrak{N}_i\tilde{\mathfrak{U}}_- \subseteq \mathfrak{M}'$. Thus $\mathfrak{N}_i\tilde{\mathfrak{U}}$ is a proper $\tilde{\mathfrak{L}}$-submodule of $\mathfrak{M} \neq 0$. This contradicts the irreducibility of $\mathfrak{M}$ and so proves that $\mathfrak{M}_i$ is $\mathfrak{L}_i$-irreducible. Now in § 3.8 we constructed a finite-dimensional irreducible $\mathfrak{L}_i$-module with a generator x' such that $x'h_i = m_ix'$, $x'e_i = 0$ and $x'f_i^{m_i+1} = 0$. It follows from the isomorphism result on irreducible e-extreme modules (Theorem 2) that this module is isomorphic to $\mathfrak{M}_i$. Hence we have $xf_i^{m_i+1} = 0$. Now suppose we have an integer $m \geqq 0$ such that $(xf_{i_1}\cdots f_{i_{r-1}})f_i^m = 0$. If $i_r = i$ this implies that $(xf_{i_1}\cdots f_{i_r})f_i^m = 0$. If $i_r = j \neq i$ we use the relation

$$f_jf_i^{m-A_{ij}} \equiv f_i^{m-A_{ij}}f_j + \binom{m - A_{ij}}{1}f_i^{m-A_{ij}-1}f_j(\operatorname{ad} f_i)$$
$$+ \cdots + \binom{m - A_{ij}}{-A_{ij}}f_i^mf_j(\operatorname{ad} f_i)^{-A_{ij}} \pmod{\theta_{ij}},$$

where θ_{ij} is as in Lemma 1 and the congruence is used in the sense of the ideal in $\tilde{\mathfrak{U}}$ generated by θ_{ij}. It follows from Lemma 1 that

$$xf_{i_1}\cdots f_{i_{r-1}}f_jf_i^{m-A_{ij}} = xf_{i_1}\cdots f_{i_{r-1}}f_i^{m-A_{ij}}f_j + \cdots + \binom{m-A_{ij}}{-A_{ij}}xf_{i_1}\cdots f_{i_{r-1}}f_i^m f_j(\operatorname{ad} f_i)^{-A_{ij}} = 0 .$$

This proves the assertion on f_i by induction on r.

LEMMA 3. *Let $\mathfrak{M}$ be as in Lemma 2. Then if M is a weight of $\mathfrak{H}$ in $\mathfrak{M}$, $M - M(h_i)\alpha_i$ is a weight for $i = 1, 2, \cdots, l$.*

Proof: Let y be a non-zero vector such that $yh = M(h)y$. If $M(h_i) \geqq 0$, then we choose q so that $z = ye_i^q \neq 0$, $ye_i^{q+1} = 0$ and m so that $zf_i^m \neq 0$, $zf_i^{m+1} = 0$. This can be done by Lemma 2. Then the determination of the finite-dimensional irreducible modules for $\mathfrak{L}_i = \Phi e_i + \Phi f_i + \Phi h_i$ (§ 3.8) shows that $\sum_{k=0}^m \Phi z f_i^k$ is such a module and $zh_i = mz$. On the other hand, $yh = M(h)y$ implies $zh = (ye_i^q)h = (M+q\alpha_i)(h)z$ and $(zf_i^k)h = (M + q\alpha_i - k\alpha_i)(h)zf_i^k$. Hence $(M+q\alpha_i)(h_i) = M(h_i) + 2q = m$ and

$$M + q\alpha_i\,, \qquad M + (q-1)\alpha_i, \cdots, \qquad M + (q-m)\alpha_i$$

are weights (corresponding to $z, zf_i, \cdots, zf_i^m$). We have $M - M(h_i)\alpha_i = M + (2q - m)\alpha_i$ and $q - m \leqq 2q - m \leqq q$ since $M(h_i) = m - 2q \geqq 0$ and $q \geqq 0$. Hence $M - M(h_i)\alpha_i$ is in the displayed sequence. If $M(h_i) \leqq 0$ we reverse the roles of e_i and f_i and argue in a similar fashion.

We can now prove

THEOREM 4. *Let $\mathfrak{M}$ be an irreducible e-extreme module for $\tilde{\mathfrak{L}}$ such that the highest weight Λ is a dominant integral linear function. Then $\mathfrak{M}$ is finite-dimensional.*

Proof: Let S_{α_i} denote the linear mapping $\xi \to \xi - \xi(h_i)\alpha_i$ in the space $\mathfrak{H}_0^*$ of rational linear combinations of the linear functions α_i such that $\alpha_i(h_j) = A_{ji}$. We have $\alpha_j S_{\alpha_i} = \alpha_j - A_{ij}\alpha_i$ so that S_{α_i} is one of the linear transformations specified in our axiom (γ) of § 1. This axiom states that the group W generated by the S_{α_i} is finite. On the other hand, Lemma 3 implies that the set of weights of $\mathfrak{H}$ in $\mathfrak{M}$ is invariant under W. We now consider the set Σ of images under W of the maximal weight Λ. This is a finite set, so it has a least element M in the lexicographic ordering we have defined in $\mathfrak{H}_0^*$ (at the beginning of this section).* Let y be a non-zero vector such that $yh = My$. Since $M - M(h_i)\alpha_i \in \Sigma$, $M \leqq M - M(h_i)\alpha_i$ so that we have $M(h_i) \leqq 0$, $i = 1, 2, \cdots, l$. One of the linear functions $M - \alpha_i$, $M + \alpha_i$ is not a weight. Thus $M = \Lambda S$, $S \in W$ and if $M - \alpha_i$, $M + \alpha_i$ are weights then $\Lambda - \alpha_i S^{-1} = (M - \alpha_i)S^{-1}$ and $\Lambda + \alpha_i S^{-1}$ are weights. However, one of these is greater than Λ in the ordering in $\mathfrak{H}_0^*$ which contradicts the maximality of Λ. If $M + \alpha_i$ is not a weight we have $ye_i = 0$ as well as $yh_i = M(h_i)y$. As in the proof of Lemma 3, y generates an irreducible $\mathfrak{L}_i$-module of dimension $M(h_i) + 1$. Hence $M(h_i) \geqq 0$. Since $M(h_i) \leqq 0$ we have $M(h_i) = 0$. If $M - \alpha_i$ is a weight we would have also that $M - \alpha_i - (M - \alpha_i)(h_i)\alpha_i = M - \alpha_i + 2\alpha_i = M + \alpha_i$ is a weight contrary to assumption. Thus $M - \alpha_i$ is not a weight if $M + \alpha_i$ is not a weight. Hence in any case, $M - \alpha_i$ is not a weight. Consequently, $yf_i = 0$, $i = 1, 2, \cdots, l$. Since $\mathfrak{M}$ is irreducible $\mathfrak{M} = y\tilde{\mathfrak{U}}$. Thus we see that $\mathfrak{M}$ is an f-extreme module. It follows that the weights of $\mathfrak{H}$ in $\mathfrak{M}$ have the form $M + \sum j_i\alpha_i$, j_i integer $\geqq 0$. If $M = \Lambda - \sum l_i\alpha_i$ it is now clear that the weights have the form $\Lambda - \sum k_i\alpha_i$ where $0 \leqq k_i \leqq l_i$. Thus there are only a finite number of different weights and since every weight space is finite-dimensional we see that $\mathfrak{M}$ is finite-dimensional.

4. *Existence theorem and isomorphism theorem for semi-simple Lie algebras*

Let F denote the collection of finite-dimensional irreducible representations of $\tilde{\mathfrak{L}}$ and let $\mathfrak{K}_0$ be the kernel of the collection F, that is, the set of elements $b \in \tilde{\mathfrak{L}}$ such that $b^R = 0$ for every $R \in F$. Then $\mathfrak{K}_0$ is an ideal in $\tilde{\mathfrak{L}}$. If $R \in F$ we can define a representation R for $\mathfrak{L} = \tilde{\mathfrak{L}}/\mathfrak{K}_0$ by setting $(l + \mathfrak{K}_0)^R = l^R$. Since the set of representing transformations in the corresponding module is the same as that furnished by the representation of $\tilde{\mathfrak{L}}$ it is clear that R is a finite-dimensional irreducible representation for $\mathfrak{L}$. We shall now show that $\mathfrak{L}$ is a split semi-simple Lie algebra and the given matrix (A_{ij}) is a Cartan matrix for $\mathfrak{L}$. Thus we have the following

THEOREM 5. *Let* $\tilde{\mathfrak{L}} = \mathfrak{FL}/\mathfrak{K}$ *where* $\mathfrak{FL}$ *is the free Lie algebra generated by* e_i, f_i, h_i, $i = 1, 2, \cdots, l$ *and* $\mathfrak{K}$ *is the ideal in* $\mathfrak{FL}$ *generated by the elements* (2). *Let* $\mathfrak{K}_0$ *be the kernel of all the finite-*

* This is the only place in our discussion in which axiom (γ) is used.

dimensional irreducible representations of $\tilde{\mathfrak{L}}$ *and set* $\mathfrak{L} = \tilde{\mathfrak{L}}/\mathfrak{K}_0$. *Then*: (1) $\mathfrak{L}$ *is a finite-dimensional split semi-simple Lie algebra.* (2) *The canonical mapping of* $\mathfrak{H} + \sum_1^l \Phi e_i + \sum_1^l \Phi f_i$ *into* $\mathfrak{L}$ *is an isomorphism so we can identify this subspace with its image.* (3) $\mathfrak{H}$ *is a splitting Cartan subalgebra, the* α_i *form a simple system of roots and the* h_i, e_i, f_i *form a set of canonical generators whose associated Cartan matrix is the given matrix* (A_{ij}). (4) *Let* $\bar{\mathfrak{L}}$ *be any finite-dimensional split semi-simple Lie algebra with l-dimensional splitting Cartan subalgebra* $\bar{\mathfrak{H}}$ *and canonical generators* $\bar{h}_i, \bar{e}_i, \bar{f}_i$ *whose associated Cartan matrix is* (A_{ij}). *Then there exists an isomorphism of* $\mathfrak{L}$ *onto* $\bar{\mathfrak{L}}$ *sending* $h_i \to \bar{h}_i$, $e_i \to \bar{e}_i$, $f_i \to \bar{f}_i$, $i = 1, 2, \cdots, l$.

Proof: Let λ_i denote the linear function on $\mathfrak{H}$ such that $\lambda_i(h_j) = \delta_{ij}$. Then λ_i is a dominant integral linear function and so it corresponds to a finite-dimensional irreducible representation R_i in a module $\mathfrak{M}_i$ for $\tilde{\mathfrak{L}}$ (and for $\mathfrak{L}$). The λ_i form a basis for $\mathfrak{H}^*$ and every dominant integral linear function Λ has the form $\Lambda = \sum m_i\lambda_i$, m_i integral $\geqq 0$. We shall show first that $\mathfrak{K}_0$ can be characterized as the kernel of the finite set of representations R_i. Thus let $\mathfrak{M}$ be any finite-dimensional irreducible module and let $\Lambda = \sum m_i\lambda_i$ be its highest weight. Let $\mathfrak{P}$ be the module which is the tensor product of m_1 copies of $\mathfrak{M}_1$, m_2 copies of $\mathfrak{M}_2, \cdots, m_l$ copies of $\mathfrak{M}_l$. Let x_j be a generator of $\mathfrak{M}_j$ such that $x_jh = \lambda_jx_j$, $h \in \mathfrak{H}$, $x_je_i = 0$, $i = 1, 2, \cdots, l$, and set

$$(14) \qquad x = \overbrace{x_1 \otimes \cdots \otimes x_1}^{m_1} \otimes \overbrace{x_2 \otimes \cdots \otimes x_2}^{m_2} \otimes \cdots \otimes \overbrace{x_l \otimes \cdots \otimes x_l}^{m_l} .$$

In this definition and formula we adopt the convention that if all the $m_i = 0$, which is the case, if and only if $\Lambda = 0$, then $\mathfrak{M}$ is the one-dimensional module with basis x and $xl = 0$, $l \in \mathfrak{L}$. Then, in any case, we have $xh = \Lambda x$, $xe_i = 0$, so $x\tilde{\mathfrak{U}}$ is an e-extreme module with highest weight Λ. Consequently, the irreducible module $\mathfrak{M}$ corresponding to Λ is a homomorphic image of $x\tilde{\mathfrak{U}}$. Suppose $l \in \tilde{\mathfrak{L}}$ and $l^{R_i} = 0$ for $i = 1, 2, \cdots, l$. Then $\mathfrak{M}_il = 0$; hence $\mathfrak{P}l = 0$ and $x\tilde{\mathfrak{U}}l = 0$. This implies that $\mathfrak{M}l = 0$ or $l^R = 0$. We have therefore proved our assertion that $\mathfrak{K}_0$ is the kernel of the finite set of representations R_i, or equally well, the kernel of the single representation S which is the direct sum of the R_i. Evidently S is finite-dimensional completely reducible and $\mathfrak{L} = \tilde{\mathfrak{L}}/\mathfrak{K}_0 \cong \mathfrak{L}^S$ a finite-dimensional completely reducible Lie algebra of linear transformations. Since $x_ih = \lambda_i(h)x_i$, $h \in \mathfrak{H}$, $h^S = 0$ implies that $\lambda_i(h) = 0$ for all i.

Since the λ_i form a basis for $\mathfrak{H}^*$ this implies that $h = 0$. Hence $\mathfrak{H}$ is mapped faithfully by S and consequently by the canonical mapping of $\tilde{\mathfrak{L}}$ onto $\mathfrak{L}$. It follows from this that the simple Lie algebras $\mathfrak{L}_i$ are mapped faithfully. Then the argument used to prove the corresponding assertion in Theorem 1 shows that (2) holds. Since $\tilde{\mathfrak{L}}$ is a direct sum of $\mathfrak{H}$ and the spaces $\tilde{\mathfrak{L}}_M$ corresponding to the roots $M = \pm(\sum k_i\alpha_i)$, k_i integral $\geqq 0$, the same holds for $\mathfrak{L}$. It follows that $\mathfrak{H}$ is a splitting Cartan subalgebra for $\mathfrak{L}$. Consequently, the center $\mathfrak{C}$ of $\mathfrak{L}$ is contained in $\mathfrak{H}$.

If $h_0 \in \mathfrak{C}$ we must have $[e_ih_0] = \alpha_i(h_0)e_i = 0$. Then $\alpha_i(h_0) = 0$ for all i and $h_0 = 0$. Thus $\mathfrak{C} = 0$. Since $\mathfrak{L}$ is isomorphic to a finite-dimensional completely reducible Lie algebra of linear transformations and has 0 center it follows that $\mathfrak{L}$ is semi-simple. Hence we have established (1). We have $[e_ih] = \alpha_ie_i$ and every root of $\mathfrak{L}$ relative to $\mathfrak{H}$ has the form $\pm(\sum k_i\alpha_i)$, k_i integral and non-negative. Hence the α_i form a simple system. We have $[e_if_j] = \delta_{ij}h_i$, $[e_ih_j] = A_{ji}e_i$, $[f_ih_j] = -A_{ji}f_i$. This means that $e_i \in \mathfrak{L}_{\alpha_i}$, $f_i \in \mathfrak{L}_{-\alpha_i}$, the generators e_i, f_i, h_i are canonical and the associated Cartan matrix is (A_{ij}) (cf. § 4.3.) This completes the proof of (3). Now let $\bar{\mathfrak{L}}$ be as in (4). Then it is clear that there exists a homomorphism of $\tilde{\mathfrak{L}}$ onto $\bar{\mathfrak{L}}$ such that $e_i \to \bar{e}_i$, $f_i \to \bar{f}_i$, $h_i \to \bar{h}_i$. Since $\bar{\mathfrak{L}}$ is semi-simple it can be identified with a completely reducible Lie algebra of linear transformations in a finite-dimensional vector space and the homomorphism of $\tilde{\mathfrak{L}}$ onto $\bar{\mathfrak{L}}$ can be considered as a representation. Since $\mathfrak{K}_0$ is mapped into 0 in any finite-dimensional irreducible representation, the homomorphism of $\tilde{\mathfrak{L}}$ maps $\mathfrak{K}_0$ into 0 and so we have an induced homomorphism of $\mathfrak{L}$ onto $\bar{\mathfrak{L}}$ such that $e_i \to \bar{e}_i$, $f_i \to \bar{f}_i$, $h_i \to \bar{h}_i$. Since $\bar{\mathfrak{H}}$ is l-dimensional the homomorphism maps $\mathfrak{H}$ isomorphically. If $\mathfrak{K}_1$ is the kernel of the homomorphism of $\mathfrak{L}$ onto $\bar{\mathfrak{L}}$, since $\mathfrak{K}_1$ is an ideal, it is invariant under $\mathrm{ad}_{\mathfrak{L}}\,\mathfrak{H}$. It follows that if $\mathfrak{K}_1 \neq 0$ then it contains a non-zero element of $\mathfrak{H}$ or it contains one of the (one-dimensional) root spaces $\mathfrak{L}_\alpha$, $\alpha \neq 0$. The first is ruled out since $\mathfrak{K}_1 \cap \mathfrak{H} = 0$. If $\mathfrak{K}_1 \supseteq \mathfrak{L}_\alpha$, $\mathfrak{K}_1 \supseteq [\mathfrak{L}_\alpha\mathfrak{L}_{-\alpha}] \neq 0$. Since $[\mathfrak{L}_\alpha\mathfrak{L}_{-\alpha}] \subseteq \mathfrak{H}$ this is ruled out too. Hence $\mathfrak{K}_1 = 0$ and the homomorphism of $\mathfrak{L}$ is an isomorphism. This proves (4).

Theorems 3 and 4 establish a $1:1$ correspondence between the isomorphism classes of finite-dimensional irreducible modules for the (infinite-dimensional) Lie algebra $\tilde{\mathfrak{L}}$ and the collection of dominant integral linear functions on $\mathfrak{H}$. Also it is clear from the definition of $\mathfrak{L}$ that any finite-dimensional irreducible $\tilde{\mathfrak{L}}$-module is an $\mathfrak{L}$-module. The converse is also clear since $\mathfrak{L}$ is a homomorphic image of $\tilde{\mathfrak{L}}$. Hence we see that Theorems 3 and 4 establish a 1 : 1 correspondence between the isomorphism classes of finite-dimensional irreducible modules for the finite-dimensional split semi-simple Lie algebra and the collection of dominant integral linear functions on a splitting Cartan subalgebra $\mathfrak{H}$ of $\mathfrak{L}$.

We recall that a set Σ of linear transformations in a vector space $\mathfrak{M}$ over Φ is called *absolutely irreducible* if the corresponding set (the set of extensions) of linear transformations is irreducible in $\mathfrak{M}_P$ for any extension P of the base field. This condition implies irreducibility since one may take $P = \Phi$. The term "absolutely irreducible" will be applied to modules and representations in the obvious way. It is clear from the definition that a module $\mathfrak{M}$ for a Lie algebra $\mathfrak{L}$ is absolutely irreducible if and only if $\mathfrak{M}_P$ is irreducible for $\mathfrak{L}_P$ for any extension P of Φ. Now let $\mathfrak{L}$ be a split semi-simple Lie algebra over Φ as in Theorem 5 and let $\mathfrak{M}$ be a finite-dimensional irreducible $\mathfrak{L}$-module. We assert that $\mathfrak{M}$ is absolutely irreducible. Thus we know that $\mathfrak{M}$ is an e-extreme module with generator x such that $xe_i = 0$, $i = 1, 2, \cdots, l$, and $xh = \Lambda(h)x$, $h \in \mathfrak{H}$ where Λ is a dominant integral linear function on $\mathfrak{H}$. We know also that the weight space $\mathfrak{M}_\Lambda$ corresponding to Λ coincides with Φx. Consider $\mathfrak{M}_P$ as module for $\mathfrak{L}_P$. Since this is finite-dimensional and $\mathfrak{L}_P$ is semi-simple, $\mathfrak{M}_P$ is completely reducible. Every irreducible submodule $\mathfrak{N}$ of $\mathfrak{M}_P$ decomposes into weight modules relative to $\mathfrak{H}_P$. Hence every weight module for $\mathfrak{H}_P$ in $\mathfrak{M}_P$ can be decomposed into submodules contained in the irreducible components of a decomposition of $\mathfrak{M}_P$ into irreducible $\mathfrak{L}_P$-modules. In particular, this holds for $(\mathfrak{M}_\Lambda)_P = Px$. Since this is one dimensional it follows that x is contained in an irreducible submodule $\mathfrak{N}$ of $\mathfrak{M}_P$. Since x generates $\mathfrak{M}_P$ we have $\mathfrak{N} = \mathfrak{M}_P$ and $\mathfrak{M}_P$ is irreducible. This proves our assertion on the absolute irreducibility of $\mathfrak{M}$.

5. *Existence of E_7 and E_8*

In § 4.6 we established the existence of the split simple Lie algebras of the types corresponding to every Dynkin diagram except E_7 and E_8. (Our method—an explicit construction for each type—admittedly was somewhat sketchy for the exceptional types G_2, F_4 and E_6.) An alternative procedure based on Theorem 5 is now available to us. This requires the verification that the Cartan matrices (A_{ij}) obtained from the Dynkin diagrams satisfy conditions (α), (β), (γ) of § 1. We shall now carry this out for the diagrams E_7 and E_8 and thereby obtain the existence of these Lie algebras. We remark first that in any such verification (α) is immediate once the matrix is written down and (β) is generally easy. To prove (γ) one displays a finite set of vectors in $\mathfrak{H}_0^*$ which span $\mathfrak{H}_0^*$ and is invariant under the Weyl reflections S_{α_i}. This will prove the finiteness of the group W generated by the S_{α_i}.

E_8. The Cartan matrix is

$$(15) \qquad (A_{ij}) = \begin{bmatrix} 2 & -1 & 0 & 0 & 0 & 0 & 0 & 0 \\ -1 & 2 & -1 & 0 & 0 & 0 & 0 & 0 \\ 0 & -1 & 2 & -1 & 0 & 0 & 0 & 0 \\ 0 & 0 & -1 & 2 & -1 & 0 & 0 & 0 \\ 0 & 0 & 0 & -1 & 2 & -1 & 0 & -1 \\ 0 & 0 & 0 & 0 & -1 & 2 & -1 & 0 \\ 0 & 0 & 0 & 0 & 0 & -1 & 2 & 0 \\ 0 & 0 & 0 & 0 & -1 & 0 & 0 & 2 \end{bmatrix}$$

One can see that det $(A_{ij}) = 1$, so (β) is clear. Also (α) is apparent. We introduce the vectors

$$\begin{aligned} \lambda_1 &= 3(\alpha_1 + \alpha_2 + \alpha_3 + \alpha_4 + \alpha_5) + 2\alpha_6 + \alpha_7 + \alpha_8 \\ \lambda_2 &= 3(\alpha_2 + \alpha_3 + \alpha_4 + \alpha_5) + 2\alpha_6 + \alpha_7 + \alpha_8 \\ \lambda_3 &= 3(\alpha_3 + \alpha_4 + \alpha_5) + 2\alpha_6 + \alpha_7 + \alpha_8 \\ \lambda_4 &= 3(\alpha_4 + \alpha_5) + 2\alpha_6 + \alpha_7 + \alpha_8 \\ \lambda_5 &= 3\alpha_5 + 2\alpha_6 + \alpha_7 + \alpha_8 \\ \lambda_6 &= 2\alpha_6 + \alpha_7 + \alpha_8 \\ \lambda_7 &= -\alpha_6 + \alpha_7 + \alpha_8 \\ \lambda_8 &= -\alpha_6 - 2\alpha_7 + \alpha_8 \end{aligned}$$

where $(\alpha_1, \alpha_2, \cdots, \alpha_8)$ is a basis for an 8-dimensional vector space over the rationals. It is immediate that the (λ_i) form a basis. We write S_i for the Weyl reflection S_{α_i} defined by (1). We can verify that S_i, $1 \leqq i \leqq 7$, permutes λ_i and λ_{i+1} and leaves the other λ_j fixed while

$$\begin{aligned} \lambda_i S_8 &= \lambda_i + \tfrac{1}{3}(\lambda_6 + \lambda_7 + \lambda_8)\,, \qquad 1 \leqq i \leqq 5 \\ \lambda_6 S_8 &= \tfrac{1}{3}(\lambda_6 - 2\lambda_7 - 2\lambda_8) \end{aligned}$$

$$\lambda_7 S_8 = \tfrac{1}{3}(-2\lambda_6 + \lambda_7 - 2\lambda_8)$$
$$\lambda_8 S_8 = \tfrac{1}{3}(-2\lambda_6 - 2\lambda_7 + \lambda_8)\,.$$

Let Σ be the following set of vectors:

$$\lambda_i - \lambda_j\,, \qquad \pm(\lambda_i + \lambda_j + \lambda_k)\,,$$
$$\pm(\lambda_i + \lambda_j + \lambda_k + \lambda_l + \lambda_m + \lambda_n)\,,$$
$$\pm(2\lambda_i + \lambda_j + \lambda_k + \lambda_l + \lambda_m + \lambda_n + \lambda_p + \lambda_q)$$

where the subscripts are all different and are in the set $(1, 2, \cdots, 8)$. It is easy to see that these vectors generate the space. Since the S_i, $i \leqq 7$, are permutation transformations of the λ's it is clear that these S_i map Σ into itself. One checks also directly that S_8 leaves Σ invariant. This implies that the group W generated by the S_i is finite.

E_7. Let $\alpha_1, \alpha_2, \cdots, \alpha_8$ be a simple system of roots of type E_8 in the split Lie algebra E_8. The matrix $(A_{jk}) = (2(\alpha_j, \alpha_k)/(\alpha_j, \alpha_j))$, $j, k = 2, \cdots, 8$ is the Cartan matrix E_7. This satisfies (α) and (β). Moreover, (γ) holds since the group generated by $S_{\alpha_2}, \cdots, S_{\alpha_8}$ is finite; hence the group generated by the restrictions of these mappings to the subspace of $\mathfrak{H}_0^*$ spanned by $\alpha_2, \cdots, \alpha_8$ is finite. This proves the existence of E_7.

A similar method applies to E_6. We remark also that it is easy to see that if e_i, f_i, h_i, $i = 1, \cdots, 8$, is a set of canonical generators for E_8, then the subalgebra generated by e_j, f_j, h_j, $j = 2, \cdots, 8$ is E_7. (Exercise 1, below.)

6. *Basic irreducible modules*

Let $\mathfrak{L}$ be a split finite-dimensional semi-simple Lie algebra over a field of characteristic 0, and, as in the proof of Theorem 5, let $\mathfrak{M}_i$ be the finite-dimensional irreducible module corresponding to the dominant integral linear function λ_i such that $\lambda_i(h_j) = \delta_{ij}$, $i, j = 1, 2, \cdots, l$. If x_i is defined as in the proof of Theorem 5 and x is as in (14), then the argument used in §4 to prove absolute irreducibility shows that the submodule of $\overbrace{\mathfrak{M}_1 \otimes \cdots \otimes \mathfrak{M}_1}^{m_1} \otimes \cdots \otimes \overbrace{\mathfrak{M}_l \otimes \cdots \otimes \mathfrak{M}_l}^{m_l}$ generated by x is the irreducible module corresponding to $\Lambda = \sum m_i \lambda_i$. The problem of explicitly determining the irreducible modules of finite-dimensionality for $\mathfrak{L}$ is therefore reduced to that of identifying the modules $\mathfrak{M}_i$, which we shall call the *basic irreducible modules* for $\mathfrak{L}$. We shall now carry this out for some of the simple Lie algebras. Others will be indicated in exercises.

A_l. As we have seen in §4.6, A_l is the Lie algebra of $(l+1) \times (l+1)$ matrices of trace 0 over Φ. If (e_{ij}) is the usual matrix basis for the matrix algebra Φ_{l+1} and the Cartan subalgebra and simple system of roots are chosen as in §4.6, then a set of canonical generators corresponding to these is

$$\begin{aligned} & e_1 = e_{21}\,, \qquad e_2 = e_{32}, \cdots, e_l = e_{l+1,l} \\ (16) \qquad & f_1 = -e_{12}\,, \qquad f_2 = -e_{23}, \cdots, f_l = -e_{l,l+1} \\ & h_1 = e_{11} - e_{22}\,, \qquad h_2 = e_{22} - e_{33}, \cdots, h_l = e_{ll} - e_{l+1,l+1}\,. \end{aligned}$$

The Cartan matrix (A_{ij}) is given by (38) of Chapter IV and we have $A_{ii} = 2$, $A_{i+1,i} = -1 = A_{i,i+1}$, $A_{ij} = 0$ otherwise. This can be checked by calculating $[e_i h_j] = A_{ji} e_i$, using (16). The simple root α_i is specified by: $\alpha_i(h_j) = A_{ji}$. The Weyl reflection S_{α_i} is $\xi \to \xi - \xi(h_i)\alpha_i$.

We consider first a representation of A_l by the Lie algebra of linear transformations of trace 0 in an $(l+1)$-dimensional vector space $\mathfrak{M}$. We can choose the basis $(u_1, u_2, \cdots, u_{l+1})$ so that $u_i h_i = u_i$, $u_{i+1} h_i = -u_{i+1}$, $u_j h_i = 0$, $j \neq i, i+1$. Then we have $u_i h = \Lambda_i(h) u_i$ where the weights Λ_i are given by the table:

$$\begin{aligned} & \Lambda_1(h_1) = 1\,, \qquad \Lambda_1(h_j) = 0\,, \qquad j \neq 1\,, \\ (17) \qquad & \Lambda_i(h_{i-1}) = -1\,, \qquad \Lambda_i(h_i) = 1\,, \qquad \Lambda_i(h_j) = 0\,, \\ & \qquad\qquad\qquad\qquad\qquad j \neq i-1, i\,, \qquad i = 2, \cdots, l\,, \\ & \Lambda_{l+1}(h_l) = -1\,, \qquad \Lambda_{l+1}(h_j) = 0\,, \qquad j \neq l\,. \end{aligned}$$

It follows from this and the table for the α_i that

$$\Lambda_i S_{\alpha_i} = \Lambda_{i+1}\,, \qquad \Lambda_{i+1} S_{\alpha_i} = \Lambda_i\,, \qquad \Lambda_j S_{\alpha_i} = \Lambda_j \quad \text{if} \quad j \neq i, i+1\,.$$

Thus S_{α_i} interchanges Λ_i and Λ_{i+1} and leaves fixed the remaining weights Λ_j. The group of transformations of the weights Λ_i generated by these mappings is the symmetric group on the $l+1$ weights.

Now let r be an integer, $1 \leqq r \leqq l$, and form the r-fold tensor product $\mathfrak{M} \otimes \mathfrak{M} \otimes \cdots \otimes \mathfrak{M}$. Let $\mathfrak{M}_r$ denote the subspace of skew symmetric tensors or *r-vectors* in $\mathfrak{M} \otimes \cdots \otimes \mathfrak{M}$. By definition, this is the space spanned by all vectors of the form

$$(18) \qquad [y_1, y_2, \cdots, y_r] \equiv \sum_P \pm y_{i_1} \otimes y_{i_2} \otimes \cdots \otimes y_{i_r}\,,$$

where the summation is taken over all permutations $P = (i_1 i_2 \cdots i_r)$ of $(1, 2, \cdots, r)$ and the sign is $+$ or $-$ according as P is even or odd. It is easy to see that if $(u_1, u_2, \cdots, u_{l+1})$ is a basis for $\mathfrak{M} = \mathfrak{M}_1$ then the $\binom{l+1}{r}$ vectors

$$[u_{j_1}, u_{j_2} \cdots, u_{j_r}] , \qquad j_1 < j_2 < \cdots < j_r = 1, \cdots, l+1 , \tag{19}$$

form a basis for $\mathfrak{M}_r$. If $a \in \mathfrak{L} = A_l$, then

$$\begin{aligned}[y_1, y_2, \cdots, y_r]a &= [y_1 a, y_2, \cdots, y_r] + [y_1, y_2 a, y_3, \cdots, y_r] \\ &\quad + \cdots + [y_1, y_2, \cdots, y_r a] ,\end{aligned} \tag{20}$$

which implies that $\mathfrak{M}_r$ is a submodule of $\mathfrak{M} \otimes \cdots \otimes \mathfrak{M}$. We may suppose that $u_j h = \Lambda_j(h) u_j$. Then (20) implies

$$\begin{aligned}&[u_{j_1}, u_{j_2} \cdots, u_{j_r}]h \\ &\quad = (\Lambda_{j_1} + \Lambda_{j_2} + \cdots + \Lambda_{j_r})[u_{j_1}, u_{j_2}, \cdots, u_{j_r}] .\end{aligned} \tag{21}$$

Hence the basis given in (19) consists of weight vectors. We have $\operatorname{tr} h = 0$, which implies that $\Lambda_{l+1} = -(\Lambda_1 + \cdots + \Lambda_l)$. On the other hand, (17) implies that $\Lambda_1 + \cdots + \Lambda_k = \lambda_k$ where $\lambda_k(h_i) = \delta_{ik}$, if $1 \leqq k \leqq l$. Hence $\Lambda_1, \cdots, \Lambda_l$ is a basis for $\mathfrak{H}_0^*$. If $j_r = l+1$, then $\Lambda_{j_1} + \cdots + \Lambda_{j_r} = -(\Lambda_{k_1} + \cdots + \Lambda_{k_{l-(r-1)}})$ where $k_1, \cdots, k_{l-(r-1)}$ is the complement of $(j_1, \cdots, j_{r-1})$ in $(1, 2, \cdots, l)$. It follows from this that if $\Lambda_{j_1} + \cdots + \Lambda_{j_r} = \Lambda_{j_1'} + \cdots + \Lambda_{j_r'}$ where $j_1 < \cdots < j_r$ and $j_1' < \cdots < j_r'$, then $j_1 = j_1'$, $j_2 = j_2'$, $\cdots, j_r = j_r'$. We now see that the weight space corresponding to the weight $\Lambda_{j_1} + \Lambda_{j_2} + \cdots + \Lambda_{j_r}$ is spanned by $[u_{j_1}, u_{j_2}, \cdots, u_{j_r}]$ and so is one-dimensional. Let $\mathfrak{N}$ be a non-zero irreducible submodule of $\mathfrak{M}_r$. Then $\mathfrak{N}$ contains one of the weight spaces corresponding to, say, $\Lambda = \Lambda_{j_1} + \cdots + \Lambda_{j_r}$. On the other hand, if $\Lambda' = \Lambda_{j_1'} + \cdots + \Lambda_{j_r'}$ is any other weight, then there exists an element S of the group generated by the S_{α_i} such that $\Lambda' = \Lambda S$. Then, by Theorem 4.1, Λ' is a weight of $\mathfrak{H}$ in $\mathfrak{N}$ also and $[u_{j_1'}, \cdots, u_{j_r'}] \in \mathfrak{N}$. Thus $\mathfrak{N} = \mathfrak{M}_r$ and so $\mathfrak{M}_r$ is irreducible. We have seen that $\Lambda_1 + \cdots + \Lambda_r = \lambda_r$ and this is a dominant integral linear function. It is clear from the definition of $[y_1, y_2, \cdots, y_r]$ that $[y_1, y_2, \cdots, y_r] = 0$ if two of the y_i are equal. This can be used to check that $[u_1, u_2, \cdots, u_r]e_i = 0$, $i = 1, 2, \cdots, l$. Since $[u_1, \cdots, u_r]h = (\Lambda_1 + \Lambda_2 + \cdots + \Lambda_r)[u_1, \cdots, u_r]$, $\mathfrak{M}_r$ is e-extreme with maximal weight λ_r. We therefore have the following

Theorem 6. *Let A_l be the Lie algebra of $(l+1) \times (l+1)$ matrices over Φ of trace 0 and let $\mathfrak{H}$ be the Cartan subalgebra of diagonal*

matrices in A_l, $h_i = e_{ii} - e_{i+1,i+1}$, $i = 1, 2, \cdots, l$. *Let* $\mathfrak{M}_r$, $1 \leqq r \leqq l$, *be the space of r-vectors. Then* $\mathfrak{M}_r$ *is an irreducible module for* A_l *and it is the irreducible module corresponding to the dominant integral linear function* λ_r *on* $\mathfrak{H}$ *such that* $\lambda_r(h_r) = 1$, $\lambda_r(h_j) = 0$ *if* $j \neq r$.

B_l, $l \geqq 2$. If we use the simple system of roots of § 4.6 for B_l we obtain the canonical generators

$$
\begin{aligned}
e_i &= e_{i+2,i+1} - e_{l+i+1,l+i+2}\,, \qquad i = 1, \cdots, l-1 \\
e_l &= e_{1,l+1} - e_{2l+1,1} \\
f_i &= e_{l+i+2,l+i+1} - e_{i+1,i+2}\,, \qquad i = 1, \cdots, l-1 \\
f_l &= 2(e_{1,2l+1} - e_{l+1,1}) \\
h_i &= e_{i+1,i+1} - e_{l+i+1,l+i+1} - e_{i+2,i+2} \\
&\qquad + e_{l+i+2,l+i+2}\,, \qquad i \leqq l-1 \\
h_l &= 2(e_{l+1,l+1} - e_{2l+1,2l+1})\,.
\end{aligned}
\tag{22}
$$

For the Cartan matrix $(A_{ij}) = (\alpha_j(h_i))$ we have the values $A_{ii} = 2$, $A_{j,j+1} = -1 = A_{j+1,j}$, $j = 1, \cdots, l-2$, $A_{l-1,l} = -1$, $A_{l,l-1} = -2$, all other $A_{ij} = 0$. In the representation of B_l by the Lie·algebra of skew symmetric linear transformations in the $n = 2l+1$ dimensional space $\mathfrak{M}$ we have a basis $(u_1, u_2, \cdots, u_n)$ such that $u_1 h = 0$, $u_2 h = \Lambda_1(h)u_2, \cdots, u_{l+1}h = \Lambda_l(h)u_{l+1}$; $u_{l+2}h = -\Lambda_1(h)u_{l+2}, \cdots, u_{2l+1}h = -\Lambda_l(h)u_{2l+1}$, where the linear functions $\Lambda_j(h)$ satisfy

$$
\begin{aligned}
&\Lambda_1(h_1) = 1\,, \quad \Lambda_1(h_i) = 0\,, \quad i \neq 1 \\
&\Lambda_j(h_{j-1}) = -1\,, \quad \Lambda_j(h_j) = 1\,, \quad \Lambda_j(h_k) = 0\,, \\
&\qquad\qquad k \neq j-1, j\,, \quad j = 2, \cdots, l-1\,. \\
&\Lambda_l(h_{l-1}) = -1\,, \quad \Lambda_l(h_l) = 2\,, \quad \Lambda_l(h_k) = 0\,, \quad k \neq l-1, l\,.
\end{aligned}
\tag{23}
$$

We have $\Lambda_1 + \cdots + \Lambda_r = \lambda_r$, $r \leqq l-1$, $\Lambda_1 + \cdots + \Lambda_l = 2\lambda_l$, where $\lambda_i(h_j) = \delta_{ij}$, $i, j = 1, \cdots, l$. Hence the Λ_i form a basis for $\mathfrak{H}_0^*$. The Weyl reflection S_{α_j}, $j = 1, \cdots, l-1$, interchanges Λ_j and Λ_{j+1} and leaves the remaining Λ_k invariant. The reflection S_{α_l} leaves fixed every Λ_j, $j = 1, \cdots, l-1$ and maps Λ_l into $-\Lambda_l$. Hence the group generated by these includes all permutations of the Λ_i and all the mappings which replace any subset of the Λ_i by their negatives leaving the remaining ones fixed. It is easy to see that the group generated by the S_{α_i} is of order $2^l l!$ and we shall show in the next chapter that this is the complete Weyl group.

Let r be an integer, $1 \leqq r \leqq l-1$, and let $\mathfrak{M}_r$ be the space of

r-vectors. It is easy to see that $[u_2, \cdots, u_{r+1}]h = \lambda_r(h)[u_2, \cdots, u_{r+1}]$ and $[u_2, \cdots, u_{r+1}]e_i = 0$, $i = 1, 2, \cdots, l$. We assert that the cyclic module generated by $[u_1, \cdots, u_{r+1}]$ is irreducible. Hence this is the required irreducible module corresponding to the dominant integral linear function λ_r. In general, we note that if $\mathfrak{N}$ is a finite-dimensional e-extreme module for a finite-dimensional split semi-simple Lie algebra $\mathfrak{L}$, then $\mathfrak{N}$ is irreducible. Thus $\mathfrak{N}$ is completely reducible and the generator x corresponding to the highest weight Λ spans the weight space of $\mathfrak{H}$ corresponding to Λ. Hence x is contained in an irreducible submodule $\mathfrak{N}'$ of $\mathfrak{N}$. Since x generates $\mathfrak{N}$ we see that $\mathfrak{N} = \mathfrak{N}'$ is irreducible.

The result we have just indicated on the determination of the irreducible module corresponding to λ_r, $r = 1, \cdots, l-1$, can be improved. As a matter of fact, this module is the complete module $\mathfrak{M}_r$ of r-vectors. To prove this one has to show that $\mathfrak{M}_r$ is irreducible. We shall not give a direct proof here but shall obtain this result later as a consequence of Weyl's formula for the dimensionality of the irreducible module with highest weight Λ, which we shall derive in Chapter VIII. For the representation with $\Lambda = \lambda_r$ we shall obtain the dimensionality: $\binom{2l+1}{r}$. Since this is the dimensionality of $\mathfrak{M}_r$ it will follow that $\mathfrak{M}_r$ is irreducible.

If we consider the space of l-vectors a similar argument will show that this is the irreducible module for B_l corresponding to the highest weight $\Lambda_1 + \cdots + \Lambda_l = 2\lambda_l$. Hence this does not provide the missing module corresponding to the linear function λ_l. In order to find this one, one has to consider the so-called *spin representation* of B_l. To give full details of this would be too lengthy. Hence we shall be content to sketch the results without giving complete proofs.

We recall first the definition of the Clifford algebra $C(\mathfrak{M}, (x, y))$ of a finite-dimensional vector space $\mathfrak{M}$ over a field Φ of characteristic not two relative to a symmetric bilinear form (x, y) on $\mathfrak{M}$. One forms the tensor algebra $\mathfrak{T} = \Phi 1 \oplus \mathfrak{M} \oplus \mathfrak{M}_2 \oplus \cdots \oplus \mathfrak{M}_i \oplus \cdots$, $\mathfrak{M}_i = \mathfrak{M} \otimes \cdots \otimes \mathfrak{M}$, i times, and one lets $\mathfrak{K}$ be the ideal in $\mathfrak{T}$ generated by all the elements of the form

$$x \otimes x - (x, x)1 \,. \tag{24}$$

Then $C(\mathfrak{M}, (x, y)) = \mathfrak{T}/\mathfrak{K}$. (For details on Clifford algebra the reader should consult Chevalley [3] pp. 36–69, or Artin [1], pp. 186–193.) It is clear that $\mathfrak{K}$ contains all the elements

$$x\otimes y+y\otimes x-2(x,y)1=(x+y)\otimes(x+y) \tag{25}$$
$$-(x+y,x+y)1-x\otimes x+(x,x)1-y\otimes y+(y,y)1\,.$$

It is known that the canonical mapping of $\mathfrak{M}$ into $C\equiv C(\mathfrak{M},(x,y))$ is an isomorphism, so that one can identify $\mathfrak{M}$ with its image in C. We do this and we write the associative product in C as ab. If $(u_1,u_2,\cdots,u_n)$ is a basis for $\mathfrak{M}$, then it is known that $\dim C=2^n$ and the set of elements

$$u_1^{\varepsilon_1}u_2^{\varepsilon_2}\cdots u_n^{\varepsilon_n}\,,\qquad \varepsilon_i=0,1 \tag{26}$$

is a basis for C. Since the elements (25) are in $\mathfrak{K}$ it follows that we have the fundamental "Jordan relation"

$$x\cdot y\equiv\tfrac{1}{2}(xy+yx)=(x,y)1 \tag{27}$$

for any $x,y\in\mathfrak{M}$.

The algebra C is not a graded algebra. However, it is a graded vector space in a natural fashion which we shall now indicate. Let $x_1,x_2,\cdots,x_r\in\mathfrak{M}$. Then we define an r-fold product $[x_1,x_2,\cdots,x_r]\in C$ inductively by the following formulas

$$\begin{aligned}&[x_1]=x_1\\ &[x_1,\cdots,x_{2k-1},x_{2k}]=[[x_1,\cdots,x_{2k-1}]x_{2k}]\\ &[x_1,\cdots,x_{2k},x_{2k+1}]=[x_1,\cdots,x_{2k}]\cdot x_{2k+1}\end{aligned} \tag{28}$$

where $[ab]=ab-ba$ and $a\cdot b=\frac{1}{2}(ab+ba)$ for $a,b\in C$.

LEMMA 4. *The product* $[x_1,\cdots,x_r]=0$ *if any two of the* x_i *are equal. Hence for any permutation* $P=(i_1,i_2,\cdots,i_r)$ *of* $(1,2,\cdots,r)$, $[x_{i_1},\cdots,x_{i_r}]=\pm[x_1,\cdots,x_r]$ *where the sign is* $+$ *or* $-$ *according to whether* P *is even or odd.*

Proof: Since $[x_1,\cdots,x_r]$ defines a multilinear function of its arguments the second statement is a consequence of the first. We prove the first by induction on r. We have $[xx]=0$. Let $r>2$ and suppose $x_k=x=x_l$ where $1\leqq k<l\leqq r$. If $l<r$ then $[x_1,\cdots,x_l]=0$ by the induction hypothesis and $[x_1,\cdots,x_r]=0$ follows from (28). Hence we may assume that $l=r$. Also the induction implies that $[x_1,\cdots,x_{r-1}]=\pm[\cdots x]$ so we have to show that $[x_1\cdots x_{r-2}xx]=0$. There are two cases: even r and odd r and these will follow by proving the following relations: $[a\cdot x,x]=0$ and $[ax]\cdot x=0$ for $a\in C$, $x\in\mathfrak{M}$. For the first of these we have

$$[\tfrac{1}{2}(ax + xa), x] = \tfrac{1}{2}(ax^2 + xax - xax - x^2a)$$
$$= \tfrac{1}{2}(ax^2 - x^2a) = \tfrac{1}{2}(x, x)a - \tfrac{1}{2}(x, x)a = 0 ,$$

and the second follows from

$$[ax]\cdot x = \tfrac{1}{2}(ax - xa)x + \tfrac{1}{2}x(ax - xa) = \tfrac{1}{2}ax^2 - \tfrac{1}{2}x^2a = 0 .$$

This completes the proof.

LEMMA 5. *Let* $\mathfrak{M}_r$ *denote the subspace of* C *spanned by all the elements* $[x_1, \cdots, x_r]$ *where* $0 \leqq r \leqq n = \dim \mathfrak{M}$ *and* $\mathfrak{M}_0 \equiv \Phi 1$. *Then* $\dim \mathfrak{M}_r = \binom{n}{r}$ *and* $C = \Phi 1 \oplus \mathfrak{M}_1 \oplus \mathfrak{M}_2 \oplus \cdots \oplus \mathfrak{M}_n$.

Proof: If $(u_1, u_2, \cdots, u_n)$ is a basis for $\mathfrak{M}$, then the skew symmetry and multilinearity of $[x_1, \cdots, x_r]$ imply that every element of $\mathfrak{M}_r$, $r \geqq 1$, is a linear combination of the elements $[u_{i_1}, u_{i_2}, \cdots, u_{i_r}]$ where $i_1 < i_2 < \cdots < i_r = 1, \cdots, n$. Now assume that the basis is orthogonal: $(u_i, u_j) = 0$ if $i \neq j$. (It is well known that such bases exist.) The condition $(u_i, u_j) = 0$ and (27) implies that $u_iu_j = -u_ju_i$. Hence $(u_{i_1} \cdots u_{i_{r-1}})u_{i_r} = \pm u_{i_r}(u_{i_1} \cdots u_{i_{r-1}})$ if $i_1 < i_2 < \cdots < i_r$ and the sign is $+$ or $-$ according as $r - 1$ is even or odd. The relation just noted and the definition of $[x_1, \cdots, x_r]$ imply that $[u_{i_1}, \cdots, u_{i_r}] = 2^k u_{i_1} \cdots u_{i_r}$ for (u_i) an orthogonal basis. Since the 2^n elements of the form $u_{i_1} \cdots u_{i_r}$, $i_1 < i_2 < \cdots < i_r$, form a basis for C it is clear that for a fixed r the elements $[u_{i_1}, \cdots, u_{i_r}]$, $i_1 < \cdots < i_r$, form a basis for $\mathfrak{M}_r$. Hence $\dim \mathfrak{M}_r = \binom{n}{r}$, and it is clear that $C = \Phi 1 \oplus \mathfrak{M}_1 \oplus \mathfrak{M}_2 \oplus \cdots \oplus \mathfrak{M}_n$.

Let $x, y, z \in \mathfrak{M}$. Then

$$(x\cdot y)\cdot z - x\cdot(y\cdot z) = (x, y)z - (y, z)x .$$

On the other hand,

$$(x\cdot y)\cdot z - x\cdot(y\cdot z) = \tfrac{1}{4}(xy + yx)z + \tfrac{1}{4}z(xy + yx)$$
$$- \tfrac{1}{4}x(yz + zy) - \tfrac{1}{4}(yz + zy)x$$
$$= \tfrac{1}{4}(xyz + yxz + zxy + zyx - xyz - xzy - yzx - zyx)$$
$$= \tfrac{1}{4}[y[xz]] .$$

Hence we have the relation

$$\tfrac{1}{4}[y[xz]] = (x, y)z - (y, z)x . \tag{29}$$

We can now prove

THEOREM 7. *The space* $\mathfrak{M}_2$ *is a subalgebra of the Lie algebra* C_L. *If* (x, y) *is non-degenerate, then this Lie algebra is isomorphic to the orthogonal Lie algebra determined by* (x, y) *in* $\mathfrak{M}$.

Proof: The space $\mathfrak{M}_2$ is the set of sums of Lie products $[xy]$, $x, y \in \mathfrak{M}$. By (29), we have

$$\begin{aligned}[[yt][xz]] &= [[y[xz]], t] + [y, [t[xz]]] \\ &= 4(x, y)[zt] - 4(y, z)[xt] + 4(x, t)[yz] - 4(t, z)[yx] \, .\end{aligned} \tag{30}$$

Hence $\mathfrak{M}_2$ is a subalgebra of C_L. This relation and (29) imply also that $\mathfrak{M} + \mathfrak{M}_2$ is a subalgebra of C_L and the restriction of the adjoint representation of $\mathfrak{M} + \mathfrak{M}_2$ to the subalgebra $\mathfrak{M}_2$ has $\mathfrak{M}$ as a submodule. If R denotes this representation, then (29) shows that $[xz]^R$ is the mapping

$$y \to 4(x, y)z - 4(y, z)x \, . \tag{31}$$

If (x, y) is non-degenerate, then (31) is in the orthogonal Lie algebra. Moreover, every element of the latter Lie algebra is a sum of mappings (31). Hence the image under R of $\mathfrak{M}_2$ is the orthogonal Lie algebra. Since both of these algebras have the same dimensionality $\binom{n}{2}$, R is an isomorphism.

The enveloping algebra of $\mathfrak{M}_2$ in $C(\mathfrak{M}, (x, y))$ will be denoted as $C^+ = C^+(\mathfrak{M}, (x, y))$. If $(u_1, \cdots, u_n)$ is an orthogonal basis for $\mathfrak{M}$, then the elements $u_i u_j$, $i < j$ constitute a basis for $\mathfrak{M}_2$. Since $u_i^2 = (u_i, u_i)1$ and $u_i u_j = -u_j u_i$ if $i \neq j$, it is easy to see that the space spanned by the elements $u_{i_1} u_{i_2} \cdots u_{i_{2r}}$, $i_1 < i_2 < \cdots < i_{2r}$, $r = 0, 1, 2, \cdots, [n/2]$ is a subalgebra of C. Since $u_{i_1} u_{i_2} \cdots u_{i_{2r}} = (u_{i_1} u_{i_2}) \cdots (u_{i_{2r-1}} u_{i_{2r}})$, this subalgebra is contained in C^+ and since it contains $\mathfrak{M}_2$ it coincides with C^+. It is now clear that C^+ is the subalgebra of even elements of C, or the so-called *second Clifford algebra* of (x, y). Its dimensionality is 2^{n-1}. The structures of C and C^+ are known. We shall state only what is needed for the representation theory of B_l and D_l. For this the symmetric bilinear form (x, y) is non-degenerate and of maximal Witt index. If $n = 2l + 1$, then C^+ is isomorphic to the complete algebra $\mathfrak{E}$ of linear transformations in a 2^l-dimensional vector space $\mathfrak{N}$ over Φ. The isomorphism can be defined explicitly in the following way.

Let $(u_1, u_2, \cdots, u_n)$ be a basis for $\mathfrak{M}$ of the type used to obtain the matrices for B_l (§ 4.6). Thus we have $(u_1, u_1) = 1$, $(u_i, u_{l+i}) = 1 = (u_{l+i}, u_i)$ if $i = 2, \cdots, l + 1$ and all other products are 0. Set $v_i = u_1 u_{i+1}$, $w_i = u_1 u_{i+l+1}$, $i = 2, \cdots, l$. Then it is easily seen that the v_i and w_i generate C^+, that $v_i v_j = -v_j v_i$, $v_i^2 = 0$, $w_i w_j = -w_j w_i$, $w_i^2 = 0$ if $i \neq j$, and that the subalgebras generated by the v's and the w's separately are isomorphic to the exterior algebra based on an l-dimensional space. A basis for C^+ is the set of elements $v_{i_1} \cdots v_{i_r} w_{j_1} \cdots w_{j_s}$ where $i_1 < i_2 < \cdots < i_r$, $j_1 < j_2 < \cdots < j_s$. We have the relations

$$v_i w_i + w_i v_i = -2\,, \qquad v_i w_j = -w_j v_i\,, \qquad \text{if } i \neq j\,. \tag{32}$$

The subspace $\mathfrak{N}$ spanned by the vectors

$$v_1 \cdots v_l w_{j_1} \cdots w_{j_s}\,, \qquad j_1 < j_2 < \cdots < j_s \tag{33}$$

is a 2^l-dimensional right ideal in C^+. The right multiplications by $a \in C^+$ in $\mathfrak{N}$ give all the linear transformations of $\mathfrak{N}$ and the correspondence between a and the restriction to $\mathfrak{N}$ of a_R is an isomorphism of C^+ onto the algebra $\mathfrak{E}$ of linear transformations of $\mathfrak{N}$.

The representation $a \to a'_R$, a'_R the restriction to $\mathfrak{N}$ of a_R, induces a representation of the subalgebra $\mathfrak{M}_2$ of C_L^+. Since $\mathfrak{M}_2$ is isomorphic to B_l this gives a representation of B_l acting in $\mathfrak{N}$. The B_l-module $\mathfrak{N}$ is irreducible and this is the irreducible module with highest weight λ_l which we require. We shall sketch the argument which can be used to establish this.

We note first that the matrix e_i, $i = 1, 2, \cdots, l-1$, in (22) can be identified with the linear transformation $x \to (x, u_{i+l+2})u_{i+1} - (x, u_{i+1})u_{i+l+2}$ relative to the basis $(u_1, u_2, \cdots, u_n)$ which we have chosen. Similarly, e_l can be identified with the transformation $x \to (x, u_1)u_{l+1} - (x, u_{l+1})u_1$, h_i, $i = 1, 2, \cdots, l-1$, with $x \to (x, u_{l+i+1})u_{i+1} - (x, u_{i+1})u_{l+i+1} - (x, u_{l+i+2})u_{i+2} + (x, u_{i+2})u_{l+i+2}$, h_l with $x \to 2(x, u_{2l+1})u_{l+1} - 2(x, u_{l+1})u_{2l+1}$. In this isomorphism with $\mathfrak{M}_2$ (cf. (29)) we have

$$\begin{aligned} e_i &\to \tfrac{1}{4}[u_{l+i+2}, u_{i+1}] = \tfrac{1}{2}v_i w_{i+1}\,, \qquad i = 1, \cdots, l-1 \\ e_l &\to \tfrac{1}{4}[u_1, u_{l+1}] = \tfrac{1}{2}v_{l+1} \\ h_i &\to \tfrac{1}{4}[u_{l+i+1}, u_{i+1}] - \tfrac{1}{4}[u_{l+i+2}, u_{i+2}] \\ &\qquad = \tfrac{1}{2}(v_i u_i - v_{i+1}u_{i+1})\,, \qquad i = 1, 2, \cdots, l-1 \\ h_l &\to \tfrac{1}{2}[u_{2l+1}, u_{l+1}] = 1 + v_l w_l\,. \end{aligned} \tag{34}$$

Let $z = v_1 \cdots v_l \in \mathfrak{N}$. Then $ze_j = 0$, $j = 1, \cdots, l$, $zh_i = 0$, $i = 1, \cdots, l-1$ and $zh_l = z$. It follows that the cyclic B_l-module generated by z is the irreducible module corresponding to λ_l. It can be shown that this is all of $\mathfrak{N}$. We shall call $\mathfrak{N}$ the *spin module* for B_l. We can summarize the results on B_l in the following

THEOREM 8. *Let B_l, $l \geqq 2$, be the orthogonal Lie algebra in a $(2l+1)$-dimensional space defined by a non-degenerate symmetric* *bilinear form of maximum Witt index. Let the basis for a Cartan subalgebra* $\mathfrak{H}$ *of* B_l *consist of the* h_i, $i = 1, 2, \cdots, l$ *of* (22) *and let* λ_j *be the linear function on* $\mathfrak{H}$ *such that* $\lambda_j(h_i) = \delta_{ij}$. *Then the irreducible module for* B_l *with highest weight* λ_j, $j = 1, \cdots, l-1$ *is the space* $\mathfrak{M}_j$ *of j-vectors. The irreducible module with highest weight* λ_l *is the spin module* $\mathfrak{N}$ *defined above.*

G_2. If (α_1, α_2) constitutes a simple system of roots for G_2, then the Cartan matrix is

$$\begin{pmatrix} 2 & -1 \\ -3 & 2 \end{pmatrix}, \tag{35}$$

so that we have

$$\alpha_1(h_1) = 2\,, \qquad \alpha_1(h_2) = -3\,, \qquad \alpha_2(h_1) = -1\,, \qquad \alpha_2(h_2) = 2\,. \tag{36}$$

We have seen in § 4.3 that the positive roots are α_1, α_2, $\alpha_1 + \alpha_2$, $\alpha_1 + 2\alpha_2$, $\alpha_1 + 3\alpha_2$, $2\alpha_1 + 3\alpha_2$. The highest of these is $2\alpha_1 + 3\alpha_2$. We have $(2\alpha_1 + 3\alpha_2)(h_1) = 1$, $(2\alpha_1 + 3\alpha_2)(h_2) = 0$ so that $2\alpha_1 + 3\alpha_2$ is the linear function λ_1 such that $\lambda_1(h_1) = 1$, $\lambda_1(h_2) = 0$. Since G_2 is simple the adjoint representation is irreducible. Thus $\mathfrak{L}$ itself is the irreducible module whose highest weight is λ_1. We shall show next that if G_2 is represented by the Lie algebra of derivations in the split Cayley algebra $\mathfrak{C}$ then the representation induced in the seven-dimensional space $\mathfrak{C}_0$ of elements of trace 0 is the irreducible representation corresponding to the linear function λ_2 such that $\lambda_2(h_1) = 0$, $\lambda_2(h_2) = 1$. We shall show this by proving that the dimensionality of any module $\mathfrak{M}$ for G_2 satisfying $\mathfrak{M}G_2 \neq 0$ is $\geqq 7$, and if the dimensionality is 7, then the module is irreducible and corresponds to λ_2.

We first express the roots in terms of the basis λ_1, λ_2. Thus we have $\alpha_1 = 2\lambda_1 - 3\lambda_2$, $\alpha_2 = -\lambda_1 + 2\lambda_2$, by (36). Hence the positive roots are

$$\begin{array}{cccc} 2\lambda_1 - 3\lambda_2\,, & -\lambda_1 + 2\lambda_2\,, & \lambda_1 - \lambda_2\,, & \lambda_2\,, \\ & -\lambda_1 + 3\lambda_2\,, & \lambda_1\,. & \end{array} \tag{37}$$

If $\Lambda = m_1\lambda_1 + m_2\lambda_2 \in \mathfrak{H}_0^*$ and $S_1 \equiv S_{\alpha_1}$, $S_2 \equiv S_{\alpha_2}$ are the Weyl reflections determined by α_1 and α_2, respectively, then $\Lambda S_1 = \Lambda - \Lambda(h_1)\alpha_1 = \Lambda - m_1(2\lambda_1 - 3\lambda_2) = -m_1\lambda_1 + (m_2 + 3m_1)\lambda_2$ and $\Lambda S_2 = (m_1 + m_2)\lambda_1 - m_2\lambda_2$. If Λ is the highest weight of $\mathfrak{H}$ in the module $\mathfrak{M}$, then the m_i are non-negative and not all 0 and the following linear functions are weights of $\mathfrak{H}$ in $\mathfrak{M}$:

$$\Lambda = m_1\lambda_1 + m_2\lambda_2,\ \Lambda S_1 = -m_1\lambda_1 + (m_2 + 3m_1)\lambda_2,$$
$$\Lambda S_2 = (m_1 + m_2)\lambda_1 - m_2\lambda_2,\ \Lambda S_1S_2 = (2m_1 + m_2)\lambda_1 - (m_2 + 3m_1)\lambda_2,$$
$$\Lambda S_2S_1 = -(m_1 + m_2)\lambda_1 + (3m_1 + 2m_2)\lambda_2,$$
$$\Lambda S_1S_2S_1 = -(2m_1 + m_2)\lambda_1 + (3m_1 + 2m_2)\lambda_2,$$
$$\Lambda S_2S_1S_2 = (2m_1 + m_2)\lambda_1 - (3m_1 + 2m_2)\lambda_2,$$
$$\Lambda(S_1S_2)^2 = (m_1 + m_2)\lambda_1 - (3m_1 + 2m_2)\lambda_2,$$
$$\Lambda(S_2S_1)^2 = -(2m_1 + m_2)\lambda_1 + (3m_1 + m_2)\lambda_2,$$
$$\Lambda(S_1S_2)^2S_1 = -(m_1 + m_2)\lambda_1 + m_2\lambda_2,$$
$$\Lambda(S_2S_1)^2S_2 = m_1\lambda_1 - (3m_1 + m_2)\lambda_2,$$
$$\Lambda(S_1S_2)^3 = -m_1\lambda_1 - m_2\lambda_2 = \Lambda(S_2S_1)^3 .$$

If $m_1 > 0$ and $m_2 > 0$, then this list gives twelve distinct weights which means that $\dim \mathfrak{M} \geqq 12$. If $m_2 = 0$ the list gives six distinct weights: $m_1\lambda_1$, $-m_1\lambda_1$, $2m_1\lambda_1 - 3m_1\lambda_2$, $-m_1\lambda_1 + 3m_1\lambda_2$, $-2m_1\lambda_1 + 3m_1\lambda_2$, $m_1\lambda_1 - 3m_1\lambda_2$. If $m_1 = 1$, then $\mathfrak{M}$ contains a submodule isomorphic to $\mathfrak{L}$ and so $\dim \mathfrak{M} \geqq 14$. If $m_1 > 1$, then since λ_1 is a root and the λ_1-string containing $m_1\lambda_1$ contains also $-m_1\lambda_1$ there are at least $2m_1 + 1 \geqq 5$ weights in this string. This adds three new weights to the list and shows that $\dim \mathfrak{M} \geqq 9$. Now let $m_1 = 0$. Then the following six weights are all different: $m_2\lambda_2$, $m_2\lambda_1 - m_2\lambda_2$, $-m_2\lambda_1 + 2m_2\lambda_2$, $m_2\lambda_1 - 2m_2\lambda_2$, $-m_2\lambda_1 + m_2\lambda_2$, $-m_2\lambda_1$. If $m_2 = 1$ the λ_2-string containing $m_2\lambda_2$ contains 0 as weight so that $\dim \mathfrak{M} \geqq 7$. If $m_2 > 1$ then the λ_2-string containing $m_2\lambda_2$ contains at least five weights and this implies that $\dim \mathfrak{M} \geqq 9$. We have therefore shown that $\dim \mathfrak{M} \geqq 7$ and $\dim \mathfrak{M} = 7$ can hold only if the highest weight $\Lambda = \lambda_2$. Since there exists a module $\mathfrak{C}_0$ for G_2 such that $\dim \mathfrak{C}_0 = 7$ it follows that $\mathfrak{C}_0$ is irreducible and its highest weight is λ_2.

THEOREM 9. *The irreducible module for $\mathfrak{L} = G_2$ with highest weight λ_1 is $\mathfrak{L}$ itself. The irreducible module for G_2 with highest weight λ_2 is the seven-dimensional module $\mathfrak{C}_0$ of elements of trace 0 in the split Cayley algebra $\mathfrak{C}$.*

Exercises

In these exercises we follow the notations of this chapter: $\mathfrak{L}$ is a finite-dimensional split semi-simple Lie algebra over a field of characteristic 0, $\mathfrak{H}$ a splitting Cartan subalgebra, e_i, f_i, h_i canonical generators for $\mathfrak{L}$, $\mathfrak{U}$ the universal enveloping algebra, etc.

1. Let e_i, f_i, h_i, $i = 1, 2, \cdots, l$, be canonical generators for $\mathfrak{L}$. Show that the subalgebra $\mathfrak{L}_1$ generated by the elements e_j, f_j, h_j, $j = 1, \cdots, k$, $k \leqq l$ is a split semi-simple Lie algebra. Show that E_6 is a subalgebra of E_7, and E_7 is a subalgebra of E_8.

2. Show that $\mathfrak{L}$ has only a finite number of inequivalent irreducible modules of dimension $\leqq N$ for N any positive integer.

3. Let $\mathfrak{X}$ be the kernel in $\mathfrak{U}$ of the representation determined by a module $\mathfrak{M}$ and let $\mathfrak{X}^{(s)}$ denote the kernel of $\mathfrak{M} \otimes \cdots \otimes \mathfrak{M}$, s-times. Show that a finite-dimensional $\mathfrak{M}$ can be chosen so that if $\mathfrak{Y}$ is any ideal in $\mathfrak{U}$ of finite co-dimension, then there exists an s such that $\mathfrak{X}^{(s)} \subseteq \mathfrak{Y}$.

4. Let $\mathfrak{K}$ be the orthogonal Lie algebra defined by a non-degenerate symmetric bilinear form in an $n \geqq 3$ dimensional space $\mathfrak{M}$. Prove that the space $\mathfrak{M}_r$ of r-vectors is $\mathfrak{K}$-irreducible for $1 \leqq r \leqq l$ if $n = 2l + 1$ and $1 \leqq r \leqq l - 1$ if $n = 2l$.

5. Determine a set of basic irreducible modules for C_l, $l \geqq 4$.

6. Show that the minimum dimensionality for an irreducible module $\mathfrak{M}$ for E_6 such that $\mathfrak{M}E_6 \neq 0$ is 27. Hence prove that E_6 and B_6, and E_6 and C_6 are not isomorphic.

7. Prove that the spin representation of D_l in the second Clifford algebra (of even elements) decomposes as a direct sum of two irreducible modules. Show that these two together with the spaces of r-vectors $1 \leqq r \leqq l - 2$ are the basic irreducible modules for D_l, $l \geqq 4$.

8. Show that a basis $(g_1, g_2, \cdots, g_l)$ can be chosen for $\mathfrak{H}$ such that $[e_i g_j] = \delta_{ij} e_i$, $[f_i g_j] = -\delta_{ij} f_i$. Let $h = 2 \sum_1^l g_i = \sum \mu_i h_i$ where the $\mu_i \in \Phi$. Let $e = \sum \gamma_i e_i$ where every $\gamma_i \neq 0$ and let $f = \sum \gamma_i^{-1} \mu_i f_i$. Prove that (e, f, h) is a canonical basis for a split three-dimensional simple Lie algebra $\mathfrak{K}$. Prove that $\mathrm{ad}_{\mathfrak{L}} \mathfrak{K}$ is a direct sum of l odd-dimensional irreducible representations for $\mathfrak{K}$. (The subalgebra $\mathfrak{K}$ is called a "principal three-dimensional subalgebra" of $\mathfrak{L}$. Such subalgebras play an important role in the cohomology theory of $\mathfrak{L}$. See Kostant [3].)

9. Determine a $\mathfrak{K}$ as in 8, for A_l. Find the characteristic roots of $\mathrm{ad}_{\mathfrak{L}} h$ and use this to obtain the dimensionalities of the irreducible components of $\mathrm{ad}_{\mathfrak{L}} \mathfrak{K}$.

10. Let $\mathfrak{M}$ be a finite-dimensional module for $\mathfrak{L}$, let $\mathfrak{L}_+$ be the (nilpotent) subalgebra of $\mathfrak{L}$ generated by the e_i and let $\mathfrak{Z}$ be the subspace of $\mathfrak{M}$ of elements z such that $zl = 0$, $l \in \mathfrak{L}_+$. Show that $\dim \mathfrak{Z}$ is the number of irreducible submodules in a direct decomposition of $\mathfrak{M}$ into irreducible submodules. (This number is independent of the particular decomposition by the Krull-Schmidt theorem. See also § 8.5.)

11. Let $\mathfrak{M}$ and $\mathfrak{N}$ be two finite-dimensional irreducible modules for $\mathfrak{L}$ and let $\mathfrak{M}^*$ be the contragredient module of $\mathfrak{M}$. Let R and S, respectively, be the representations in $\mathfrak{M}$ and $\mathfrak{N}$, R^* the representation in $\mathfrak{M}^*$. Show that the number of submodules in a decomposition of $\mathfrak{M}^* \otimes \mathfrak{N}$ as direct sum of irreducible submodules is dim $\mathfrak{Z}$ where $\mathfrak{Z}$ is the subspace of the space $\mathfrak{E}(\mathfrak{M}, \mathfrak{N})$ ($\cong \mathfrak{M}^* \otimes \mathfrak{N}$ by §1) of linear mappings of $\mathfrak{M}$ into $\mathfrak{N}$ defined by

$$\mathfrak{Z} = \{Z \mid Z \in \mathfrak{E}(\mathfrak{M}, \mathfrak{N}), \, l^R Z = Z l^S\} \ .$$

12. (Dynkin). If $\mathfrak{M}_1$ and $\mathfrak{M}_2$ are finite-dimensional irreducible $\mathfrak{L}$-modules with highest weights Λ_1 and Λ_2 and canonical generators x_1 and x_2, respectively, then $\mathfrak{M}_2$ is said to be *subordinate* to $\mathfrak{M}_1$ if $x_1 u = 0$ implies $x_2 u = 0$ for every u in the universal enveloping algebra $\mathfrak{U}_-$ of the subalgebra $\mathfrak{L}_-$ of $\mathfrak{L}$ generated by the f_i. Prove that $\mathfrak{M}_2$ is subordinate to $\mathfrak{M}_1$ if and only if $\Lambda_1 = \Lambda_2 + M$ where M is a dominant integral linear function on $\mathfrak{H}^*$. (*Hint*: Note that if $\mathfrak{N}$ is the finite-dimensional irreducible module with highest weight M and y is a canonical generator then $\mathfrak{M}_1$ can be taken to be the submodule of $\mathfrak{M}_2 \otimes \mathfrak{N}$ generated by $x_2 \otimes y$.)

13. Note that the definition of subordinate in 12 is equivalent to the following: there exists a $\mathfrak{U}$-homomorphism of $\mathfrak{M}_1$ onto $\mathfrak{M}_2$ mapping x_1 onto x_2. Use this to prove that if a finite-dimensional irreducible module $\mathfrak{M}$ with $\mathfrak{M}\mathfrak{L} \neq 0$ has minimal dimensionality (for such modules), then $\mathfrak{M}$ is basic.

CHAPTER VIII

Characters of the Irreducible Modules

The main result of this chapter is the formula, due to Weyl, for the character of any finite-dimensional irreducible module for a split semi-simple Lie algebra $\mathfrak{L}$ over a field of characteristic 0. If the base field is the field of complex numbers and R is a finite-dimensional irreducible representation, then the character of R is the function on a Cartan subalgebra $\mathfrak{H}$ defined by

$$(1) \qquad \chi(h) = \operatorname{tr} \exp h^R ,$$

where, as usual, $\exp a = 1 + a + a^2/2! + \cdots$. If $\mathfrak{G}$ is a connected semi-simple compact Lie group, then (1) gives the character in the ordinary sense of an irreducible representation of $\mathfrak{G}$. Thus in this case $\mathfrak{H}$ corresponds to a maximal torus and it is known that any element of $\mathfrak{G}$ is conjugate to an element in this torus. Then (1) gives the values of the characters for elements of the torus. We know that h^R acts diagonally, that is, a basis can be chosen so that the matrix of h^R is

$$(2) \qquad \operatorname{diag}\{\Lambda(h), M(h), \cdots\} ,$$

where $\Lambda(h), M(h), \cdots$ are the weights of $\mathfrak{H}$ in the representation. Then the matrix $\exp h^R$ is $\operatorname{diag}\{\exp \Lambda(h), \exp M(h), \cdots\}$; hence

$$(3) \qquad \chi(h) = \sum n_M \exp M(h)$$

where n_M is the multiplicity of the weight $M(h)$, that is, the dimensionality of the weight space $\mathfrak{M}_M$.

One obtains a purely algebraic form of the definition of the character $\chi(h)$ by replacing the exponentials $\exp M(h)$ of (3) by formal exponentials which are elements of a certain group algebra. Weyl's formula gives an expression for the character χ_Λ of the finite-dimensional irreducible module with highest weight Λ as a quotient of two quite simple elementary alternating expressions in the exponentials.

Weyl derived his formula originally by using integration on

compact groups. An elementary purely algebraic method for obtaining the result is due to Freudenthal and we shall follow this in our discussion. A preliminary result of Freudenthal's gives a recursion formula for the multiplicities n_M. Weyl's formula can be used to derive a formula for the dimensionality of the irreducible module with highest weight Λ. It can also be used to obtain the irreducible constituents of the tensor product of two irreducible modules.

1. *Some properties of the Weyl group*

In this section we derive some properties of the Weyl group which are needed for the proof of Weyl's formula and for the determination of the automorphisms of semi-simple Lie algebras over an algebraically closed field of characteristic 0 (Chapter IX). As usual, $\mathfrak{L}$ denotes a finite-dimensional split semi-simple Lie algebra over a field Φ of characteristic zero, e_i, f_i, h_i, $i = 1, 2, \cdots, l$, are canonical generators for $\mathfrak{L}$ as in Chapters IV and VII, and $\mathfrak{H}$ is the splitting Cartan subalgebra spanned by the h_i. Let $\mathfrak{H}^*$ be the conjugate space of $\mathfrak{H}$, $\mathfrak{H}_0^*$ the rational vector space spanned by the roots of $\mathfrak{H}$ in $\mathfrak{L}$. If α is a non-zero root, then the Weyl reflection is the mapping

$$S_\alpha\colon \xi \to \xi - \frac{2(\xi, \alpha)}{(\alpha, \alpha)}\alpha \tag{4}$$

in $\mathfrak{H}_0^*$. Here (ξ, η) is the positive definite scalar product in $\mathfrak{H}_0^*$ defined as in § 4.1. The mapping S_α is characterized by the properties that it is a linear transformation in $\mathfrak{H}_0^*$ which maps α into $-\alpha$ and leaves fixed every vector orthogonal to α. The reflection S_α is an orthogonal transformation relative to (ξ, η) and S_α permutes the weights of $\mathfrak{H}$ in any finite-dimensional module for $\mathfrak{L}$. The S_α generate the Weyl group W which is a finite group. If T is an orthogonal transformation such that αT is a root for some root $\alpha \neq 0$, then a direct calculation shows that

$$T^{-1}S_\alpha T = S_{\alpha T}\ . \tag{5}$$

In particular, this holds for every $T \in W$. We note also that if α is a root then $-\alpha$ is a root and it is clear from (4) that $S_\alpha = S_{-\alpha}$.

The roots α_i, $i = 1, 2, \cdots, l$, such that $\alpha_i(h_j) = A_{ji}$, (A_{ij}) the Cartan matrix, form a simple system. The notion of a simple system of roots is that given in § 4.3 and depends on a lexicographic ordering of the rational vector space $\mathfrak{H}_0^*$. We recall that l roots form a simple system if and only if every root $\alpha = \pm(\sum k_i\alpha_i)$ where the k_i are non-negative integers. This criterion implies that if $\pi = \{\alpha_1, \alpha_2, \cdots, \alpha_l\}$ is a simple system and T is a linear transformation in $\mathfrak{H}_0^*$ which leaves the set of roots invariant, then $\pi T = \{\alpha_1 T, \cdots, \alpha_l T\}$ is a simple system of roots. The set P of positive roots in the lexicographic ordering which gives rise to the simple system π is the set of non-zero roots of the form $\sum k_i\alpha_i$, $k_i \geqq 0$. On the other hand, if the ordering is given, so that P is known, then π is the set of elements of P which cannot be written in the form $\beta + \gamma$, $\beta, \gamma \in P$. Thus P and π determine each other. Any simple system of roots defines a set of canonical generators e_i, f_i, h_i (cf. § 4.3).

LEMMA 1. *Let* $\pi = \{\alpha_1, \alpha_2, \cdots, \alpha_l\}$ *be a simple system of roots and let* α *be a positive* (*negative*) *root. Then* $\alpha S_{\alpha_i} > 0$ (< 0) *if* $\alpha \neq -\alpha_i$ *and* $\alpha S_{\alpha_i} < 0$ (> 0) *if* $\alpha = \alpha$ $(\alpha = -\alpha_i)$.

Proof: If α is a negative root, then $-\alpha$ is a positive root and the assertion on α will follow from that on $-\alpha$. Hence it suffices to assume $\alpha > 0$. We have $\alpha = \sum k_j\alpha_j$, $k_j \geqq 0$ and

$$\alpha S_{\alpha_i} = \sum_{j \neq i} k_j\alpha_j + \left(k_i - \frac{2(\alpha, \alpha_i)}{(\alpha_i, \alpha_i)}\right)\alpha_i \ . \tag{6}$$

If $\alpha \neq \alpha_i$, then since no multiple except $0, \pm\alpha$ of a root is a root, $\alpha \neq k_i\alpha_i$ and so $k_j \neq 0$ for some $j \neq i$. Then the expression for αS_{α_i} has a positive coefficient for some α_j and hence $\alpha S_{\alpha_i} > 0$. If $\alpha = \alpha_i$, then $\alpha S_{\alpha_i} = -\alpha_i < 0$. This completes the proof.

Our results on W will be given in two theorems. The first of these is

THEOREM 1. *If* $\pi = \{\alpha_1, \alpha_2, \cdots, \alpha_l\}$ *is a simple system of roots, then the Weyl group* W (*relative to* $\mathfrak{H}$) *is generated by the reflections* S_{α_i}, $\alpha_i \in \pi$. *If* α *is any non-zero root, then there exists* $\alpha_j \in \pi$ *and* $S \in W$ *such that* $\alpha_j S = \alpha$.

Proof: Let W' be the subgroup of W generated by the S_{α_i}. We show first that any positive root α has the form $\alpha_j S$, $\alpha_j \in \pi$, $S \in W'$. We recall that if $\alpha = \sum k_i\alpha_i$ then $\sum k_i$ is the level of α and we shall prove the result by induction on the level. The result is clear if the level is one since this condition is equivalent to $\alpha \in \pi$. If $\sum k_i > 1$, $\alpha \notin \pi$, and there exists an α_i such that $(\alpha, \alpha_i) > 0$. Otherwise $(\alpha, \alpha_i) \leqq 0$ for all i and since $\alpha = \sum k_i \alpha_i$, $k_i \geqq 0$ this gives $(\alpha, \alpha) \leqq 0$ contrary to $(\alpha, \alpha) > 0$. Choose α_i so that $(\alpha, \alpha_i) > 0$. Then $\beta = \alpha S_{\alpha_i} > 0$ by Lemma 1 and we have $\beta = \sum_{j \neq i} k_j \alpha_j + (k_i - [2(\alpha, \alpha_i)/(\alpha_i, \alpha_i)])\alpha_i$. Since $(\alpha, \alpha_i) > 0$ it is clear that the level of β is lower than that of α. Hence the induction shows that $\beta = \alpha_j S'$, $\alpha_j \in \pi$, $S' \in W'$. Then $\alpha = \alpha_j S' S_{\alpha_i}$ as required. Since $\alpha = \alpha_j S$ where $S \in W'$, $S_\alpha = S^{-1} S_{\alpha_j} S \in W'$. Since $S_\alpha = S_{-\alpha}$ this shows that every $S_\alpha \in W'$ and so $W' = W$. It remains to show that if α is any non-zero root then $\alpha = \alpha_i S$, $\alpha_i \in \pi$, $S \in W$. This has been shown for $\alpha > 0$. Since $\alpha S_\alpha = -\alpha$ the result is clear also for $\alpha < 0$.

THEOREM 2. *Let π and π' be any two simple systems of roots. Then there exists one and only one element $S \in W$ such that $\pi S = \pi'$.*

Proof: Let P and P', respectively, denote the sets of positive roots determined by π and π'. It is clear that P and P' contain the same number q of elements, since this is half the number of non-zero roots. It is clear also that $P = P'$ if and only if $\pi = \pi'$ and if $P \neq P'$, then $\pi \nsubseteq P'$ and $\pi' \nsubseteq P$. Let r be the number of elements in the intersection $P \cap P'$. If $r = q$, $P = P'$ and $S = 1$ satisfies $\pi S = \pi'$, so the result holds in this case. We now employ induction on $q - r$ and we may assume $r < q$, or $P \neq P'$. Then there exists $\alpha_i \in \pi$ such that $\alpha_i \notin P'$ and hence $-\alpha_i = \alpha_i S_{\alpha_i} \in P'$. If $\beta \in P \cap P'$, $\beta S_{\alpha_i} \in P$, by Lemma 1. Hence $\beta S_{\alpha_i} \in P \cap P' S_{\alpha_i}$. Also $\alpha_i = (-\alpha_i) S_{\alpha_i} \in P \cap P' S_{\alpha_i}$. Hence $P \cap P' S_{\alpha_i}$ contains at least $r + 1$ elements. The simple system corresponding to $P' S_{\alpha_i}$ is $\pi' S_{\alpha_i}$ so the induction hypothesis permits us to conclude that there exists a $T \in W$ such that $\pi T = \pi' S_{\alpha_i}$. Then $S = T S_{\alpha_i}$ satisfies $\pi S = \pi'$. This proves the existence of S. To prove uniqueness it suffices to show that if $S \in W$ satisfies $\pi S = \pi$, or equivalently, $PS = P$, then $S = 1$. We give an elementary but somewhat long proof of this here and in an exercise (Exercise 2, below) we shall indicate a short proof which is based on the existence theorem for irreducible modules. We now write $S_i = S_{\alpha_i}$ and by Theorem 1, $S = S_{i_1} S_{i_2} \cdots S_{i_m}$, $i_j = 1, 2, \cdots, l$. We cannot have $m = 1$ since $\alpha_{i_1} S_{i_1} = -\alpha_{i_1} \notin \pi$ and if $m = 2$, then $\alpha_{i_1} S_{i_1} = -\alpha_{i_1}$ and since $(-\alpha_{i_1}) S_{i_2} = \alpha_{i_1} S_{i_1} S_{i_2} > 0$, $i_2 = i_1$ by Lemma 1. Thus $S = S_{i_1}^2 = 1$. We now suppose $m > 2$ and we may assume that if $T = S_{j_1} S_{j_2} \cdots S_{j_r}$, $r < m$, and $\pi T = \pi$, then $T = 1$. Let $S' = S_{i_1}S_{i_2}\cdots S_{i_{m-1}}$, so that $S = S'S_{i_m}$. Since $S \neq S_{i_m}$, $S' \neq 1$ and since S' is a product of $m-1$ S_j, $PS' \neq P$ by the induction hypothesis. Then there exists an $\alpha_j \in \pi$ such that $\alpha_j S' < 0$. Since $\alpha_j S' S_{i_m} > 0$, Lemma 1 implies that $\alpha_j S' = -\alpha_{i_m}$. Similarly, if β is any positive root such that $\beta S' < 0$, then $\beta S' S_{i_m} > 0$ implies that $\beta S' = -\alpha_{i_m} = \alpha_j S'$; hence $\beta = \alpha_j$. Thus S' has the following two properties: $\alpha_j S' = -\alpha_{i_m}$ and $\beta S' > 0$ for every positive $\beta \neq \alpha_j$. Set $S_{i_0} = 1$. Then $\alpha_j S_{i_0} > 0$ and

$$\alpha_j S_{i_0} S_{i_1} \cdots S_{i_{m-1}} = \alpha_j S' = -\alpha_{i_m} < 0\,.$$

Hence there exists a k, $1 \leqq k \leqq m-1$, such that $\alpha_j S_{i_0} S_{i_1} \cdots S_{i_{k-1}} > 0$ and $\alpha_j S_{i_0} S_{i_1} \cdots S_{i_k} < 0$. Since $\alpha_j S_{i_0} \cdots S_{i_{k-1}} > 0$ and $(\alpha_j S_{i_0} \cdots S_{i_{k-1}}) S_{i_k} < 0$, $\alpha_j S_{i_0} \cdots S_{i_{k-1}} = \alpha_{i_k}$. Hence if we set $T = S_{i_0} S_{i_1} \cdots S_{i_{k-1}}$, then $T^{-1} S_j T = S_{i_k}$ or $S_j T = T S_{i_k}$. Then, if $T' = S_{i_{k+1}} \cdots S_{i_{m-1}}$ for $k < m-1$ and $T' = 1$ for $k = m-1$, $S' = T S_{i_k} T' = S_j T T'$. Hence $TT' = S_j S'$. If α is a positive root $\neq \alpha_j$, then $\beta = \alpha S_j$ is a positive root $\neq \alpha_j$. Hence $\alpha T T' = \alpha S_j S' = \beta S' > 0$. Also $\alpha_j T T' = \alpha_j S_j S' = (-\alpha_j) S' = \alpha_{i_m} > 0$. Hence $\pi T T' = \pi$ and since TT' is a product of $m-1$ S_j's the induction hypothesis gives $TT' = 1$. Then $S_j S' = 1$ and so $S' = S_j$ and $S = S_{i_1} S_j$. Hence $S = 1$ by the case $m = 2$ which we considered before.

2. *Freudenthal's formula*

Let $\mathfrak{M}$ be a finite-dimensional irreducible module for $\mathfrak{L}$ with highest weight $\Lambda = \Lambda(h)$ a dominant integral linear function on $\mathfrak{H}$. We know that $\mathfrak{M}$ is a direct sum of weight spaces relative to $\mathfrak{H}$ and that the weights are integral linear functions on $\mathfrak{H}$ of the form $\Lambda - \sum k_i \alpha_i$, k_i a non-negative integer. If $M = M(h)$ is any integral linear function on $\mathfrak{H}$ we define the *multiplicity* n_M *of* M in $\mathfrak{M}$ to be 0 if M is not a weight and otherwise, define $n_M = \dim \mathfrak{M}_M$, where $\mathfrak{M}_M$ is the weight space in $\mathfrak{M}$ of $\mathfrak{H}$ corresponding to the weight M. We recall that for the highest weight Λ we have $n_\Lambda = 1$. We shall now derive a recursion formula due to Freudenthal which expresses n_M in terms of the $n_{M'}$ for $M' > M$ in the lexicographic ordering of $\mathfrak{H}_0^*$ determined by the simple system of roots $\pi = \{\alpha_1, \alpha_2, \cdots, \alpha_l\}$.

Let α be a non-zero root of $\mathfrak{H}$ and choose $e_\alpha \in \mathfrak{L}_\alpha$, $e_{-\alpha} \in \mathfrak{L}_{-\alpha}$ so that $(e_\alpha, e_{-\alpha}) = -1$ where $(x, y) = \mathrm{tr}\, \mathrm{ad}\, x\, \mathrm{ad}\, y$, the Killing form on $\mathfrak{L}$. Then we know (§ 4.1) that $[e_\alpha e_{-\alpha}] = h_\alpha$ where, in general for $\rho \in \mathfrak{H}^*$, h_ρ is the element of $\mathfrak{H}$ such that $(h, h_\rho) = \rho(h)$. As in § 4.2, let $\mathfrak{L}^{(\alpha)}$ be the subalgebra $\mathfrak{H} + \Phi e_\alpha + \Phi e_{-\alpha}$. Then we know that $\mathfrak{M}$ is completely reducible as $\mathfrak{L}^{(\alpha)}$-module (Theorem 4.1). We consider a particular decomposition of $\mathfrak{M}$ as a direct sum of irreducible $\mathfrak{L}^{(\alpha)}$-submodules and we shall refer to these as *the irreducible* $\mathfrak{L}^{(\alpha)}$*-constituents* of $\mathfrak{M}$. Let $\mathfrak{N}$ be one of these. Then we know that $\mathfrak{N}$ has a basis $(y_0, y_1, \cdots, y_m)$ such that

$$\begin{aligned} &y_i h = (M - i\alpha)y_i\,, \qquad i = 0, 1, \cdots, m \\ (7)\qquad &y_i e_{-\alpha} = y_{i+1}\,, \qquad y_m e_{-\alpha} = 0\,, \qquad i = 0, 1, \cdots, m-1 \\ &y_0 e_\alpha = 0\,, \qquad y_i e_\alpha = \frac{i(i-1-m)}{2}(\alpha, \alpha)y_{i-1}\,, \qquad i = 1, 2, \cdots, m\,. \end{aligned}$$

We know also that

$$(8)\qquad m = 2(M, \alpha)/(\alpha, \alpha)$$

where $(\rho, \sigma) = \rho(h_\sigma) = \sigma(h_\rho)$, as in § 4.1. Equations (7) imply that

$$(9)\qquad y_i e_{-\alpha}^R e_\alpha^R = \frac{(i+1)(i-m)}{2}(\alpha, \alpha)y_i\,, \qquad i = 0, 1, \cdots, m\,.$$

We note also that $\mathfrak{N}$ is a direct sum of weight spaces, that the weights are $M, M-\alpha, M-2\alpha, \cdots, M-m\alpha$ and the weight spaces $\mathfrak{N}_{M-p\alpha} = \mathfrak{M}_{M-p\alpha} \cap \mathfrak{N}$ are one-dimensional.

Let M be a weight of $\mathfrak{H}$ in $\mathfrak{M}$ such that $M+\alpha$ is not a weight. Then the α-string of weights in $\mathfrak{M}$ containing M is $M, M-\alpha, \cdots, M-m\alpha$ where $m = 2(M, \alpha)/(\alpha, \alpha)$. Let $0 \leqq p \leqq m$. Then $M-p\alpha$ is a weight and $\mathfrak{M}_{M-p\alpha}$ is a direct sum of the $M-p\alpha$ weight spaces of those irreducible $\mathfrak{L}^{(\alpha)}$-constituents which have this as weight. If $0 \leqq p \leqq (M, \alpha)/(\alpha, \alpha)$, then these are the irreducible $\mathfrak{L}^{(\alpha)}$-constituents having maximal weight $M, M-\alpha, \cdots, M-p\alpha$. Let m_j, $0 \leqq j \leqq (M, \alpha)/(\alpha, \alpha)$, denote the number of irreducible $\mathfrak{L}^{(\alpha)}$-constituents of highest weight $M-j\alpha$. Then it is now clear that

$$(10)\qquad n_{M-j\alpha} = m_0 + m_1 + \cdots + m_j\,,$$

so that,

$$(11)\qquad m_j = n_{M-j\alpha} - n_{M-(j-1)\alpha}\,.$$

If $0 \leqq j \leqq p \leqq (M, \alpha)/(\alpha, \alpha)$, then the weight space corresponding to $M-p\alpha$ in an irreducible $\mathfrak{L}^{(\alpha)}$-constituent having highest weight $M-j\alpha$ is spanned by the vector y_{p-j} in the notation of (7). The dimensionality of this module is $2(M-j\alpha, \alpha)/(\alpha, \alpha)+1=(m-2j)+1$.

Hence we may replace i by $p - j$ and m by $m - 2j$ in (9) to obtain

$$y_{p-j}e^R_{-\alpha}e^R_\alpha = \frac{(p-j+1)(p-m+j)}{2}(\alpha, \alpha)y_{p-j} . \tag{12}$$

We can use this to compute $\mathrm{tr}_{\mathfrak{M}_{M-p\alpha}} e^R_{-\alpha}e^R_\alpha$, the trace of the restriction to $\mathfrak{M}_{M-p\alpha}$ of $e^R_{-\alpha}e^R_\alpha$. We remark that it is clear that

$$\mathfrak{M}_{M-p\alpha}e^R_{-\alpha}e^R_\alpha \subseteq \mathfrak{M}_{M-(p+1)\alpha}e^R_\alpha \subseteq \mathfrak{M}_{M-p\alpha} ,$$

so that $\mathfrak{M}_{M-p\alpha}$ is invariant under $e^R_{-\alpha}e^R_\alpha$. By (12), the contribution to $\mathrm{tr}_{\mathfrak{M}_{M-p\alpha}} e^R_{-\alpha}e^R_\alpha$ of the m_j irreducible $\mathfrak{L}^{(\alpha)}$-constituents with highest weight $M - j\alpha$ is

$$m_j \frac{(p-j+1)(p-m+j)}{2}(\alpha, \alpha) ;$$

hence,

$$\begin{aligned}
\mathrm{tr}_{\mathfrak{M}_{M-p\alpha}} e^R_{-\alpha}e^R_\alpha &= \sum_{j=0}^{p} m_j \frac{(p-j+1)(p-m+j)}{2}(\alpha, \alpha) \\
&= \sum_{j=0}^{p} (n_{M-j\alpha} - n_{M-(j-1)\alpha}) \frac{(p-j+1)(p-m+j)}{2}(\alpha, \alpha) \\
&= \sum_{j=0}^{p} n_{M-j\alpha} \frac{(2j-m)}{2}(\alpha, \alpha) .
\end{aligned}$$

Since $m/2 = (M, \alpha)/(\alpha, \alpha)$ we have the formula

$$\mathrm{tr}_{\mathfrak{M}_{M-p\alpha}} e^R_{-\alpha}e^R_\alpha = -\sum_{j=0}^{p} n_{M-j\alpha}(M - j\alpha, \alpha) , \tag{13}$$

for $0 \leqq p \leqq (M, \alpha)/(\alpha, \alpha)$.

Next let $(M, \alpha)/(\alpha, \alpha) < p \leqq m$. If we apply the Weyl reflection S_α to $M - p\alpha$ we obtain $M - p\alpha - 2((M - p\alpha, \alpha)/(\alpha, \alpha))\alpha = M - (m - p)\alpha$, since $m = 2(M, \alpha)/(\alpha, \alpha)$. We have $0 \leqq m - p \leqq (M, \alpha)/(\alpha, \alpha)$ and we see that $M - p\alpha$ is a weight for the irreducible $\mathfrak{L}^{(\alpha)}$-constituents having the following highest weights: $M, M - \alpha, \cdots, M - (m - p)\alpha$ and only these. The reasoning used to establish (13) now gives

$$\mathrm{tr}_{\mathfrak{M}_{M-p\alpha}} e^R_{-\alpha}e^R_\alpha = -\sum_{j=0}^{m-p-1} n_{M-j\alpha}(M - j\alpha, \alpha) \tag{14}$$

for $(M, \alpha)/(\alpha, \alpha) < p \leqq m$. We recall that if M is a weight, then

MS_α is a weight and $n_M = n_{MS_\alpha}$ (Theorem 4.1). Hence $n_{M-j\alpha} = n_{M-(m-j)\alpha}$. On the other hand, $(M - j\alpha, \alpha) + (M - (m-j)\alpha, \alpha) = (2M - m\alpha, \alpha) = 0$. Hence

$$n_{M-j\alpha}(M - j\alpha, \alpha) + n_{M-(m-j)\alpha}(M - (m-j)\alpha, \alpha) = 0 . \tag{15}$$

This and (14) imply that (13) is valid also for $p > (M, \alpha)/(\alpha, \alpha)$. We recall that M was any weight of $\mathfrak{H}$ in $\mathfrak{M}$ such that $M + \alpha$ is not a weight. Hence $M - p\alpha$ can represent any weight of $\mathfrak{H}$ in $\mathfrak{M}$. We now change our notation and write M for $M - p\alpha$. If we recall that $n_{M'} = 0$ if M' is not a weight, then we can re-write (13) as

$$\mathrm{tr}_{\mathfrak{M}_M} e_{-\alpha}^R e_\alpha^R = -\sum_{j=0}^{\infty} n_{M+j\alpha}(M + j\alpha, \alpha) \tag{16}$$

for any weight M and any root $\alpha \neq 0$.

We consider next the Casimir element defined by the Killing form. This is $\Gamma = \sum_i u_i^R u^{iR}$ where (u_i), (u^i) are dual bases for $\mathfrak{L}$ relative to (x, y). We know that $[a^R\Gamma] = 0$ for all $a \in \mathfrak{L}$. Since R is absolutely irreducible it follows from Schur's lemma that $\Gamma = \gamma 1$, $\gamma \in \Phi$ (cf. Jacobson, *Lectures in Abstract Algebra* II, p. 276). We choose dual bases in the following way: $(h_1, \cdots, h_l)$, $(h^1, \cdots, h^l)$ are dual bases in $\mathfrak{H}$ relative to (x, y). Then since $(e_\alpha, e_{-\alpha}) = -1$ and e_α is orthogonal to $\mathfrak{H}$ and to every e_β, $\beta \neq -\alpha$, the following are dual:

$$(h_1, \cdots, h_l, e_{-\alpha}, e_{-\beta}, \cdots)$$
$$(h^1, \cdots, h^l, -e_\alpha, -e_\beta, \cdots) .$$

Then

$$\Gamma = \sum_{i=1}^{l} h_i^R h^{iR} - \sum_{\alpha \neq 0} e_{-\alpha}^R e_\alpha^R . \tag{17}$$

If we take the trace of the induced mappings in $\mathfrak{M}_M$, M a weight, we obtain

$$\gamma n_M = \sum_i \mathrm{tr}_{\mathfrak{M}_M} h_i^R h^{iR} - \sum_{\alpha \neq 0} \mathrm{tr}_{\mathfrak{M}_M} e_{-\alpha}^R e_\alpha^R . \tag{18}$$

Since h^R is the scalar multiplication by $M(h)$ in $\mathfrak{M}_M$ we have $\mathrm{tr}_{\mathfrak{M}_M} h_i^R h^{iR} = n_M M(h_i) M(h^i)$, so the first term on the right hand side of (18) is $n_M \sum_i M(h_i) M(h^i)$. We proceed to show that $\sum_i M(h_i) M(h^i) = (M, M)$. Thus we write $h_M = \sum \mu_i h_i$. Then $(M, M) = \sum (h_i, h_j)\mu_i \mu_j$ and $M(h_i) = (h_M, h_i) = \sum_j \mu_j (h_j, h_i)$ and $M(h^i) = (h_M, h^i) = \sum \mu_j (h_j, h^i) = \mu_i$.

Hence $\sum M(h_i)M(h^i) = \sum(h_i, h_j)\mu_i\mu_j = (M, M)$. It follows that $\sum_i \mathrm{tr}_{\mathfrak{M}_M} h_i^R h^{iR} = n_M(M, M)$. If we use this result and (16) in (18), we obtain

$$(19) \qquad \gamma n_M = (M, M)n_M + \sum_{\alpha\neq 0} \sum_{j=0}^{\infty} n_{M+j\alpha}(M + j\alpha, \alpha)\,.$$

The terms $n_M(M, \alpha)$ and $n_M(M, -\alpha)$ cancel in this formula so we may replace $\sum_{j=0}^{\infty}$ by $\sum_{j=1}^{\infty}$. The resulting formula is valid also if M is not a weight. In this case $n_M = 0$. If, for a particular α, no $M + j\alpha$, $j \geqq 1$, is a weight then $n_{M+j\alpha} = 0$ and the particular sum $\sum_{j=1}^{\infty} n_{M+j\alpha}(M + j\alpha, \alpha) = 0$. If $M + j\alpha$ is a weight for some $j \geqq 1$, then no $M - k\alpha$ is a weight for $k \geqq 1$ since the α-string containing $M+j\alpha$ does not contain M. Hence the set $\{M+k\alpha \mid k\geqq 1\}$ contains the complete α-string containing $M + j\alpha$. Then $\sum_{j=1}^{\infty} n_{M+j\alpha}(M + j\alpha, \alpha) = 0$, since

$$n_{M+j\alpha}(M + j\alpha, \alpha) = -n_{(M+j\alpha)S_\alpha}((M + j\alpha)S_\alpha, \alpha)\,.$$

Thus in all cases $\sum_{j=1}^{\infty} n_{M+j\alpha}(M + j\alpha, \alpha) = 0$ and (19) holds. The argument shows also that

$$(20) \qquad \sum_{k=-\infty}^{\infty} n_{M+k\alpha}(M + k\alpha, \alpha) = 0$$

holds for any integral linear function M on $\mathfrak{H}$. This implies that $\sum_{k=1}^{\infty} n_{M-k\alpha}(M - k\alpha, -\alpha) = n_M(M, \alpha) + \sum_{k=1}^{\infty} n_{M+k\alpha}(M + k\alpha, \alpha)$, so if we substitute in (19) with $\sum_{j=1}^{\infty}$, we obtain

$$\gamma n_M = (M, M)n_M + \sum_{\alpha>0} n_M(M, \alpha) + 2\sum_{\substack{k=1\\ \alpha>0}}^{\infty} n_{M+k\alpha}(M + k\alpha, \alpha)\,.$$

Setting $\delta = (1/2)(\sum_{\alpha>0}\alpha)$ this gives the following formula:

$$(21) \qquad \gamma n_M = (M, M + 2\delta)n_M + 2\sum_{\substack{k=1\\ \alpha>0}}^{\infty} n_{M+k\alpha}(M + k\alpha, \alpha)\,.$$

If $M = \Lambda$ the highest weight of $\mathfrak{H}$ in $\mathfrak{M}$, then $n_M = 1$ and $n_{M+k\alpha} = 0$ for $\alpha > 0$, $k \geqq 1$. Hence (21) gives

$$\gamma = (\Lambda, \Lambda + 2\delta) = (\Lambda + \delta, \Lambda + \delta) - (\delta, \delta)$$

and substitution in (21) gives *Freudenthal's recursion formula*:

$$(22) \; ((\Lambda + \delta, \Lambda + \delta) - (M + \delta, M + \delta))n_M = 2\sum_{\substack{k=1\\ \alpha>0}}^{\infty} n_{M+k\alpha}(M + k\alpha, \alpha)\,.$$

We shall now show that this gives an effective procedure for calculating n_M beginning with $n_\Lambda = 1$. For this we require a couple of lemmas.

LEMMA 2. *Let* $\delta = (1/2)\sum_{\alpha>0}\alpha$. *Then* $\delta(h_i) = 1$, $i = 1, 2, \cdots, l$, *if* e_i, f_i, h_i *is a set of canonical generators for* $\mathfrak{L}$. *Also if* $S \neq 1$ *is in* W, *then* $\delta - \delta S$ *is a non-zero sum of distinct positive roots.*

Proof: If α is a positive root we know (Lemma 1) that $\alpha S_i \equiv \alpha S_{\alpha_i} = \alpha - \alpha(h_i)\alpha_i > 0$ unless $\alpha = \alpha_i$, in which case $\alpha_i S_i = -\alpha_i$. Hence

$$\delta S_i = \left(\frac{1}{2}\sum_{\alpha>0}\alpha\right)S_i = \left(\frac{1}{2}\sum_{\substack{\alpha>0\\ \alpha\neq\alpha_i}}\alpha\right) - \frac{1}{2}\alpha_i = \delta - \alpha_i\,.$$

Also, $(\delta S_i, \alpha_i) = (\delta, \alpha_i S_i) = (\delta, -\alpha_i)$. Hence $(\delta - \alpha_i, \alpha_i) = (\delta, -\alpha_i)$ and $2(\delta, \alpha_i) = (\alpha_i, \alpha_i)$. Thus $\delta(h_i) = 2(\delta, \alpha_i)/(\alpha_i, \alpha_i) = 1$, $i = 1, 2, \cdots, l$. Since any $S \in W$ maps roots into roots we evidently have $\delta S = \delta - \sum\beta$, where the summation is taken over the $\beta = -\alpha S > 0$. If there are no such β, then $\alpha S > 0$ for all $\alpha > 0$. This implies $S = 1$, by Theorem 2, contrary to hypothesis.

LEMMA 3. *Let* Λ *be the highest weight of* $\mathfrak{H}$ *in* $\mathfrak{M}$. *Then*

$$(M + \delta, M + \delta) < (\Lambda + \delta, \Lambda + \delta) \tag{23}$$

for any weight $M \neq \Lambda$ *of* $\mathfrak{H}$ *in* $\mathfrak{M}$.

Proof: We shall prove that there exists a weight $M' > M$ such that $(M' + \delta, M' + \delta) > (M + \delta, M + \delta)$. This will prove the result by an evident ascending chain argument. Assume first there exists an i such that $M(h_i) < 0$, h_i in the set of canonical generators. Then we take $M' = MS_i = M - M(h_i)\alpha_i > M$ and we have

$$\begin{aligned}(M' + \delta, M' + \delta) &- (M + \delta, M + \delta)\\ &= -2M(h_i)(M + \delta, \alpha_i) + M(h_i)^2(\alpha_i, \alpha_i)\\ &= -2M(h_i)/(\delta, \alpha_i)\,,\end{aligned}$$

since $M(h_i) = 2(M, \alpha_i)/(\alpha_i, \alpha_i)$. Since $(\delta, \alpha_i) > 0$ this shows that $(M' + \delta, M' + \delta) > (M + \delta, M + \delta)$. Next suppose $M(h_i) \geqq 0$, $i = 1, 2, \cdots, l$. If no $M + \alpha_i$ is a root and $x \neq 0$ satisfies $xh = Mx$, then $xe_i = 0$ for all i and x is canonical generator of an e-extreme $\mathfrak{L}$-module. This must coincide with $\mathfrak{M}$, so M is the highest weight, contrary to hypothesis. Now let α_i be one of the simple roots such that $M' = M + \alpha_i$ is a weight. Then $M' > M$ and

$$(M' + \delta, M' + \delta) - (M + \delta, M + \delta) = 2(M + \delta, \alpha_i) + (\alpha_i, \alpha_i) .$$

Since $M(h_i) \geqq 0$, $(M, \alpha_i) \geqq 0$. Also $(\delta, \alpha_i) > 0$ and $(\alpha_i, \alpha_i) > 0$, so again we have $(M' + \delta, M' + \delta) > (M + \delta, M + \delta)$.

It is now clear that (22) can be used to determine the weights and their multiplicities. Thus we begin with $n_\Lambda = 1$. Suppose that for a given $M = \Lambda - \sum k_i\alpha_i$, k_i non-negative integers with at least one $k_i > 0$, we already know $n_{M'}$ for every $M' = \Lambda - \sum k_i'\alpha_i$, k_i' integers, $0 \leqq k_i' \leqq k_i$, $M' \neq M$. Then every term on the right hand side of (22) is known. Moreover, if $(\Lambda + \delta, \Lambda + \delta) = (M + \delta, M + \delta)$, then, by Lemma 3, M is not a weight and $n_M = 0$. Otherwise, the coefficient of n_M in (22) is not zero so we can solve for n_M using the formula.

3. *Weyl's character formula*

In order to obtain a general formulation of Weyl's formula valid for any field it is necessary to replace the exponentials which appear in this formula by "formal" exponentials. This notion can be made precise by introducing the group algebra over the base field of the group of integral linear functions on $\mathfrak{H}$. We recall that an element $M \in \mathfrak{H}^*$ is called integral if $M(h_i)$ is an integer for $i = 1, 2, \cdots, l$, where e_i, f_i, h_i are canonical generators for $\mathfrak{L}$. The set $\mathfrak{J}$ of integral linear functions is a group under addition, and it is clear that $\mathfrak{J}$ is the direct sum of the cyclic groups generated by elements λ_i of $\mathfrak{J}$ such that $\lambda_i(h_j) = \delta_{ij}$. We now introduce an algebra $\mathfrak{A}$ over Φ with basis $\{e(M) \mid M \in \mathfrak{J}\}$ in 1 : 1 correspondence with the elements of $\mathfrak{J}$ in which the multiplication table is

$$e(M)e(M') = e(M + M') . \tag{24}$$

Then $\mathfrak{A}$ is the group algebra over Φ of $\mathfrak{J}$ and $e(0) = 1$ is the identity element of $\mathfrak{A}$. The elements of $\mathfrak{A}$ are the formal exponentials to which we alluded before. We now define *the character* χ of a finite-dimensional module $\mathfrak{M}$ to be the formal exponential

$$\chi = \sum_M n_M e(M) , \tag{25}$$

where n_M is the multiplicity of $M \in \mathfrak{J}$ as defined before: $n_M = 0$ if M is not a weight and $n_M = \dim \mathfrak{M}_M$ the dimensionality of the weight space $\mathfrak{M}_M$ if M is a weight. The summation in (25) is taken over all $M \in \mathfrak{J}$. This is a finite sum since $n_M \neq 0$ for only a finite number of M.

Let $x_i = e(\lambda_i)$, $\lambda_i(h_j) = \delta_{ij}$. Since any integral linear function can be written in one and only one way as $M = \sum m_i\lambda_i$ where the m_i are integers, the base element $e(M) = e(\sum m_i\lambda_i) = e(m_1\lambda_1)\cdots e(m_l\lambda_l) = e(\lambda_1)^{m_1}\cdots e(\lambda_l)^{m_l} = x_1^{m_1}\cdots x_l^{m_l}$. We have called M dominant if $M(h_i) \geqq 0$ for $i = 1, 2, \cdots, l$. This is equivalent to $m_i \geqq 0$. The set of linear combinations of the $e(M)$, M dominant, is a subalgebra of $\mathfrak{A}$ which is the same as the set of linear combinations of the monomials $x_1^{m_1}\cdots x_l^{m_l}$, $m_i \geqq 0$. The set of these monomials obtained from the sequences of integers $(m_1, \cdots, m_l)$ with $m_i \geqq 0$ are linearly independent. Hence the subalgebra we have indicated can be identified with the commutative polynomial algebra $\Phi[x_1, x_2, \cdots, x_l]$ in the algebraically independent elements $x_1, \cdots, x_l$. Every element of $\mathfrak{A}$ has the form $(x_1^{r_1}\cdots x_l^{r_l})^{-1}f$ where the $r_i \geqq 0$ and $f \in \Phi[x_1, \cdots, x_l]$. It is well known that $\Phi[x_1, \cdots, x_l]$ is an integral domain. It follows that $\mathfrak{A}$ is a commutative integral domain.

With each element S in the Weyl group W we associate the linear mapping in $\mathfrak{A}$ such that $e(M)S = e(MS)$. Since

$$(e(M)e(M'))S = (e(M + M'))S = e((M + M')S)$$
$$= e(MS + M'S) = e(MS)e(M'S) = e(M)Se(M')S\,,$$

it is clear that S is an automorphism in $\mathfrak{A}$. The set of S in $\mathfrak{A}$ is a group of automorphisms isomorphic to W. An element $a \in \mathfrak{A}$ is called *symmetric* if $aS = a$, $S \in W$, and *alternating* if $aS = (\det S)a$, $S \in W$ where $\det S$ is the determinant of the orthogonal transformation S in $\mathfrak{H}_0^*$. Thus $\det S = \pm 1$ and if $S = S_\alpha$ the Weyl reflection determined by the root α then $\det S_\alpha = -1$. In particular, $\det S_i = -1$ for $S_i = S_{\alpha_i}$. Since the S_i generate W, a is symmetric if and only if $aS_i = a$, $i = 1, 2, \cdots, l$, and a is alternating if and only if $aS_i = -a$ for all i. The set of symmetric elements is a subalgebra; the set of alternating elements is a subspace. The product of two alternating elements is symmetric and the product of an alternating element and a symmetric element is alternating. Since $n_{MS} = n_M$ for $S \in W$, it follows that the character $\chi = \sum n_M e(M)$ is a symmetric element. We note next that the element

$$Q = e(-\delta)\prod_{\alpha>0}(e(\alpha) - 1) = e(\delta)\prod_{\alpha>0}(1 - e(-\alpha))\,, \tag{26}$$

where $\delta = (1/2)(\sum_{\alpha>0}\alpha)$ the integral linear function defined in § 2, is an alternating element of $\mathfrak{A}$. Thus we have seen (proof of Lemma 2) that $\delta S_i = \delta - \alpha_i$ so $(-\delta)S_i = -\delta + \alpha_i$ and $e(-\delta)S_i = e(-\delta)e(\alpha_i)$. Also we know that S_i permutes the positive roots $\neq \alpha_i$ and sends α_i into $-\alpha_i$. Hence

$$\prod_{\alpha>0}(e(\alpha)-1)S_i = \Big(\prod_{\substack{\alpha>0\\ \alpha\neq\alpha_i}}(e(\alpha)-1)\Big)(e(-\alpha_i)-1)$$

and

$$\begin{aligned} QS_i &= \Big(\prod_{\substack{\alpha>0\\ \alpha\neq\alpha_i}}(e(\alpha)-1)\Big)(e(-\alpha_i)-1)e(-\delta)e(\alpha_i) \\ &= e(-\delta)\prod_{\substack{\alpha>0\\ \alpha\neq\alpha_i}}(e(\alpha)-1)(1-e(\alpha_i)) \\ &= -Q\,. \end{aligned}$$

We set

$$f = Q\chi\,. \tag{27}$$

This element is alternating. It turns out that we can obtain a formula for f and this will give the desired formula for χ.

Let $\sigma = \sum_{S\in W}(\det S)S$ as linear operator in $\mathfrak{A}$. We have for any $T \in W$ that $\sigma T = \sum_S(\det S)ST = (\det T)^{-1}\sum_S(\det ST)ST = (\det T)\sigma$ and similarly $T\sigma = (\det T)\sigma$. If a is any element of $\mathfrak{A}$, then $(a\sigma)T = (\det T)(a\sigma)$. Hence $a\sigma$ is an alternating element. If a is alternating to begin with, then $aS = (\det S)a$ and so $a\sigma = wa$ where w is the order of the Weyl group. It follows that $(1/w)\sigma$ is a projection operator of $\mathfrak{A}$ onto the space of alternating elements. Hence any alternating element has the form $a\sigma$, $a \in \mathfrak{A}$, and consequently such an element is a linear combination of the elements $e(M)\sigma$, M an integral linear function. Since $S\sigma = \pm\sigma$ it is clear that $e(M)\sigma$ can be replaced by $e(MS)\sigma$ and so we may express any alternating element as a linear combination of elements $e(M)\sigma$ where M is highest among its conjugates MS under the Weyl group. In particular, we may suppose that $M \geqq MS_i = M - M(h_i)\alpha_i$, that is, we may suppose that $M(h_i) \geqq 0$ for $i = 1, 2, \cdots, l$. Suppose $M(h_i) = 0$ for some i, or equivalently $MS_i = M$. Then $e(M)\sigma = -e(M)S_i\sigma = -e(MS_i)\sigma = -e(M)\sigma$ and $e(M)\sigma = 0$. We therefore conclude that every alternating element is a linear combination of the elements $e(M)\sigma$ where $M(h_i) > 0$, $i = 1, 2, \cdots, l$.

We now apply this argument to the element Q defined in (26). If we multiply out the product in (26) we see that Q is a linear combination of elements $e(M)$ where $M = \delta - \rho$ and ρ is a sum of a subset of positive roots. Thus $M = (1/2)\sum_{\alpha>0}\varepsilon_\alpha\alpha$ where $\varepsilon_\alpha = \pm 1$ and any conjugate MS of M under the Weyl group again has the form $\delta - \rho' = (1/2)\sum_{\alpha>0}\varepsilon'_\alpha\alpha$ where ρ' is a sum of positive roots and $\varepsilon'_\alpha = \pm 1$. If we apply the projection operator $(1/w)\sigma$ to the expression indicated for Q we obtain an expression for Q as linear combination of elements $e(M)\sigma$ where M is of the form $\delta - \rho$, ρ a sum of positive roots. We have seen also that we may assume that M satisfies $M(h_i) > 0$, $i = 1, 2, \cdots, l$. These conditions imply that $M = \delta$. Thus, if $M = \delta - \rho$, $\rho = \sum k_i\alpha_i$ where the k_i are non-negative integers, then $0 < M(h_i) = \delta(h_i) - \rho(h_i) = 1 - \rho(h_i)$. Since $\rho \in \mathfrak{I}$, $\rho(h_i)$ is an integer. Hence we must have $\rho(h_i) \leqq 0$. On the other hand, $0 \leqq (\rho, \rho) = \sum k_i(\alpha_i, \rho) = (1/2)\sum k_i(\alpha_i, \alpha_i)\rho(h_i) \leqq 0$. Hence $(\rho, \rho) = 0$ and $\rho = 0$. Thus we have shown that $Q = \eta e(\delta)\sigma$, $\eta \in \Phi$. Now $e(\delta)\sigma = \sum_S(\det S)e(\delta S)$. By Lemma 2, $\delta S \neq \delta$ if $S \neq 1$. Hence $e(\delta)\sigma = e(\delta) + \sum \pm e(M)$ where $M = \delta - \rho$, $\rho \neq 0$ and a sum of positive roots. Also it is clear from (26) that $Q = e(\delta) + \sum \pm e(M)$. Since $Q = \eta e(\delta)\sigma$ it follows that $\eta = 1$. We have therefore proved the following

LEMMA 4. *Let* $Q = e(-\delta)\Pi_{\alpha>0}(e(\alpha) - 1)$. *Then*

$$Q = e(\delta)\sigma = \sum_{S\in W}(\det S)e(\delta S)\,. \tag{28}$$

We shall next introduce the vector space $\mathfrak{H}^* \otimes_\Phi \mathfrak{A}$ and we shall define certain linear mappings of this space and of the algebra $\mathfrak{A}$. The elements of $\mathfrak{H}^* \otimes \mathfrak{A}$ have the form $\sum\rho_i \otimes a_i$, $\rho_i \in \mathfrak{H}^*$, $a_i \in \mathfrak{A}$. The algebra composition in $\mathfrak{A}$ provides a linear mapping of $\mathfrak{A} \otimes_\Phi \mathfrak{A}$ into $\mathfrak{A}$ such that $a \otimes b \rightarrow ab$. This gives a linear mapping of $(\mathfrak{H}^* \otimes \mathfrak{A}) \otimes \mathfrak{A} = \mathfrak{H}^* \otimes (\mathfrak{A} \otimes \mathfrak{A})$ into $\mathfrak{H}^* \otimes \mathfrak{A}$ so that $(\rho \otimes a) \otimes b \rightarrow \rho \otimes ab$. It follows that if we set $(\sum\rho_i \otimes a_i)b = \sum\rho_i \otimes a_ib$, then this product of $\sum\rho_i \otimes a_i$ and b is single-valued and coincides with the image of $\sum\rho_i \otimes a_i \otimes b$ in $\mathfrak{H}^* \otimes \mathfrak{A}$. It is clear that the product $(\sum\rho_i \otimes a_i)b = \sum\rho_i \otimes a_ib$ turns $\mathfrak{H}^* \otimes \mathfrak{A}$ into a right $\mathfrak{A}$-module. We recall that we have the bilinear form $(\rho, \sigma) = (h_\rho, h_\sigma)$ on $\mathfrak{H}^*$ which defines the linear mapping of $\mathfrak{H}^* \otimes_\Phi \mathfrak{H}^*$ into Φ so that $\rho \otimes \tau \rightarrow (\rho, \tau)$. If we combine this with the linear mapping of $\mathfrak{A} \otimes \mathfrak{A}$ into $\mathfrak{A}$ we obtain the linear mapping of $(\mathfrak{H}^* \otimes \mathfrak{A}) \otimes (\mathfrak{H}^* \otimes \mathfrak{A})$ into $\Phi \otimes \mathfrak{A} = \mathfrak{A}$ such that $(\rho \otimes a) \otimes (\tau \otimes b) \rightarrow (\rho, \tau)ab$. This defines a Φ-bilinear mapping of $\mathfrak{H}^* \otimes \mathfrak{A}$ such that the value $(\rho \otimes a, \tau \otimes b) = (\rho, \tau)ab$. Since (ρ, σ) is symmetric and $\mathfrak{A}$ is commutative this is a symmetric bilinear form. Also if

$c \in \mathfrak{A}$, then $((\rho \otimes a)c, \tau \otimes b) = (\rho \otimes ac, \tau \otimes b) = (\rho, \tau)acb$ and $(\rho \otimes a, \tau \otimes b)c = (\rho, \tau)abc$ and similarly $(\rho \otimes a, (\tau \otimes b)c) = (\rho \otimes a, \tau \otimes b)c$ which implies that (x, y), $x, y \in \mathfrak{H}^* \otimes \mathfrak{A}$, is $\mathfrak{A}$-bilinear.

Next we define a linear mapping, the *gradient*, of $\mathfrak{A}$ into $\mathfrak{H}^* \otimes \mathfrak{A}$: $a \to aG$ such that $e(M)G = M \otimes e(M)$ and a linear mapping, the *Laplacian*, of $\mathfrak{A}$ into $\mathfrak{A}$: $a \to a\Delta$ such that $e(M)\Delta = (M, M)e(M)$. We have

$$\begin{aligned}(e(M)e(M'))G &= e(M + M')G \\ &= (M + M') \otimes e(M + M') = (M + M') \otimes e(M)e(M') \\ &= M \otimes e(M)e(M') + M' \otimes e(M')e(M) \\ &= (e(M)G)e(M') + (e(M')G)e(M) .\end{aligned}$$

The linearity then implies that

$$(ab)G = (aG)b + (bG)a , \qquad a, b \in \mathfrak{A} . \tag{29}$$

We have

$$\begin{aligned}(e(M)e(M'))\Delta &= e(M + M')\Delta = (M + M', M + M')e(M)e(M') \\ &= (M, M)e(M)e(M') + 2(M, M')e(M)e(M') \\ &\quad + (M', M')e(M)e(M') \\ &= (e(M)\Delta)e(M') + 2(M \otimes e(M), M' \otimes e(M')) \\ &\quad + (e(M')\Delta)e(M) \\ &= (e(M)\Delta)e(M') + 2(e(M)G, e(M')G) + (e(M')\Delta)e(M) .\end{aligned}$$

This implies that

$$(ab)\Delta = (a\Delta)b + 2(aG, bG) + (b\Delta)a . \tag{30}$$

We now return to the formulas which we developed for the multiplicity n_M of the integral linear function M in a finite-dimensional irreducible module of highest weight Λ, γ the element of Φ (rational number) determined by the Casimir operator. We consider again (19):

$$\gamma n_M = (M, M)n_M + \sum_{\alpha \neq 0} \sum_{j=0}^{\infty} n_{M+j\alpha}(M + j\alpha, \alpha) .$$

We multiply both sides by $e(M)$ and sum on M. This gives

$$\gamma \chi = \chi \Delta + \sum_{M} \sum_{\alpha \neq 0} \sum_{j=0}^{\infty} n_{M+j\alpha}(M + j\alpha, \alpha)e(M) ,$$

which we multiply through by

$$\prod_{\alpha\neq 0}(e(\alpha)-1)=\prod_{\alpha>0}(e(\alpha)-1)\prod_{\alpha>0}(e(-\alpha)-1)=\pm Q^2\,,$$

by (26). This gives

$$\begin{aligned}(31)\qquad &\pm\gamma\chi Q^2\mp(\chi\Delta)Q^2\\ &=\sum_M\sum_{\alpha\neq 0}\sum_{j=0}^{\infty}\prod_{\beta\neq\alpha}(e(\beta)-1)n_{M+j\alpha}(M+j\alpha,\alpha)(e(M+\alpha)-e(M))\,.\end{aligned}$$

The coefficient of $e(M+\alpha)$ in the right-hand side of (31) is

$$\begin{aligned}&\prod_{\beta\neq\alpha}(e(\beta)-1)\left(\sum_{j=0}^{\infty}n_{M+j\alpha}(M+j\alpha,\alpha)-\sum_{j=0}^{\infty}n_{M+(j+1)\alpha}(M+(j+1)\alpha,\alpha)\right)\\ &\qquad=\prod_{\beta\neq\alpha}(e(\beta)-1)n_M(M,\alpha)\,.\end{aligned}$$

Hence (31) can be written in the form

$$\begin{aligned}(32)\qquad &\pm\gamma\chi Q^2\mp(\chi\Delta)Q^2\\ &=\sum_{\alpha\neq 0}\prod_{\beta\neq\alpha}(e(\beta)-1)e(\alpha)\sum_M n_M(M,\alpha)e(M)\\ &=\left(\sum_{\alpha\neq 0}\alpha\otimes e(\alpha)\prod_{\beta\neq\alpha}(e(\beta)-1),\ \sum_M M\otimes n_M e(M)\right)\\ &=(\pm Q^2G,\chi G)=\pm 2((QG)Q,\chi G)\,.\end{aligned}$$

Hence we have

$$\gamma\chi Q^2-(\chi\Delta)Q^2=2((QG)Q,\chi G)=2(QG,\chi G)Q$$

and canceling $Q\neq 0$ in the integral domain $\mathfrak{A}$, we obtain

$$\begin{aligned}\gamma\chi Q-(\chi\Delta)Q&=2(QG,\chi G)\\ &=(\chi Q)\Delta-(\chi\Delta)Q-(Q\Delta)\chi,\ \text{by (30)}\,.\end{aligned}$$

If we set $f=\chi Q$, as before, we obtain

$$(33)\qquad \gamma f=f\Delta-(Q\Delta)\chi\,.$$

Since $Q=\sum_{S\in W}\det S(e(\delta S))$ and $(\delta S,\delta S)=(\delta,\delta)$ we have $Q\Delta=(\delta,\delta)Q$. Since $\gamma=(\Lambda+\delta,\Lambda+\delta)-(\delta,\delta)$ (eq. below (21)) these substitutions convert (33) to the following fundamental equation for f

$$(34)\qquad f\Delta=(\Lambda+\delta,\Lambda+\delta)f\,.$$

The element $f=\chi Q$ is an alternating element and consequently this element is a linear combination of elements of the form $e(M)\sigma$. Moreover, we can limit the M which are needed here by looking at the form of χ and Q. Thus we have $\chi=\sum n_M e(M)$

where we now consider the summation as taken just over the weights M of the representation. Also we have seen that $Q = \sum_{S \in W}(\det S)e(\delta S)$. If we multiply we obtain χQ as a linear combination of terms $e(M+\delta S)$ where M is a weight: $f = \sum \eta_{M+\delta S} e(M+\delta S)$, M a weight, $S \in W$. Now

$$\begin{aligned} e(M + \delta S)\Delta &= (M + \delta S, M + \delta S)e(M + \delta S) \\ &= (MS^{-1} + \delta, MS^{-1} + \delta)e(M + \delta S) \end{aligned}$$

so that $e(M + \delta S)$ is in the characteristic space of the characteristic root $(MS^{-1}+\delta, MS^{-1}+\delta)$ of Δ. Since f belongs to the characteristic root $(\Lambda + \delta, \Lambda + \delta)$ for Δ and characteristic spaces belonging to distinct roots are linearly independent it follows that f is a linear combination of $e(M + \delta S)$ such that

$$(MS^{-1} + \delta, MS^{-1} + \delta) = (\Lambda + \delta, \Lambda + \delta) .$$

By Lemma 3, $(MS^{-1} + \delta, MS^{-1} + \delta) = (\Lambda + \delta, \Lambda + \delta)$ for the weight MS^{-1} implies that $MS^{-1} = \Lambda$. Hence we see that f is a linear combination of the terms $e(\Lambda + \delta)S$. If we apply the projection operator $(1/w)\sigma$ to f we see that $f = \eta e(\Lambda + \delta)\sigma = \eta \sum_S (\det S)e(\Lambda + \delta)S$. Since $\delta S < \delta$ if $S \neq 1$, $(\Lambda + \delta)S < \Lambda + \delta$ if $S \neq 1$. Hence the coefficient of $e(\Lambda + \delta)$ in our expression for f is η. On the other hand, the coefficient of this term in χQ is $n_\Lambda = 1$. Hence $\eta = 1$ and we have proved

Weyl's Theorem. Let $\mathfrak{M}$ be the irreducible module for $\mathfrak{L}$ with highest weight Λ. Then the character $\chi_\Lambda = \sum n_M e(M)$ of $\mathfrak{L}$ in $\mathfrak{M}$ is given by the formula

$$\chi_\Lambda \sum_{S \in W} (\det S)e(\delta S) = \sum_{S \in W} \det S(e(\Lambda + \delta)S) , \tag{35}$$

where $\delta = (1/2)\sum_{\alpha>0}\alpha$, α a root.

This theorem means that the expression on the right is divisible by $Q = \sum_S (\det S)e(\delta S)$ in $\mathfrak{A}$ and the quotient is the character χ of the representation.

It is easy to see that this result gives Weyl's original formula

$$\chi_\Lambda(h) = \frac{\sum_{S \in W} \det S \exp((\Lambda + \delta)S)(h)}{\sum_{S \in W} \det \exp(\delta S)(h)} \tag{36}$$

in the complex case. Weyl employed his result to obtain by a limiting process the dimensionality of $\mathfrak{M} = \sum n_M = \chi_\Lambda(0)$. We proceed to obtain the same result by a somewhat similar device.

We introduce the algebra $\Phi\langle t\rangle$ of formal power series in an indeterminate t with coefficients in Φ. We recall that the mapping of power series into their constant terms is a homomorphism ζ of $\Phi\langle t\rangle$ onto Φ. We can also define homomorphisms of $\mathfrak{A}$ into $\Phi\langle t\rangle$ by employing exponentials: $\exp z = 1 + z + (z^2/2!) + \cdots$ which is defined for any $z \in \Phi\langle t\rangle$ with zero constant term. We have the relation $\exp(z_1 + z_2) = (\exp z_1)(\exp z_2)$; hence, if $\lambda, \mu, \rho \in \mathfrak{H}^*$, then $\exp(\lambda, \rho)t \exp(\mu, \rho)t = \exp(\lambda + \mu, \rho)t$. In particular, this holds for $\lambda = M$, $\mu = M'$, integral linear functions on $\mathfrak{H}$, which implies, in view of (24), that we have a homomorphism ζ_ρ of $\mathfrak{A}$ into $\Phi\langle t\rangle$ such that $e(M)\zeta_\rho = \exp(M, \rho)t$. Now consider $\chi_\Lambda\zeta_\rho\zeta$. Since $\chi_\Lambda = \sum n_M e(M)$ we have $\chi_\Lambda\zeta_\rho = \sum n_M \exp(M, \rho)t$ and since the constant term of an exponential is 1, $\chi_\Lambda\zeta_\rho\zeta = \sum n_M = \dim \mathfrak{M}$. We shall obtain the formula for $\dim \mathfrak{M}$ by applying $\zeta_\rho\zeta$ to (35), taking $\rho = \delta = (1/2)\sum_{\alpha>0}\alpha$.

Let $\sigma = \sum_{S \in W}(\det S)S$ as before and let M, M' be integral linear functions. Then

$$\begin{aligned} e(M)\sigma\zeta_{M'} &= \sum_S \det S \exp(MS, M')t \\ &= \sum_S \det S \exp(M'S^{-1}, M)t = e(M')\sigma\zeta_M . \end{aligned}$$

Hence

$$\begin{aligned} (37)\quad e(M)\sigma\zeta_\delta &= e(\delta)\sigma\zeta_M = e(-\delta)\zeta_M \prod_{\alpha>0}(e(\alpha)\zeta_M - 1) \qquad \text{(Lemma 4)} \\ &= \exp(-\delta, M)t \prod_{\alpha>0}(\exp(\alpha, M)t - 1) \\ &= \prod_{\alpha>0}\left(\exp\frac{1}{2}(\alpha, M)t - \exp\frac{1}{2}(-\alpha, M)t\right), \end{aligned}$$

since $\delta = (1/2)\sum_{\alpha>0}\alpha$. Applying this to (35) we obtain

$$\begin{aligned} (38)\quad (\chi_\Lambda\zeta_\delta) &\prod_{\alpha>0}\left(\exp\frac{1}{2}(\alpha, \delta)t - \exp\frac{1}{2}(-\alpha, \delta)t\right) \\ &= \prod_{\alpha>0}\left(\exp\frac{1}{2}(\alpha, \Lambda + \delta)t - \exp\frac{1}{2}(-\alpha, \Lambda + \delta)t\right). \end{aligned}$$

Now

$$\begin{aligned} (39)\quad &\prod_{\alpha>0}\left(\exp\frac{1}{2}(\alpha, M)t - \exp\frac{1}{2}(-\alpha, M)t\right) \\ &\qquad \equiv \prod_{\alpha>0}(\alpha, M)t^k \qquad (\text{mod } t^{k+1}), \end{aligned}$$

where k is the number of positive roots. Hence if we divide both sides of (38) by t^k and then apply the homomorphism ζ which picks out the constant term, we obtain

$$(\dim \mathfrak{M}) \prod_{\alpha>0} (\alpha, \delta) = \prod_{\alpha>0} (\alpha, \delta + \Lambda) ,$$

or

$$\text{(40)} \qquad \dim \mathfrak{M} = \frac{\prod_{\alpha>0}(\alpha, \delta + \Lambda)}{\prod_{\alpha>0} (\alpha, \delta)} .$$

4. Some examples

We now indicate how $\dim \mathfrak{M} = \Pi_{\alpha>0}(\Lambda + \delta, \alpha)/\Pi_{\alpha>0}(\delta, \alpha)$ can be calculated from the Dynkin diagram for $\mathfrak{L}$. Let $w_i = 1, 2$ or 3 be the weight of the vertex α_i in the diagram and let α_r be one of the roots in the simple system such that $w_r = 1$. We replace the scalar product (α, β) by $(\alpha, \beta)' = 2(\alpha, \beta)/(\alpha_r, \alpha_r)$. Then we evidently have $\dim \mathfrak{M} = \Pi_{\alpha>0}(\Lambda + \delta, \alpha)'/\Pi_{\alpha>0}(\delta, \alpha)'$. We write $\Lambda = \sum m_i\lambda_i$, $m_i \geqq 0$, $\alpha = \sum k_j\alpha_j$, $k_j \geqq 0$. Then since $(\sum\lambda_i)(h_j) = 1$ and $\delta(h_j) = 1$, $j = 1, \cdots, l$, $\delta = \sum\lambda_i$. Hence we require $(\sum(m_i + 1)\lambda_i, \sum k_j\alpha_j)' = \sum_{i,j}(m_i + 1)k_j(\lambda_i, \alpha_j)'$ and $\sum_{i,j}k_j(\lambda_i, \alpha_j)'$. Now

$$(\lambda_i, \alpha_j)' = \frac{2(\lambda_i, \alpha_j)}{(\alpha_r, \alpha_r)} = \frac{2(\lambda_i, \alpha_j)}{(\alpha_j, \alpha_j)} \cdot \frac{(\alpha_j, \alpha_j)}{(\alpha_r, \alpha_r)} = \delta_{ij}w_j$$

Hence $(\Lambda + \delta, \alpha)' = \sum(m_i + 1)w_ik_i$ and $(\delta, \alpha)' = \sum w_ik_i$ and

$$\text{(41)} \qquad \dim \mathfrak{M} = \prod_{(k_i)} \frac{\sum_{i=1}^{l}(m_i + 1)w_ik_i}{\sum_{i=1}^{l} w_ik_i}$$

where the product is taken over all the sequences $(k_1, k_2, \cdots, k_l)$, $k_i \geqq 0$, such that $\sum k_j\alpha_j$ is a root. We have seen that this set as well as the w_i can be determined from the Dynkin diagram.

G_2. Here $w_1 = 3$, $w_2 = 1$ and the roots are α_1, α_2, $\alpha_1 + \alpha_2$, $\alpha_1 + 2\alpha_2$, $\alpha_1 + 3\alpha_2$, $2\alpha_1 + 3\alpha_2$. Then (41) gives

$$\text{(42)} \qquad \dim \mathfrak{M} = \frac{1}{120} [(m_1 + 1)(m_2 + 1)(3m_1 + m_2 + 4) \cdot(3m_1 + 2m_2 + 5)(m_1 + m_2 + 2)(2m_1 + m_2 + 3)] ,$$

if the highest weight of $\mathfrak{M}$ is $m_1\lambda_1 + m_2\lambda_2$. For $\Lambda = \lambda_1$ and $\Lambda = \lambda_2$ we obtain, respectively, 14 and 7, which we had obtained previously.

B_l, $l \geqq 2$. Here $w_1 = w_2 = \cdots = w_{l-1} = 2$, $w_l = 1$ and the weights are

$$\begin{aligned} &\alpha_i + \alpha_{i+1} + \cdots + \alpha_j\,, \qquad 1 \leqq i \leqq j \leqq l-1 \\ &\alpha_i + \alpha_{i+1} + \cdots + \alpha_l\,, \qquad 1 \leqq i \leqq l \\ &\alpha_i + \cdots + \alpha_{j-1} + 2(\alpha_j + \cdots + \alpha_l)\,, \qquad 1 \leqq i < j \leqq l\,. \end{aligned} \tag{43}$$

These contribute the following factors to the dimensionality formula:

$$\begin{aligned} &\frac{m_i + m_{i+1} + \cdots + m_j + j - i + 1}{j - i + 1}\,, \qquad 1 \leqq i \leqq j \leqq l-1 \\ &\frac{2(m_i + m_{i+1} + \cdots + m_{l-1}) + m_l + 2(l-i) + 1}{2l - j - i + 1}\,, \qquad 1 \leqq i \leqq l \\ &\frac{(m_i + \cdots + m_{j-1} + m_l) + 2(m_j + \cdots + m_{l-1}) + 2l - j - i + 1}{2l - j - i + 1}\,, \\ &\qquad\qquad 1 \leqq i < j \leqq l. \end{aligned} \tag{44}$$

Let $\Lambda = \lambda_k$, $1 \leqq k \leqq l-1$, so that $m_k = 1$, $m_i = 0$ if $i \neq k$. The product taken over the first set of factors in (44) is

$$\prod_{i=1}^{k} \prod_{j=k}^{l-1} \frac{j-i+2}{j-i+1} = \binom{l}{k}.$$

The second set of factors gives

$$\prod_{i=1}^{k} \frac{2l-2i+3}{2l-2i+1} = \frac{2l+1}{2l-2k+1}$$

and the last is

$$\begin{aligned} &\prod_{i=1}^{k} \prod_{j=k+1}^{l} \frac{2l-j-i+2}{2l-j-i+2} \prod_{1 \leqq i < j \leqq k} \frac{2l-j-i+3}{2l-j-i+1} \\ &\qquad = \frac{\binom{2l-k}{k}}{\binom{l}{k}} \frac{\binom{2l}{k}}{\binom{2l-k+1}{k}}. \end{aligned}$$

Multiplication of these results gives

$$\dim \mathfrak{M} = \binom{2l+1}{k}. \tag{45}$$

We recall that this result was what was required to complete our proof (§ 7.6) of the irreducibility of the module of k-vectors relative to B_l.

5. *Applications and further results*

We shall now call a character of a finite-dimensional irreducible representation of $\mathfrak{L}$ a *primitive* character. Such a character has the form $\chi_\Lambda = e(\Lambda) + \sum n_M e(M)$, where the summation is taken over $M < \Lambda$ in the lexicographic ordering in $\mathfrak{H}_0^*$ or in $\mathfrak{J}$. Since the $e(\Lambda)$ constitute a basis for the group algebra $\mathfrak{A}$ of $\mathfrak{J}$ it is clear that $\chi_{\Lambda_1} = \chi_{\Lambda_2}$ implies $\Lambda_1 = \Lambda_2$ and this implies that the associated representations are equivalent. Conversely, equivalence of finite-dimensional irreducible representations implies equality of the characters. It is clear also from the expression we have indicated for a primitive character that distinct primitive characters are linearly independent. If $\mathfrak{M}$ is any finite-dimensional module for $\mathfrak{L}$, $\mathfrak{M}$ is a direct sum of, say, m_1 irreducible modules with character χ_{Λ_1}, m_2 with character $\chi_{\Lambda_2} \neq \chi_{\Lambda_1}$, etc. Then the character χ of $\mathfrak{M}$ has the form

$$\chi = m_1\chi_{\Lambda_1} + m_2\chi_{\Lambda_2} + \cdots + m_k\chi_{\Lambda_k} . \tag{46}$$

Since this expression is unique we see that if χ and the primitive characters are known then the m_i can be determined. This gives the isomorphism classes and multiplicities of the irreducible constituents of $\mathfrak{M}$. As a corollary, we see that these classes and multiplicities are independent of the particular decomposition of $\mathfrak{M}$ into irreducible constituents.

The characters can be used to determine the isomorphism classes and multiplicities of the tensor product of two irreducible modules. Suppose $(x_1, \cdots, x_m)$ is a basis for a module $\mathfrak{M}$ such that $x_i h = \Lambda_i(h)x_i$ and $(y_1, \cdots, y_n)$ is a basis for a module $\mathfrak{N}$ such that $y_j h = M_j(h)y_j$. Then the mn products $x_i \otimes y_j$ form a basis for $\mathfrak{M} \otimes \mathfrak{N}$ and we have

$$(x_i \otimes y_j)h = (\Lambda_i + M_j)(h)(x_i \otimes y_j) .$$

Hence we have the following expressions for the characters of $\mathfrak{M}$, $\mathfrak{N}$ and $\mathfrak{M} \otimes \mathfrak{N}$: $\sum_i e(\Lambda_i)$, $\sum_j e(M_j)$, $\sum_{i,j} e(\Lambda_i + M_j) = \sum_{i,j} e(\Lambda_i)e(M_j)$. Thus we see that the character of $\mathfrak{M} \otimes \mathfrak{N}$ is the product in $\mathfrak{A}$ of the characters of $\mathfrak{M}$ and $\mathfrak{N}$. One obtains the structure of $\mathfrak{M} \otimes \mathfrak{N}$

by writing the product of two characters as a sum of primitive characters. In the case of the Lie algebra A_1 there is a classical formula for this called the Clebsch-Gordon formula. We shall give an extension of this which is due to R. Steinberg. This is based on an explicit formula for the multiplicity n_M of the weight M in the irreducible module with highest weight Λ. This formula is due to Kostant; the simple proof we shall give is due to Cartier and to Steinberg (independently).

We introduce first a partition function $P(M)$ for $M \in \mathfrak{Z}$, which is analogous to the (unordered) partition function of number theory. If $M \in \mathfrak{Z}$, $P(M)$ is the number of ways of writing M as a sum of positive roots, that is, $P(M)$ is the number of solution $(k_\alpha, k_\beta, \cdots, k_\rho)$, of $\sum_{\alpha>0} k_\alpha \alpha = M$ where the k_α are non-negative integers and $\{\alpha, \beta, \cdots, \rho\}$ is the set of positive roots. We have $P(0) = 1$ and $P(M) = 0$ unless $M = \sum m_i \alpha_i$, m_i non-negative integer and α_i defined as before. Then $e(M) = x_1^{m_1} \cdots x_l^{m_l}$, $x_j = e(\lambda_j)$. It is convenient to replace the group algebra $\mathfrak{A}$ by the field $\tilde{\mathfrak{A}}$ of power series of the form $(x_1^{m_1} \cdots x_l^{m_l})^{-1} \tilde{f}$ where $\tilde{f}$ is an infinite series with coefficients in Φ in the elements $x_1^{n_1} \cdots x_l^{n_l}$, n_i non-negative integral. In $\tilde{\mathfrak{A}}$ we can consider the "generating function" $\sum_{M \in \mathfrak{Z}} P(M) e(M)$ which is defined since $P(M) = 0$ unless $M = \sum m\alpha_i$, $m_i \geqq 0$. It is clear that we have the identity

$$\sum_{M \in \mathfrak{Z}} P(M) e(M) = \prod_{\alpha>0} (1 + e(\alpha) + e(2\alpha) + \cdots) .$$

Since $(1 - e(\alpha))^{-1} = 1 + e(\alpha) + e(2\alpha) + \cdots$, we have the identity

$$\left(\sum_{M \in \mathfrak{Z}} P(M) e(M)\right) \prod_{\alpha>0} (1 - e(\alpha)) = 1 . \tag{47}$$

We re-write Weyl's formula (35) as

$$\left(\sum n_M e(M)\right) \sum_{S \in W} (\det S) e(\delta S) = \sum_{S \in W} (\det S) e(\Lambda + \delta) S .$$

We replace M by $-M$ in this and multiply the result through by $e(\delta)$ to obtain

$$\sum n_M e(-M)) \sum_{S \in W} (\det S) e(\delta - \delta S) = \sum_{S \in W} (\det S) e(\delta - (\Lambda + \delta) S) . \tag{48}$$

By Lemma 4,

$$Q = \sum (\det S) e(\delta S) = e(\delta) \prod_{\alpha>0} (1 - e(-\alpha)) .$$

Hence

$$\sum_{S\in W}(\det S)e(\delta-\delta S)=\prod_{\alpha>0}(1-e(\alpha))\,.$$

Hence if we multiply both sides of (48) by $\sum_{M\in\mathfrak{Z}}P(M)e(M)$, we obtain, by (47), that

$$\begin{aligned}\sum n_M e(-M) &= \Big(\sum_{S\in W}(\det S)e(\delta-(\Lambda+\delta)S)\Big)\Big(\sum_{M\in\mathfrak{Z}}P(M)e(M)\Big)\\ &= \sum_{\substack{S\in W\\ M\in\mathfrak{Z}}}(\det S)P(M)e(M+\delta-(\Lambda+\delta)S)\,.\end{aligned}\tag{49}$$

Comparison of coefficients of $e(-M)$ gives Kostant's formula for the multiplicity n_M in the irreducible module with highest weight Λ, namely,

$$n_M=\sum_{S\in W}(\det S)P((\Lambda+\delta)S-(M+\delta))\,.\tag{50}$$

We now consider the formula for $\chi_{\Lambda_1}\chi_{\Lambda_2}$ where $\chi_{\Lambda_i}=\sum n_M^{(i)}e(M)$ is the character of the irreducible module $\mathfrak{M}_i$ with highest weight Λ_i. We have seen that $\chi_{\Lambda_1}\chi_{\Lambda_2}=\sum m_\Lambda\chi_\Lambda$, where m_Λ is the multiplicity in $\mathfrak{M}_1\otimes\mathfrak{M}_2$ of the irreducible module with highest weight Λ. The summation is taken over all the dominant Λ. If we multiply through by $\sum_{S\in W}(\det S)e(\delta S)$ and apply Weyl's formula we get

$$\begin{aligned}&\Big(\sum_{M\in\mathfrak{Z}}n_M^{(1)}e(M)\Big)\Big(\sum_{T\in W}(\det T)e((\Lambda_2+\delta)T)\Big)\\ &\qquad\qquad=\sum_\Lambda m_\Lambda\sum_{S\in W}(\det S)e((\Lambda+\delta)S)\,.\end{aligned}$$

The summation on the right-hand side can be taken for all $\Lambda\in\mathfrak{Z}$ if we define $m_\Lambda=0$ for non-dominant Λ. Applying the formula (50) for $n_M^{(1)}$ we get

$$\begin{aligned}&\Big\{\sum_{M\in\mathfrak{Z}}\Big(\sum_{S\in W}(\det S)P((\Lambda_1+\delta)S-(M+\delta))e(M)\Big)\Big\}\\ &\qquad\cdot\Big\{\sum_{T\in W}(\det T)e((\Lambda_2+\delta)T)\Big\}\\ &\qquad\qquad=\sum_{\Lambda\in\mathfrak{Z}}m_\Lambda\sum_{S\in W}(\det S)e((\Lambda+\delta)S)\,.\end{aligned}$$

Hence

$$\sum_{M\in\mathfrak{Z}}\sum_{S,T\in W}(\det ST)P((\Lambda_1+\delta)S-(M+\delta))e((\Lambda_2+\delta)T+M)$$
$$=\sum_{\Lambda\in\mathfrak{Z}}m_\Lambda\sum_{S\in W}(\det S)e((\Lambda+\delta)S)\,.$$

If we put $(\Lambda_2+\delta)T+M=\Lambda+\delta$ on the left we get

$$\sum_{M\in\mathfrak{Z}}\sum_{S,T\in W}(\det ST)P((\Lambda_1+\delta)S+(\Lambda_2+\delta)T-(\Lambda+2\delta))e(\Lambda+\delta)$$
$$=\sum_{\Lambda\in\mathfrak{Z}}\Bigl(\sum_{\substack{S\in W\\ M\in\mathfrak{Z}\\ (M+\delta)S=\Lambda+\delta}}(\det S)m_M\Bigr)e(\Lambda+\delta)$$
$$=\sum_{\Lambda\in\mathfrak{Z}}\Bigl(\sum_{S\in W}(\det S)m_{(\Lambda+\delta)S^{-1}-\delta}\Bigr)e(\Lambda+\delta)$$
$$=\sum_{\Lambda\in\mathfrak{Z}}\Bigl(\sum_{S\in W}(\det S)m_{(\Lambda+\delta)S-\delta}\Bigr)e(\Lambda+\delta)\,.$$

Hence

$$\sum_{S,T\in W}(\det ST)P((\Lambda_1+\delta)S+(\Lambda_2+\delta)T-(\Lambda+2\delta))$$
$$=\sum_{S\in W}(\det S)m_{(\Lambda+\delta)S-\delta}$$
$$=m_\Lambda+\sum_{\substack{S\in W\\ S\neq 1}}(\det S)m_{(\Lambda+\delta)S-\delta}\,.$$

It is easy to see that if Λ is dominant, then $(\Lambda+\delta)S$ and hence $(\Lambda+\delta)S-\delta$ is not dominant if $S\neq 1$. Hence if Λ is dominant then $m_{(\Lambda+\delta)S-\delta}=0$ if $S\neq 1$ and so we obtain the formula

$$m_\Lambda=\sum_{S,T\in W}(\det ST)P((\Lambda_1+\delta)S+(\Lambda_2+\delta)T-(\Lambda+2\delta)) \tag{51}$$

for the multiplicity of the module with highest weight Λ in $\mathfrak{M}_1\otimes\mathfrak{M}_2$.

Exercises

The notations and conventions are as in Chapters VII and VIII.

1. Let $\rho\in\mathfrak{H}_0^*$. Give a direct proof that $\rho\geqq\rho S$ for every $S\in W$ if and only if $\rho(h_i)\geqq 0$, $i=1,2,\cdots,l$.

2. (Seligman). Prove the uniqueness assertion in Theorem 2 by using the fact that there exists a finite-dimensional irreducible module $\mathfrak{M}$ with highest weight Λ satisfying: $\Lambda(h_i)$ are distinct and positive. Note that $\Lambda S=\Lambda$ if $\pi S=S$, so that $2(\Lambda,\alpha_i)/(\alpha_i,\alpha_i)=2(\Lambda S,\alpha_i S)/(\alpha_i S,\alpha_i S)=2(\Lambda,\alpha_i S)/(\alpha_i S,\alpha_i S)$. This leads to $\Lambda(h_i)=\Lambda(h_j)$ if $\alpha_i S=\alpha_j$.

3. Call a weight Λ in $\mathfrak{M}$ a *frontier* weight if for every root $\alpha \neq 0$ either $\Lambda + \alpha$ or $\Lambda - \alpha$ is not a weight. Show that if $\mathfrak{M}$ is finite-dimensional irreducible then any two frontier weights are conjugate under the Weyl group.

4. Let the base field be the field of real numbers. If α is a root let P_α be the hyperplane in $\mathfrak{H}$ defined by $\alpha(h) = 0$. A *chamber* is defined to be a connected component (maximal connected subset) of the complement of $\cup_{\alpha \neq 0} P_\alpha$ in $\mathfrak{H}$. Show that every chamber C is a convex set. A set of roots $\sum$ is a *defining system* for C if C is the set of elements h satisfying $\alpha(h) > 0$ for all $\alpha \in \sum$. Defining systems which are minimal are called fundamental systems. Show that these are just simple systems of roots determined by the lexicographic orderings in $\mathfrak{H}^*$.

5. Show that the group algebra $\mathfrak{A}$ (of the group $\mathfrak{J}$ of integral functions on $\mathfrak{H}$) is a domain with unique factorization (into elements).

6. Prove that if $P \in \mathfrak{A}$ is alternating, then P is divisible by $Q = \sum_S (\det S) e(\delta S)$.

7. Let η be the automorphism of $\mathfrak{A}$ such that $e(\Lambda)^\eta = e(-\Lambda)$. Show that if χ is the character of a finite-dimensional module $\mathfrak{M}$ then χ^η is the character of the contragredient module $\mathfrak{M}^*$.

8. Let $\mathfrak{M}$ be a finite-dimensional irreducible module whose character satisfies $\chi^\eta = \chi$. Assume the base field algebraically closed. Show that the image $\mathfrak{L}^R$ under the representation R in $\mathfrak{M}$ is a subalgebra of an orthogonal or a symplectic Lie algebra of linear transformations in $\mathfrak{M}$.

9. Let $\mathfrak{K}$ be the split three-dimensional simple Lie algebra with canonical basis e, f, h. Show that the character of the $(m + 1)$-dimensional irreducible module $\mathfrak{M}_{m+1}$ for $\mathfrak{K}$ is $x^m + x^{m-1} + \cdots + x^{-m}$ where $x = e(\lambda)$, $\lambda(h) = 1$. Use this to obtain the irreducible constituents of $\mathfrak{M}_{m+1} \otimes \mathfrak{M}_{n+1}$.

10 (Dynkin). Let $\mathfrak{M}$ and $\mathfrak{N}$ be finite-dimensional irreducible modules for $\mathfrak{L}$. Prove that $\mathfrak{M} \otimes \mathfrak{N}$ is irreducible if and only if for every l in any simple ideal of $\mathfrak{L}$ either $\mathfrak{M}l = 0$ or $\mathfrak{N}l = 0$.

11. Use Weyl's formula to show that the dimensionalities of the four basic irreducible modules for F_4 are: 26, 52, 273, 1274. Use Freudenthal's formula to obtain the character of the 26-dimensional basic module.

12. Use Weyl's formula to prove $\dim \mathfrak{M} = 2^l$ if $\mathfrak{M}$ has highest weight λ_l for B_l.

13 (Steinberg). Let $\mathfrak{M}$ be the finite-dimensional irreducible module with highest weight Λ. Show that M is a weight of $\mathfrak{M}$ if $\Lambda - MS$ is a sum of positive roots for every $S \in W$.

14 (Kostant). Prove the following recursion formula for the partition function $P(M)$:

$$P(M) = -\sum_{\substack{S \in W \\ S \neq 1}} (\det S) P(M - (\delta - \delta S)).$$

CHAPTER IX

Automorphisms

In this chapter we shall study the groups of automorphisms of semi-simple Lie algebras over an algebraically closed field of characteristic 0. If z is an element of a Lie algebra $\mathfrak{L}$ of characteristic 0 such that $\operatorname{ad} z$ is nilpotent, then we know that $\exp(\operatorname{ad} z)$ is an automorphism. Products of automorphisms of this type will be called invariant automorphisms. These constitute a subgroup $G_0(\mathfrak{L})$ of the group of automorphisms $G(\mathfrak{L})$ of $\mathfrak{L}$. If τ is any automorphism, then $\tau^{-1}(\exp \operatorname{ad} z)\tau = \exp \operatorname{ad} z^{\tau}$ and $\operatorname{ad} z^{\tau}$ is nilpotent. It is clear from this that G_0 is an invariant subgroup of G.

The main problem we shall consider in this chapter is the determination of the index of G_0 in G for $\mathfrak{L}$ finite dimensional simple over an algebraically closed field Φ of characteristic zero. We show first that if $\mathfrak{L}$ is finite-dimensional over Φ (not necessarily simple), then G_0 acts transitively on the set of Cartan subalgebras, that is, if $\mathfrak{H}_1$ and $\mathfrak{H}_2$ are Cartan subalgebras then there exists a $\sigma \in G_0$ such that $\mathfrak{H}_1^{\sigma} = \mathfrak{H}_2$. A conjugacy theorem of this type was first noted by Cartan in the case of $\mathfrak{L}$ semi-simple over the field of complex numbers and it was applied by him to the study of the automorphisms of these algebras. The extension and rigorous proof of the conjugacy theorem is due to Chevalley. This result reduces the study of the position of G_0 in G to the study of automorphisms which map a Cartan subalgebra into itself.

If $\mathfrak{L}$ is semi-simple, then the Weyl group plays an important role in our considerations. We shall require also some explicit calculations of invariant automorphisms due to Seligman. The final results we derive give the group of automorphisms for the Lie algebras $A_l, B_l, C_l, D_l,\ l > 4,\ G_2$ and F_4. It is noteworthy that similar results can be obtained in the characteristic p case (see Jacobson [8] and Seligman [4]). As usual, for the sake of simplic ity we stick to the characteristic 0 case. In the next chapter we shall extend our final results for non-exceptional simple Lie algebras over algebraically closed fields to algebras of this type over arbitrary base fields (of characteristic 0).

1. *Lemmas from algebraic geometry*

Let $\mathfrak{M}$ be a finite-dimensional vector space with basis $(u_1, u_2, \cdots, u_m)$ over an infinite field Φ. Any x has a unique representation as $x = \sum \xi_i u_i$ and the ξ_i are the coordinates of x relative to the basis (u_i). Let $f(\lambda_1, \cdots, \lambda_m)$ be an element of the polynomial ring $\Phi[\lambda_1, \cdots, \lambda_m]$ in the m indeterminates λ_i with coefficients in Φ. Then $f(\lambda_1, \cdots, \lambda_m)$ and the basis (u_i) define a mapping f of $\mathfrak{M}$ into Φ by the rule that the image $f(x) = f(\xi) = f(\xi_1, \cdots, \xi_m)$. (In this chapter we shall often use the notation $f(x)$ rather than xf or x^f.) We call f a *polynomial function* on $\mathfrak{M}$. If $(u_1', \cdots, u_m')$ is a second basis, then it is readily seen that the function f can be defined by another polynomial in $\Phi[\lambda_1, \cdots, \lambda_m]$ with respect to (u_i'). In this sense the notion of a polynomial function is independent of the choice of the basis for $\mathfrak{M}$. The set of polynomial functions is an algebra $\Phi[\mathfrak{M}]$ relative to the usual compositions of functions. We recall that if f and g are functions on $\mathfrak{M}$ with values in Φ, then $(f+g)(x) = f(x) + g(x)$, $(\alpha f)(x) = \alpha f(x)$ for α in Φ, $(fg)(x) = f(x)g(x)$. The mapping $f(\lambda_1, \cdots, \lambda_m) \rightarrow f \in \Phi[\mathfrak{M}]$ is a homomorphism of $\Phi[\lambda_1, \cdots, \lambda_m]$ into $\Phi[\mathfrak{M}]$. Since Φ is infinite, $f(\xi_1, \cdots, \xi_m) = 0$ for all ξ_i in Φ implies $f(\lambda_1, \cdots, \lambda_m) = 0$. Hence the homomorphism is an isomorphism. The isomorphism maps λ_i into the projection function π_i such that $\pi_i(x) = \xi_i$. Hence it is clear that the π_i generate the algebra $\Phi[\mathfrak{M}]$.

Let $\mathfrak{N}$ be a second finite-dimensional space over Φ with the basis $(v_1, v_2, \cdots, v_n)$. A *polynomial mapping* P of $\mathfrak{M}$ into $\mathfrak{N}$ is a mapping of the form $x = \sum_1^m \xi_i u_i \rightarrow y = \sum_1^n \eta_j v_j$ where $\eta_j = p_j(\xi_1, \cdots, \xi_m)$, $p_j(\lambda_1, \cdots, \lambda_m) \in \Phi[\lambda_1, \cdots, \lambda_m]$. This notion is independent of the basis. The set of polynomial mappings of $\mathfrak{M}$ into $\mathfrak{N}$ is a vector space under the usual addition and scalar multiplication of mappings. The resultant of a polynomial mapping P of $\mathfrak{M}$ to $\mathfrak{N}$ and a polynomial mapping Q of $\mathfrak{N}$ to $\mathfrak{P}$ is a polynomial mapping PQ of $\mathfrak{M}$ into $\mathfrak{P}$. The notion of a polynomial function is the special case of a polynomial mapping in which the image space is the one-dimensional space Φ. Hence if P is a polynomial mapping of $\mathfrak{M}$ into $\mathfrak{N}$ and f is a polynomial function on $\mathfrak{N}$, then Pf is a polynomial function on $\mathfrak{M}$. The mapping $f \rightarrow Pf$ is a mapping of the algebra $\Phi[\mathfrak{N}]$ of polynomial functions on $\mathfrak{N}$ into $\Phi[\mathfrak{M}]$. We now look at the form of this mapping. Thus we have $P(x) = y$ where $x = \sum \xi_i u_i$, $y = \sum \eta_j v_j$ and

$$\eta_j = p_j(\xi_1, \cdots, \xi_m), \tag{1}$$

$p_j(\lambda_1, \cdots, \lambda_m) \in \Phi[\lambda_1, \cdots, \lambda_m]$. Also we have $f(y) = f(\eta_1, \cdots, \eta_n)$, $f(\mu_1, \cdots, \mu_n) \in \Phi[\mu_1, \cdots, \mu_n]$, μ_j indeterminates. Hence Pf maps x into

$$f(p_1(\xi), p_2(\xi), \cdots, p_n(\xi)) \tag{2}$$

where $p_j(\xi) \equiv p_j(\xi_1, \xi_2, \cdots, \xi_m)$. If $f, g \in \Phi[\mathfrak{N}]$, then

$$(f+g)(p_1(\xi), \cdots, p_n(\xi)) = f(p_1(\xi), \cdots, p_n(\xi)) + g(p_1(\xi), \cdots, p_n(\xi))$$
$$\alpha f(p_1(\xi), \cdots, p_n(\xi)) = \alpha(f(p_1(\xi), \cdots, p_n(\xi))$$
$$(fg)(p_1(\xi), \cdots, p_n(\xi)) = f(p_1(\xi), \cdots, p_n(\xi))g(p_1(\xi), \cdots, p_n(\xi)) .$$

This shows that $f \to Pf$ is a homomorphism σ_P of $\Phi[\mathfrak{N}]$ into $\Phi[\mathfrak{M}]$.

Conversely, let σ be an algebra homomorphism of $\Phi[\mathfrak{N}]$ into $\Phi[\mathfrak{M}]$. Consider the projection ρ_j: $\sum \eta_k v_k \to \eta_j$. Suppose ρ_j^σ is the mapping $\sum \xi_i u_i \to f_j(\xi_1, \xi_2, \cdots, \xi_m)$ and let P be the polynomial mapping of $\mathfrak{M}$ into $\mathfrak{N}$ such that $\sum \xi_i u_i \to \sum \eta_j v_j$ where $\eta_j = f_j(\xi)$. Then $P\rho_j$ maps $\sum \xi_i u_i$ into $f_j(\xi)$ so that $P\rho_j = \rho_j^\sigma$. Since the ρ_j generate $\Phi[\mathfrak{N}]$ it follows that σ coincides with the homomorphism σ_P determined by P. Thus every homomorphism of $\Phi[\mathfrak{N}]$ into $\Phi[\mathfrak{M}]$ (sending 1 into 1) is realized by a polynomial mapping of $\mathfrak{M}$ into $\mathfrak{N}$. It is easy to see that if P and Q are two such polynomial mappings, then the homomorphism $\sigma_P = \sigma_Q$ if and only if $P = Q$. Hence we have a 1 : 1 correspondence between the polynomial mappings of $\mathfrak{M}$ into $\mathfrak{N}$ and the homomorphisms of $\Phi[\mathfrak{N}]$ into $\Phi[\mathfrak{M}]$.

Of particular importance for us is the set of algebra homomorphisms of $\Phi[\mathfrak{N}]$ into the base field Φ. This can be obtained in a somewhat devious manner by identifying Φ with the algebra $\Phi[\mathfrak{M}]$ of polynomial functions on $\mathfrak{M} = 0$ and applying the foregoing result. However, it is more straightforward to look at this directly. We note first that if $y \in \mathfrak{N}$, then the mapping σ_y: $f \to f(y)$, the specialization of f at y, is a homomorphism of $\Phi[\mathfrak{N}]$ into Φ. Conversely, if σ is any homomorphism of $\Phi[\mathfrak{N}]$ into Φ, we let $\eta_j = \rho_j^\sigma$, as before. Then it is immediate that $\sigma = \sigma_y$ where $y = \sum \eta_j v_j$. It is clear also that if $y_1 \neq y_2$ in $\mathfrak{N}$, then $\sigma_{y_1} \neq \sigma_{y_2}$.

Let $f \in \Phi[\mathfrak{M}]$ and let $a = \sum \alpha_i u_i \in \mathfrak{M}$. Then we can define a linear function $d_a f$ on $\mathfrak{M}$ by

$$(3) \qquad (d_a f)(\sum \xi_i u_i) = \sum_k \left(\frac{\partial f}{\partial \lambda_k}\right)_{\lambda_l = \alpha_l} \xi_k \,.$$

It is easy to see that $d_a f$ is independent of the basis used to define this mapping. The linear mapping $d_a f$ is called *the differential of f at a*. We have the following properties

$$(4) \qquad \begin{aligned} d_a(f+g) &= d_a f + d_a g \\ d_a(\alpha f) &= \alpha(d_a f) \\ d_a(fg) &= f(a)(d_a g) + g(a)(d_a f) \,. \end{aligned}$$

If t is an indeterminate and f is extended to $\mathfrak{M}_{\Phi(t)}$ in the obvious way, then we have the following relation in the algebra $\Phi[t]$, which is a consequence of Taylor's formula

$$(5) \qquad f(a + tx) \equiv f(a) + t(d_a f)(x) \pmod{t^2} \,.$$

More generally, let P be a polynomial mapping $x = \sum_1^m \xi_i u_i \rightarrow \sum_1^n p_j(\xi) v_j$ of $\mathfrak{M}$ into $\mathfrak{N}$. Then we define a linear mapping $d_a P$, *the differential of P at a* by

$$(6) \qquad (d_a P)(x) = \sum_{j=1}^{n} \left(\sum_{k=1}^{m} \left(\frac{\partial p_j}{\partial \lambda_k}\right)_{\lambda_l = \alpha_l} \xi_k \right) v_j \,.$$

Again, one can verify that this is independent of the bases chosen in $\mathfrak{M}$ and $\mathfrak{N}$. Also one has the useful generalization of (5):

$$(7) \qquad P(a + tx) \equiv P(a) + t(d_a P)(x) \pmod{t^2} \,.$$

A set of generators for the image space $\mathfrak{M}(d_a P)$ is the set of vectors

$$(8) \qquad (d_a P)(u_i) = \sum_j \left(\frac{\partial p_j}{\partial \lambda_i}\right)_{\lambda_k = \alpha_k} v_j \,.$$

Hence $d_a P$ is surjective if and only if the Jacobian matrix

$$(9) \qquad \left(\left(\frac{\partial p_j}{\partial \lambda_i}\right)_{\lambda_k = \alpha_k}\right), \qquad i = 1, \cdots, m \,, \qquad j = 1, 2, \cdots, n$$

has rank n.

LEMMA 1. *Assume Φ perfect. Then if $d_a P$ is surjective for some a, the homomorphism σ_P is an isomorphism of $\Phi[\mathfrak{N}]$ into $\Phi[\mathfrak{M}]$.*

Proof: Our hypothesis is that one of the n-rowed minors of the matrix $(\partial p_j / \partial \lambda_i)$ is not zero. If σ_P is not an isomorphism, then there exists a polynomial $f(\mu_1, \cdots, \mu_n) \neq 0$ such that $f(p_1(\xi), p_2(\xi), \cdots, p_n(\xi)) =$

0 for all ξ_i in Φ. This implies that $f(p_1(\lambda), \cdots, p_n(\lambda)) = 0$, that is, the polynomials $p_1(\lambda_1, \cdots, \lambda_m), \cdots, p_n(\lambda_1, \cdots, \lambda_m)$ are algebraically dependent. If this is the case, then we may assume that the polynomial $f \neq 0$ giving the dependence is of least degree. The relation $f(p_1(\lambda), \cdots, p_n(\lambda)) = 0$ gives

$$(10) \qquad 0 = \frac{\partial f}{\partial \lambda_i} = \sum_{j=1}^{n} \left(\frac{\partial f}{\partial \mu_j}\right)_{\mu_j = p_j(\lambda)} \frac{\partial p_j}{\partial \lambda_i}, \qquad i = 1, \cdots, m .$$

This contradicts the hypothesis on the Jacobian matrix unless $(\partial f/\partial \mu_j)(p_1(\lambda), \cdots, p_n(\lambda)) = 0$. Since the degree of f is minimal for algebraic relations in the $p_1(\lambda), \cdots, p_m(\lambda)$ we must have $(\partial f/\partial \mu_j) = 0$, $j = 1, \cdots, n$. This implies that f is a non-zero element of Φ—which is absurd—if the characteristic is 0. If the characteristic is $p \neq 0$, we obtain that f is a polynomial in $\mu_1^p, \mu_2^p, \cdots, \mu_n^p$. Since Φ is perfect this implies that $f = g^p$, g a polynomial in the μ's. Then $g(p_1(\lambda), \cdots, p_m(\lambda)) = 0$ which again contradicts the minimality of the degree of f.

The main result we shall require for the conjugacy theorem for Cartan subalgebras is the following

THEOREM 1. *Let Φ be algebraically closed and let P be a polynomial mapping of $\mathfrak{M}$ into $\mathfrak{N}$ such that $d_a P$ is surjective for some $a \in \mathfrak{M}$. If f is a non-zero polynomial function on $\mathfrak{M}$, then there exists a non-zero polynomial function g on $\mathfrak{N}$ such that if y is any element of $\mathfrak{N}$ satisfying $g(y) \neq 0$, then there exists an x in $\mathfrak{M}$ such that $f(x) \neq 0$ and $P(x) = y$.*

In geometric form this result has the following meaning: Given an "open" set in $\mathfrak{M}$ defined to be the set of elements x such that $f(x) \neq 0$ for the non-zero polynomial f, then there exists an open set in $\mathfrak{N}$ defined by $g(y) \neq 0$, g a non-zero polynomial which is completely contained in the image under P of the given open set in $\mathfrak{M}$. (*Suggestion*: Draw a figure for this.) We shall see that Theorem 1 is an easy consequence of the following theorem on extensions of homomorphisms.

THEOREM 2. *Let Φ be an algebraically closed field and let Γ be an extension field of Φ, $\mathfrak{A}$ a subalgebra of Γ and $\mathfrak{A}'$ an extension algebra of $\mathfrak{A}$ of the form $\mathfrak{A}' = \mathfrak{A}[u_1, u_2, \cdots, u_r]$, $u_i \in \Gamma$. Let f be a non-zero element of $\mathfrak{A}'$. Then there exists a non-zero element g in $\mathfrak{A}$ such that if σ is any homomorphism of $\mathfrak{A}$ into Φ such that $g^\sigma \neq 0$,* *then* σ *has an extension homomorphism* τ *of* $\mathfrak{A}'$ *into* Φ *such that* $f^\tau \neq 0$.

Proof: Suppose first that $r = 1$, so that $\mathfrak{A}' = \mathfrak{A}[u]$. Case I: u is transcendental over the subfield Σ of Γ generated by $\mathfrak{A}$. We write $f = f_0 + f_1u + \cdots + f_mu^r$, $f_i \in \mathfrak{A}$, $f_r \neq 0$. Let $g = f_r$ and let σ be a homomorphism of $\mathfrak{A}$ into Φ such that $g^\sigma \neq 0$. Consider the polynomial $f_0^\sigma + f_1^\sigma\lambda + \cdots + f_m^\sigma\lambda^r$ in $\Phi[\lambda]$, λ an indeterminate. Since $f_m^\sigma \neq 0$, this polynomial has at most m roots in Φ so we can choose c in Φ so that $\sum_0^r f_i^\sigma c^i \neq 0$. Let τ be the homomorphism of $\mathfrak{A}' = \mathfrak{A}[u]$ into Φ such that $\sum a_iu^i \to \sum a_i^\sigma c^i$. This is an extension of σ and $f^\tau \neq 0$ as required. Case II: u is algebraic over Σ. The canonical homomorphism of $\mathfrak{A}[\lambda]$, λ an indeterminate, onto $\mathfrak{A}' = \mathfrak{A}[u]$ (identity on $\mathfrak{A}$, $\lambda \to u$) has a non-zero kernel $\mathfrak{P}$. Since $f \in \mathfrak{A}[u]$, f is algebraic over Σ also. Let $p(\lambda)$, $q(\lambda)$ be non-zero polynomials in $\mathfrak{A}[\lambda]$ of least degree such that $p(u) = 0$ and $q(f) = 0$. Then these polynomials are also of least degree in $\Sigma[\lambda]$ such that $p(u) = 0$, $q(f) = 0$. Hence they are irreducible in $\Sigma[\lambda]$. Let g_1 be the leading coefficient of $p(\lambda)$ and $g_2 = q(0)$ and choose $g = g_1g_2$. We shall show that g has the required property for f. Thus suppose σ is a homomorphism of $\mathfrak{A}$ into Φ satisfying $g^\sigma = g_1^\sigma g_2^\sigma \neq 0$ and suppose τ is any extension of σ to a homomorphism of $\mathfrak{A}' = \mathfrak{A}[u]$ into Φ. Since $g(f) = 0$ we shall have $g^\sigma(f^\tau) = 0$, which implies that $f^\tau \neq 0$ since $g^\sigma(0) \neq 0$. Thus we need to show only that the homomorphism σ can be extended to a homomorphism of $\mathfrak{A}'$. For this purpose let c be a root of $p^\sigma(\lambda) = 0$ and consider the homomorphism τ' of $\mathfrak{A}[\lambda]$ which coincides with σ on $\mathfrak{A}$ and maps λ into c. Let $h(\lambda)$ be any element in the ideal $\mathfrak{P}$. Then the minimality of the degree of $p(\lambda)$ implies that there exists a non-negative integer k such that $g_1^kh(\lambda)$ is divisible in $\mathfrak{A}[\lambda]$ by $p(\lambda)$. Since $p^\sigma(c) = 0$, $(g_1^\sigma)^kh^\sigma(c) = 0$ and since $g_1^\sigma \neq 0$, $h^\sigma(c) = 0$. Hence $h(\lambda)^{\tau'} = 0$ and so $\mathfrak{P}$ is mapped into 0 by τ'. It follows that τ' induces a homomorphism τ of $\mathfrak{A}' = \mathfrak{A}[u] \cong \mathfrak{A}[\lambda]/\mathfrak{P}$ which is an extension of σ.

Now assume the result holds for $r - 1$. Let $\mathfrak{B} = \mathfrak{A}[u_r]$ so that $\mathfrak{A}' = \mathfrak{B}[u_1, \cdots, u_{r-1}]$. Then there exists an element $h \in \mathfrak{B}$ such that any homomorphism ρ of $\mathfrak{B}$ such that $h^\rho \neq 0$ has an extension τ to $\mathfrak{A}'$ such that $f^\tau \neq 0$. By the case $r = 1$, there exists $g \in \mathfrak{A}$ such that any homomorphism σ of $\mathfrak{A}$ into Φ such that $g^\sigma \neq 0$ has an extension ρ to $\mathfrak{B} = \mathfrak{A}[u_r]$ such that $h^\rho \neq 0$. Hence τ is an extension of σ such that $f^\tau \neq 0$ as required.

We now give the

Proof of Theorem 1: The hypothesis on $d_a P$ implies that σ_P is an isomorphism of $\Phi[\mathfrak{N}]$ into $\Phi[\mathfrak{M}]$. Let $\mathfrak{A} = \Phi[\mathfrak{N}]^{\sigma_P} \subseteq \mathfrak{A}' = \Phi[\mathfrak{M}]$. If $\pi_1, \cdots, \pi_m$ are the projection mappings of $\mathfrak{M}$ into Φ we have $\mathfrak{A}' = \Phi[\pi_1, \cdots, \pi_m] = \mathfrak{A}[\pi_1, \cdots, \pi_m]$. Since $\mathfrak{A}'$ is an integral domain we may suppose it imbedded in its field of quotients Γ. Hence we can apply Theorem 2 to $\mathfrak{A}$ and $\mathfrak{A}'$. Let f be a non-zero element of $\mathfrak{A}' = \Phi[\mathfrak{M}]$. Then there exists a non-zero element $g \in \Phi[\mathfrak{N}]$ such that if y is an element of $\mathfrak{N}$ such that $g(y) \neq 0$ then the homomorphism σ: $h^{\sigma_P} \rightarrow h(y)$ of $\mathfrak{A} = \Phi[\mathfrak{N}]^{\sigma_P}$ into Φ, which satisfies $g^{\sigma_P \sigma} = g(y) \neq 0$, can be extended to a homomorphism τ of $\Phi[\mathfrak{M}]$ into Φ satisfying $f^{\tau} \neq 0$. We have seen that τ has the form $k \rightarrow k(x)$ where x is an element of $\mathfrak{M}$. Then $f(x) \neq 0$ and for every $h \in \mathfrak{N}$, $h^{\sigma_P} = h^{\sigma_P \sigma}$. This means that $h(P(x)) = h(y)$. Hence $P(x) = y$ and the theorem is proved.

2. *Conjugacy of Cartan subalgebras*

Let $\mathfrak{L}$ be a finite-dimensional Lie algebra over an algebraically closed field of characteristic 0 and let $\mathfrak{H}$ be a Cartan subalgebra of $\mathfrak{L}$. Let

$$\mathfrak{L} = \mathfrak{H} + \sum \mathfrak{L}_\alpha \tag{11}$$

be the decomposition of $\mathfrak{L}$ into root spaces corresponding to the roots $0, \alpha, \beta, \cdots$ of $\mathfrak{H}$ acting in $\mathfrak{L}$. If $h \in \mathfrak{H}$ and $e_\alpha \in \mathfrak{L}_\alpha$, then there exists an integer r such that $e_\alpha(\operatorname{ad} h - \alpha(h)1)^r = 0$. This is equivalent to the condition that $\alpha(h)$ is the only characteristic root of the restriction of $\operatorname{ad} h$ to $\mathfrak{L}_\alpha$. The α are linear functions on $\mathfrak{H}$. If $x \in \mathfrak{L}_\rho$ ($\rho = 0$ or $\rho \neq 0$) then $[xe_\alpha] = 0$ or $\rho + \alpha$ is a root and $[xe_\alpha] \in \mathfrak{L}_{\rho+\alpha}$. In the latter case either $[[xe_\alpha]e_\alpha] = 0$ or $\rho + 2\alpha$ is a root and $[[xe_\alpha]e_\alpha] \in \mathfrak{L}_{\rho+2\alpha}$. If we continue in this way and we take into account the fact that there are only a finite number of distinct roots we see that $x(\operatorname{ad} e_\alpha)^k = 0$ for sufficiently high k. This implies that $\operatorname{ad} e_\alpha$ is nilpotent for every $e_\alpha \in \mathfrak{L}_\alpha$, $\alpha \neq 0$. It follows that if $e_{\alpha_1} \in \mathfrak{L}_{\alpha_1}, \cdots, e_{\alpha_k} \in \mathfrak{L}_{\alpha_k}$, $\alpha_1, \alpha_2, \cdots, \alpha_k$ non-zero roots, then

$$\eta = \exp(\operatorname{ad} e_{\alpha_1}) \exp(\operatorname{ad} e_{\alpha_2}) \cdots \exp(\operatorname{ad} e_{\alpha_k}) \tag{12}$$

is an invariant automorphism of $\mathfrak{L}$.

Now let $(h_1, h_2, \cdots, h_l, e_{l+1}, \cdots, e_n)$ be a basis for $\mathfrak{L}$ such that $(h_1, h_2, \cdots, h_l)$ is a basis for $\mathfrak{H}$ and the elements $e_{l+1}, \cdots, e_n$ are in root spaces $\mathfrak{L}_\alpha$, $\alpha \neq 0$. Let $\lambda_1, \cdots, \lambda_n$ be indeterminates, $P =$

$\Phi(\lambda_1, \cdots, \lambda_n)$ and form the element

$$\left(\sum_1^l \lambda_i h_i\right) \exp(\operatorname{ad} \lambda_{l+1} e_{l+1}) \cdots \exp(\operatorname{ad} \lambda_n e_n) = \sum_1^l p_i(\lambda_1, \cdots, \lambda_n) h_i + \sum_{j=l+1}^n p_j(\lambda_1, \cdots, \lambda_n) e_j \tag{13}$$

where the p_i and p_j are polynomials in the λ's. These determine a polynomial mapping

$$P\colon \sum_1^l \xi_i h_i + \sum_{l+1}^n \xi_j e_j \to \sum p_i(\xi) h_i + \sum p_j(\xi) e_j \tag{14}$$

in $\mathfrak{L}$.

The product $\alpha\beta \cdots \rho$ of the non-zero roots is a non-zero polynomial function. It follows that there exist $\xi_i^0 \in \Phi$ such that if $h^0 = \sum_1^l \xi_i^0 h_i$, then $\alpha(h^0)\beta(h^0) \cdots \rho(h^0) \neq 0$. Then the characteristic roots of the restriction of $\operatorname{ad} h^0$ to $\mathfrak{L}_\alpha + \mathfrak{L}_\beta + \cdots + \mathfrak{L}_\rho$ are all different from 0 and so this restriction of $\operatorname{ad} h^0$ is non-singular.

We shall now calculate the differential $d_{h^0} P$ of P at h^0. For this purpose we let $x = h + e$, $h = \sum_1^l \xi_i h_i$, $e = \sum_{l+1}^n \xi_j e_j$, let t be an indeterminate and we consider

$$\begin{aligned} P(h^0 + t(h + e)) &= (h^0 + th) \exp(\operatorname{ad} t\xi_{l+1} e_{l+1}) \exp(\operatorname{ad} t\xi_{l+2} e_{l+2}) \cdots \exp(\operatorname{ad} t\xi_n e_n) \\ &\equiv (h^0 + th)(1 + \operatorname{ad} t\xi_{l+1} e_{l+1}) \cdots (1 + \operatorname{ad} t\xi_n e_n) \pmod{t^2} \\ &\equiv h^0 + th + h^0 \operatorname{ad} t\xi_{l+1} e_{l+1} + \cdots + h^0 \operatorname{ad} t\xi_n e_n \pmod{t^2} \\ &= h^0 + th + t\xi_{l+1}[h^0 e_{l+1}] + \cdots + t\xi_n [h^0 e_n] \pmod{t^2} \\ &\equiv h^0 + th + t[h^0 e] \pmod{t^2}. \end{aligned} \tag{15}$$

If we compare this with (7) we we see that $d_{h^0} P$ is the mapping

$$h + e \to h + [h^0 e]. \tag{16}$$

Since $h \to h$ and $e \to [h^0 e]$ are non-singular it follows that $d_{h^0} P$ is surjective. We are therefore in a position to apply Theorem 1. Accordingly, we have the following result: If f is a polynomial function $\neq 0$ on $\mathfrak{L}$, then there exists a polynomial function $g \neq 0$ on $\mathfrak{L}$ such that if $y \in \mathfrak{L}$ and $g(y) \neq 0$, then there is an x in $\mathfrak{L}$ such that $P(x) = y$ and $f(x) \neq 0$.

We recall the definition of a regular element a of $\mathfrak{L}$ as an element such that ad a has the minimum number l' of 0 characteristic roots. We recall also that if a is regular, then the set of vectors h belonging to the characteristic root 0 of ad a is a Cartan subalgebra. It follows from this that if $\mathfrak{H}$ is a Cartan subalgebra and $\mathfrak{H}$ contains a regular element a, then $\mathfrak{H}$ is just the collection of elements $h \in \mathfrak{L}$ such that $h(\text{ad}\, a)^r = 0$ for some integer r. We shall need also the characterization given in § 3.1 of regular elements. For this we take the element $u = \sum_1^l \lambda_i h_i + \sum_{l+1}^n \lambda_j e_j$ in $\mathfrak{L}_P$, $P = \Phi(\lambda_1, \cdots, \lambda_n)$ and we consider the characteristic polynomial

$$
\begin{aligned}
f_u(\lambda) &= \det(\lambda 1 - \text{ad}\, u) \\
&= \lambda^n - \tau_1(\lambda_i)\lambda^{n-1} + \cdots + (-1)^{n-l'}\tau_{n-l'}(\lambda_i)\lambda^{l'}
\end{aligned}
\tag{17}
$$

Then $\tau_{n-l'}(\lambda_1, \cdots, \lambda_n)$ is a non-zero homogeneous polynomial of degree $n - l'$ in the λ's and if $x = \sum \xi_i h_i + \sum \xi_j e_j$, then $x \to \tau_{n-l'}(x) \equiv \tau_{n-l'}(\xi_1, \cdots, \xi_n)$ is a polynomial function on $\mathfrak{L}$. The element x is regular in $\mathfrak{L}$ if and only if $\tau_{n-l'}(x) \neq 0$. (It will be a consequence of the theorem we are going to prove that $l' = l$.)

We now consider again the Cartan subalgebra $\mathfrak{H}$ and the basis $(h_1, \cdots, h_l, e_{l+1}, \cdots, e_n)$ for $\mathfrak{L}$. We apply Theorem 1 to the polynomial function $f = \tau_{n-l'}^{\sigma_P}$ which is $\neq 0$ since $\tau_{n-l'} \neq 0$ and σ_P is an isomorphism. Accordingly we see that there is a non-zero polynomial function g on $\mathfrak{L}$ such that every $y \in \mathfrak{L}$ satisfying $g(y) \neq 0$ has the form $P(x)$ where $f(x) = \tau_{n-l'}(P(x)) = \tau_{n-l'}(y) \neq 0$. Hence every y such that $g(y) \neq 0$ is regular and if $x = \sum_1^l \xi_i h_i + \sum_{l+1}^n \xi_j e_j$, then $y = P(x) = (\sum \xi_i h_i)(\exp \text{ad}\, \xi_{l+1} e_{l+1}) \cdots (\exp \text{ad}\, \xi_n e_n) = h^\eta$ where $h = \sum \xi_i h_i$ and η is an invariant automorphism. Thus y is the image of an element $h \in \mathfrak{H}$ under an invariant automorphism. It follows that $h = y^{\eta^{-1}}$ is regular. It is now easy to prove the conjugacy theorem for Cartan subalgebras.

THEOREM 3. *If* $\mathfrak{H}_1$ *and* $\mathfrak{H}_2$ *are Cartan subalgebras of a finite-dimensional Lie algebra over an algebraically closed field of characteristic* 0, *then there exists an invariant automorphism* η *such that* $\mathfrak{H}_1^\eta = \mathfrak{H}_2$.

Proof: There exists a non-zero polynomial function g_i such that if y is an element satisfying $g_i(y) \neq 0$, then $y = h_i^{\eta_i}$, h_i a regular element in $\mathfrak{H}_i$ and η_i an invariant automorphism. Since $g_1 g_2 \neq 0$ we can choose y so that $g_1(y) \neq 0$ and $g_2(y) \neq 0$. Then $y = h_1^{\eta_1} = h_2^{\eta_2}$, h_i a regular element of $\mathfrak{H}_i$, η_i an invariant automorphism. Then $h_2 = h_1^\eta$, $\eta = \eta_1\eta_2^{-1}$. Since h_i is regular and is contained in $\mathfrak{H}_i$ it follows that $\mathfrak{H}_2 = \mathfrak{H}_1^\eta$.

Remarks. It is a consequence of the theorem that every Cartan subalgebra contains regular elements and that all Cartan subalgebras have the same dimensionality l which is the same as the number l' indicated above. It is easy to see also that if the notations are as before, then the regular elements of $\mathfrak{L}$ belonging to $\mathfrak{H}$ are just the elements h^0 such that $\alpha(h^0)\beta(h^0) \cdots \rho(h^0) \neq 0$.

3. *Non-isomorphism of the split simple Lie algebras*

We shall apply the conjugacy theorem for Cartan subalgebras first to settle a point which has been left open hitherto, namely, that the split simple Lie algebras which were listed in §§ 4.5–4.6 are distinct in the sense of isomorphism. We recall that these were: A_l, $l \geqq 1$, B_l, $l \geqq 2$, C_l, $l \geqq 3$, D_l, $l \geqq 4$, G_2, F_4, E_6, E_7 and E_8. The dimensionalities of these algebras are given in the following table:

type	*dimensionality*
A_l	$l(l+2)$
B_l	$l(2l+1)$
C_l	$l(2l+1)$
D_l	$l(2l-1)$
G_2	14
F_4	52
E_6	78
E_7	133
E_8	248

For the classical types and for G_2, F_4 and E_6 this was derived in § 4.6. We have proved the existence of E_7 and E_8 in § 7.5. The dimensionalities of these Lie algebras can be derived by determining the positive roots directly from the Cartan matrices. We shall not carry this out but we shall assume the result for these two Lie algebras.

To prove that no two of the Lie algebras we have listed are isomorphic it suffices to assume the base field algebraically closed. This is clear since $\mathfrak{L}_1 \cong \mathfrak{L}_2$ implies $\mathfrak{L}_{1P} \cong \mathfrak{L}_{2P}$ for any extension P of the base field. We therefore assume Φ algebraically closed. The subscript l in the designation X_l (e.g., A_l, E_6) for our Lie algebras is the dimensionality of a Cartan subalgebra. The conjugacy theorem shows that this is an invariant. Hence necessary conditions for isomorphism of X_l and $Y_{l'}$ are $l = l'$ and $\dim X_l = \dim Y_{l'}$. A glance at the list of dimensionalities shows that the only possible isomorphisms which we may have are between B_l and C_l, $l \geqq 3$ and between B_6 and E_6 and between C_6 and E_6. The latter two have been ruled out in an exercise (Exercise 7.6). It therefore remains to show that $B_l \not\cong C_l$, $l \geqq 3$.

Since C_l has an irreducible module in a $2l$-dimensional space it will suffice to show that if $\mathfrak{M}$ is an irreducible module for B_l such that $\mathfrak{M}B_l \neq 0$, then $\dim \mathfrak{M} \geqq 2l + 1$. We could establish this, as in a similar discussion for G_2 (§ 7.6), by using the fact that the set of weights is invariant under the Weyl group. However, we can now obtain the result more quickly by using Weyl's dimensionality formula. We observe first that (8.41) shows that if $\mathfrak{M}$ is an irreducible module of least dimension satisfying $\mathfrak{M}B_l \neq 0$, then $\mathfrak{M}$ is a basic module (cf. also Exercise 7.13). The dimensionalities of these are $\binom{2l+1}{k}$, $k = 1, 2, \cdots, l-1$ and 2^l (cf. (8.45) and Exercise 8.12). Since $l \geqq 3$, these numbers exceed $2l$.

This proves our assertion and completes the proof that B_l and C_l are not isomorphic if $l \geqq 3$.

4. *Automorphisms of semi-simple Lie algebras over an algebraically closed field*

Let $\mathfrak{L}$ be a finite-dimensional semi-simple Lie algebra over an algebraically closed field of characteristic 0, $\mathfrak{H}$ a Cartan subalgebra, $\pi = \{\alpha_1, \cdots, \alpha_l\}$ a simple system of roots relative to $\mathfrak{H}$, e_i, f_i, h_i, $i = 1, 2, \cdots, l$, a set of canonical generators for $\mathfrak{L}$ determined by π. Thus the h_i form a basis for $\mathfrak{H}$, $e_i \in \mathfrak{L}_{\alpha_i}$, $f_i \in \mathfrak{L}_{-\alpha_i}$ and we have the following relations:

$$(18)\qquad \begin{aligned} [h_i h_j] &= 0 \\ [e_i f_j] &= \delta_{ij} h_i \\ [e_i h_j] &= A_{ji} e_i \\ [f_i h_j] &= -A_{ji} f_i \end{aligned}$$

where (A_{ij}) is the Cartan matrix determined by π.

Let τ be an automorphism. Then $\mathfrak{H}^\tau$ is a second Cartan subalgebra. Hence there exists an invariant automorphism σ such that $\mathfrak{H}^\sigma = \mathfrak{H}^\tau$. Then the automorphism $\tau' = \tau\sigma^{-1}$ maps $\mathfrak{H}$ into itself. We now consider an automorphism $\tau(=\tau')$ which maps the Cartan subalgebra $\mathfrak{H}$ into itself. If $e_\alpha \in \mathfrak{L}_\alpha$ we have $[e_\alpha h] = \alpha(h) e_\alpha$. Hence

$[e_\alpha^\tau h^\tau] = \alpha(h)e_\alpha^\tau$. It follows that $\mathfrak{L}_\alpha^\tau = \mathfrak{L}_\beta$ where β is a root. In this way we obtain a mapping $\alpha \to \beta$ of the set of roots. Since $e_\alpha{}^\tau \in \mathfrak{L}_\beta$ we have $[e_\alpha^\tau h^\tau] = \beta(h^\tau)e_\alpha^\tau$. Hence we see that

$$\alpha(h) = \beta(h^\tau)\,. \tag{19}$$

Let τ^* denote the transpose in $\mathfrak{H}^*$ of the restriction of τ to $\mathfrak{H}$. By definition, if $\xi \in \mathfrak{H}^*$, then $\xi^{\tau^*}(h) = \xi(h^\tau)$. Then (19) implies that $\beta^{\tau^*} = \alpha$ or $\beta = \alpha^{(\tau^*)^{-1}}$. Hence we have the following

PROPOSITION 1. *Let τ be an automorphism of $\mathfrak{L}$ such that $\mathfrak{H}^\tau = \mathfrak{H}$ for a Cartan subalgebra $\mathfrak{H}$ of $\mathfrak{L}$. Then if α is any root of $\mathfrak{H}$ in $\mathfrak{L}$,*

$$\mathfrak{L}_\alpha^\tau = \mathfrak{L}_{\alpha^{(\tau^*)^{-1}}} \tag{20}$$

where τ^* is the transpose in $\mathfrak{H}^*$ of the restriction of τ to $\mathfrak{H}$.

We note next that if $\pi = \{\alpha_1, \cdots, \alpha_l\}$ and $\beta_i = \alpha_i^{(\tau^*)^{-1}}$, then $\pi' = \{\beta_1, \cdots, \beta_l\}$ is a simple system of roots. Thus every root has the form $\pm \sum k_i\alpha_i$ where the k_i are non-negative integers. If we apply $(\tau^*)^{-1}$ we see that every root also has the form $\pm \sum k_i\beta_i$. This guarantees that π' is a simple system. We prove next

PROPOSITION 2. *Let π and π' be simple systems of roots. Then there exists an invariant automorphism σ such that $\mathfrak{H}^\sigma = \mathfrak{H}$ and $\pi^{(\sigma^*)^{-1}} = \pi'$.*

Proof: Let $\pi = \{\alpha_1, \alpha_2, \cdots, \alpha_l\}$, $\pi' = \{\beta_1, \cdots, \beta_l\}$. Then we know (Theorems 8.1, 8.2) that $\pi' = \pi S_{\alpha_{i_1}} S_{\alpha_{i_2}} \cdots S_{\alpha_{i_r}}$ for suitable Weyl reflections S_{α_i}. It is therefore clear that it suffices to prove the result for $\pi' = \pi S_{\alpha_i}$. For this purpose we introduce the invariant automorphisms $\exp \operatorname{ad} \xi f_i$ and $\exp \operatorname{ad} \xi e_i$, $\xi \in \Phi$. In our calculation we shall use the formulas for the irreducible representations of the three-dimensional split simple Lie algebra given in (36) of § 3.8. We note that the matrices of the restrictions of $\operatorname{ad} f_i$, $\operatorname{ad} e_i$ to $\mathfrak{L}_i = \Phi e_i + \Phi h_i + \Phi f_i$, using the basis $(e_i, h_i, [h_i f_i])$ are, respectively,

$$\operatorname{ad} f_i \to \begin{pmatrix} 0 & 1 & 0 \\ 0 & 0 & 1 \\ 0 & 0 & 0 \end{pmatrix}, \qquad \operatorname{ad} e_i \to \begin{pmatrix} 0 & 0 & 0 \\ -2 & 0 & 0 \\ 0 & -2 & 0 \end{pmatrix}. \tag{21}$$

Hence for $\exp \operatorname{ad} \xi f_i$, $\exp \operatorname{ad} \xi e_i$ in $\mathfrak{L}_i$ we have the matrices

$$A_i(\xi) = \begin{pmatrix} 1 & \xi & \xi^2/2 \\ 0 & 1 & \xi \\ 0 & 0 & 1 \end{pmatrix}, \qquad B_i(\xi) = \begin{pmatrix} 1 & 0 & 0 \\ -2\xi & 1 & 0 \\ 2\xi^2 & -2\xi & 1 \end{pmatrix}. \tag{22}$$

It follows that the matrix of the restriction to $\mathfrak{L}_i$ of

$$\sigma_i(\xi) \equiv \exp \operatorname{ad} \xi f_i \exp \operatorname{ad} \xi^{-1} e_i \exp \operatorname{ad} \xi f_i \tag{23}$$

is

$$A_i(\xi)B_i(\xi^{-1})A_i(\xi) = \begin{pmatrix} 0 & 0 & \xi^2/2 \\ 0 & -1 & 0 \\ 2\xi^{-2} & 0 & 0 \end{pmatrix}. \tag{24}$$

In particular, $h_i^{\sigma_i(\xi)} = -h_i$. Next let $h \in \mathfrak{H}$ satisfy $\alpha_i(h) = 0$. Then $[he_i] = 0 = [hf_i]$ and consequently $h^{\sigma_i(\xi)} = h$. Since $\mathfrak{H}$ is the direct sum of Φh_i and the subspace of vectors h such that $\alpha_i(h) = 0$ it is clear that $\mathfrak{H}^{\sigma_i(\xi)} \subseteq \mathfrak{H}$. Moreover, the restriction of $\sigma_i(\xi)$ to $\mathfrak{H}$ is the reflection determined by h_i: $h \to h - [2(h, h_i)/(h_i, h_i)]h_i = h - [2(h, h_{\alpha_i})/(h_{\alpha_i}, h_{\alpha_i})]h_{\alpha_i} = h - [2\alpha_i(h)/\alpha_i(h_{\alpha_i})]h_{\alpha_i}$. If $\rho \in \mathfrak{H}^*$, then

$$\rho\left(h - \frac{2\alpha_i(h)}{\alpha_i(h_{\alpha_i})} h_{\alpha_i}\right) = \rho(h) - \frac{2\rho(h_{\alpha_i})}{\alpha_i(h_{\alpha_i})}\alpha_i(h) = \rho(h) - \frac{2(\rho, \alpha_i)}{(\alpha_i, \alpha_i)}\alpha_i(h) .$$

This shows that the transpose inverse of the restriction to $\mathfrak{H}$ of $\sigma_i(\xi)$ is the Weyl reflection S_{α_i} in $\mathfrak{H}^*$. Hence the invariant automorphism $\sigma_i(\xi)$ satisfies $\pi^{(\sigma_i(\xi)^*)^{-1}} = \pi S_{\alpha_i} = \pi'$, as required.

Proposition 2 and the considerations preceding it show that if τ is an automorphism such that $\mathfrak{H}^\tau = \mathfrak{H}$, then there exists an invariant automorphism σ such that $\mathfrak{H}^\sigma = \mathfrak{H}$ and if $\tau' = \tau\sigma^{-1}$, then $\pi((\tau')^*)^{-1} = \pi$ for the simple system π. We simplify our notation again by writing τ for τ'. Then we have a permutation $i \to i'$ of $i = 1, 2, \cdots, l$ such that $\alpha_i^{(\tau^*)^{-1}} = \alpha_{i'}$. Also we have $[e_i^\tau h_i^\tau] = 2e_i^\tau$, $[f_i^\tau h_i^\tau] = -2f_i^\tau$, $[e_i^\tau f_i^\tau] = h_i^\tau$. Since $e_i^\tau \in \mathfrak{L}_{\alpha_{i'}}$, $f_i^\tau \in \mathfrak{L}_{-\alpha_{i'}}$ we have $e_i^\tau = \mu_i e_{i'}$, $f_i^\tau = \nu_i f_{i'}$. Then $h_i^\tau = \mu_i \nu_i h_{i'}$. Since $[e_{i'} h_{i'}] = 2e_{i'}$ and $[e_i^\tau h_i^\tau] = 2e_i^\tau$ so that $\mu_i^2 \nu_i [e_{i'} h_{i'}] = 2\mu_i e_{i'}$, we have $\nu_i = \mu_i^{-1}$ and $h_i^\tau = h_{i'}$. Since $[e_i^\tau h_j^\tau] = A_{ji} e_i^\tau$, $[e_{i'} h_{j'}] = A_{j'i'} e_{i'}$. Hence we have

$$A_{i'j'} = A_{ij}, \qquad i, j = 1, 2, \cdots, l . \tag{25}$$

The subgroup of the symmetric group on $1, 2, \cdots, l$ of the permutations $i \to i'$ satisfying (25) will be called the *group of automorphisms of the Cartan matrix* (A_{ij}). If we recall the definition of the Dynkin diagram of the Cartan matrix, it is clear that if $\alpha_1, \alpha_2, \cdots, \alpha_l$ are the vertices of the diagram, then any element $i \to i'$ of the group of automorphisms of (A_{ij}) defines an automorphism of the Dynkin diagram, that is, a $1:1$ mapping $\alpha_i \to \alpha_{i'}$ such that $(\alpha_i, \alpha_i) = (\alpha_{i'}, \alpha_{i'})$ and for any i, j, the number of lines connecting α_i to α_j

is the same as that connecting $\alpha_{i'}$ and $\alpha_{j'}$ (cf. § 4.5). The argument used to show that the Dynkin diagram determines the Cartan matrix shows also that the converse holds, that is, if $\alpha_i \rightarrow \alpha_{i'}$ is an automorphism of the Dynkin diagram, then $i \rightarrow i'$ is an automorphism of the Cartan matrix.

Now suppose that τ is an automorphism such that $\mathfrak{L}_{\alpha_i}^{\tau} = \mathfrak{L}_{\alpha_i}$, $i = 1, 2, \cdots, l$. Then we have the identity mapping $i \rightarrow i' = i$ in the foregoing argument. This shows that $e_i^{\tau} = \mu_i e_i$, $f_i^{\tau} = \mu_i^{-1} f_i$, $h_i^{\tau} = h_i$. Hence τ acts as the identity in $\mathfrak{H}$. Conversely, Prop. 1 shows that if the restriction of τ to $\mathfrak{H}$ is the identity, then $\mathfrak{L}_{\alpha}^{\tau} = \mathfrak{L}_{\alpha}$ for every α so $e_i^{\tau} = \mu_i e_i$, $f_i^{\tau} = \mu_i^{-1} f_i$. For these automorphisms we have the following

Proposition 3. *If τ is an automorphism such that $h^{\tau} = h$ for every h of a Cartan subalgebra $\mathfrak{H}$ of $\mathfrak{L}$, then τ is an invariant automorphism.*

Proof: We have seen that $e_i^{\tau} = \mu_i e_i$ and $f_i^{\tau} = \mu_i^{-1} f_i$. Let $\sigma_i(\xi)$ be the invariant automorphism defined by (23) and let

$$\omega_i(\xi) = \sigma_i(\xi)\sigma_i(1) \, . \tag{26}$$

We have seen that $h_i^{\sigma_i(\xi)} = -h_i$, $h^{\sigma_i(\xi)} = h$ if $\alpha_i(h) = 0$. It follows that $h^{\omega_i(\xi)} = h$ for all h. The matrix relative to $(e_i, h_i, [h_i f_i])$ of the restriction of $\omega_i(\xi)$ to $\mathfrak{L}_i$ is the product of the matrix in (24) by the matrix obtained from this by taking $\xi = 1$. The result is

$$\operatorname{diag}\{\xi^2, 1, \xi^{-2}\} \, . \tag{27}$$

Hence we have

$$e_i^{\omega_i(\xi)} = \xi^2 e_i = \xi^{A_{ii}} e_i \, . \tag{28}$$

We wish to calculate next $e_j^{\omega_i(\xi)}$ for $j \neq i$. We have $[f_j e_i] = 0$ and $[f_j h_i] = -A_{ij} f_j$. Hence f_j generates an irreducible module for $\mathfrak{L}_i$. There are four possibilities $-A_{ij} = 0, 1, 2$ or 3. If $A_{ij} = -2$ the module is equivalent to $\mathfrak{L}_i$ and the argument just used shows that $f_j^{\omega_i(\xi)} = \xi^{-A_{ij}} f_j$ and this implies that $e_j^{\omega_i(\xi)} = (-\xi)^{A_{ij}} e_j$. If $A_{ij} = 0$, $[f_j \mathfrak{L}_i] = 0$ and this implies that $e_j^{\omega_i(\xi)} = e_j = (-\xi)^{A_{ij}} e_j$. Next let $A_{ij} = -1$. Then the matrices of f_i, e_i acting in the module generated by f_j are, respectively

$$\begin{pmatrix} 0 & 1 \\ 0 & 0 \end{pmatrix}, \quad \begin{pmatrix} 0 & 0 \\ -1 & 0 \end{pmatrix}. \tag{29}$$

It follows that the matrix of $\sigma_i(\xi)$ is

$$\begin{pmatrix} 0 & \xi \\ -\xi^{-1} & 0 \end{pmatrix} \tag{30}$$

and that of $\omega_i(\xi)$ is $\mathrm{diag}\{-\xi, -\xi^{-1}\}$. Hence we have $f_j^{\omega_i(\xi)} = -\xi f_j$ and $e_j^{\omega_i(\xi)} = -\xi^{-1}e_j = (-\xi)^{A_{ij}}e_j$. Finally, let $A_{ij} = -3$. Then the representing matrices for f_i and e_i are, according to to (36) of § 3.8:

$$\begin{pmatrix} 0 & 1 & 0 & 0 \\ 0 & 0 & 1 & 0 \\ 0 & 0 & 0 & 1 \\ 0 & 0 & 0 & 0 \end{pmatrix}, \quad \begin{pmatrix} 0 & 0 & 0 & 0 \\ -3 & 0 & 0 & 0 \\ 0 & -4 & 0 & 0 \\ 0 & 0 & -3 & 0 \end{pmatrix}. \tag{31}$$

It follows that the matrix of $\sigma_i(\xi)$ as defined in (23) is

$$\begin{pmatrix} 0 & 0 & 0 & \xi^3/6 \\ 0 & 0 & -\frac{1}{2}\xi & 0 \\ 0 & 2\xi^{-1} & 0 & 0 \\ -6\xi^{-3} & 0 & 0 & 0 \end{pmatrix}. \tag{32}$$

Consequently, the matrix of $\omega_i(\xi)$ in the module generated by f_j is $\mathrm{diag}\{-\xi^3, -\xi, -\xi^{-1}, -\xi^{-3}\}$. Hence $f_j^{\omega_i(\xi)} = -\xi^3 f_j$ and $e_j^{\omega_i(\xi)} = (-\xi)^{-3}e_j = (-\xi)^{A_{ij}}e_j$. Thus in all cases we have

$$e_j^{\omega_i(\xi)} = (-\xi)^{A_{ij}}e_j \tag{33}$$

($j = i$ or $j \neq i$). Now set

$$\Omega \equiv \Omega(\xi_1, \xi_2, \cdots, \xi_l) = \omega_1(-\xi_1)\omega_2(-\xi_2) \cdots \omega_l(-\xi_l)\,. \tag{34}$$

Then (33) implies that

$$e_j^{\Omega} = \xi_1^{A_{1j}}\xi_2^{A_{2j}} \cdots \xi_l^{A_{3j}}e_j\,, \qquad j = 1, 2, \cdots, l\,. \tag{35}$$

We recall that the matrix (A_{ij}) is non-singular and its determinant is clearly an integer d. Hence we have an integral matrix (B_{ij}) such that $(A_{ij})(B_{ij}) = d1$. Let j be fixed and set $\xi_k = (\mu_j^{1/d})^{B_{jk}}$, $k = 1, \cdots, l$. Then

$$\begin{aligned} e_j^{\Omega(\xi_1,\cdots,\xi_l)} &= (\mu_j^{1/d})^{B_{j1}A_{1j}+\cdots+B_{jl}A_{lj}}e_j = (\mu_j^{1/d})^d e_j = \mu_j e_j \\ e_i^{\Omega(\xi_1,\cdots,\xi_l)} &= (\mu_j^{1/d})^{B_{j1}A_{1i}+\cdots+B_{jl}A_{li}}e_i = e_i \qquad \text{if} \qquad i \neq j\,. \end{aligned} \tag{36}$$

It is clear that a suitable product of the invariant automorphisms defined here for $j = 1, 2, \cdots, l$ coincides with the given automorphism τ in its action on the e_i, f_i, h_i. Since these are generators it is clear that τ is an invariant automorphism.

It is now easy to see that the index of the invariant subgroup

G_0 in G does not exceed the order of the group of automorphisms of the Cartan matrix. In fact, we can display a subgroup K of G which is isomorphic to the group of automorphisms of the Cartan matrix such that every $\tau \in G$ is congruent modulo G_0 to a $\tau_P \in K$. Let P: $i \rightarrow i'$ be an automorphism of the Cartan matrix. Then we have the relations

$$[h_{i'}h_{j'}] = 0\,, \qquad [e_{i'}f_{j'}] = \delta_{ij}h_{i'}$$
$$[e_{i'}h_{j'}] = A_{ji}e_{i'}\,, \qquad [f_{i'}h_{j'}] = -A_{ji}f_{i'}\,.$$

Hence the isomorphism theorem for split semi-simple Lie algebras (Theorem 4.3) implies that there exists a (unique) automorphism τ_P of $\mathfrak{L}$ such that $e_i^{\tau_P} = e_{i'}$, $f_i^{\tau_P} = f_{i'}$. The set of these automorphisms is evidently a subgroup K of G isomorphic to the group of automorphisms of the Cartan matrix.

Now let τ be any automorphism of $\mathfrak{L}$. We have seen that τ is congruent modulo G_0 to an automorphism τ_1 such that $\mathfrak{H}^{\tau_1} = \mathfrak{H}$. Also we know that τ_1 is congruent modulo G_0 to an automorphism τ_2 such that $\mathfrak{H}^{\tau_2} = \mathfrak{H}$ and $\pi^{(\tau_2^*)^{-1}} = \pi$ for the simple system of roots $\pi = \{\alpha_1, \alpha_2, \cdots, \alpha_l\}$ relative to $\mathfrak{H}$. We have $\mathfrak{L}_{\alpha_i}^{\tau_2} = \mathfrak{L}_{\alpha_{i'}}$, $i = 1, 2, \cdots, l$ and $i \rightarrow i'$ is an automorphism P of the Cartan matrix. Let τ_P be the corresponding element of K. Then $\sigma = \tau_2\tau_P^{-1}$ satisfies $\mathfrak{L}_{\alpha_i}^{\sigma} = \mathfrak{L}_{\alpha_i}$, $\mathfrak{L}_{-\alpha_i}^{\sigma} = \mathfrak{L}_{-\alpha_i}$, $i = 1, 2, \cdots, l$. Hence σ is an invariant automorphism, by Prop. 3. Thus τ is congruent modulo G_0 to τ_P.

It can be shown that no element of K is in G_0, which means that G is the semi-direct product of K and G_0. This is equivalent to showing that the index of G_0 in G is the order of the group of automorphisms of the Cartan matrix. A proof of this will be indicated in the exercises. We shall now restrict our attention to $\mathfrak{L}$ simple. The result stated will follow for the Lie algebras A_l, B_l, C_l, D_l, $l \neq 4$, G_2 and F_4 from the explicit determination of the groups of automorphisms for these Lie algebras which we shall give in the next section. The result will also be clear for E_7 and E_8.

We now examine the groups of automorphisms of the connected Dynkin diagrams. We recall that these are the ones which correspond to the simple Lie algebras $\mathfrak{L}$. The types are A_l, $l \geqq 1$, B_l, $l \geqq 2$, C , $l \geqq 3$, D_l, $l \geqq 4$, G_2, F_4, E_6, E_7, E_8. If we look at these diagrams as given in § 4.5 we see that for $A_1, B_l, C_l, G_2, F_4, E_7$ and E_8 the group of automorphisms of the diagram is the identity.

For A_l, $l \geqq 2$: 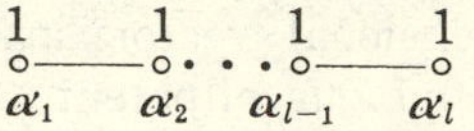

we have in addition to the identity mapping the automorphism $\alpha_i \rightarrow \alpha_{l+1-i}$. For

D_l : 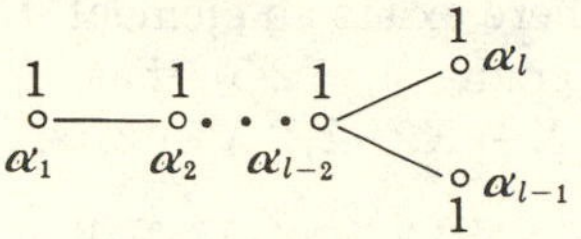

we have the identity automorphism and the mapping $\alpha_i \rightarrow \alpha_i$ $i \leqq l-2$, $\alpha_{l-1} \rightarrow \alpha_l$, $\alpha_l \rightarrow \alpha_{l-1}$ which is an automorphism. These are the only automorphisms if $l \geqq 5$. For $l = 4$ the diagram

D_4 : 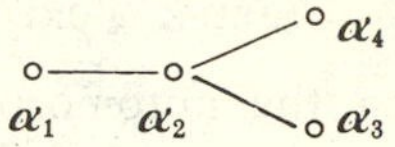

has the group of automorphisms which permute $\alpha_1, \alpha_3, \alpha_4$ and leave α_2 fixed. This is isomorphic to the symmetric group on three elements. For

E_6 : 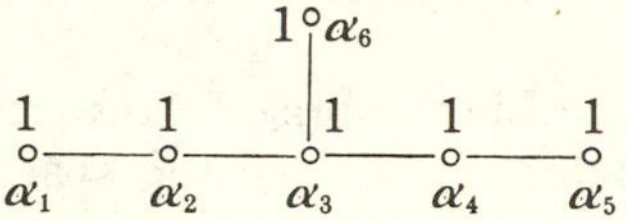

the group of automorphisms consists of the identity mapping and the mapping $\alpha_6 \rightarrow \alpha_6$, $\alpha_i \rightarrow \alpha_{6-i}$, $i \leqq 5$.

It is now clear that we have the following

THEOREM 4. *Let $\mathfrak{L}$ be finite-dimensional simple over an algebraically closed field of characteristic* 0, *G the group of automorphisms of $\mathfrak{L}$, G_0 the invariant subgroup of invariant automorphisms. Then $G = G_0$ unless $\mathfrak{L}$ is of one of the following types: A_l, $l > 1$, D_l or E_6. In all of these cases except D_4, the index $[G : G_0] \leqq 2$ and for D_4, $[G : G_0] \leqq 6$.*

Remark. The group G is an algebraic linear group (cf. Chevalley [2]). It is easy to see that G_0 is the algebraic component of the identity element of G. If Φ is the field of complex numbers, then G is a topological group and G_0 is the connected component of 1 in G.

5. *Explicit determination of the automorphisms for the simple Lie algebras*

Let $\mathfrak{L}$ be a semi-simple subalgebra of $\mathfrak{E}_L$, $\mathfrak{E}$ the algebra of linear transformations of a finite-dimensional vector space over Φ. Let Z be an element of $\mathfrak{L}$ such that $\operatorname{ad} Z$ is nilpotent. Since the algebra of linear transformations $\operatorname{ad}_{\mathfrak{L}} \mathfrak{L}$ is semi-simple it follows from Theorem 3.17 that there exists an element $H \in \mathfrak{L}$ such that $[ZH] = Z$. This implies (Lemma 4, § 2.5) that Z is nilpotent. We have $\operatorname{ad} Z = Z_R - Z_L$ (Z_R: $X \to XZ$, Z_L: $X \to ZX$). Hence

$$\exp \operatorname{ad} Z = \exp (Z_R - Z_L) = \exp Z_R \exp (-Z_L) ,$$

since $[Z_R Z_L] = 0$. Then

$$\begin{aligned} \exp \operatorname{ad} Z &= (\exp Z)_R (\exp (-Z))_L \\ &= (\exp Z)_R (\exp Z)_L^{-1} . \end{aligned} \tag{37}$$

If we set $A = \exp Z$, then the automorphism $\exp \operatorname{ad} Z$ of $\mathfrak{L}$ is the mapping

$$X \to A^{-1} X A . \tag{38}$$

We now consider the simple Lie algebras of types A_l, B_l, C_l, D_l, G_2, and F_4.

A_l. Here $\mathfrak{L}$ is the Lie algebra of linear transformations of trace zero in an $(l + 1)$-dimensional vector space. We can identify $\mathfrak{L}$ with the Lie algebra of matrices of trace 0 in Φ_{l+1}. If A is a non-singular matrix, then $X \to A^{-1}XA$ is an automorphism of $\mathfrak{L}$. Since the only matrices which commute with all matrices of trace 0 are the scalar matrices it is clear that the automorphisms $X \to A^{-1}XA$, $X \to B^{-1}XB$ are identical if and only if $B = \rho A$, $\rho \in \Phi$. Besides the automorphisms $X \to A^{-1}XA$ we have the automorphism $X \to -X'$ of $\mathfrak{L}$ and, more generally, we have the set of automorphisms $X \to -A^{-1}X'A$. Suppose the automorphism $X \to -X'$ coincides with one of the automorphisms $X \to A^{-1}XA$. Then we have

$$A^{-1} X A = -X' \tag{39}$$

for all X of trace 0. This implies that

$$-X = A'X'(A')^{-1} = - A'A^{-1}XA(A')^{-1}$$

so that $A(A')^{-1}$ commutes with all X. It follows that $A' = \rho A$; hence $A' = \pm A$. The condition (39) can be rewritten as $A^{-1}X'A = -X$. The set of X's satisfying this condition is either the symplectic Lie algebra or the orthogonal Lie algebra. Accordingly the dimensionality is either $(l + 1)(l + 2)/2$ (odd l only) or $l(l + 1)/2$ (l even or odd). Since the dimensionality of the space of $(l + 1) \times (l + 1)$

matrices of trace 0 is $l^2 + 2l$, we must have either $l^2 + 2l = (l+1)(l+2)/2$ or $l^2 + 2l = l(l+1)/2$. The only solution is $l = 1$ in which case $l^2 + 2l = (l+1)(l+2)/2$. Thus we see that $X \to -X'$ coincides with an automorphism of the form $X \to A^{-1}XA$ only if $l = 1$. If $l = 1$, then we have $-X' = A^{-1}XA$ for $A = \begin{pmatrix} 0 & 1 \\ -1 & 0 \end{pmatrix}$ and all X of trace 0.

The result obtained at the beginning of this section shows that every invariant automorphism of $\mathfrak{L}$ has the form $X \to A^{-1}XA$ where A is a product of exponentials of nilpotent matrices. For the Lie algebra A_1 this is the complete automorphism group. For A_l, $l > 1$, we have the automorphism $X \to -X'$ which is not invariant. Hence $[G : G_0] = 2$ and the automorphisms are the mappings $X \to A^{-1}XA$ and $X \to -A^{-1}X'A$.

THEOREM 5. *The group of automorphisms of the Lie algebra of* 2×2 *matrices of trace* 0 *is the set of mappings* $X \to A^{-1}XA$. *The group of automorphisms of the Lie algebra of* $n \times n$ *matrices of trace* 0, $n > 2$, *is the set of mappings* $X \to A^{-1}XA$ *and* $X \to -A^{-1}X'A$. (*The base field* Φ *is algebraically closed of characteristic* 0.)

B_l, C_l, D_l. These algebras are the sets of matrices X satisfying $S^{-1}X'S = -X$ where $S = 1$ for B_l and D_l and $S' = -S$ for C_l. The size of the matrices is $2l$ for C_l and D_l and $2l + 1$ for B_l. We take $l \geqq 2$ for B_l, $l \geqq 3$ for C_l and $l \geqq 4$ for D_l. Let O be a matrix such that

$$O'SO = \rho S \tag{40}$$

where $\rho \neq 0$ is in Φ. Then O is non-singular and if $X \in \mathfrak{L}$, then $Y = O^{-1}XO$ satisfies

$$\begin{aligned} S^{-1}Y'S &= S^{-1}(O^{-1}XO)'S = S^{-1}O'X'(O')^{-1}S \\ &= -S^{-1}O'SXS^{-1}(O')^{-1}S = -\rho O^{-1}X(\rho^{-1}O) \\ &= -O^{-1}XO = -Y\,. \end{aligned} \tag{41}$$

Hence $Y \in \mathfrak{L}$ and consequently $X \to Y = O^{-1}XO$ is an automorphism of $\mathfrak{L}$. Since the base field is algebraically closed we can replace O by $\rho^{-\frac{1}{2}}O = O_1$ and we obtain $O_1^{-1}XO_1 = Y$, $O_1'SO_1 = S$. If we write O for O_1, then we see that for B_l and D_l, O is an orthogonal matrix ($S = 1$) and for C_l, O is a symplectic matrix. It is easy to verify that the enveloping associative algebra of $\mathfrak{L}$ is the complete matrix algebra. We leave it to the reader to prove this. It follows that the only matrices which commute with all the elements of $\mathfrak{L}$ are the scalars. Hence if $O_1^{-1}XO_1 = O_2^{-1}XO_2$ for all $X \in \mathfrak{L}$, then $O_2 = \rho O_1$, $\rho \in \Phi$.

Let Z be an element of $\mathfrak{L}$ such that $\operatorname{ad} Z$ is nilpotent. Then we have seen that Z is nilpotent and the automorphism $\exp \operatorname{ad} Z$ has the form $X \to A^{-1}XA$ where $A = \exp Z$. The nilpotence of Z implies that $\exp Z$ is unimodular (Exercise 5.4). Also, we have

$$\begin{aligned} S^{-1}(\exp Z)'S &= S^{-1}(\exp Z')S \\ &= \exp S^{-1}Z'S = \exp(-Z) = (\exp Z)^{-1} . \end{aligned}$$

Hence $A = \exp Z$ is orthogonal for B_l and D_l and symplectic for C_l. For B_l and C_l every automorphism is an invariant automorphism. Hence in these cases the automorphisms of $\mathfrak{L}$ have the form $X \to O^{-1}XO$ where O is unimodular and satisfies $O'SO = S$. For B_l this states that O is a proper orthogonal matrix (corresponding to a rotation). For C_l it is known that if O is symplectic ($O'SO = S$) then O is necessarily unimodular (see Artin [1], p. 139, or Exercise 12 below). Hence this condition can be dropped. Now consider D_l. In this case there exist orthogonal matrices O of determinant -1 and we cannot have $O = \rho O_1$ where O_1 is a proper orthogonal matrix. Thus if $O = \rho O_1$, then $\rho = \pm 1$ and in either case $\det O = \det \rho O_1 = \det O_1 = 1$. This contradiction implies that the automorphism $X \to O^{-1}XO$, where O is improper orthogonal, is not invariant and we see that $G \supset G_0$. If $l > 4$, then the index of G_0 in G is $\leqq 2$; hence this index is 2 and every automorphism of D_l has the form $X \to O^{-1}XO$ where O is orthogonal. We therefore have the following

THEOREM 6. *Let Φ be algebraically closed of characteristic* 0 *and let $\mathfrak{L}$ be the Lie algebra of skew matrices or the "symplectic" Lie algebra of matrices X such that $S^{-1}X'S = -X$ where $S' = -S$. Assume the number of rows $n \geqq 5$ in the odd-dimensional skew case, $n \geqq 6$ in the symplectic case and $n \geqq 10$ in the even-dimensional skew case. Then the groups of automorphisms of $\mathfrak{L}$ in the skew cases consist of the mappings $X \to O^{-1}XO$ where O is orthogonal. In the odd-dimensional case one can add the condition that O is proper. In the symplectic case the group of automorphisms is the set of mappings $X \to O^{-1}XO$ where $O'SO = S$.*

The case of D_4 is not covered by this theorem. It can be shown that in this case the group G/G_0 is isomorphic to the symmetric group on three letters and the group G can be determined. This will be indicated in some exercises below.

We consider next the Lie algebra G_2. Here we use the representation of $\mathfrak{L}$ as the algebra of derivations of a Cayley algebra $\mathfrak{C}$ (§ 4.6). Now, in general, if $\mathfrak{C}$ is a non-associative algebra, D a derivation of $\mathfrak{C}$, A an automorphism of $\mathfrak{C}$, then $A^{-1}DA$ is a derivation. It follows that the mapping $X \to A^{-1}XA$ determined by an automorphism of $\mathfrak{C}$ is an automorphism of the derivation algebra $\mathfrak{L} = \mathfrak{D}(\mathfrak{C})$. In the case of $\mathfrak{C}$, the Cayley algebra over an algebraically closed field of characteristic 0, we know that every automorphism of $\mathfrak{L}$ is invariant and so it is a product of mappings of the form $X \to A^{-1}XA$ where $A = \exp Z$, Z in $\mathfrak{L}$ and Z a nilpotent linear transformation in $\mathfrak{C}$. Since $Z \in \mathfrak{L}$, Z is a derivation of $\mathfrak{C}$. It follows that $A = \exp Z$ is an automorphism of $\mathfrak{C}$. We therefore see that every automorphism of $\mathfrak{L}$ has the form $X \to A^{-1}XA$ where A is an automorphism in $\mathfrak{C}$.

The same reasoning applies to the Lie algebra F_4. Here we represent $\mathfrak{L}$ as the derivation algebra of the exceptional Jordan algebra M_3^8 and we obtain the result that the group of automorphisms of $\mathfrak{L}$ is the set of mappings $X \to A^{-1}XA$ where A is an automorphism of M_3^8.

THEOREM 7. *Let $\mathfrak{L}$ be the Lie algebra of derivations of the Cayley algebra $\mathfrak{C}$ or of the Jordan algebra M_3^8 over an algebraically closed field of characteristic* 0. *Then the automorphisms of $\mathfrak{L}$ have the form $X \to A^{-1}XA$ where A is an automorphism of $\mathfrak{C}$ or of M_3^8.*

Exercises

The base field in all of these exercises will be of characteristic 0 and all spaces are finite-dimensional.

1. Show that any non-singular linear transformation A can be written in the form A_uA_s where A_u is *unipotent*, that is, $A_u = 1 + N$, N nilpotent, and A_s is semi-simple and A_u and A_s are polynomials in A. Prove that if $A = B_sB_u$ and B_u and B_s commute where B_u is unipotent and B_s is semi-simple then $A_u = B_u$ and $A_s = B_s$. (*Hint*: Use Theorem 3.16.)

2. Let τ be an automorphism of a non-associative algebra $\mathfrak{A}$ and let $\tau = \tau_u\tau_s$ as in 1. Prove that τ_s and τ_u are automorphisms. (*Hint*: Assume the base field is algebraically closed and use Exercise 2.5. Extend the field to obtain the result in the general case.)

3. Let $\mathfrak{L}$ be a finite-dimensional simple Lie algebra over an algebraically closed field and let G_0 be the group of invariant automorphisms of $\mathfrak{L}$. Prove that $\mathfrak{L}$ is irreducible relative to G_0.

4. Let $\mathfrak{C}$ be the split Cayley algebra over an algebraically closed field, $\mathfrak{C}_0$ the subspace of elements of trace 0 in $\mathfrak{C}$: $a + \bar{a} = 0$. We have seen that the derivation algebra $\mathfrak{T}(=G_2)$ acts irreducibly in $\mathfrak{C}_0$ (§ 7.6). Use this result to prove that $\mathfrak{C}_0$ is irreducible relative to the group of automorphisms G of $\mathfrak{C}$.

5. Same as 4. with $\mathfrak{C}$ replaced by M_3^8 (cf. § 4.6).

Exercises 6 through 9 are designed to prove that if $\mathfrak{L}$ is semi-simple over an algebraically closed field, then G/G_0 is isomorphic to the group of automorphisms of the Cartan matrix or Dynkin diagram determined by a Cartan subalgebra $\mathfrak{H}$ of $\mathfrak{L}$. In all of these exercises $\mathfrak{L}$, $\mathfrak{H}$, G, G_0, etc., are as indicated in the text.

6. If $a \in \mathfrak{L}$ and $\mathfrak{C}_a = \{z \mid [za] = 0\}$, then $\dim \mathfrak{C}_a \geqq l$, the dimensionality of a Cartan subalgebra of $\mathfrak{L}$. *Sketch of proof*: Note that $\dim \mathfrak{C}_a = \dim \mathfrak{L} -$ rank (ad a). Show that if a is a regular element, that $\mathfrak{C}_a = \mathfrak{H}$ a Cartan subalgebra so $\dim \mathfrak{C}_a = l$. If $(u_1, \cdots, u_n)$ is a basis for $\mathfrak{L}$ over Φ and $(\xi_1, \cdots, \xi_n)$ are indeterminates, then $x = \sum \xi_i u_i$ is a regular element of $\mathfrak{L}_P$, $P = \Phi(\xi_i)$ so rank (ad x) $= n - l$. If $a = \sum \alpha_i u_i$ the specialization $\xi_i = \alpha_i$ shows that rank (ad a) $\leqq n - l$. Hence $\dim \mathfrak{C}_a \geqq l$.

7. If τ is an automorphism in $\mathfrak{L}$ let I_τ be the set of fixed points of τ: $b^\tau = b$. Prove that if τ is invariant, then $\dim I_\tau \geqq l$. *Sketch of proof*: We have $\tau = \exp Z_1 \exp Z_2 \cdots \exp Z_r$ where $Z_i = \operatorname{ad} z_i$ is nilpotent and we have to show that rank $(\tau - 1) \leqq n - l$. Let $\xi_1, \cdots, \xi_r$ be indeterminates and let P be the field of formal power series in the ξ_i, that is, the quotient field of the algebra $\Phi < \xi_1, \cdots, \xi_r >$ of formal power series in the ξ_i with coefficients in Φ. Then $\tau(\xi) \equiv \exp \xi_1 Z_1 \exp \xi_2 Z_2 \cdots \exp \xi_r Z_r$ is an invariant automorphism of $\mathfrak{L}_P$. The matrix of $\tau(\xi)$ relative to the basis $(u_1, \cdots, u_n)$ of $\mathfrak{L}$ has entries which are polynomials in the ξ_i and the specialization $\xi_i = 1$ gives the matrix of τ relative to this basis. Hence if $\dim I_{\tau(\xi)} \geqq l$, a specialization argument will show that $\dim I_\tau \geqq l$ (semi-simplicity and the rank are unchanged in passing from $\mathfrak{L}$ to $\mathfrak{L}_P$). Now the exponential formula can be used to show that $\tau(\xi) = \exp \xi_1 Z_1 \cdots \exp \xi_r Z_r = \exp Z$ where $Z = \operatorname{ad} z$, $z \in \mathfrak{L}_P$ (Exercise 5.12). Then $I_{\tau(\xi)} \supset \mathfrak{C}_z$ so $\dim I_{\tau(\xi)} \geqq l$ by 6.

8. Let τ be an invariant automorphism such that $\mathfrak{H}^\tau = \mathfrak{H}$ and $\pi^{(\tau^*)^{-1}} = \pi$ for a simple system of roots π. Then $h^\tau = h$ for every $h \in \mathfrak{H}$. *Sketch of proof*: $(\tau^*)^{-1}$ induces a permutation of the set of roots which we can write as a product of cycles $(\beta_1, \cdots, \beta_r)(\gamma_1, \cdots, \gamma_s) \cdots$. Since $\pi^{(\tau^*)^{-1}} = \pi$, $(\tau^*)^{-1}$ leaves the set of positive roots invariant; hence the β_i in a cycle $(\beta_1, \cdots, \beta_r)$ are either all positive or all negative. $\mathfrak{L}_{\beta_1} + \cdots + \mathfrak{L}_{\beta_r}$ is invariant under τ. If e_{β_i} is a base element for $\mathfrak{L}_{\beta_i}$, then we have $e_{\beta_1}^\tau = \nu_1 e_{\beta_2}$, $e_{\beta_2}^\tau = \nu_2 e_{\beta_3}, \cdots, e_{\beta_r}^\tau = \nu_r e_{\beta_1}$ so that the characteristic polynomial of the restriction of τ to $\mathfrak{L}_{\beta_1} + \cdots + \mathfrak{L}_{\beta_r}$ is $\lambda^r - \nu_1 \cdots \nu_r$. If $\nu_1 \cdots \nu_r \neq 1$ then $\lambda - 1$ is not a factor of this and consequently $(\mathfrak{L}_{\beta_1} + \cdots + \mathfrak{L}_{\beta_r}) \cap I_\tau = 0$. If this holds for every cycle, then $I_\tau \subset \mathfrak{H}$ since $\mathfrak{L} = \mathfrak{H} \oplus (\mathfrak{L}_{\beta_1} \oplus \cdots \oplus \mathfrak{L}_{\beta_r}) \oplus (\mathfrak{L}_{\gamma_1} \oplus \cdots \oplus \mathfrak{L}_{\gamma_s}) + \cdots$. Since $\dim I_\tau \geqq l$ by 7. it will follow that $I_\tau = \mathfrak{H}$ and $h^\tau = h$ for all $h \in \mathfrak{H}$. Now let σ be the automorphism in $\mathfrak{L}$ such that $h^\sigma = h$, $h \in \mathfrak{H}$ and $e_i^\sigma = \mu_i e_i \neq 0$, $f_i^\sigma = \mu_i^{-1} f_i$ (e_i, f_i, h_i canonical generators). We have shown that σ is invariant. If α is a positive root then $\alpha = \sum k_i \alpha_i$ and $e_\alpha^\sigma = \mu_1^{k_1} \mu_2^{k_2} \cdots \mu_l^{k_l} e_\alpha$ and $e_{-\alpha}^\sigma = \mu_1^{-k_1} h_2^{-k_2} \cdots \mu_l^{-k_l} e_{-\alpha}$. The automorphism $\tau' = \sigma\tau$ satisfies the same conditions as τ and $e_{\beta_1}^{\tau'} = \nu_1' e_{\beta_2}, \cdots, e_{\beta_r}^{\tau'} = \nu_r' e_{\beta_1}$ where $\nu_1' \cdots \nu_l' = \nu_1 \cdots \nu_l \mu_1^{s_1} \cdots \mu_l^{s_l}$ if the β_i are all positive and $\beta_1 + \cdots + \beta_r = \sum_1^l s_i \alpha_i \neq 0$ s_i non-negative integers, and $\nu_1 \cdots \nu_l' = \nu_1 \cdots \nu_l \mu_1^{-s_1} \cdots h_l^{-s_l}$ if the β_i are all negative and $\beta_1 + \cdots + \beta_r = -\sum_1^l s_i \alpha_i$. Since $s_i \neq 0$ for some i we can choose the μ's so that $\nu_1' \cdots \nu_l' \neq 1$ for every cycle. Then the argument used before shows that $h^{\tau'} = h$, $h \in \mathfrak{H}$. Hence $h^\tau = h$, $h \in \mathfrak{H}$.

9. Prove that G/G_0 is isomorphic to the group of automorphisms of the Cartan matrix.

10. Let $\mathfrak{L} = \Phi_{l+1}'$, $l + 1 = 2r$ and let τ be an automorphism of the form $X \to -A^{-1}X'A$ in $\mathfrak{L}$. Show that $\dim I_\tau \geqq r$ and that there exist τ such that $\dim I_\tau = r$.

11. Let $\mathfrak{L}$ be semi-simple over an algebraically closed field and let H be the subgroup of G_0 of η such that $\mathfrak{H}^\eta \subseteq \mathfrak{H}$, K the subgroup of G_0 of ξ such that $h^\xi = h$ for all $h \in \mathfrak{H}$. Show that K is an invariant subgroup of H and that $H/K \cong W$. (This gives a conceptual description of the Weyl group.)

12. Use the proof of Theorem 6 to prove that if O is a symplectic matrix with entries in a field of characteristic 0, then $\det O = 1$.

13. Let $\mathfrak{C}$ be a split Cayley algebra and let $\bar{a} = \alpha 1 - a_0$ if $a = \alpha 1 + a_0$, $a_0 \in \mathfrak{C}_0$. Set $N(a) = a\bar{a} = \bar{a}a$ and $(a, b) = \frac{1}{2}[N(a + b) - N(a) - N(b)]$. Verify that (a, b) is a non-degenerate symmetric bilinear form of maximal Witt index. Prove that $N(ac, b) = N(a, b\bar{c})$ and $N(ca, b) = N(a, \bar{c}b)$. Hence show that if $c \in \mathfrak{C}_0$, then c_R $(x \to xc)$, c_L $(x \to cx)$ and $R_c = \frac{1}{2}(c_L + c_R)$ are in the orthogonal Lie algebra (D_4) of the space $\mathfrak{C}$ relative to the form (a, b). Show that every element of this Lie algebra has the form $R_c + \sum [R_{c_i} R_{d_i}]$ where $c, c_i, d_i \in \mathfrak{C}_0$.

14. Use the alternative law (eq. (4.79)) to prove the following identity in $\mathfrak{C}$

$$c(ab) + (ab)c = (ca)b + a(bc)$$

or $2(ab)R_c = (ac_L)b + a(bc_R)$. Use this to prove the *principle of local triality*: For every linear transformation A in $\mathfrak{C}$ which is skew relative to (a, b) there exist a unique pair (B, C) of skew linear transformations such that

$$(ab)A = (aB)b + a(bC) .$$

15. Show that the mappings $A \to B$ and $A \to C$ determined in 14 are automorphisms of the orthogonal Lie algebra. Prove that every automorphism of this Lie algebra is of the form $X \to O^{-1}XO$ where O is orthogonal or is the product of one of the automorphisms defined in 14. by an automorphism $X \to O^{-1}XO$.

16. Let $\mathfrak{L}_1$ and $\mathfrak{L}_2$ be two subalgebras of D_4 isomorphic to D_3. Prove that there exists an automorphism of D_4 mapping $\mathfrak{L}_1$ onto $\mathfrak{L}_2$.

17. Show that the automorphism in A_l, $l > 1$, such that $e_i \rightarrow f_i$, $f_i \rightarrow e_i$ (cf. p.127) is not invariant. Show that for D_l, $l \geqq 4$, this automorphism is invariant if and only if l is even.

18. (Steinberg). If $\mathfrak{H}$ is a Cartan subalgebra let $G_0(\mathfrak{H})$ be group of automorphisms generated by all exp ad e where e belongs to a root space of $\mathfrak{L}$ relative to $\mathfrak{H}$ (cf. Chevalley [7]). Let $\mathfrak{H}_1$ be a second Cartan subalgebra and let $G_0(\mathfrak{H}_1)$ be the corresponding group of automorphisms. Show that there exists $\eta \in G_0(\mathfrak{H})$, $\eta_1 \in G_0(\mathfrak{H}_1)$ such that $\mathfrak{H}^{\eta} = \mathfrak{H}_1^{\eta_1}$. Then $\mathfrak{H} = \mathfrak{H}_1^{\eta_1\eta^{-1}}$. This implies that $G_0(\mathfrak{H}) = \eta\eta_1^{-1}G_0(\mathfrak{H}_1)\eta_1\eta^{-1}$ so that

$$G_0(\mathfrak{H}) = \eta^{-1}G_0(\mathfrak{H})\eta = \eta_1^{-1}G_0(\mathfrak{H}_1)\eta_1 = G_0(\mathfrak{H}_1) \,.$$

19. Prove that $G_0(\mathfrak{H}) = G_0$, the group of invariant automorphisms.

CHAPTER X

Simple Lie Algebras over an Arbitrary Field

In this chapter we study the problem of classifying the finite-dimensional simple Lie algebras over an arbitrary field of characteristic 0. The known methods for handling this involve reductions to the problem treated in Chapter IV of classifying the simple Lie algebras over an algebraically closed field of characteristic 0. One first defines a certain extension field Γ called the centroid of the simple Lie algebra which has the property that $\mathfrak{L}$ can be considered as an algebra over Γ and $(\mathfrak{L} \text{ over } \Gamma)_\Omega \equiv \Omega \otimes_\Gamma \mathfrak{L}$ is simple for every extension field Ω of Γ. It is natural to replace the given base field Φ by the field Γ. In this way the classification problem is reduced to the special case of classifying the Lie algebras $\mathfrak{L}$ such that $\mathfrak{L}_\Omega$ is simple for every extension field Ω of the base field. If $\mathfrak{L}$ is of this type and Ω is the algebraic closure, then the possibilities for $\mathfrak{L}_\Omega$ are known (A_l, B_l, C_l, etc.) and one now has the problem of determining all the $\mathfrak{L}$ such that $\mathfrak{L}_\Omega$ is one of the known simple Lie algebras over the algebraically closed field Ω.

This problem can be transferred to the analogous one in which the algebraically closed field Ω is replaced by a finite-dimensional Galois extension P of Γ and $\mathfrak{L}_P$ is one of the split simple Lie algebras over P. This is equivalent to the problem of determining the finite groups of automorphisms of $\mathfrak{L}_P$ considered as an algebra over Φ which are semi-linear transformations in $\mathfrak{L}_P$ over P.

We shall consider this problem in detail for $\mathfrak{L}_P$ one of the classical types $A_l - D_l$ except D_4. Our results will not give complete classifications even in these cases but will amount to a reduction of the problem to fairly standard questions on associative algebras. For certain base fields (e.g., the field of real numbers) complete solutions of these problems are known, so in these cases the classification problem can be solved.

The classification problem for the field of real numbers is quite old. A complete solution was given by Cartan in 1914. Simplifications of the treatment are due to Lardy and to Gantmacher. For an arbitrary base field of characteristic 0 the results for the classical types are due to Landherr and to the present author, for G_2 to the author and for F_4 to Tomber. It is worth noting that most of the results carry over also to the characteristic $p \neq 0$ case. This has been shown by the author. References to the literature can be found in the bibliography.

1. Multiplication algebra and centroid of a non-associative algebra

Let $\mathfrak{A}$ be an arbitrary non-associative algebra over a field Φ (cf. § 1.1). If $a \in \mathfrak{A}$, the right (left) multiplication a_R (a_L) is the linear mapping $x \to xa$ ($x \to ax$). We define the *multiplication algebra* $\mathfrak{T}(\mathfrak{A})$ to be the enveloping algebra of all the a_R and a_L, $a \in \mathfrak{A}$. Thus $\mathfrak{T}(\mathfrak{A})$ is the algebra (associative with 1) generated by the a_L and a_R. If $\mathfrak{A}$ is a Lie algebra, then $\mathfrak{T}(\mathfrak{A})$ is the enveloping algebra of the Lie algebra ad $\mathfrak{A}$. We define the *centroid* $\Gamma(\mathfrak{A})$ of $\mathfrak{A}$ to be the centralizer of $\mathfrak{T}(\mathfrak{A})$ in the algebra $\mathfrak{E}(\mathfrak{A})$ of all linear transformations in $\mathfrak{A}$. Thus the elements of $\Gamma = \Gamma(\mathfrak{A})$ are the linear transformations γ such that $[\gamma, A] = 0$ for all $A \in \mathfrak{T}(\mathfrak{A})$. Evidently, $\gamma \in \Gamma$ if and only if $[\gamma a_R] = 0 = [\gamma a_L]$ for all $a \in \mathfrak{A}$, and these conditions can be written in the form

$$(1) \qquad (ab)\gamma = (a\gamma)b = a(b\gamma)\,, \qquad a, b \in \mathfrak{A}\,.$$

LEMMA 1. *If $\mathfrak{A}^2 = \mathfrak{A}$, then Γ is commutative.*

Proof: Let $\gamma, \delta \in \Gamma$, $a, b \in \mathfrak{A}$. Then $(ab)\gamma\delta = ((a\gamma)b)\delta = (a\gamma)(b\delta)$ and $(ab)\gamma\delta = (a(b\gamma))\delta = (a\delta)(b\gamma)$. If we interchange γ and δ we obtain $(ab)\delta\gamma = (a\delta)(b\gamma) = (a\gamma)(b\delta)$. Hence $(ab)(\gamma\delta - \delta\gamma) = 0$. Since $\mathfrak{A}^2 = \mathfrak{A}$, any element c of $\mathfrak{A}$ has the form $c = \sum a_i b_i$. It follows that $c(\gamma\delta - \delta\gamma) = 0$ for all c and Γ is commutative.

A non-associative algebra $\mathfrak{A}$ is *simple* if $\mathfrak{A}$ has no (two-sided) ideals $\neq 0$, $\neq \mathfrak{A}$ and $\mathfrak{A}^2 \neq 0$. Since $\mathfrak{A}^2$ is an ideal it follows that $\mathfrak{A}^2 = \mathfrak{A}$ for $\mathfrak{A}$ simple. Hence the lemma implies that Γ is commutative.

The ideals of a non-associative algebra $\mathfrak{A}$ are just the subspaces which are invariant relative to the right and left multiplications. These are the same as the subspaces which are invariant relative to the multiplication algebra $\mathfrak{T} \equiv \mathfrak{T}(\mathfrak{A})$. It follows that $\mathfrak{A}$ is simple if and only if $\mathfrak{T}$ is an irreducible algebra of linear transformations. If $x \in \mathfrak{A}$, then the smallest $\mathfrak{T}$-invariant subspace containing x is $x\mathfrak{T}$.

Hence if $x \neq 0$ and $\mathfrak{A}$ is simple, then $x\mathfrak{T} = \mathfrak{A}$. The converse is easily seen also: if $\mathfrak{A}^2 \neq 0$ and $x\mathfrak{T} = \mathfrak{A}$ for every $x \neq 0$, then $\mathfrak{A}$ is simple. We recall the well-known lemma of Schur: If $\mathfrak{T}$ is an irreducible algebra of linear transformations, then the centralizer of $\mathfrak{T}$ is a division algebra (for proof see, for example, Jacobson [2], vol. II, p. 271). In the special case of $\mathfrak{T} = \mathfrak{T}(\mathfrak{A})$ for $\mathfrak{A}$ simple this and Lemma 1 give

THEOREM 1. *The centroid Γ of a simple non-associative algebra is a field.*

Since the centroid Γ is a field we can consider $\mathfrak{A}$ as a (left) vector space over Γ by setting $\gamma a = a\gamma$, $a \in \mathfrak{A}$, $\gamma \in \Gamma$. Then condition (1) can be re-written as

$$\gamma(ab) = (\gamma a)b = a(\gamma b)\,, \tag{1'}$$

which is just the condition that $\mathfrak{A}$ as vector space over Γ be a non-associative algebra over Γ relative to the product ab defined in $\mathfrak{A}$ over Φ.

A non-associative algebra $\mathfrak{A}$ will be called *central* if its centroid Γ coincides with the base field. If $\mathfrak{A}$ is simple with centroid Γ, then $\mathfrak{A}$ is central simple over Γ; for we have

THEOREM 2. *Let $\mathfrak{A}$ be a simple non-associative algebra over a field Φ and let $\Gamma(\supseteq \Phi)$ be the centroid. Consider $\mathfrak{A}$ as algebra over Γ by defining $\gamma a = a\gamma$, $a \in \mathfrak{A}$, $\gamma \in \Gamma$. Then $\mathfrak{A}$ is simple and central over Γ and the multiplication algebra of $\mathfrak{A}$ over Γ is the same set of transformations as the multiplication algebra of $\mathfrak{A}$ over Φ.*

Proof: Since $\Gamma \supseteq \Phi$ it is clear that a Γ-ideal of $\mathfrak{A}$ is a Φ-ideal so $\mathfrak{A}$ is Γ-simple. Similarly, the centroid $\tilde{\Gamma}$ of $\mathfrak{A}$ over Γ is contained in Γ; hence $\tilde{\Gamma} = \Gamma$ and $\mathfrak{A}$ is central. Let $\tilde{\mathfrak{T}}$ denote the multiplication algebra of $\mathfrak{A}$ over Γ. Then it is clear that $\tilde{\mathfrak{T}} = \Gamma\mathfrak{T}$ the set of Γ-linear combinations of the elements of $\mathfrak{T}$. Now let $\mathfrak{T}_0$ be the subset of $\mathfrak{T}$ of elements A such that $\gamma A \in \mathfrak{T}$ for all γ in Γ. It is clear that $\mathfrak{T}_0$ is a Φ-subalgebra of $\mathfrak{T}$. If $a, x \in \mathfrak{A}$, then $\gamma(ax) = a(\gamma x) = (\gamma a)x$ which means that $\gamma a_L = a_L\gamma = (\gamma a)_L$. Hence $a_L \in \mathfrak{T}_0$ and similarly, $a_R \in \mathfrak{T}_0$. Thus $\mathfrak{T}_0$ contains all the left and the right multiplications and consequently $\mathfrak{T}_0 = \mathfrak{T}$. We therefore have $\tilde{\mathfrak{T}} = \Gamma\mathfrak{T} = \mathfrak{T}$.

We consider next the question of extension of the base field of a non-associative simple algebra. The fundamental result in this connection is the following

THEOREM 3. *If $\mathfrak{A}$ is a non-associative central simple algebra over Φ and P is any extension field of Φ, then $\mathfrak{A}_P$ is central simple over P. Next let $\mathfrak{A}$ be an arbitrary non-associative algebra over Φ, let Δ be a subfield (over Φ) of the centroid and suppose the Δ-algebra $\Gamma \otimes_\Delta \mathfrak{A}$ is simple. Then $\mathfrak{A}$ is simple over Φ and $\Delta = \Gamma$.*

Proof: For the proof of the first assertion we shall need a well-known density theorem on irreducible algebras of linear transformations. (See, for example, Jacobson [2], vol. II, p. 272.) A special case of this result states that if $\mathfrak{T}$ is a non-zero irreducible algebra of linear transformations in a vector space $\mathfrak{A}$ over Φ and the centralizer of $\mathfrak{T}$ in $\mathfrak{E}(\mathfrak{A})$ is the set Φ of scalar multiplications, then $\mathfrak{T}$ is a dense algebra of linear transformations in $\mathfrak{A}$ over Φ. This means that if $\{x_1, x_2, \cdots, x_n\}$ is an ordered finite set of linearly independent elements of $\mathfrak{A}$ and $y_1, y_2, \cdots, y_n$ are n arbitrary elements of $\mathfrak{A}$, then there exists a $T \in \mathfrak{T}$ such that $x_iT = y_i$, $i = 1, 2, \cdots, n$. The density theorem is applicable to the multiplication algebra $\mathfrak{T}$ of a central simple non-associative algebra. Now let P be an extension field of Φ and consider the extension algebra $\mathfrak{A}_P$, for which we choose a basis $\{u_\alpha \mid \alpha \in I\}$ consisting of elements $u_\alpha \in \mathfrak{A}$. Let x and y be any elements of $\mathfrak{A}_P$ with $x \neq 0$. Then we can write $x = \sum_1^n \xi_i x_i$, $y = \sum_1^n \eta_i x_i$ where $\{x_1, x_2, \cdots, x_n\}$ is a suitable subset of the basis $\{u_\alpha\}$ and the ξ's and η's are in P. We may assume also that $\xi_1 \neq 0$. The extension to $\mathfrak{A}_P$ of any element of $\mathfrak{T}$ is in the multiplication algebra $\tilde{\mathfrak{T}}$ of $\mathfrak{A}_P$ over P. Hence there exists an $A_i \in \tilde{\mathfrak{T}}$ such that $x_1A_i = x_i$ and $x_jA_i = 0$ if $j \neq 1$, $i = 1, 2, \cdots, n$. Then $xA_i = \xi_1 x_i$ and $A = \sum_{i=1}^n \eta_i \xi_1^{-1} A_i \in \tilde{\mathfrak{T}}$ and satisfies $xA = y$. Thus $x\tilde{\mathfrak{T}} \ni y$ and since y is arbitrary, $x\tilde{\mathfrak{T}} = \mathfrak{A}_P$. This implies that $\mathfrak{A}_P$ is simple by the criterion we noted before. Let C be a linear transformation in $\mathfrak{A}_P$ such that $[CA] = 0$ for all $A \in \tilde{\mathfrak{T}}$. Let u_α be one of the base elements and apply the previous considerations to $x = u_\alpha = x_1$, $y = u_\alpha C$. Let $A_1 \in \tilde{\mathfrak{T}}$ satisfy $x_1A_1 = x_1$, $x_iA_1 = 0$, $i \neq 1$. Then $u_\alpha C = x_1C = x_1A_1C = x_1CA_1 = yA_1 = \eta_1 x_1 = \eta_1 u_\alpha$. Since η_1 depends on u_α we write $\eta_1 = \rho_\alpha$, and so we have $u_\alpha C = \rho_\alpha u_\alpha$, $\rho_\alpha \in P$. Next we note that if u_α and u_β are any two base elements, then there exists a $B_{\alpha\beta} \in \tilde{\mathfrak{T}}$ such that $u_\alpha B_{\alpha\beta} = u_\beta$. This is clear because of the density of $\mathfrak{T}$. Then $\rho_\beta u_\beta = u_\beta C = u_\alpha B_{\alpha\beta} C = u_\alpha C B_{\alpha\beta} = \rho_\alpha u_\alpha B_{\alpha\beta} = \rho_\alpha u_\beta$. Hence $\rho_\alpha = \rho_\beta = \rho$ and C is the scalar multiplication by the element ρ. This shows that the centroid is P, so $\mathfrak{A}_P$ is central over P. This completes the proof of the first assertion.

Now suppose $\mathfrak{A}$ is a non-associative algebra over Φ such that $\Gamma \otimes_\Delta \mathfrak{A}$ is simple over Δ, Δ a subfield containing Φ of the centroid Γ of $\mathfrak{A}$. We can consider $\mathfrak{A}$ as algebra over Δ ($\delta a = a\delta$, $a \in \mathfrak{A}$, $\delta \in \Delta$) and we show first that $\mathfrak{A}$ is simple over Δ. Thus let $\mathfrak{B}$ be a Δ-ideal in $\mathfrak{A}$. Then the subset of elements of the form $\sum \gamma_i \otimes b_i$, $\gamma_i \in \Gamma$, $b_i \in \mathfrak{B}$ is a Δ-ideal in $\Gamma \otimes_\Delta \mathfrak{A}$. The properties of tensor products over fields imply that if $\mathfrak{B} \neq 0$, $\mathfrak{A}$, then the ideal indicated is a proper non-zero Δ-ideal in $\Gamma \otimes_\Delta \mathfrak{A}$. Hence $\mathfrak{B} = 0$ or $\mathfrak{B} = \mathfrak{A}$. Also $\mathfrak{A}^2 \neq 0$ since $(\Gamma \otimes_\Delta \mathfrak{A})^2 \neq 0$. This shows that $\mathfrak{A}$ is simple over Δ. The proof of Theorem 2 shows that the multiplication algebra $\mathfrak{T}$ of $\mathfrak{A}$ over Φ is the same set as the multiplication algebra of $\mathfrak{A}$ over Δ. Since we have $x\mathfrak{T} = \mathfrak{A}$ for all $x \neq 0$ it is now clear that $\mathfrak{A}$ is simple over Φ also and Γ is a field. Consider $\mathfrak{A}$ again as algebra over Δ. Then one checks that the mapping

$$\sum \gamma_i \otimes a_i \rightarrow \sum \gamma_i a_i (\equiv \sum a_i \gamma_i) \tag{2}$$

$\gamma_i \in \Gamma$, $a_i \in \mathfrak{A}$ is a Δ-algebra homomorphism of $\Gamma \otimes_\Delta \mathfrak{A}$ onto $\mathfrak{A}$. Suppose $\Gamma \supset \Delta$ and let $\gamma \in \Gamma$, $\notin \Delta$ so that 1, γ are Δ-independent. Then a relation $\gamma \otimes a_1 + 1 \otimes a_2 = 0$, $a_i \in \mathfrak{A}$, implies that the $a_i = 0$. Choose $a_1 = a \neq 0$, $a_2 = -\gamma a$. Then $\gamma \otimes a_1 + 1 \otimes a_2 \neq 0$ and the image of this element under (2) is $\gamma a - \gamma a = 0$. Hence if $\Gamma \neq \Delta$, then (2) has a non-zero kernel as well as a non-zero image. This contradicts the assumption that $\Gamma \otimes_\Delta \mathfrak{A}$ is simple over Δ. Hence $\Gamma = \Delta$ and the proof is complete.

It is immediate that a dense algebra $\mathfrak{T}$ of linear transformations in a finite-dimensional vector space is the complete algebra $\mathfrak{E}(\mathfrak{A})$. Thus if $(x_1, \cdots, x_n)$ is a basis and A is any linear transformation, then $\mathfrak{T}$ contains T such that $x_i T = x_i A$. Hence $A = T \in \mathfrak{T}$. If $\mathfrak{A}$ is non-associative simple with centroid Γ, then we have seen that the multiplication algebra $\mathfrak{T}$ is a dense algebra of linear transformations in $\mathfrak{A}$ considered as a vector space over Γ. Hence if $\mathfrak{A}$ is finite-dimensional over Φ and consequently over Γ, then $\mathfrak{T}$ is the complete algebra of linear transformations in $\mathfrak{A}$ over Γ. We therefore have the following

THEOREM 4. *Let $\mathfrak{A}$ be a finite-dimensional simple algebra with centroid Γ and multiplication algebra $\mathfrak{T}$. Then $\mathfrak{T}$ is the complete set of linear transformations in $\mathfrak{A}$ considered as a vector space over Γ.*

Let the dimensionality $[\mathfrak{A} : \Phi]$ of $\mathfrak{A}$ over Φ be n and let $[\Gamma : \Phi] =$

r, $[\mathfrak{A}:\Gamma] = m$. Then it is well known that $n = rm$ and we now see that $[\mathfrak{T}:\Gamma] = m^2$ and $[\mathfrak{T}:\Phi] = m^2 r$.

Again let $\mathfrak{A}$ be arbitrary and let $a \to a^\theta$ be an isomorphism of $\mathfrak{A}$ onto a second non-associative algebra $\bar{\mathfrak{A}}$. Then θ is a $1:1$ linear mapping of $\mathfrak{A}$ onto $\bar{\mathfrak{A}}$ and $(ab)^\theta = a^\theta b^\theta$, which implies that

$$(3) \qquad a_L\theta = \theta(a^\theta)_L\,, \qquad b_R\theta = \theta(b^\theta)_R\,.$$

Thus

$$(3') \qquad \theta^{-1}a_L\theta = (a^\theta)_L\,, \qquad \theta^{-1}b_R\theta = (b^\theta)_R\,.$$

This implies that the mapping $X \to \theta^{-1}X\theta$ is an isomorphism of the multiplication algebra $\mathfrak{T}(\mathfrak{A})$ onto $\mathfrak{T}(\bar{\mathfrak{A}})$. It is clear also that $\gamma \to \gamma^\theta \equiv \theta^{-1}\gamma\theta$ is an isomorphism of the centroid $\Gamma(\mathfrak{A})$ onto $\Gamma(\bar{\mathfrak{A}})$. If $a \in \mathfrak{A}$ we have $(a\gamma)^\theta = (a^{\theta\theta^{-1}}\gamma)^\theta = a^\theta\gamma^\theta$. In particular, if $\mathfrak{A}$ is simple and $\mathfrak{A}$ is considered as a vector space over Γ, then we have

$$(4) \qquad (\gamma a)^\theta = \gamma^\theta a^\theta\,.$$

Next let D be a derivation in $\mathfrak{A}$. Then D is linear and $(ab)D = (aD)b + a(bD)$, which gives

$$(5) \qquad [a_L, D] = (aD)_L\,, \qquad [b_R, D] = (bD)_R\,.$$

This implies that the inner derivation $X \to [X, D]$ in $\mathfrak{E}(\mathfrak{A})$ maps $\mathfrak{T}(\mathfrak{A})$ into itself. Consequently, this induces a derivation $\gamma \to \gamma^d \equiv [\gamma D]$ in $\Gamma(\mathfrak{A})$. By definition of γ^d we have $(a\gamma)D = (aD)\gamma + a\gamma^d$ so that for simple $\mathfrak{A}$ we have

$$(6) \qquad (\gamma a)D = \gamma^d a + \gamma(aD)\,.$$

We can state the results we have just noted in the following convenient form.

THEOREM 5. *Let $\mathfrak{A}$ and $\bar{\mathfrak{A}}$ be central simple non-associative algebras over a field Γ and let Φ be a subfield of Γ. Then any isomorphism θ of $\mathfrak{A}$ over Φ onto $\bar{\mathfrak{A}}$ over Φ is a semi-linear transformation of $\mathfrak{A}$ over Γ onto $\bar{\mathfrak{A}}$ over Γ. Any derivation D in $\mathfrak{A}$ over Φ satisfies $(\gamma a)D = \gamma^d a + \gamma(aD)$, $\gamma \in \Gamma$, $a \in \mathfrak{A}$ where d is a derivation in Γ as a field over Φ.*

We add one further remark to this discussion. Suppose $(u_1, \cdots, u_n)$ is a basis for $\mathfrak{A}$ over Γ and that $u_iu_j = \sum \gamma_{ijk}u_k$, $\gamma_{ijk} \in \Gamma$, gives the multiplication table. A $1:1$ semi-linear transformation θ of $\mathfrak{A}$ onto $\bar{\mathfrak{A}}$ maps the u_i into the basis u_i^θ, $i = 1, \cdots, n$ for $\bar{\mathfrak{A}}$ over Γ. If θ is a Φ-isomorphism we have $u_i^\theta u_j^\theta = \sum \gamma_{ijk}^\theta u_k^\theta$ where θ acting on γ_{ijk}

is the automorphism in Γ associated with the semi-linear mapping θ of $\mathfrak{A}$. In many cases which are of interest one can choose a basis so that the $\gamma_{ijk} \in \Phi$, that is, $\mathfrak{A} = \mathfrak{A}_{0P}$ where $\mathfrak{A}_0$ is a non-associative algebra over Φ. Then we have $u_i^\theta u_j^\theta = \sum \gamma_{ijk} u_k^\theta$ which implies that $\mathfrak{A}$ and $\bar{\mathfrak{A}}$ are isomorphic as algebras over Γ.

2. *Isomorphism of extension algebras*

Let $\mathfrak{A}$ be a non-associative algebra over Φ and let P be a finite-dimensional Galois extension field of Φ, $G = \{1, s, \cdots, u\}$ the Galois group of P over Φ. In this section we shall obtain a survey of the isomorphism classes of non-associative algebras $\mathfrak{B}$ such that $\mathfrak{B}_P \cong \mathfrak{A}_P$. We recall that if $\mathfrak{A}$ is a vector space over Φ then $\mathfrak{A}$ can be identified with a subset of $\mathfrak{A}_P$. This subset is a Φ-subspace of $\mathfrak{A}_P$ which generates $\mathfrak{A}_P$ as space over P and has the property that any set of elements of $\mathfrak{A}$ which is Φ-independent is necessarily P-independent. Moreover, these properties are characteristic. If $(\rho_1, \rho_2, \cdots, \rho_n)$ is a basis for P over Φ, then any element of $\mathfrak{A}_P$ has a unique representation in the form $\sum \rho_i x_i$, $x_i \in \mathfrak{A}$. If $s \in G$, then s defines a semi-linear transformation U_s in $\mathfrak{A}_P$ by the rule

$$(\textstyle\sum \rho_i x_i) U_s = \sum \rho_i^s x_i \,. \tag{7}$$

It is easy to check that U_s is independent of the choice of the basis in P over Φ, that the automorphism in P associated with U_s is s (that is, $(\xi x)U_s = \xi^s(xU_s)$, $\xi \in P$, $x \in \mathfrak{A}_P$) and

$$U_s U_t = U_{st} \,, \qquad U_1 = 1 \,. \tag{8}$$

Hence the U_s form a group isomorphic to G. If we take $\rho_1 = 1$, then the elements of $\mathfrak{A}$ have the form $\rho_1 x_1$, $x_1 \in \mathfrak{A}$ and it is clear from (7) that these elements are fixed points for every U_s, $s \in G$. It is easy to show directly that $\mathfrak{A}$ is just the set of fixed points relative to the U_s. This will follow also from the following basic lemma:

LEMMA 2. *Let $\tilde{\mathfrak{A}}$ be a vector space over a field P which is a finite-dimensional Galois extension of a field Φ. Suppose that for each s in the Galois group G of P over Φ there is associated a semi-linear transformation U_s in $\tilde{\mathfrak{A}}$ with associated automorphism s in P such that $U_1 = 1$, $U_s U_t = U_{st}$. Let $\mathfrak{A}$ be the set of fixed points relative to the U_s, $s \in G$. Then $\mathfrak{A}$ is a Φ-subspace of $\tilde{\mathfrak{A}}$ such that $\tilde{\mathfrak{A}} = \mathfrak{A}_P$.*

Proof: One verifies directly that $\mathfrak{A}$ is a Φ-subspace of $\tilde{\mathfrak{A}}$. If $x \in \tilde{\mathfrak{A}}$ and $\rho \in P$, then $\sum_{s \in G} \rho^s(xU_s) \in \mathfrak{A}$ since $(\sum_s \rho^s(xU_s))U_t = \sum_s \rho^{st}(xU_sU_t) = \sum_s \rho^{st} xU_{st} = \sum_s \rho^s(xU_s)$. Let $(\rho_1, \rho_2, \cdots, \rho_n)$ be a basis for P over Φ. Then it is well known that the $n \times n$ matrix whose rows are $(\rho_1^s, \rho_2^s, \cdots, \rho_n^s)$, $s \in G$, is non-singular. It follows from this that x is a P-linear combination of the n elements $\sum_s \rho_i^s(xU_s)$ which belong to $\mathfrak{A}$. Thus the P-space $P\mathfrak{A}$ spanned by $\mathfrak{A}$ is $\tilde{\mathfrak{A}}$. Next let $x_1, \cdots, x_r$ be elements of $\mathfrak{A}$ which are Φ-independent. Assume there exist $\xi_i \in P$, not all 0, such that $\sum_1^r \xi_j x_j = 0$. Then we may take r minimal for such relations and we may suppose $\xi_1 = 1$. Evidently we have $r > 1$ and we may assume that $\xi_2 \notin \Phi$. Then $0 = (\sum_j \xi_j x_j)U_s = \sum_j \xi_j^s x_j$ and we can choose s in G so that $\xi_2^s \neq \xi_2$. We can therefore obtain a relation $0 = \sum_1^r \xi_j x_j - \sum_1^r \xi_j^s x_j = \sum_{k=2}^r (\xi_k - \xi_k^s)x_s$ and this is non-trivial and shorter than the relation $\sum_1^r \xi_j x_j = 0$. This contradiction shows that the x_j's are P-independent. Hence $\tilde{\mathfrak{A}} = \mathfrak{A}_P$.

If $x = \sum \xi_j x_j$, $\xi_j \in P$, $x_j \in \mathfrak{A}$, then $xU_s = \sum \xi_j^s x_j$. Thus the U_s are the transformations we constructed before for $\mathfrak{A}_P$. We return to the situation considered before in which we were given $\tilde{\mathfrak{A}} = \mathfrak{A}_P$ and we defined the U_s by (7). We saw that $\mathfrak{A} \subseteq \mathfrak{B}$, the Φ-subspace of fixed elements relative to the U_s. On the other hand, we have $\tilde{\mathfrak{A}} = \mathfrak{A}_P = \mathfrak{B}_P$. Let $\{e_\alpha\}$ be a basis for $\mathfrak{A}$ over Φ and let $b \in \mathfrak{B}$. Then $b = \sum \xi_i x_i$, $\xi_i \in P$, $x_i \in \{e_\alpha\}$. Since the x_i and b are in $\mathfrak{B}$ this relation implies that these elements are Φ-dependent and since the x_i are Φ-independent we have $b = \sum \alpha_i x_i$, $\alpha_i \in \Phi$. Hence $b \in \mathfrak{A}$ and $\mathfrak{B} = \mathfrak{A}$ so that $\mathfrak{A}$ is just the set of fixed elements relative to the U_s. Our results establish the following

THEOREM 6. *Let $\tilde{\mathfrak{A}}$ be a vector space over P, a finite-dimensional Galois extension of the field Φ. Let $\{U_s \mid s \in G\}$ be a* (*finite*) *set of semi-linear transformations in $\tilde{\mathfrak{A}}$ over P such that*: (1) $U_1 = 1$, $U_sU_t = U_{st}$, (2) *the automorphism in P associated with U_s is s. Let $\mathfrak{A}$ be the Φ-space of fixed elements relative to the U_s. Then $\tilde{\mathfrak{A}} = \mathfrak{A}_P$ and the correspondence $\{U_s\} \rightarrow \mathfrak{A}$ is a bijection of the set of finite sets of semi-linear transformations satisfying* (1) *and* (2) *and the set of Φ-subspaces $\mathfrak{A}$ of $\tilde{\mathfrak{A}}$ satisfying $\tilde{\mathfrak{A}} = \mathfrak{A}_P$.*

Now assume that $\tilde{\mathfrak{A}}$ is a non-associative algebra over Φ. Then it is easy to check that the U_s defined by (7) are automorphisms of $\tilde{\mathfrak{A}} = \mathfrak{A}_P$ considered as an algebra over Φ. Conversely, if the group $\{U_s\}$ of semi-linear transformations in $\tilde{\mathfrak{A}}$ is given and every

U_s is an automorphism of $\tilde{\mathfrak{A}}$ over Φ then the set $\mathfrak{A}$ of fixed elements is an algebra over Φ. Hence the correspondence of Theorem 6 induces a bijection of the set of groups $\{U_s\}$, U_s an automorphism of $\tilde{\mathfrak{A}}$ over Φ and the set $\{\mathfrak{A}\}$ of Φ-subalgebras $\mathfrak{A}$ such that $\mathfrak{A}_P = \tilde{\mathfrak{A}}$. Let $\mathfrak{A}$ and $\mathfrak{B}$ be two of the Φ-algebras such that $\mathfrak{A}_P = \mathfrak{B}_P$ and suppose A is an isomorphism of $\mathfrak{A}$ onto $\mathfrak{B}$. Let $\{V_s\}$ be the group of semi-linear transformations associated with $\mathfrak{B}$, $\{U_s\}$ that with $\mathfrak{A}$. The isomorphism A has a unique extension to an automorphism A of $\tilde{\mathfrak{A}}$ over P. The mapping $A^{-1}U_sA$ is a semi-linear transformation in $\tilde{\mathfrak{A}}$ with associated automorphism s in P. Also $A^{-1}U_1A = 1$ and $(A^{-1}U_sA)(A^{-1}U_tA) = A^{-1}U_{st}A$. The set of fixed points relative to the $A^{-1}U_sA$ is the space $\mathfrak{B}$. Hence $A^{-1}U_sA = V_s$, $s \in G$. Conversely, let $\{U_s\}$, $\{V_s\}$ be groups of semi-linear mappings as in Theorem 6 such that the U_s, V_s are automorphisms of $\tilde{\mathfrak{A}}$ as non-associative algebra over Φ and assume there exists a P-automorphism A of $\tilde{\mathfrak{A}}$ such that

$$V_s = A^{-1}U_sA\,, \qquad s \in G\,. \tag{9}$$

Then $\mathfrak{A}A = \mathfrak{B}$ for the associated non-associative algebras $\mathfrak{A}$ and $\mathfrak{B}$. Hence $\mathfrak{A}$ and $\mathfrak{B}$ are isomorphic as algebras over Φ.

We therefore have the following

THEOREM 7. *Let $\tilde{\mathfrak{A}}$ be a non-associative algebra over a finite dimensional Galois extension P of a field Φ. Then the correspondence of Theorem* 6 *between sets of semi-linear transformations $\{U_s\}$ in $\tilde{\mathfrak{A}}$ satisfying* (1) *and* (2) *and Φ-subspaces $\mathfrak{A}$ such that $\mathfrak{A}_P = \tilde{\mathfrak{A}}$ induces a bijection of the set of $\{U_s\}$ such that every U_s is a Φ-algebra isomorphism and the $\mathfrak{A}$ which are Φ-subalgebras of $\tilde{\mathfrak{A}}$. The corresponding Φ-subalgebras are isomorphic if and only if there exists an automorphism A of $\tilde{\mathfrak{A}}$ over P such that* (9) *holds for the associated groups.*

If $\mathfrak{B}$ is a non-associative algebra over Φ such that $\mathfrak{B}_P \cong \mathfrak{A}_P$, then $\mathfrak{B}$ can be identified with its image in $\tilde{\mathfrak{A}} = \mathfrak{A}_P$. In this way we see that Theorem 7 gives a survey of the isomorphism classes of algebras $\mathfrak{B}$ such that $\mathfrak{B}_P \cong \tilde{\mathfrak{A}}_P$ via certain similarity classes of groups of automorphisms of $\tilde{\mathfrak{A}}$ over Φ. This is the type of result we wished to establish.

In the sequel we shall be concerned with finite-dimensional non-associative algebras over a field Φ of characteristic 0 and we shall be concerned with the question of equality $\mathfrak{A}_\Omega = \mathfrak{B}_\Omega$ for two such algebras, Ω being the algebraic closure of the base field Φ. Let $(e_1, \cdots, e_m)$, $(f_1, \cdots, f_m)$ be bases for $\mathfrak{A}$ over Φ and $\mathfrak{B}$ over Φ respectively. Then the equality $\mathfrak{A}_\Omega = \mathfrak{B}_\Omega$ implies that $f_i = \sum \rho_{ij} e_j$, $\rho_{ij} \in \Omega$, $i, j = 1, \cdots, m$. Let Σ be the subfield over Φ generated by the ρ_{ij}. Since every ρ_{ij} is algebraic over Φ and there are only a finite number of these, Σ is a finite-dimensional extension of Φ. Since Ω is algebraic and of characteristic zero, Σ is contained in a subfield P of Ω which is a finite-dimensional Galois extension of Φ. Since the $\rho_{ij} \in P$ it is clear that $\sum_1^m Pf_i \subseteqq \sum_1^m Pe_i$ and the converse inequality holds since $e_j = \sum \rho'_{ij} f_j$, $(\rho'_{ij}) = (\rho_{ij})^{-1}$. Also it is clear that the f_i and the e_i both form linearly independent sets over P. Hence $\sum Pe_i = \mathfrak{A}_P$ and $\sum Pf_i = \mathfrak{B}_P$, so $\mathfrak{A}_\Omega = \mathfrak{B}_\Omega$ for Ω the algebraic closure of the base field Φ implies $\mathfrak{A}_P = \mathfrak{B}_P$ for P a suitable finite-dimensional Galois extension of Φ.

3. *Simple Lie algebras of types A–D*

We now take up the main problem of this chapter: the classification of the finite-dimensional simple Lie algebras over any field Φ of characteristic 0. If $\mathfrak{L}$ is such an algebra and Γ is its centroid, then Γ is a finite-dimensional field extension of Φ and $\mathfrak{L}$ is finite-dimensional central simple over Γ. Conversely, if $\mathfrak{L}$ is finite-dimensional central simple over any finite-dimensional extension field Γ of Φ, then $\mathfrak{L}$ is finite-dimensional simple over Φ. If $\mathfrak{L}_1$ and $\mathfrak{L}_2$ are two isomorphic simple Lie algebras over Φ, then the respective centroids Γ_1 and Γ_2 are isomorphic so both algebras may be considered as central simple over the same field $\Gamma \cong \Gamma_1 \cong \Gamma_2$. Moreover, Theorem 5 shows that we have a semi-linear transformation θ of $\mathfrak{L}_1$ over Γ onto $\mathfrak{L}_2$ over Γ such that θ is a Φ-isomorphism of $\mathfrak{L}_1$ onto $\mathfrak{L}_2$ as non-associative algebras over Φ. These results reduce the classification problem for simple Lie algebras over Φ to the following problems: (1) the classification of the finite-dimensional field extensions Γ of Φ, (2) the classification of the finite-dimensional central simple Lie algebras over the Γ in (1), and (3) determination of conditions for the existence of a semi-linear transformation θ of two central simple Lie algebras over Γ such that θ is a Φ-algebra isomorphism. To make this concrete we consider the important special case: Φ the field of real numbers. Here any algebraic extension field Γ of Φ is either Φ itself or is isomorphic to the field of complex numbers. If Γ is the field of complex numbers, then Γ is algebraically closed and we know the classification of the finite-dimensional simple Lie algebras over Γ. We recall that the algebras in this classification (A_l, B_l etc.) all have the form $\mathfrak{L}_{0\Gamma}$, $\mathfrak{L}_0$ an algebra over Φ (Theorem 4.2). It follows from the remark following Theorem 5 that the algebras in our list (which are not Γ-isomorphic) are not isomorphic over Φ. Thus, to complete the classification over the reals it remains to classify the central simple algebras over this field.

Now suppose $\mathfrak{L}$ is a finite-dimensional central simple Lie algebra over Γ which is any field of characteristic 0. If Ω is the algebraic closure of Γ, then $\mathfrak{L}_\Omega$ is simple. Conversely, if $\mathfrak{L}$ is finite-dimensional over Γ and $\mathfrak{L}_\Omega$ is simple, then $\mathfrak{L}$ is evidently simple so its centroid Γ' is a finite-dimensional field extension of Γ. This can be considered as a subfield of Ω. Then $\mathfrak{L}_{\Gamma'}$ is simple since $(\mathfrak{L}_{\Gamma'})_\Omega = \mathfrak{L}_\Omega$ is simple. Hence $\mathfrak{L}$ is central by Theorem 3. Thus $\mathfrak{L}$ is central simple over Γ if and only if $\mathfrak{L}_\Omega$ is simple over Ω.

Since Ω is algebraically closed of characteristic 0 we know the possibilities for $\mathfrak{L}_\Omega$. They are the Lie algebras A_l, $l \geqq 1$, B_l, $l \geqq 2$, C_l, $l \geqq 3$, D_l, $l \geqq 4$, G_2, F_4, E_6, E_7, E_8 in the Killing Cartan list. If $\mathfrak{L}_\Omega$ is the Lie algebra X in this list then we shall say that $\mathfrak{L}$ is of *type* X. Usually the subscript l will be dropped and we shall speak simply of $\mathfrak{L}$ of *type* A, *type* B, etc. For each type X we shall choose a fixed Lie algebra $\mathfrak{L}_0$ of this type. For example, we can take $\mathfrak{L}_0$ to be the split Lie algebra over Γ of type X. Then our problem is to classify the Lie algebras $\mathfrak{L}$ such that $\mathfrak{L}_\Omega = \mathfrak{L}_{0\Omega}$. For a particular $\mathfrak{L}$ there exists a finite-dimensional Galois extension P such that $\mathfrak{L}_P = \mathfrak{L}_{0P}$ and we have seen that to determine the $\mathfrak{L}$ which satisfy this condition for a particular P, then we have to look at the automorphisms of $\mathfrak{L}_{0P}$ over Γ. We shall study the cases of $\mathfrak{L}_0$ of types, A_l, B_l, C_l, D_l, $l > 4$. In this section we shall give some constructions of Lie algebras of types A to D. In the next section we give the conditions for isomorphism of these Lie algebras and in § 5 we shall prove that every Lie algebra of type A_l, B_l, C_l, D_l, $l > 4$ can be obtained in the manner given here.

The starting point of our constructions is the fundamental Wedderburn structure theorem on simple associative algebras: Any finite-dimensional simple associative algebra $\mathfrak{A}$ is isomorphic to an algebra $\mathfrak{E}$ of all the linear transformations of a finite-dimensional vector space $\mathfrak{M}$ over a finite-dimensional division algebra Δ. An equivalent formulation is that $\mathfrak{A} \cong \Delta_n$ the algebra of $n \times n$ matrices over a finite-dimensional division algebra Δ. (For a proof see: Jacobson, *Structure of Rings*, p. 39, or Artin, Nesbitt, and Thrall, *Rings with Minimum Condition*, p. 32.) If the base field Γ is algebraically closed then the only finite-dimensional division algebra over Γ is Γ itself, so in this case $\mathfrak{A} \cong \Gamma_n$ for some n. If the base field is the field of real numbers, then it is known that there are just three possibilities for Δ: $\Delta = \Gamma$, $\Delta = \Gamma(i)$ the field of complex numbers or Δ the division algebra of quaternions. (Theorem of Frobenius, cf. Dickson [1], p. 62, or Pontrjagin [1], p. 175.) The center $\mathfrak{C}$ of a finite-dimensional simple associative algebra is a field and the centroid consists of the mappings $c_L = c_R$, $c \in \mathfrak{C}$. This can be identified with the center. If $\mathfrak{A}$ is central ($\mathfrak{C} = \Gamma$) and Ω is the algebraic closure of Γ, then $\mathfrak{A}_\Omega$ is finite-dimensional simple over Ω. Hence $\mathfrak{A}_\Omega \cong \Omega_n$ for some n. Since $[\mathfrak{A} : \Gamma] = [\mathfrak{A}_\Omega : \Omega] = [\Omega_n : \Omega] = n^2$ this shows that *the dimensionality of any finite-dimensional central simple associative algebra is a square.*

Let $\mathfrak{A}$ be a finite-dimensional central simple associative algebra over Γ. Consider the derived algebra $\mathfrak{L} = \mathfrak{A}'_L$ of the Lie algebra $\mathfrak{A}_L$. If Ω is the algebraic closure of Γ, then $\mathfrak{L}_\Omega = (\mathfrak{A}'_L)_\Omega = (\mathfrak{A}_\Omega)'_L \cong \Omega'_{nL}$, since $\mathfrak{A}_\Omega \cong \Omega_n$. On the other hand, we know that if $n = l + 1$, $l \geqq 1$, then Ω'_{nL} is the Lie algebra of $(l + 1) \times (l + 1)$ matrices of trace 0 and this is the simple Lie algebra A_l over Ω. Since $\mathfrak{L}_\Omega \cong \Omega'_{nL}$ it follows that $\mathfrak{L} = \mathfrak{A}'_L$ is a central simple Lie algebra of type A_l.

Next let $\mathfrak{A}$ be a finite-dimensional simple associative algebra with an *involution* J. By definition, J is an anti-automorphism of period two in $\mathfrak{A}$. Hence $-J$ is an automorphism of $\mathfrak{A}_L$. The set $\mathfrak{S}(\mathfrak{A}, J)$ of J-skew elements ($a^J = -a$) is the subset of fixed elements relative to the automorphism $-J$. Hence this is a subalgebra of $\mathfrak{A}_L$. The anti-automorphism J induces an automorphism in the center $\mathfrak{C}$ of $\mathfrak{A}$ which is either the identity or is of period two. In the first case J is of *first kind* and in the second J is an involution of *second kind.*

We assume first that J is of second kind. More precisely, we assume that $\mathfrak{C} = P = \Gamma(q)$ a quadratic extension of the base field Γ and that $q^J = -q$. Let $\mathfrak{H}(\mathfrak{A}, J)$ be the space of J-symmetric elements ($a^J = a$). Any $a \in \mathfrak{A}$ has the form $a = b + c$, $b = \frac{1}{2}(a + a^J) \in \mathfrak{H}$ and $c = \frac{1}{2}(a - a^J) \in \mathfrak{S}$. If $b \in \mathfrak{H}$, $qb \in \mathfrak{S}$ and if $c \in \mathfrak{S}$ then $qc \in \mathfrak{H}$. Hence the mapping $x \rightarrow qx$ is a $1:1$ linear mapping of $\mathfrak{S}$ onto $\mathfrak{H}$ so we have the dimensionality relation: $[\mathfrak{S} : \Gamma] = [\mathfrak{H} : \Gamma]$. Since $\mathfrak{S} \cap \mathfrak{H} = 0$ and $\mathfrak{A} = \mathfrak{S} + \mathfrak{H}$, $\mathfrak{A} = \mathfrak{S} \oplus \mathfrak{H}$ and $[\mathfrak{A} : \Gamma] = 2[\mathfrak{S} : \Gamma]$. Thus $[\mathfrak{A} : P] =$

$\frac{1}{2}[\mathfrak{A} : \Gamma] = [\mathfrak{S} : \Gamma]$. We recall also that $[\mathfrak{A} : P]$ is a square n^2, so $[\mathfrak{S} : \Gamma] = n^2$. Let $(a_1, \cdots, a_{n^2})$ be a basis for $\mathfrak{S}$ over Γ. Then every element of $\mathfrak{A}$ has the form $\sum_1^{n^2} \alpha_i a_i + \sum_1^{n^2} \beta_i (q a_i) = \sum \rho_i a_i$, $\rho_i = \alpha_i + \beta_i q$, $\alpha_i, \beta_i \in \Gamma$. It follows that the a_i form a basis for $\mathfrak{A}$ (or $\mathfrak{A}_L$) over P. This implies that $\mathfrak{S}_P = \mathfrak{A}_L$ over P. Let Ω be the algebraic closure of Γ which we may assume to be an extension of the field P. Then $\mathfrak{S}_\Omega = (\mathfrak{S}_P)_\Omega = (\mathfrak{A}_L \text{ over } P)_\Omega \cong \Omega_{nL}$ since $(\mathfrak{A} \text{ over } P)_\Omega \cong \Omega_n$. Let $\mathfrak{L}$ be the derived algebra $\mathfrak{S}' = \mathfrak{S}(\mathfrak{A}, J)'$. Then $\mathfrak{L}_\Omega \cong \Omega'_{nL}$ so $[\mathfrak{L}_\Omega : \Omega] = n^2 - 1$ and $[\mathfrak{S}(\mathfrak{A}, J)' : \Gamma] = n^2 - 1$. If $n = l + 1$, $l \geqq 1$, $\mathfrak{L}_\Omega$ is the simple Lie algebra A_l. Hence $\mathfrak{L}$ is central simple of type A_l.

We have now given two constructions of Lie algebras of type A. We summarize our results in the following

THEOREM 8. *Let $\mathfrak{A}$ be a finite-dimensional central simple associative algebra over Γ, $\mathfrak{A} \neq \Gamma$. Then the derived algebra $\mathfrak{A}'_L$ is a central simple Lie algebra of type A_l, $l \geqq 1$. Let $\mathfrak{A}$ be a finite-dimensional simple associative algebra with center P a quadratic extension of the base field Γ and suppose $\mathfrak{A}$ possesses an involution J of second kind. Suppose also that $\mathfrak{A} \neq P$. Let $\mathfrak{S}(\mathfrak{A}, J)$ be the Lie algebra of skew elements of $\mathfrak{A}$. Then the derived algebra $\mathfrak{S}(\mathfrak{A}, J)'$ is a central simple Lie algebra of type A_l, $l \geqq 1$.*

Before proceeding to the discussion of the Lie algebras $\mathfrak{S}(\mathfrak{A}, J)$, J of first kind we quote some well-known results on involutions in algebras of linear transformations (cf. Jacobson [3], pp. 80–83). Let $\mathfrak{M}$ be a finite-dimensional vector space over a division algebra Δ and let $\mathfrak{E}$ be the algebra of linear transformations in $\mathfrak{M}$ over Δ. Then it is known that $\mathfrak{E}$ has an involution $A \to A^J$ if and only if Δ has an anti-automorphism $d \to \bar{d}$ of period one or two. If the period is one, then $\bar{d} \equiv d$ and $\overline{d_1 d_2} = \bar{d}_2 \bar{d}_1$ implies that Δ is commutative. Conversely if Δ is commutative, then one can take $\bar{d} \equiv d$. If $d \to \bar{d}$ is given, then one can define a non-degenerate hermitian or skew hermitian form (x, y) in $\mathfrak{M}$ over Δ relative to $d \to \bar{d}$. Such a form is defined by the conditions

$$(10) \qquad \begin{aligned} (x_1 + x_2, y) &= (x_1, y) + (x_2, y)\,, \qquad (x, y_1 + y_2) = (x, y_1) + (x, y_2) \\ (dx, y) &= d(x, y)\,, \qquad (x, dy) = (x, y)\bar{d} \end{aligned}$$

for $x, x_1, x_2, y, y_1, y_2 \in \mathfrak{M}$, $d \in \Delta$,

$$(11) \qquad (x, y) = \overline{(y, x)} \quad \text{or} \quad (x, y) = -\overline{(y, x)}\,,$$

according as the form is hermitian or skew-hermitian, and non-degeneracy means that $(x, z) = 0$ for all x implies $z = 0$. If $\bar{d} \equiv d$, then we obtain a symmetric or skew bilinear form. If $(u_1, u_2, \cdots, u_n)$ is a basis for $\mathfrak{M}$ over Δ, then $(x, y) = \sum \xi_i \bar{\eta}_i$ for $x = \sum \xi_i u_i$, $y = \sum \eta_i u_i$ defines a non-degenerate hermitian form and $(x, y) = \sum \xi_i \rho \bar{\eta}_i$ is a non-degenerate skew hermitian form if $\bar{\rho} = -\rho \neq 0$. If $\bar{d} \equiv d$, then non-degenerate skew bilinear forms exist for $\mathfrak{M}$ if and only if $\mathfrak{M}$ is of even dimensionality over Γ. Let (x, y) be any non-degenerate hermitian or skew hermitian form associated with $d \to \bar{d}$ in Δ. If $A \in \mathfrak{E}$ we let A^J denote the adjoint of A relative to (x, y), that is, A^J is the linear transformation in $\mathfrak{M}$ such that $(xA, y) = (x, yA^J)$ for $x, y \in \mathfrak{M}$. Then it is easily seen that $A \to A^J$ is an involution in $\mathfrak{E}$. Moreover, one has the fundamental theorem that every involution J of $\mathfrak{E}$ is obtained in this way.

In particular, suppose $\Delta = \Gamma$ is an algebraically closed field. Since $\Delta = \Gamma$, the only anti-automorphism of Δ as algebra over Γ is the identity mapping. Hence (x, y) is either a non-degenerate symmetric bilinear form or is a non-degenerate skew bilinear form. The latter can occur only in the even-dimensional case. The Lie algebra $\mathfrak{S}(\mathfrak{E}, J)$ determined by the form is the Lie algebra B_l if the form is symmetric and $\dim \mathfrak{M} = 2l + 1$. The Lie algebra $\mathfrak{S}(\mathfrak{E}, J)$ is C_l if the form is skew and $\dim \mathfrak{M} = 2l$ and it is D_l if the form is symmetric and $\dim \mathfrak{M} = 2l$.

Now let $\mathfrak{A}$ be a finite-dimensional central simple associative algebra over Γ which has an involution J of first kind. If Ω is the algebraic closure of Γ, then $\mathfrak{A}_\Omega \cong \Omega_n$. The extension J of J to a linear transformation in $\mathfrak{A}_\Omega$ is an involution in $\mathfrak{A}_\Omega$ and $\mathfrak{S}(\mathfrak{A}_\Omega, J) = \mathfrak{S}(\mathfrak{A}, J)_\Omega$. Since $\mathfrak{A}_\Omega \cong \Omega_n$ and J is an involution in Ω_n the result above shows that $\mathfrak{S}(\mathfrak{A}_\Omega, J)$ is one of the Lie algebras B_l, C_l or D_l. We assume now that $l \geqq 2$, $l \geqq 3$ or $l \geqq 4$ according as $\mathfrak{S}(\mathfrak{A}_\Omega, J)$ is B_l, C_l or D_l. Then the algebras $\mathfrak{S}(\mathfrak{A}_\Omega, J)$ are simple and we see that $\mathfrak{S}(\mathfrak{A}, J)$ is central simple over Φ of types B_l, C_l or D_l. We recall that n is the dimensionality of the space $\mathfrak{M}$ considered before and that $n^2 = [\Omega_n : \Omega] = [\mathfrak{A} : \Gamma]$. For B_l we have $n = 2l + 1$ and $[\mathfrak{S}(\mathfrak{A}, J) : \Gamma] = [\mathfrak{S}(\mathfrak{A}_\Omega, J) : \Omega] = l(2l + 1)$. For C_l, $n = 2l$ and $[\mathfrak{S} : \Gamma] = l(2l + 1)$ and for D_l, $n = 2l$ and $[\mathfrak{S} : \Gamma] = l(2l - 1)$. We can now state the following

THEOREM 9. *Let $\mathfrak{A}$ be a finite-dimensional central simple associative algebra over Γ of dimensionality n^2 and suppose $\mathfrak{A}$ has an involution*

J of first kind. Let $\mathfrak{S}(\mathfrak{A}, J)$ be the Lie algebra of J-skew elements of $\mathfrak{A}$. If $n = 2l + 1$, then $[\mathfrak{S}(\mathfrak{A}, J) : \Gamma] = l(2l + 1)$ and $\mathfrak{S}$ is central simple of type B_l for $l \geqq 2$. If $n = 2l$, then $[\mathfrak{S} : \Gamma] = l(2l + 1)$ or $l(2l - 1)$. In the former case assume $l \geqq 3$. Then $\mathfrak{S}$ is central simple of type C_l. If $n = 2l$ and $[\mathfrak{S} : \Gamma] = l(2l - 1)$ then we assume $l \geqq 4$. Then $\mathfrak{S}$ is central simple of type D_l.

4. Conditions for isomorphism

Let $\mathfrak{L}_1$ and $\mathfrak{L}_2$ be finite-dimensional central simple Lie algebras over the field Γ of characteristic 0. Suppose Φ is a subfield of Γ and that $\mathfrak{L}_1$ and $\mathfrak{L}_2$ are isomorphic as algebras over Φ: $\mathfrak{L}_1 \cong_\Phi \mathfrak{L}_2$. Then we know that a Φ-isomorphism θ of $\mathfrak{L}_1$ onto $\mathfrak{L}_2$ is a semi-linear transformation of $\mathfrak{L}_1$ over Γ onto $\mathfrak{L}_2$ over Γ. We denote the associated automorphism in Γ by θ also so that we have $(\gamma a)^\theta = \gamma^\theta a^\theta$ if $\gamma \in \Gamma$, $a \in \mathfrak{L}_1$. Let $(a_1, \cdots, a_m)$ be a basis for $\mathfrak{L}_1$ over Γ with the multiplication table $[a_i a_j] = \sum \gamma_{ijk} a_k$. Then $(a_1^\theta, \cdots, a_m^\theta)$ is a basis for $\mathfrak{L}_2$ over Γ and $[a_i^\theta a_j^\theta] = \sum \gamma_{ijk}^\theta a_k^\theta$. Let Ω be the algebraic closure of Γ. Then it is a well-known result of Galois theory that the automorphism θ in Γ has an extension to an automorphism θ in Ω. The a_i form a basis for $\mathfrak{L}_{1\Omega}$ over Ω and the a_i^θ form a basis for $\mathfrak{L}_{2\Omega}$ over Ω. It follows that the mapping $\sum \omega_i a_i \to \sum \omega_i^\theta a_i^\theta$, $\omega_i \in \Omega$, is a Φ-isomorphism of $\mathfrak{L}_{1\Omega}$ onto $\mathfrak{L}_{2\Omega}$. Thus $\mathfrak{L}_{1\Omega} \cong_\Phi \mathfrak{L}_{2\Omega}$. On the other hand, $\mathfrak{L}_{1\Omega}$ is simple over the algebraically closed field Ω. Hence it has a basis over Ω whose multiplication coefficients are in the prime field and so are in Φ. This implies (remark following Theorem 5) that $\mathfrak{L}_{1\Omega} \cong_\Omega \mathfrak{L}_{2\Omega}$. We have therefore proved the following

LEMMA 3. *Let $\mathfrak{L}_1$ and $\mathfrak{L}_2$ be finite-dimensional central simple Lie algebras over a field Γ of characteristic zero. Let Φ be a subfield of Γ and Ω the algebraic closure of Γ. Then $\mathfrak{L}_1 \cong_\Phi \mathfrak{L}_2$ implies $\mathfrak{L}_{1\Omega} \cong_\Omega \mathfrak{L}_{2\Omega}$.*

This result implies that the only Φ-isomorphisms which can exist for the Lie algebras of Theorems 8 and 9 are those between the algebras defined in Theorem 8 and between algebras of the same type (B_l, C_l, D_l) of Theorem 9. We shall call the Lie algebras of the form $\mathfrak{A}_L'$ of Theorem 8 Lie algebras of *type* A_{I}, those of the form $\mathfrak{S}(\mathfrak{A}, J)'$, J of second kind, Lie algebras of *type* A_{II}. For the latter class we assume $l > 1$ if the type is A_l. This amounts to assuming that $[\mathfrak{A} : P] = n^2 > 4$. We shall suppose also from now on that $l > 4$ for the algebras of type D_l. We consider next the enveloping associative algebras of the Lie algebras of Theorems 8, 9.

LEMMA 4. *Let* Φ, Γ, Ω *be as in Lemma* 3. (1) *Let* $\mathfrak{A}$ *be a finite-dimensional central simple associative algebra over* Γ *of dimensionality* $n^2 > 1$. *Then the enveloping associative algebra of* $\mathfrak{L} = \mathfrak{A}'_L$ *over* Φ *is* $\mathfrak{A}$. (2) *Let* $\mathfrak{A}$ *be a finite-dimensional simple associative algebra with center* P *a quadratic extension of* Γ *and with an involution* J *of second kind. Assume* $[\mathfrak{A} : P] = n^2 > 4$. *Then the enveloping associative algebra of* $\mathfrak{L} = \mathfrak{S}(\mathfrak{A}, J)'$ *over* Φ *is* $\mathfrak{A}$. (3) *Let* $\mathfrak{A}$ *be finite-dimensional central simple over* Γ *with an involution* J *of first kind and* $[\mathfrak{A} : \Gamma] = n^2 > 1$. *Then the enveloping associative algebra of* $\mathfrak{L} = \mathfrak{S}(\mathfrak{A}, J)$ *over* Φ *is* $\mathfrak{A}$.

Proof: We note first that in all cases the enveloping associative algebra of $\mathfrak{L}$ over Φ is the same as that of $\mathfrak{L}$ over its centroid Γ. Thus since $\mathfrak{L}$ is a vector space over Γ, $\gamma l \in \mathfrak{L}$ if $\gamma \in \Gamma$ and $l \in \mathfrak{L}$. Hence the two enveloping associative algebras indicated coincide with the set of sums of products $l_1 l_2 \cdots l_k$, $l_i \in \mathfrak{L}$. This remark shows that we may as well assume the base field $\Phi = \Gamma$ and we shall now do this. In the cases 1 and 3 we introduce the algebra $\mathfrak{A}_\Omega = \Omega_n$, $n > 1$, and we consider $\mathfrak{L}_\Omega$ which is a subalgebra of Ω_{nL}. We know that $\mathfrak{L}_\Omega$ is the Lie algebra of matrices A_l, B_l, C_l or D_l as defined in § 4.6. In all cases an elementary direct calculation with the bases given in § 4.6 shows that the Ω-subalgebra generated by $\mathfrak{L}_\Omega$ is Ω_n, that is, $(\mathfrak{L}_\Omega)^* = \Omega_n$ where the * denotes the enveloping associative algebra. If $\mathfrak{L}^*$ denotes the enveloping associative algebra over Φ of $\mathfrak{L}$ (in $\mathfrak{A}$), then it clear that the Ω-subspace of Ω_n spanned by $\mathfrak{L}^*$ is $(\mathfrak{L}_\Omega)^* = \Omega_n$. Hence $\mathfrak{L}^*$ contains a basis for Ω_n over Ω. Since the elements of this basis are contained in $\mathfrak{A}$ it follows that they constitute a basis for $\mathfrak{A}$. Thus we have $\mathfrak{L}^* = \mathfrak{A}$. The argument just used cannot be applied readily to the case 2 (Lie algebras of type A_{II}) since in this case $\mathfrak{A}_\Omega = \Omega_n \oplus \Omega_n$. We therefore proceed in a somewhat different manner. Let $P = \Phi(q)$ where $q^J = -q$, as before. Then $q \in \mathfrak{S}(\mathfrak{A}, J)$ and $\mathfrak{A} = \mathfrak{S}(\mathfrak{A}, J) + q\mathfrak{S}(\mathfrak{A}, J) = \mathfrak{S}(\mathfrak{A}, J)^*$. Hence it suffices to show that $\mathfrak{L}^* \supseteq \mathfrak{S}(\mathfrak{A}, J)$. Since $\Phi = \Gamma$ and $\mathfrak{L} = \mathfrak{S}'$ we have $[\mathfrak{S} : \Phi] = n^2$ and $[\mathfrak{L} : \Phi] = n^2 - 1$. Hence, it suffices to show that $\mathfrak{L}^*$ contains an element $a \in \mathfrak{S}$, $\notin \mathfrak{S}'$. Assume the contrary that $\mathfrak{L}^* \cap \mathfrak{S} = \mathfrak{S}'$. Let b_i, $i = 1, 2, 3$, be skew elements. Then it is immediate that $\{b_1 b_2 b_3\} \equiv b_1 b_2 b_3 + b_3 b_2 b_1$ is skew. If the $b_i \in \mathfrak{S}'$, then $\{b_1 b_2 b_3\} \in \mathfrak{S}' = \mathfrak{L}^* \cap \mathfrak{S}$.

We now consider the algebra $(\mathfrak{A}$ over $P)_\Omega = \Omega_n$. We have seen that this algebra has a basis consisting of elements of $\mathfrak{S}$ and that Ω'_{nL} has a basis of elements in $\mathfrak{L} = \mathfrak{S}'$. The multilinear character of $\{b_1b_2b_3\}$ now implies that $\{b_1b_2b_3\} \in \Omega'_{nL}$ for all $b_i \in \Omega'_{nL}$. If we take $b_i = b$ this implies that $b^3 \in \Omega'_{nL}$ if $b \in \Omega'_{nL}$. Thus we must have $\operatorname{tr} b^3 = 0$ for all b satisfying $\operatorname{tr} b = 0$. This is impossible since $n > 2$. For example, we can take $b = 2e_{11} - e_{22} - e_{33}$ so that $\operatorname{tr} b = 0$, $\operatorname{tr} b^3 = 6 \neq 0$. This contradiction shows that $\mathfrak{L}^* \supseteq \mathfrak{S}$ and $\mathfrak{L}^* = \mathfrak{A}$.

We are now ready to prove our main isomorphism theorems. In all of these Φ, Γ and Ω are as in the foregoing discussion.

THEOREM 10. *Let $\mathfrak{A}$ and $\mathfrak{B}$ be finite-dimensional central simple associative algebras over Γ and let θ be an isomorphism of $\mathfrak{A}'_L$ over Φ onto $\mathfrak{B}'_L$ over Φ. Assume $[\mathfrak{A}:\Gamma] = n^2 > 1$ and $[\mathfrak{B}:\Gamma] > 1$. Then if $n = 2$, θ can be extended in one and only one way to an isomorphism of $\mathfrak{A}$ over Φ onto $\mathfrak{B}$ over Φ and if $n > 2$ then θ can be extended in one and only one way to either an isomorphism or the negative of an anti-isomorphism of $\mathfrak{A}$ over Φ onto $\mathfrak{B}$ over Φ.*

Proof: We have $\mathfrak{A}_\Omega \cong \Omega_n$ and $\mathfrak{B}_\Omega \cong \Omega_m$. By Lemma 3, $\mathfrak{A}'_{L\Omega} \cong_\Omega \mathfrak{B}'_{L\Omega}$; hence $\Omega'_{nL} \cong \Omega'_{mL}$ and $m = n$. Thus we may assume that $\mathfrak{A}_\Omega = \Omega_n = \mathfrak{B}_\Omega$ so that Ω_n has a basis $(a_1, \cdots, a_{n^2})$ such that the a_i form a basis for $\mathfrak{A}$ over Γ. We may assume also that $(a_1, \cdots, a_{n^2-1})$ is a basis for $\mathfrak{A}'_L$ over Γ. Now θ is a semi-linear mapping of $\mathfrak{A}'_L$ over Γ onto $\mathfrak{B}'_L$ over Γ whose automorphism in Γ we denote by θ. Then $(a_1^\theta, \cdots, a_{n^2-1}^\theta)$ is a basis for $\mathfrak{B}'_L$ over Γ. If θ is extended to the automorphism θ in Ω, then $\sum_1^{n^2-1} \omega_i a_i \rightarrow \sum \omega_i^\theta a_i^\theta$ is an automorphism θ' in Ω'_{nL} over Φ which is a semi-linear transformation with associated automorphism θ in Ω. Let (e_{ij}), $i, j = 1, \cdots, n$, be the usual matrix units for Ω_n. Then the mapping $\sum \omega_{ij} e_{ij} \rightarrow \sum \omega_{ij}^\theta e_{ij}$, $\omega_{ij} \in \Omega$, is an automorphism θ'' of the associative algebra Ω_n over Φ whose associated automorphism in Ω is θ. θ'' induces an automorphism θ'' in the Lie algebra Ω'_{nL} and since θ' and θ'' have the same associated automorphism in Ω, $\eta = (\theta'')^{-1}\theta'$ is an automorphism in Ω'_{nL} over Ω. By Theorem 9.5 this has the form $X \rightarrow M^{-1}XM$ if $n = 2$ and it either has this form or the form $X \rightarrow -M^{-1}X'M$ if $n > 2$. Since the mapping $X \rightarrow M^{-1}XM$ is an automorphism of Ω_n and $X \rightarrow M^{-1}X'M$ is an anti-automorphism of Ω_n, η can be realized by an automorphism of Ω_n over Ω if $n = 2$ and either by an automorphism or by the negative of an anti-automorphism of Ω_n over Ω for $n > 2$. Then $\theta' = \theta''\eta$ can be extended to an automorphism

ζ of Ω_n over Φ if $n = 2$, or to an automorphism ζ or the negative of an anti-automorphism ζ of Ω_n over Φ if $n > 2$. Since θ' is the mapping $\sum \omega_i a_i \to \sum \omega_i^\theta a_i^\theta$ it is clear that θ' coincides with the given θ on $\mathfrak{A}'_L$. Since the enveloping associative algebra over Φ of $\mathfrak{A}'_L$ is $\mathfrak{A}$ and that of $\mathfrak{B}'_L$ is $\mathfrak{B}$ it follows that ζ maps $\mathfrak{A}$ onto $\mathfrak{B}$ and consequently θ can be extended to an isomorphism or the negative of an anti-isomorphism of $\mathfrak{A}$ over Φ and $\mathfrak{B}$ over Φ. Since $\mathfrak{A}'_L$ generates $\mathfrak{A}$ over Φ it is clear that the extension is unique.

This result shows that if $\mathfrak{A}'_L$ and $\mathfrak{B}'_L$ are isomorphic as algebras over Φ, then $\mathfrak{A}$ and $\mathfrak{B}$ are either isomorphic or anti-isomorphic as Φ-algebras. The converse is clear since any isomorphism of $\mathfrak{A}$ onto $\mathfrak{B}$ induces an isomorphism of $\mathfrak{A}'_L$ into $\mathfrak{B}'_L$ and if θ is an anti-isomorphism of $\mathfrak{A}$ onto $\mathfrak{B}$ then $-\theta$ induces an isomorphism of $\mathfrak{A}'_L$ onto $\mathfrak{B}'_L$. Our result also gives a description of the group of automorphisms of $\mathfrak{A}'_L$. If $n = 2$ or if $\mathfrak{A}$ has no anti-automorphism, then the group of automorphisms of $\mathfrak{A}'_L$ over Φ can be identified with the group of automorphisms of $\mathfrak{A}$ over Φ. If $n > 2$ and $\mathfrak{A}$ has an anti-automorphism J, then it is easy to prove by a field extension argument that the automorphism $a \to -a^J$ in $\mathfrak{A}'_L$ is not of the form $a \to a^\theta$, θ an automorphism of $\mathfrak{A}$. It follows that the group of automorphisms of $\mathfrak{A}$ over Φ is isomorphic to a subgroup of index two of the group of automorphisms of $\mathfrak{A}'_L$ over Φ.

If we take $\Phi = \Gamma$, then it is a known result of the associative theory that every automorphism of $\mathfrak{A}$ over Φ is inner. This could also be deduced from the form of the automorphisms of Ω'_{nL} by a field extension argument. Then it follows that the automorphisms of $\mathfrak{A}'_L$ over Φ are of the form $x \to m^{-1}xm$ or of the form $x \to -m^{-1}x^Jm$ where J is a fixed anti-automorphism in $\mathfrak{A}$.

THEOREM 11. *Let $\mathfrak{A}_i$, $i = 1, 2$, be a finite-dimensional simple associative algebra over Γ with center a quadratic field $P_i = \Gamma(q_i)$ and involution J_i of second kind such that $q_i^{J_i} = -q_i$. Assume $[\mathfrak{A}_i : P_i] = n_i^2 > 4$ and let Φ be a subfield of Γ. Then any Φ-isomorphism θ of $\mathfrak{S}(\mathfrak{A}_1, J_1)'$ onto $\mathfrak{S}(\mathfrak{A}_2, J_2)'$ can be extended in one and only one way to a Φ-isomorphism of $\mathfrak{A}_1$ onto $\mathfrak{A}_2$. The Lie algebra $\mathfrak{S}(\mathfrak{A}_1, J_1)'$ is not isomorphic to any Lie algebra $\mathfrak{B}'_L$ of type A_I.*

Proof: We may choose a basis $(a_1, \cdots, a_{n^2})$ for $\mathfrak{A}_1$ over P_1 so that $(a_1, \cdots, a_{n^2-1})$ is a basis for $\mathfrak{S}(\mathfrak{A}_1, J_1)'$ over Γ and for $\mathfrak{A}'_{1L}$ over P_1. If Ω is the algebraic closure of Γ (chosen to contain P_1 and P_2), then $(\mathfrak{A}_1 \text{ over } P_1)_\Omega = \Omega_{n_1}$ so $\mathfrak{S}(\mathfrak{A}_1, J_1)_\Omega = \Omega_{n_1L}$ and $\mathfrak{S}(\mathfrak{A}_1, J_1)' = \Omega'_{n_1L}$.

Similarly, we have $\mathfrak{S}(\mathfrak{A}_2, J_2)'_\Omega = \Omega'_{n_2L}$. Assume there exists a Φ-isomorphism of $\mathfrak{S}(\mathfrak{A}_1, J_1)'$ onto $\mathfrak{S}(\mathfrak{A}_2, J_2)'$. Then we know that $\Omega'_{n_1L} \cong_\Omega \Omega'_{n_2L}$ so $n_1 = n_2 = n$ and we may assume that $\mathfrak{A}_1$ and $\mathfrak{A}_2$ are Φ-subalgebras of Ω_n such that any P_i-basis for $\mathfrak{A}_i$ is a basis for Ω_n. Let θ be a Φ-isomorphism of $\mathfrak{L}_1 = \mathfrak{S}(\mathfrak{A}_1, J_1)'$ onto $\mathfrak{L}_2 = \mathfrak{S}(\mathfrak{A}_2, J_2)'$. Then θ is semi-linear in $\mathfrak{L}_1$ over Γ onto $\mathfrak{L}_2$ over Γ with associated automorphism θ in Γ. If $(a_1, \cdots, a_{n^2-1})$ is a basis for $\mathfrak{L}_1$ over Γ, then $(a_1^\theta, \cdots, a_{n^2-1}^\theta)$ is a basis for $\mathfrak{L}_2$ over Γ and both of these are bases for Ω'_{nL} over Ω. If the automorphism θ in Γ is extended to an automorphism θ in Ω, then the mapping $\sum_1^{n^2-1} \omega_i a_i \rightarrow \sum \omega_i^\theta a_i^\theta$ is an automorphism in Ω'_{nL} over Φ whose associated automorphism in Ω is θ. The proof of Theorem 10 shows that this can be extended to an automorphism or the negative of an anti-automorphism of the associative algebra Ω_n over Φ. Since the enveloping associative algebra of $\mathfrak{S}(\mathfrak{A}_i, J_i)'$ over Φ is $\mathfrak{A}_i$ it follows that the isomorphism θ on $\mathfrak{L}_1$ over Φ can be extended to an isomorphism or the negative of an anti-isomorphism of $\mathfrak{A}_1$ over Φ onto $\mathfrak{A}_2$ over Φ. If the second possibility holds let ζ denote the anti-isomorphism. Then $J_1\zeta$ is an isomorphism of $\mathfrak{A}_1$ onto $\mathfrak{A}_2$ and if $a \in \mathfrak{S}(\mathfrak{A}_1, J_1)'$ then $a^\theta = -a^\zeta = a^{J_1\zeta}$ since $a^{J_1} = -a$ for $a \in \mathfrak{S}(\mathfrak{A}_1, J_1)'$. Hence $J_1\zeta$ is an isomorphism of $\mathfrak{A}_1$ which coincides with θ on $\mathfrak{S}(\mathfrak{A}_1, J_1)'$. Thus in every case we can extend θ to an isomorphism of $\mathfrak{A}_1$ onto $\mathfrak{A}_2$. This extension is unique since $\mathfrak{A}_1$ is the enveloping associative algebra of $\mathfrak{S}(\mathfrak{A}_1, J_1)'$. This proves our first assertion. Next let $\mathfrak{L}_1 = \mathfrak{S}(\mathfrak{A}_1, J_1)'$ and suppose we have a Φ-isomorphism θ of $\mathfrak{L}_1$ onto $\mathfrak{L}_2 = \mathfrak{B}'_L$ where $\mathfrak{B}$ is central simple associative over Γ. The argument just used for $\mathfrak{A}_1$ and $\mathfrak{A}_2$ shows that θ can be extended to an isomorphism θ of $\mathfrak{A}_1$ onto $\mathfrak{B}$. Since the centroids of the Lie algebras $\mathfrak{S}(\mathfrak{A}_1, J_1)'$ and $\mathfrak{B}'_L$ consist of the multiplications $(x \rightarrow \gamma x)$ in these Lie algebras by the elements $\gamma \in \Gamma$ it follows that θ maps Γ into itself. Hence $P_1 = \Gamma(q_1)$ is mapped into a subfield of $\mathfrak{B}$ which properly contains Γ. On the other hand, P_1 is the center of $\mathfrak{A}_1$ so this is mapped into the center Γ of $\mathfrak{B}$. This contradiction proves that $\mathfrak{L}_1 = \mathfrak{S}(\mathfrak{A}_1, J_1)'$ cannot be isomorphic to $\mathfrak{B}'_L$.

We consider next the Lie algebras of types B, C and D which we defined in Theorem 9. The main isomorphism result on these is the following

THEOREM 12. *Let $\mathfrak{A}_i$, $i = 1, 2$, be a finite-dimensional central simple associative algebra over Γ such that $\mathfrak{A}_i$ has an involution J_i of first* *kind. If* $[\mathfrak{A}_i : \Gamma] = n_i^2$, *then* $[\mathfrak{S}(\mathfrak{A}_i, J_i) : \Gamma] = l_i(2l_i + 1)$ *if* $n_i = 2l_i + 1$ *and* $[\mathfrak{S}(\mathfrak{A}_i, J_i) : \Gamma] = l_i(2l_i + 1)$ *or* $l_i(2l_i - 1)$ *if* $n_i = 2l_i$. *In the respective cases indicated we assume that* $l_i \geqq 2$, $l_i \geqq 3$, $l_i > 4$. *Then if* Φ *is a subfield of* Γ, *any* Φ*-isomorphism* θ *of* $\mathfrak{S}(\mathfrak{A}_1, J_1)$ *onto* $\mathfrak{S}(\mathfrak{A}_2, J_2)$ *can be extended in one and only one way to a* Φ*-isomorphism of* $\mathfrak{A}_1$ *onto* $\mathfrak{A}_2$.

Proof: If Ω is the algebraic closure of Φ, then $\mathfrak{A}_{i\Omega} \cong \Omega_{n_i}$ and $\mathfrak{S}(\mathfrak{A}_i, J_i)_\Omega$ is the Lie algebra B_{l_i}, C_{l_i} or D_{l_i} according as $n_i = 2l_i + 1$, $n_i = 2l_i$ and $[\mathfrak{S}(\mathfrak{A}_i, J_i) : \Gamma] = l_i(2l_i + 1)$ or $n_i = 2l_i$ and $[\mathfrak{S}(\mathfrak{A}_i, J_i) : \Gamma] = l_i(2l_i - 1)$. It follows that if $\mathfrak{S}(\mathfrak{A}_1, J_1) \cong_\Phi \mathfrak{S}(\mathfrak{A}_2, J_2)$ then $n_1 = n_2 = n$ and we may suppose that $\mathfrak{A}_{1\Omega} = \Omega_n = \mathfrak{A}_{2\Omega}$ and $\mathfrak{S}(\mathfrak{A}_1, J_1)_\Omega = \mathfrak{S}(\mathfrak{A}_2, J_2)_\Omega$ is either the Lie algebra of skew symmetric matrices in Ω_n (types B or D) or the Lie algebra of matrices satisfying $Q^{-1}A'Q = -A$, Q a skew symmetric matrix with entries in the prime field (type C). Let $(a_1, \cdots, a_m)$ $(m = l(2l + 1)$ or $l(2l - 1))$ be a basis for $\mathfrak{S}(\mathfrak{A}_1, J_1)$ over Γ and hence for $\mathfrak{S}_\Omega$ over Ω. If θ is a Φ-isomorphism of $\mathfrak{S}(\mathfrak{A}_1, J_1)$ onto $\mathfrak{S}(\mathfrak{A}_2, J_2)$, θ is semi-linear with associated automorphism θ in Γ and $(a_1^\theta, \cdots, a_n^\theta)$ is a basis for $\mathfrak{S}(\mathfrak{A}_2, J_2)$ over Γ and for $\mathfrak{S}_\Omega$ over Ω. If θ is extended to the automorphism θ in Ω then $\sum \omega_i a_i \rightarrow \sum \omega_i^\theta a_i^\theta$, $\omega_i \in \Omega$, is an automorphism θ' in $\mathfrak{S}_\Omega$ over Φ with θ as its associated automorphism in Ω. If (e_{ij}) is a usual matrix basis for Ω_n over Ω then $\sum \omega_{ij} e_{ij} \rightarrow \sum \omega_{ij}^\theta e_{ij}^\theta$ is an automorphism θ'' of Ω_n over Ω whose associated automorphism in Ω is θ. Moreover, since Q has entries in Φ, θ'' maps $\mathfrak{S}_\Omega$ into itself and so it induces an automorphism θ'' in $\mathfrak{S}_\Omega$ over Φ. It follows that $\theta'(\theta'')^{-1}$ is an automorphism of $\mathfrak{S}_\Omega$ over Ω. In all cases this has the form $X \rightarrow M^{-1}XM$ and so it can be extended to an inner automorphism of Ω_n over Ω. It follows that θ' can be extended to an automorphism ζ of Ω_n over Ω. Since the restriction of ζ to $\mathfrak{S}(\mathfrak{A}_1, J_1)$ is the given θ and since the enveloping associative algebra of $\mathfrak{S}(\mathfrak{A}_i, J_i)$ is $\mathfrak{A}_i$, ζ maps $\mathfrak{A}_1$ isomorphically on $\mathfrak{A}_2$ and coincides with θ on $\mathfrak{S}(\mathfrak{A}_1, J_1)$. Thus θ can be extended to a Φ-isomorphism of $\mathfrak{A}_1$ onto $\mathfrak{A}_2$. Since $\mathfrak{S}(\mathfrak{A}_1, J_1)$ generates $\mathfrak{A}_1$ this extension is unique.

5. Completeness theorems

Let $\mathfrak{L}$ be a finite-dimensional central simple Lie algebra of type A over Γ. Then there exists a finite-dimensional Galois extension field P of Γ such that $\mathfrak{L}_P \cong P'_{nL}$, $n \geqq 2$. Thus we may suppose that $\mathfrak{L}$ is a Γ-subalgebra of P'_{nL} such that the P-space spanned by $\mathfrak{L}$ is

P'_{nL} and elements of $\mathfrak{L}$ which are Γ-independent are P-independent. We have seen also that for each s in the Galois group G of P over Γ there corresponds an automorphism U_s of P'_{nL} over Γ such that $(\rho x)U_s = \rho^s(xU_s)$. If $(a_1, \cdots, a_{n^2-1})$ is a basis for $\mathfrak{L}$ over Γ, hence for P'_{nL} over P, then U_s is the mapping $\sum_1^{n^2-1} \rho_i a_i \rightarrow \sum \overset{s}{\rho_i} a_i$. We have $U_1 = 1$, $U_s U_t = U_{st}$ and $\mathfrak{L}$ which is the set of elements $\sum \gamma_i a_i$, $\gamma_i \in \Gamma$, is the set of fixed elements for the U_s, $s \in G$. We have seen in the last section that U_s has an extension U_s in the enveloping associative algebra P_n of P'_{nL} such that U_s is an automorphism of P_n over Γ if $n = 2$ and U_s is either an automorphism of P_n over Γ or the negative of an anti-automorphism of P_n over Γ if $n > 2$. Every $y \in P_n$ is a sum of products $x_1 x_2 \cdots x_r$, $x_i \in P'_{nL}$, and if $\rho \in P$ and U_s is an automorphism of P_n, then $(\rho x_1 \cdots x_r)U_s = (\rho x_1 U_s)(x_2 U_s) \cdots (x_r U_s) = \rho^s(x_1 U_s) \cdots (x_r U_s) = \rho^s((x_1 \cdots x_r)U_s)$. If U_s is the negative of an anti-automorphism, then $(\rho x_1 \cdots x_r)U_s = (-1)^r(x_r U_s) \cdots (x_2 U_s)(\rho x_1 U_s) = (-1)^r \rho^s(x_r U_s) \cdots (x_1 U_s) = \rho^s((x_1 \cdots x_r)U_s)$. Hence in either case U_s is semi-linear with automorphism s in P. Since $U_s U_t$ and U_{st} have the same effect on the set of generators P'_{nL} of P_n we have $U_s U_t = U_{st}$. Similarly $U_1 = 1$ is valid in P_n. If U_s and U_t are the negatives of anti-automorphisms then $U_{st^{-1}} = U_s U_t^{-1}$ is an automorphism. It follows immediately from this that the subset H of elements $s \in G$ such that U_s is an automorphism of P_n over Γ is a subgroup of index one or two in G.

Case I. $H = G$. The subset of P_n of fixed elements under the U_s is a Γ-subalgebra $\mathfrak{A}$ of P_n such that $\mathfrak{A}_P = P_n$ (Theorem 7). Hence $\mathfrak{A}$ is finite-dimensional central simple over Γ. Evidently $\mathfrak{L} \subseteq \mathfrak{A}$ and so $\mathfrak{L}' = \mathfrak{L} \subseteq \mathfrak{A}'_L$. On the other hand, $[\mathfrak{L} : \Gamma] = n^2 - 1$ and $[\mathfrak{A}'_L : \Gamma] = n^2 - 1$. Hence $\mathfrak{L} = \mathfrak{A}'_L$ is a Lie algebra of type A_I as defined on p. 303.

Case II. $H \neq G$. Then H has index two in G. We know also that $n > 2$ in this case. The subset of P of elements ξ such that $\xi^s = \xi$, $s \in H$, is a quadratic subfield $\Gamma(q)$ over Γ. It is clear from the form of the U_s that $x = \sum_1^{n^2-1} \rho_i a_i$ satisfies $xU_s = x$, $s \in H$, if and only if all the $\rho_i \in \Gamma(q)$. Thus $\mathfrak{L}_{\Gamma(q)}$, the set of $\Gamma(q)$-linear combinations of the a_i, is the set of elements of P'_{nL} which are fixed for all the U_s, $s \in H$, and H is the Galois group of P over $\Gamma(q)$. It follows from case I that $\mathfrak{L}_{\Gamma(q)} = \mathfrak{A}'_L$ where $\mathfrak{A}$ is central simple over $\Gamma(q)$ and is the enveloping associative algebra of $\mathfrak{L}_{\Gamma(q)}$. Now let $t \in G$, $\notin H$. Since H is of index two in G, an element $\sum_1^{n^2-1} \rho_i a_i$ of $\mathfrak{L}_{\Gamma(q)}$ is in $\mathfrak{L}$ if and only if $(\sum \rho_i a_i)U_t = \sum \bar{\rho}_i a_i$. Now

$U_t = -J$ where J is an anti-automorphism of P_n over Γ. Since $(\sum \rho_i a_i)U_t = \sum \bar{\rho}_i a_i$ for $\rho_i \in \Gamma(q)$ and $\bar{\rho}$ the conjugate of ρ under the automorphism $\neq 1$ of $\Gamma(q)$ over Γ, $J = -U_t$ maps $\mathfrak{L}_{\Gamma(q)}$ into itself. Hence J induces an anti-automorphism in the enveloping algebra $\mathfrak{A}$ of $\mathfrak{L}_{\Gamma(q)}$. If $\rho \in \Gamma(q)$, $y \in \mathfrak{A}$, then $(\rho y)^J = \bar{\rho} y^J$. Hence J induces the automorphism $\rho \to \bar{\rho}$ in $\Gamma(q)$, so J is of second kind. Since $aU_t = a$ for $a \in \mathfrak{L}$, $a^J = -a$ and $a \in \mathfrak{S}(\mathfrak{A}, J)$. Thus $\mathfrak{L} \subseteq \mathfrak{S}(\mathfrak{A}, J)$ and $\mathfrak{L}' = \mathfrak{L} \subseteq \mathfrak{S}(\mathfrak{A}, J)'$. Comparison of dimensionalities over Γ shows that $\mathfrak{L} = \mathfrak{S}(\mathfrak{A}, J)'$. Hence $\mathfrak{L}$ is a simple Lie algebra of type A_{II}.

We have therefore proved the following

THEOREM 13. *Any central simple Lie algebra of type A_l is isomorphic either to a Lie algebra $\mathfrak{A}'_L$, $\mathfrak{A}$ a finite-dimensional central simple associative algebra or to an algebra $\mathfrak{S}(\mathfrak{A}, J)'$ where $\mathfrak{A}$ is finite-dimensional simple associative with an involution J of second kind.*

We consider next the Lie algebras of types B, C and D in the following

THEOREM 14. *Let $\mathfrak{L}$ be a central simple Lie algebra of type B_l, $l \geqq 2$, C_l, $l \geqq 3$ or D_l, $l \geqq 5$. Then $\mathfrak{L}$ is isomorphic to a Lie algebra $\mathfrak{S}(\mathfrak{A}, J)$ where $\mathfrak{A}$ is a finite-dimensional central simple associative algebra, J an involution of first kind in $\mathfrak{A}$.*

Proof: There exists a finite-dimensional Galois extension P of Γ such that $\mathfrak{L}_P$ is the Lie algebra $\mathfrak{S}(P_n, J)$ of J-skew matrices in P_n where J is the involution $X \to X'$, the transpose of X in P_n, or the involution $X \to Q^{-1}X'Q$ where $Q' = -Q$ and the entries of Q are in the prime field. Also we have $n \geqq 5$ if $n = 2l + 1$, $n \geqq 6$ if $n = 2l$ and the involution is $X \to Q^{-1}X'Q$, and $n \geqq 10$ in the remaining case. For each s in the Galois group G of P over Γ we have the automorphism U_s of $\mathfrak{L}_P$ over Γ: $\sum_1^m \rho_i a_i \to \sum \rho_i^s a_i$ where $(a_1, \cdots, a_m)$ is a basis for $\mathfrak{L}$ over Γ and for $\mathfrak{L}_P$ over P. $\mathfrak{L}$ is the subset of $\mathfrak{L}_P$ of elements which are fixed relative to the U_s. The conditions on n insure that U_s can be extended to an automorphism U_s of the enveloping associative algebra P_n of $\mathfrak{L}_P$ (Theorem 12). The extension U_s is semi-linear in P_n with associated automorphism s, $U_1 = 1$ and $U_sU_t = U_{st}$ hold in P_n. Hence the subset of P_n of elements which are fixed relative to the U_s, $s \in G$, is a subalgebra $\mathfrak{A}$ of P_n such that $\mathfrak{A}_P = P_n$. Hence $\mathfrak{A}$ is finite-dimensional central simple over Γ. If $X \in \mathfrak{S}_P$, then $X^J = -X$ and $XU_s \in \mathfrak{S}_P$. Hence $X^J U_s = -XU_s = (XU_s)^J$. Thus $JU_s = U_sJ$ holds in $\mathfrak{S}_P$. Since P_n is the enveloping associative algebra of $\mathfrak{S}_P = \mathfrak{S}(P_n, J)$ it follows that $JU_s = U_sJ$ in P_n also. This implies that J maps $\mathfrak{A}$ into itself and hence J induces an involution in $\mathfrak{A}$ over Γ which is of first kind since Γ is the center of $\mathfrak{A}$. If $a \in \mathfrak{L}$ then $a^J = -a$ so $a \in \mathfrak{S}(\mathfrak{A}, J)$ and $\mathfrak{L} \subseteq \mathfrak{S}(\mathfrak{A}, J)$. On the other hand, $\mathfrak{S}(\mathfrak{A}, J)_P \subseteq \mathfrak{S}(P_n, J) = \mathfrak{L}_P$. Hence $\mathfrak{L} = \mathfrak{S}(\mathfrak{A}, J)$. This completes the proof.

6. *A closer look at the isomorphism conditions*

We have seen in § 4, that if J_i, $i = 1, 2$, is an involution (either kind) in a finite-dimensional simple associative algebra $\mathfrak{A}_i$ over Φ, then $\mathfrak{S}(\mathfrak{A}_1, J_1)' \cong_\Phi \mathfrak{S}(\mathfrak{A}_2, J_2)'$ implies that $\mathfrak{A}_1$ and $\mathfrak{A}_2$ are isomorphic. It therefore suffices to consider one algebra $\mathfrak{A} = \mathfrak{A}_1 = \mathfrak{A}_2$ and consider the condition for isomorphism of $\mathfrak{S}(\mathfrak{A}, J)'$ and $\mathfrak{S}(\mathfrak{A}, K)'$ where J and K are involutions in $\mathfrak{A}$. We have seen that any isomorphism θ of $\mathfrak{S}(\mathfrak{A}, J)'$ onto $\mathfrak{S}(\mathfrak{A}, K)'$ can be realized by an automorphism of $\mathfrak{A}$. If $a \in \mathfrak{S}(\mathfrak{A}, J)'$ then $a^{J\theta} = -a^\theta = a^{\theta K}$. Thus $J\theta = \theta K$ holds in $\mathfrak{S}(\mathfrak{A}, J)'$. Since the enveloping Φ-algebras of $\mathfrak{S}(\mathfrak{A}, J)'$ and $\mathfrak{S}(\mathfrak{A}, K)'$ are $\mathfrak{A}$ we have $J\theta = \theta K$ in $\mathfrak{A}$ or $K = \theta^{-1}J\theta$. We shall call the involutions J and K of $\mathfrak{A}$ *cogredient* if there exists an automorphism θ of $\mathfrak{A}$ such that $K = \theta^{-1}J\theta$. We have seen that cogredience is a necessary condition for isomorphism of $\mathfrak{S}(\mathfrak{A}, J)'$ and $\mathfrak{S}(\mathfrak{A}, K)'$. Conversely, if J and K are cogredient and $K = \theta^{-1}J\theta$ where θ is an automorphism then θ maps $\mathfrak{S}(\mathfrak{A}, J)$ onto $\mathfrak{S}(\mathfrak{A}, K)$ and $\mathfrak{S}(\mathfrak{A}, J)'$ onto $\mathfrak{S}(\mathfrak{A}, K)'$. Hence θ induces an isomorphism of the Lie algebra $\mathfrak{S}(\mathfrak{A}, J)'$ onto $\mathfrak{S}(\mathfrak{A}, K)'$. Thus $\mathfrak{S}(\mathfrak{A}, J)' \cong_\Phi \mathfrak{S}(\mathfrak{A}, K)'$ if and only if J and K are cogredient. We see also that the group of automorphisms of the Lie algebra $\mathfrak{S}(\mathfrak{A}, J)'$ can be identified with the subgroup of the group of automorphisms θ of the algebra $\mathfrak{A}$ such that $\theta J = J\theta$.

Now let $\mathfrak{A} = \mathfrak{E}$ the algebra of linear transformations in the finite-dimensional vector space $\mathfrak{M}$ over the finite dimensional division algebra Δ. Let $\alpha \to \bar{\alpha}$ be an involution in Δ and let (x, y) be a hermitian or skew hermitian form corresponding to this involution. Then the mapping $A \to A^*$, $A \in \mathfrak{E}$, A^* the adjoint of A, is an involution in $\mathfrak{E}$. We recall that A^* is the unique linear transformation such that $(xA, y) = (x, yA^*)$ and we have $(y, x) = \varepsilon\overline{(x, y)}$ where $\varepsilon = 1$ or -1 according as the form is hermitian or skew hermitian. Suppose $\alpha \to \alpha'$ is a second involution in Δ and $(x, y)_1$

a second hermitian or skew hermitian form corresponding to this so we have $(y, x)_1 = \varepsilon_1(x, y)_1'$ where $\varepsilon_1 = \pm 1$. Suppose the involution determined by this form is the same as that given by (x, y). Thus if $(x, yA^*) = (xA, y)$ then $(x, yA^*)_1 = (xA, y)_1$. Let u and v be arbitrary vectors in $\mathfrak{M}$. Then $x \to (x, u)v$ is a linear transformation in $\mathfrak{M}$ and one checks that its adjoint relative to (x, y) is $x \to \varepsilon(x, v)u$. Hence we have

$$((x, u)v, y)_1 = (x, \varepsilon(y, v)u)_1$$
$$(x, u)(v, y)_1 = \varepsilon(x, u)_1(y, v)' .$$

Since x, y, u, v are arbitrary this shows that $(x, y)_1 = (x, y)\rho$, $\rho \neq 0$ in Δ. If $\alpha \in \Delta$, $(x, \alpha y)_1 = (x, y)_1\alpha'$ gives $(x, y)\bar{\alpha}\rho = (x, y)\rho\alpha'$; hence $\alpha' = \rho^{-1}\bar{\alpha}\rho$. Also we have $(y, x)\rho = \varepsilon_1((x, y)\rho)' = \varepsilon_1\rho'(x, y)' = \varepsilon_1\rho^{-1}\bar{\rho}\overline{(x, y)}\rho = \varepsilon_1\varepsilon\rho^{-1}\bar{\rho}(y, x)\rho$. Hence $\rho = \varepsilon_1\varepsilon\bar{\rho}$. Conversely, let ρ be any element of Δ satisfying $\bar{\rho} = \pm\rho \neq 0$. Then a direct verification shows that $\alpha \to \alpha' = \rho^{-1}\bar{\alpha}\rho$ is an involution in Δ and $(x, y)_1 \equiv (x, y)\rho$ is hermitian or skew hermitian relative to this involution. If (x, y) is skew hermitian and $\bar{\rho} = -\rho$, then $(x, y)\rho$ is hermitian. Hence if Δ contains a skew element $\neq 0$, then a skew hermitian form can be replaced by a hermitian one which gives the same involution $A \to A^*$ in $\mathfrak{E}$. If Δ contains no such elements then $\bar{\rho} = \rho$ for all $\rho \in \Delta$ and this implies that Δ is a field. Hence we may restrict our attention to hermitian forms and to alternate forms (Δ a field). Two such forms $(x, y)_1$ and (x, y) give the same involution in $\mathfrak{E}$ if and only if $(x, y)_1 = (x, y)\rho$, $\bar{\rho} = \rho$ if $\alpha \to \bar{\alpha}$ is the involution of (x, y).

It is known that any automorphism of $\mathfrak{E}$ has the form $A \to S^{-1}AS$ where S is a semi-linear transformation in $\mathfrak{M}$ over Δ (Jacobson [3], p. 45). If θ is the automorphism in Δ associated with S, then one checks that $(x, y)_1 \equiv (xS, yS)^{\theta^{-1}}$ is hermitian or alternate with involution $\alpha \to (\bar{\alpha^\theta})^{\theta^{-1}}$. Moreover, if $A \in \mathfrak{E}$ then

$$\begin{aligned}(xAS, yS)^{\theta^{-1}} &= (xSS^{-1}AS, yS)^{\theta^{-1}} \\ &= (xS, yS(S^{-1}AS)^*)^{\theta^{-1}} \\ &= (xS, y(S(S^{-1}AS)^*S^{-1})S)^{\theta^{-1}} .\end{aligned}$$

Thus $(xA, y)_1 = (x, y(S(S^{-1}AS)^*S^{-1})_1$ and the involution in $\mathfrak{E}$ determined by $(x, y)_1$ is $A \to S(S^{-1}AS)^*S^{-1}$. Thus if we call $A \to A^*$, J, and $A \to S^{-1}AS$, θ, then the new involution is $K = \theta J\theta^{-1}$ which is cogredient to J. Because of this relation it is natural to extend the usual notion of equivalence of forms in the following manner:

Two hermitian forms (x, y) and $(x, y)_1$ are said to be S-equivalent if there exists a $1:1$ semi-linear transformation S with associated isomorphism θ such that $(x, y)_1 = (xS, yS)^{\theta^{-1}}$.

It is well known that any two non-degenerate alternate forms are equivalent in the ordinary sense. Hence the involutions in $\mathfrak{E}$ determined by any two such forms are cogredient. Moreover, these are not cogredient to any involution determined by a hermitian form. Our results imply also that the non-degenerate hermitian forms (x, y) and $(x, y)_1$ define cogredient involutions in $\mathfrak{E}$ if and only if $(x, y)_1 = (xS, yS)^{\theta^{-1}}\rho$ where S is semi-linear with automorphism θ and ρ is symmetric relative to $\alpha \to \alpha' \equiv (\overline{\alpha^{\theta}})^{\theta^{-1}}$.

7. *Central simple real Lie algebras*

We shall now apply our results and known results on associative algebras to classify the central simple Lie algebras of types A–D (except D_4) over the field Φ of real numbers. By Frobenius' theorem, the finite-dimensional division algebras over Φ are: Φ; the complex field $P = \Phi(i)$, $i^2 = -1$; the quaternion division algebra Δ with basis $1, i, j, k$ such that

$$
\begin{aligned}
i^2 = j^2 = k^2 = -1\,, \qquad ij = -ji = k\,, \\
jk = -kj = i\,, \qquad ki = -ik = j\,.
\end{aligned}
\tag{12}
$$

Δ has the standard involution $a = \alpha + \beta i + \gamma j + \delta k \to \bar{a} = \alpha - \beta i - \gamma j - \delta k$. Since the automorphisms in Δ are all inner, every involution in Δ is either standard or it has the form $a \to q^{-1}\bar{a}q$ where $\bar{q} = -q$. The dimensionality of the space of skew elements under the standard involution is three; under $a \to q^{-1}\bar{a}q$ the dimensionality is one. We denote the automorphism $\neq 1$ in P over Φ by $\rho \to \bar{\rho}$.

By the Wedderburn theorem, the finite-dimensional simple associative algebras over Φ are the full matrix algebras Φ_n, P_n and Δ_n. These can be identified with the algebras $\mathfrak{E}(\Phi, n)$, $\mathfrak{E}(P, n)$ and $\mathfrak{E}(\Delta, n)$ of linear transformations in $\mathfrak{M}$ over Φ, P and Δ respectively. The algebras $\Phi_n \cong \mathfrak{E}(\Phi, n)$ and $\Delta_n \cong \mathfrak{E}(\Delta, n)$ are central and have only inner automorphisms. The center of $\mathfrak{E}(P, n) \cong P_n$ is P. In addition to inner automorphisms, $\mathfrak{E}(P, n)$ has the outer automorphisms $X \to S^{-1}XS$ where S is a semi-linear transformation with associated automorphism $\rho \to \bar{\rho}$ in P.

All our algebras have involutions and $\mathfrak{E}(P, n)$ has involutions of second kind. The involutions of $\mathfrak{E}(\Phi, n)$ have the form $X \to X^*$ where X^* is the adjoint of X relative to a non-degenerate symmetric or skew bilinear form in $\mathfrak{M}$ over Φ. Involutions determined by skew forms are not cogredient to any determined by a symmetric form. Any two non-degenerate skew bilinear forms are equivalent so these give a single cogredience class of involutions. If (x, y) and $(x, y)_1$ are non-degenerate the criterion of the last section shows that the involution determined by (x, y) is cogredient to that of $(x, y)_1$ if and only if (x, y) is equivalent to a multiple of $(x, y)_1$. Since (x, y) is equivalent to any positive multiple of (x, y) it follows that the involution given by (x, y) is cogredient to that of $(x, y)_1$ if and only if (x, y) is equivalent to $\pm(x, y)_1$. If (x, y) is non-degenerate symmetric it is well known that there exists a basis $(u_1, u_2, \cdots, u_n)$ for $\mathfrak{M}$ such that

$$(13) \qquad \begin{aligned} (u_i, u_i) &= 1\,, && 1 \leqq i \leqq p \\ (u_j, u_j) &= -1\,, && p < j \leqq n \\ (u_i, u_j) &= 0\,, && i \neq j\,. \end{aligned}$$

The number p is an invariant by Sylvester's theorem. Accordingly, we obtain $[n/2] + 1$ cogredience classes of involutions corresponding to the values $p = 0, 1, \cdots, [n/2]$. A simple calculation using the canonical basis (13) shows that the dimensionality over Φ of the space of skew elements determined by (x, y) is $n(n-1)/2$. This implies that the Lie algebra $\mathfrak{S}(\mathfrak{E}, J)$ of these elements is of type B or D according as n is odd or even. The Lie algebra determined by a skew bilinear form (x, y) is of type C.

We consider next the involutions of second kind in $\mathfrak{E}(P, n)$. Such an involution is the adjoint mapping relative to a non-degenerate hermitian form (x, y) in $\mathfrak{M}$ over P. A basis $(u_1, u_2, \cdots, u_n)$ can be chosen so that (13) holds. If (x, y) is S-equivalent to $(x, y)_1$ in the sense that there exists a semi-linear transformation S with automorphism $\rho \to \bar{\rho}$ in P such that $(x, y) = (xS, yS)_1$, then we have $(u_iS, u_iS)_1 = 1$, $1 \leqq i \leqq p$, $(u_jS, u_jS)_1 = -1$, $p < j \leqq n$, $(u_iS, u_jS)_1 = 0$, $i \neq j$. Then (x, y) and $(x, y)_1$ are equivalent in the usual sense. Since (x, y) is equivalent to $\gamma(x, y)$, γ real and positive, it follows that (x, y) and $(x, y)_1$ determine cogredient involutions if and only if (x, y) is equivalent to $\pm(x, y)_1$. This and Sylvester's theorem for hermitian forms implies again that the number of non-cogredient involutions of second kind in $\mathfrak{E}(P, n)$ is $[n/2] + 1$.

The discussion we have just given carries over verbatim to hermitian forms in $\mathfrak{M}$ over Δ for which the involution in Δ is the standard one. A basis satisfying (13) can be chosen (cf. Jacobson [2], vol. II, p. 159). The number of cogredience classes of involutions given by the standard hermitian forms is $[n/2]+1$. If we use a canonical basis for which (13) holds it is easy to calculate that the dimensionality of $\mathfrak{S}(\mathfrak{E}(\Delta, n), J)$ determined by the associated involution J is $n(2n+1)$. Since $\Delta_P \cong P_2$, $\mathfrak{E}(\Delta, n)_P \cong P_{2n}$. Since $\mathfrak{S}(\mathfrak{E}(\Delta, n), J)_P$ has dimensionality $n(2n+1)$ over P it follows that $\mathfrak{S}$ is central simple of type C_n.

It remains to consider the hermitian forms (x, y) in $\mathfrak{M}$ over Δ whose involutions in Δ are of the form $a \to q^{-1}\bar{a}q$, $\bar{q} = -q$. If we replace (x, y) by $(x, y)q^{-1}$ we obtain a skew hermitian form relative to the standard involution. We prefer to treat these.

LEMMA 5. *If q_1 and q_2 are non-zero skew in Δ (relative to the standard involution), then there exists a non-zero a in Δ such that $q_2 = \bar{a}q_1a$.*

Proof: We note first that if b_1 and b_2 are elements of Δ not in Φ which have equal traces and norms, then there exists an isomorphism of $\Phi(b_1)$ onto $\Phi(b_2)$ mapping b_1 into b_2. This can be extended to an inner automorphism of Δ (Jacobson, *Structure of Rings*, p. 162). It follows that b_1 and b_2 are similar, that is, $b_2 = cb_1c^{-1}$ for some c in Δ. In particular, if $N(q_2) = N(q_1)$, then since the traces $T(q_1) = 0 = T(q_2)$, there exists a c such that $q_2 = cq_1c^{-1} = N(c)^{-1}\bar{c}q_1c$. We note also that since the norm of any non-zero element is positive, q_1 is necessarily similar to a suitable positive multiple of q_2. We now see that it suffices to prove the lemma for $q_2 = \gamma q_1$, $\gamma > 0$. Then we consider the quadratic field $\Phi(q_1) = \Phi(q_2)$. We have $\gamma = N(c)$ for some c in this field. Hence $q_2 = N(c)q_1 = \bar{c}q_1c$ as required.

The usual method of obtaining a diagonal matrix for a hermitian or skew hermitian form now gives the following

LEMMA 6. *If (x, y) is a non-degenerate skew hermitian bilinear form relative to the standard involution in Δ, then there exists a basis $(u_1, u_2, \cdots, u_n)$ for $\mathfrak{M}$ such that the matrix $((u_i, u_j)) =$*

$$\operatorname{diag}\{q, q, \cdots, q\} \tag{14}$$

where q is any selected non-zero skew element of Δ.

This result implies that there is just one cogredience class of involutions in $\mathfrak{E}(\Delta, n)$ given by skew hermitian forms in $\mathfrak{M}$ over Δ.

If we use (14) we can calculate the dimensionality for the Lie algebra defined by the involution. This is $n(2n-1)$ which implies that the Lie algebra is central simple of type D_n.

We can now list a set of representatives for the isomorphism classes of central simple Lie algebras of types A–D over Φ, as follows.

Type $A_{\rm I}$. Φ'_{nL}, $n > 1$; Δ'_{nL}, $n \geqq 1$.

Type $A_{\rm II}$. Suppose $n > 1$ and let

$$S_p = \mathrm{diag}\,\{\overbrace{1, \cdots, 1}^{p}, \overbrace{-1, \cdots, -1}^{n-p}\}. \tag{15}$$

Let $\mathfrak{S}(P, n, p)$ denote the Lie algebra of matrices $X \in P_n$ such that $S_p^{-1}X'S_p = -X$. Then the Lie algebras $\mathfrak{S}(P, n, p)$ for $p = 0, 1, \cdots, [n/2]$ constitute our list.

Type B. Let $\mathfrak{S}(\Phi, n, p)$ be the Lie algebra of matrices $X \in \Phi_n$ such that $S_p^{-1}X'S_p = -X$. Then our list is the set of Lie algebras $\mathfrak{S}(\Phi, n, p)$ with n odd, $n \geqq 5$.

Type C. Let $\mathfrak{S}(\Delta, n, p)$ be the Lie algebra of matrices $X \in \Delta_n$ satisfying $S_p^{-1}X'S_p = -X$ and let $\mathfrak{S}(\Phi, 2n, Q)$ the Lie algebra of matrices in Φ_{2n} such that $Q^{-1}X'Q = -X$ where Q is any skew symmetric matrix. Then the list is: $\mathfrak{S}(\Delta, n, p)$, $n \geqq 3$, $p = 0, \cdots, [n/2]$ and $\mathfrak{S}(\Phi, 2n, Q)$, $n \geqq 3$.

Type D. Let $\mathfrak{S}(\Delta, n, Q)$ be the Lie algebra of matrices in Δ_n satisfying $Q^{-1}X'Q = -X$ where Q is a skew hermitian matrix. The list is; $\mathfrak{S}(\Delta, n, Q)$, $n > 5$, and $\mathfrak{S}(\Phi, 2n, S_p)$, $n > 5$, and $p = 0, 1, \cdots, n$.

Exercises

1. Determine the groups of automorphisms of the simple real Lie algebras of types A–D, except D_4.

2. Determine an invariant non-degenerate symmetric bilinear form for every central simple real Lie algebra of type A–D. Use this to enumerate the compact Lie algebras in the list (cf. § 4.7).

3. Show that the Lie algebras $S(\Delta, 4, Q)$ and $S(\Phi, 8, S_1)$ in the notation of § 7 are isomorphic, (This shows that Theorem 12 is not valid for Lie algebras of type D_4, since Δ_4 and Φ_8 are not isomorphic.)

4. A (generalized) *quaternion algebra* over an arbitrary field Φ is defined to be an algebra with basis $1, i, j, k$ such that 1 is the identity and the multiplication table for i, j, k is

$$i^2 = \alpha 1, \quad j^2 = \beta 1, \quad k^2 = -\alpha\beta 1 \neq 0$$

$$ij = -ji = k, \quad jk = -kj = -\beta i, \quad ki = -ik = -\alpha j.$$

Show that an algebra Δ is a quaternion algebra over Φ if and only if $\Delta_\Omega \cong \Omega_2$ for Ω the algebraic closure of Φ.

5. A non-associative algebra $\mathfrak{C}$ over an arbitrary field Φ is called a *Cayley algebra* if $\mathfrak{C}$ has an identity 1, $\mathfrak{C}$ contains a quaternion subalgebra Δ containing 1 and every element of $\mathfrak{C}$ can be writen in one and only one way in the form $a + bu$, a, b in Δ and u an element of $\mathfrak{C}$ such that

$$a(bu) = (ba)u\ , \qquad (bu)a = (b\bar{a})u\ , \qquad (au)(bu) = \mu\overline{b}a\ ,$$

where μ is a non-zero element of Φ. Show that $\mathfrak{C}$ is alternative (cf. § 4.6) and that the mapping $x = a + bu \to \bar{a} - bu$ is an involution in $\mathfrak{C}$. Prove that $x\bar{x} = N(x)1 = \bar{x}x$ where $N(x) \in \Phi$ and satisfies $N(xy) = N(x)N(y)$. Let $(x, y) = \frac{1}{2}[N(x + y) - N(x) - N(y)]$. Show that (x, y) is a non-degenerate symmetric bilinear form and that two Cayley algebras are isomorphic if and only if their forms (x, y) are equivalent. Prove that a non-associative algebra $\mathfrak{C}$ is a Cayley algebra if and only if $\mathfrak{C}_\Omega$ is the split Cayley algebra of § 4.6.

6. Use Theorem 9.7 and Exercise 5 to prove that two Cayley algebras over a field of characteristic 0 are isomorphic if and only if their derivation algebras are isomorphic.

7. Prove that a central simple Lie algebra over a field of characteristic 0 is of type G if and only if it is isomorphic to the derivation algebra of a Cayley algebra.

BIBLIOGRAPHY

ADO, I. D.

[1] The representation of Lie algebras by matrices. Uspehi Mat. Nauk (N.S.) 2, No. 6(22) (1947), pp. 159-173. Am. Math. Soc. Transl. No. 2 (1949).

ALBERT, A. A., and FRANK, M. S.

[1] Simple Lie algebras of characteristic p. Univ. e Politecnico Torino. Rend. Sem. Mat. Fis. 14 (1954-1955), pp. 117-139.

ARTIN, E.

[1] Geometric Algebra. Interscience, 1957.

BLOCK, R.

[1] New Simple Lie algebras of prime characteristic. Trans. Am. Math. Soc. 89 (1958), pp. 421-449.

[2] On Lie algebras of classical type. Proc. Am. Math. Soc. 11 (1960), pp. 377-379.

BOREL, A.

[1] Topology of Lie groups and characteristic classes. Bull. Am. Math. Soc. 61 (1955), pp. 397-432.

[2] Groupes linéaires algébriques. Ann. Math. 64 (1956), pp. 20-82.

BOREL A., and CHEVALLEY, C.

[1] The Betti numbers of the exceptional groups. Mem. Am. Math. Soc. No. 14 (1955), pp. 1-9.

BOREL, A., and MOSTOW, G. D.

[1] On semi-simple automorphisms of Lie algebras. Ann. Math. 61 (1955), pp. 389-504.

BOREL, A., and SERRE, J. P.

[1] Sur certains sous-groupes des groupes de Lie compacts. Comment. Math. Helv. 27 (1953), pp. 128-139.

CARTAN, E.

[1] Thèse. Paris, 1894. 2nd ed., Vuibert, Paris, 1933.

[2] Les groupes réels simples, finis et continus. Ann. Sci. École Norm. Sup. 31 (1914), pp. 263-355.

[3] Les groupes projectifs qui ne laissent invariante aucune multiplicité plane. Bull. Soc. Math. France 41 (1913), pp. 53-96.

[4] Les groupes projectifs continus réels qui ne laissent invariante aucune multiplicité plane. J. Math. Pures Appl. 10, (1914), pp. 149-186.
[5] La géométrie des groupes de transformations. J. Math. Pures Appl. 6, Ser. 9 (1927), pp. 1-119.

CARTAN, H., and EILENBERG, S.
[1] Homological Algebra. Princeton Univ. Press, 1956.

CARTIER, P.
[1] Séminaire Sophus Lie, Vol. II. (Hyperalgèbres et groupes de Lie formels) (1957) École Norm. Sup.
[2] Dualité de Tannaka des groupes et des algèbres de Lie. Comptes Rendus 242 (1956), pp. 322-325.

CARTIER, P. et al.
[1] Séminaire Sophus Lie. (1954-1955) École Norm. Sup.

CHANG, HO-JUI.
[1] Über Wittsche Lie-Ringe. Hamburg Abhandl. 14 (1941), pp. 151-184.

CHEVALLEY, C.
[1] Theory of Lie Groups: Vol. I. Princeton Univ. Press, 1946.
[2] Théorie des Groupes de Lie: Tome II, Groupes Algébriques. Actualités Sci. Ind. No. 1152. Paris, 1951.
[3] The Algebraic Theory of Spinors. Columbia Univ. Press, 1954.
[4] Théorie des Groupes de Lie: Tome III, Théorèmes Généraux sur les Algébres de Lie. Actualitès. Sci. Ind. No. 1226. Paris, 1955.
[5] An algebraic proof of a property of Lie groups. Am. J. Math. 63 (1941), pp. 785-793.
[6] A new kind of relationship between matrices. Am. J. Math. 65 (1943), pp. 521-531.
[7] Sur certains groupes simples. Tôhoku Math. J. 7-8, Ser. 2, (1955-1956), pp. 14-66.
[8] Séminaire Chevalley, Vols. I and II. (Classification des groupes de Lie algébriques) (1956-1958) École Norm. Sup.

CHEVALLEY, C. and EILENBERG, S.
[1] Cohomology theory of Lie groups and Lie algebras. Trans. Am. Math. Soc. 63 (1948), pp. 85-124.

CHEVALLEY, C. and SCHAFER, R. D.
[1] The exceptional simple Lie algebras F_4 and E_6. Proc. Nat. Acad. Sci. U.S. 36 (1950), pp. 137-141.

COHN, P. M.
[1] Lie Groups. Cambridge Tracts in Math. and Math. Phys. 46. Cambridge, 1957.

COXETER, H. S. M.

[1] The product of the generators of a finite group generated by reflections. Duke Math. J. 18 (1951), pp. 765-782.

CURTIS, C. W.

[1] Modular Lie algebras, I and II. I: Trans. Am. Math. Soc. 82 (1956), pp. 160-179. II: Trans. Am. Math. Soc. 86 (1957), pp. 91-108.

[2] On the dimensions of the irreducible modules of Lie algebras of classical type. Trans. Am. Math. Soc. 96 (1960), pp. 135-142.

[3] Representations of Lie algebras of classical type with applications to linear groups. J. Math. Mech. 9 (1960), pp. 307-326.

DICKSON, L. E.

[1] Algebras and their Arithmetics, Chicago Univ. Press, 1923.

DIEUDONNÉ, J.

[1] La Géométrie des Groupes Classiques. Ergeb. Math. Berlin, 1955.

[2] Sur les groupes de Lie algébriques sur un corps de caractéristique $p > 0$. Rend. Circ. Mat. Palermo (2) 1 (1952), pp. 380-402.

[3] Sur quelques groupes de Lie abéliens sur un corps de caractéristique $p > 0$. Arch. Math. 5 (1954), pp. 274-281.

[4] Lie groups and Lie hyperalgebras over a field of characteristic $p > 0$, I-VI. I: Comment. Math. Helv. 28 (1954), pp. 87-118. II: Am. J. Math. 77 (1955), pp. 218-244. III: Math. Z. 62 (1955), pp. 53-75. IV: Am. J. Math. 77 (1955), pp. 429-452. V: Bull. Soc. Math. France 84 (1956), pp. 207-239. VI: Am. J. Math. 79 (1957), pp. 331-388.

[5] Witt groups and hyperexponential groups. Mathematika. 2 (1955), pp. 21-31.

[6] On simple groups of type B_n. Am. J. Math. 79 (1957), pp. 922-923.

[7] Les algèbres de Lie simples associées aux groupes, simple algébriques sur un corps de caractéristique $p > 0$. Rend. Circ. Mat. Palermo (2), 6 (1957), pp. 198-204.

DIXMIER, J.

[1] Sur un théorème d'Harish-Chandra. Acta Sci. Math. Szeged. 14 (1952), pp. 145-156.

[2] Sur les algèbres dérivées d'une algèbre de Lie. Proc. Cambridge Phil. Soc. 51 (1955), pp. 541-544.

[3] Sous-algèbres de Cartan et decompositions de Levi dans les algèbres de Lie. Trans. Roy. Soc. Canada, Sec. III, (3) 50 (1956), pp. 17-21.

[4] Certaines factorizations canoniques dans l'homologie et la cohomologie des algèbres de Lie. J. Math. Pures Appl. (9) 35 (1956), pp. 77-86.

DIXMIER, J., and LISTER, W. G.

[1] Derivations of nilpotent Lie algebras. Proc. Am. Math. Soc. 8 (1957), pp. 155-158.

DYNKIN, E.

[1] The structure of semi-simple Lie algebras. Uspehi Mat. Nauk (N.S.) 2 (1947), pp. 59-127. Am. Math. Soc. Transl. No. 17 (1950).

[2] On the representation of the series $\log(e^x e^y)$ for non-commutative x and y by commutators. Mat. Sbornik 25 (1949), pp. 155-162.

[3] Semi-simple subalgebras of semi-simple Lie algebras. Mat. Sbornik 30 (1952), pp. 349-462. Am. Math. Soc. Transl, Ser. 2, 6 (1957), pp. 111-244.

[4] Maximal subgroups of the classical groups. Trudy Moskov. Mat. Obshchestva. 1 (1952), pp. 39-166. Am. Math. Soc. Transl., Ser. 2, 6 (1957), pp. 245-378.

FRANK, M. S.

[1] A new class of simple Lie algebras. Proc. Nat. Acad. Sci. U.S. 40 (1954), pp. 713-719.

FREUDENTHAL, H.

[1] Oktaven, Ausnahmegruppen und Oktavengeometrie. Mimeographed, Utrecht, 1951.

[2] Sur le groupe exceptionnel E_7. Indag. Math. 15 (1953), pp. 81-89.

[3] Sur les invariants caractéristiques des groupes semi-simples. Indag. Math. 15 (1953), pp. 90-94.

[4] Sur le groupe exceptionnel E_8. Indag. Math. 15 (1953), pp. 95-98.

[5] Zur ebenen Oktavengeometrie. Indag Math. 15 (1953), pp. 195-200.

[6] Zur Berechnung der Charaktere der halbeinfachen Lieschen Gruppen, I II and III. I: Indag. Math. 16 (1954), pp. 369-376. II: ibid. pp. 487-491. III: ibid. 18 (1956), pp. 511-514.

[7] Beziehungen der E_7 und E_8 zur Oktavenebene. I-IX. I: Indag. Math. 16 (1954), pp. 218-230. II: Indag. Math. 16 (1954), pp. 363-368. III: Indag. Math. 17 (1955), pp. 151-157. IV: Indag. Math. 17 (1955), pp. 277-285. V: Indag. Math. 21 (1959), pp. 165-179. VI: Indag. Math. 21 (1959), pp. 180-191. VII: Indag. Math. 21 (1959), pp. 192-201. VIII: Indag. Math. 21 (1959), pp. 447-465. IX: Indag. Math. 21 (1959), pp. 466-474.

GANTMACHER, F.

[1] Canonical representation of automorphisms of a complex semi-simple Lie group. Mat. Sbornik 5 (1939), pp. 101-146.

[2] On the classification of real simple Lie groups. Mat. Sbornik 5 (1939), pp. 217-249.

GUREVIĈ, G. B.

[1] Standard Lie algebras. Mat. Sbornik 35 (1954), pp. 437-460.

HARISH-CHANDRA.

[1] On representations of Lie algebras. Ann. Math. 50 (1949), pp. 900-915.

[2] Lie algebras and the Tannaka duality theorem. Ann. Math. 51 (1950), pp. 299-330.

[3] Some applications of the universal enveloping algebra of a semi-simple Lie algebra. Trans. Am. Math. Soc. 70 (1951), pp. 28-99.

HIGMAN, G.

[1] Lie ring methods in the theory of finite nilpotent groups. Proceedings of the International Congress of Mathematicians. Edinburgh, 1958. pp. 307-312.

HOCHSCHILD, G.

[1] Representation Theory of Lie Algebras. Technical Report of the Office of Naval Research. University of Chicago Press, 1959.

[2] Cohomology of restricted Lie algebras. Am. J. Math. 76 (1954), pp. 555-580.

[3] Lie algebra kernels and cohomology. Am. J. Math. 76 (1954), pp. 698-716.

[4] Representations of restricted Lie algebras of characteristic p. Proc. Am. Math. Soc. 5 (1954), pp. 603-605.

[5] On the algebraic hull of a Lie algebra. Proc. Am. Math. Soc. 11 (1960), pp. 195-199.

HOCHSCHILD, G., and MOSTOW, G. D.

[1] Extensions of representations of Lie groups and Lie algebras, I. Am. J. Math. 79 (1957), pp. 924-942.

[2] Representations and representative functions of Lie groups, I and II. I: Ann. Math. 66 (1957), pp. 495-542. II: Ann. Math. 68 (1958), pp. 295-313.

HOCHSCHILD, G., and SERRE, J. P.

[1] Cohomology of Lie algebras. Ann. Math. 57 (1953), pp. 591-603.

HOOKE, R.

[1] Linear p-adic groups and their Lie algebras. Ann. Math. 43 (1942), 641-655.

IWAHORI, N.

[1] On some matrix operators. J. Math. Soc. Japan 6 (1954), pp. 76-105.

[2] On real irreducible representations of Lie algebras. Nagoya Math. J. 14 (1959), pp. 59-83.

IWASAWA, K.

[1] On the representations of Lie algebras. Japan. J. Math. 19 (1948), pp. 405-426.

JACOBSON, N.

[1] Theory of Rings. American Mathematical Society, 1943.

[2] Lectures in Abstract Algebra. Vols. I and II. I: Van Nostrand, 1951. II: Van Nostrand, 1953.

[3] Structure of Rings. American Mathematical Society, 1956.

[4] Abstract derivation and Lie algebras. Trans. Am. Math. Soc. 42 (1937), pp. 206-224.

[5] Simple Lie algebras over a field of characteristic zero. Duke Math. J. 4 (1938), pp. 534-551.

[6] Cayley numbers and simple Lie algebras of type G. Duke Math. J. 5 (1939), pp. 775-783.

[7] Restricted Lie algebras of characteristic p. Trans. Am. Math. Soc. 50 (1941), pp. 15-25.

[8] Classes of restricted Lie algebras of characteristic p, I and II. I: Am. J. Math. 63 (1941), pp. 481-515. II: Duke Math. J. 10 (1943), pp. 107-121.

[9] Commutative restricted Lie algebras. Proc. Am. Math. Soc. 6 (1955), pp. 476-481.

[10] Composition algebras and their automorphisms. Rend. Circ. Mat. Palermo (2) 7 (1958), pp. 55-80.

[11] Exceptional Lie Algebras. Yale mimeographed notes, 1957.

[12] Some groups of transformations defined by Jordan algebras, I, II and III. I: J. Reine Angew. Math. 201 (1959), pp. 178-195. II: J. Reine Angew. Math. 204 (1960), pp. 74-98. III: J. Reine Angew. Math. 207 (1961), pp. 61-85.

[13] A note on automorphisms of Lie algebras. Pacific J. Math. 12 (1962).

JENNINGS, S. H., and REE, R.

[1] On a family of Lie algebras of characteristic p. Trans. Am. Math. Soc. 84 (1957), pp. 192-207.

KAPLANSKY, I.

[1] Lie algebras of characteristic p. Trans. Am. Math. Soc. 89 (1958), pp. 149-183.

KILLING, W.

[1] Die Zusammensetzung der stetigen endlichen Transformationsgruppen. I, II, III, and IV. I: Math. Ann. 31 (1888), pp. 252-290. II: Math. Ann. 33 (1889), pp. 1-48. III: Math. Ann. 34 (1889), pp. 57-122. IV: Math. Ann. 36 (1890), pp. 161-189.

KOSTANT, B.

[1] On the conjugacy of real Cartan subalgebras, I. Proc. Nat. Acad. Sci. U.S. 41 (1955), pp. 967-970.

[2] A theorem of Frobenius, a theorem of Amitsur-Levitski and cohomology theory. J. Math. Mech. 7 (1958), pp. 237-264.

[3] The principal three-dimensional subgroups and the Betti numbers of a complex simple Lie group. Am. J. Math. 81 (1959), pp. 973-1032.

[4] A formula for the multiplicity of a weight. Trans. Am. Math. Soc. 93 (1959), pp. 53-73.

KOSTRIKIN, A. I.

[1] On Lie rings satisfying the Engel condition. Doklady Akad. Nauk SSSR (N.S.) 108 (1956), pp. 580-582.

[2] On the connection between periodic groups and Lie rings. Izvest. Akad. Nauk S.S.R., Ser, Mat. 21 (1957), pp. 289-310.

[3] Lie rings satisfying the Engel condition. Izvest. Akad. Nuk S.S.R., Ser. Mat. 21 (1957), pp. 515-540.

[4] On local nilpotency of Lie rings that satisfy Engel's condition. Doklady Akad. Nauk S.S.R. (N.S.) 118 (1958), pp. 1074-1077.

[5] On a problem of Burnside. Izvest. Akad. Nauk S.S.R., Ser. Mat. 23 (1959), pp. 3-34.

KOSZUL, J. L.

[1] Homologie et cohomologie des algèbres de Lie. Bull. Soc. Math. France 78 (1950), pp. 65-127.

LANDHERR, W.

[1] Über einfache Liesche Ringe. Abhandl. Math. Sem. Univ. Hamburg. 11 (1935), pp. 41-64.

[2] Liesche Ringe vom Typus A. Abhandl. Math. Sem. Univ. Hamburg. 12 (1938), pp. 200-241.

LARDY, P.

[1] Sur la détermination des structures réelles de groupes simples, finis et continus, au moyen des isomorphies involutives. Comment. Math. Helv. 8 (1935), pp. 189-234.

LAZARD, M.

[1] Sur les algèbres enveloppantes universelles de certaines algèbres de Lie. Comptes Rendus 234 (1952), pp. 788-791.

[2] Sur les groupes nilpotents et les anneaux de Lie. Ann. Sci. École Norm Sup., Ser. 3., 71 (1954), pp. 101-190.

[3] Sur les algèbres enveloppantes universelles de certaines algèbres de Lie. Publ. Sci. Univ. Algérie, Ser. A., 1 (1954), pp. 281-294 (1955).

[4] Lois de groupes et analyseurs. Ann. Sci. Ecole Norm. Sup. (3) 72 (1955), pp. 299-400.

LEGER, G. F.

[1] A note on the derivations of Lie algebras. Proc. Am. Math. Soc. 4 (1953), pp. 511-514.

MALCEV, A.

[1] On semi-simple subgroups of Lie groups. Izvest. Akad. Nauk. Ser. Mat. 8 (1944), pp. 143-174. Am. Math. Soc. Transl. No. 33 (1950).

[2] On solvable Lie algebras. Izvest. Akad. Nauk S.S.R., Ser. Mat. 9 (1945), pp. 329-356. Am. Math. Soc. Transl No. 27 (1950).

[3] Commutative subalgebras of semi-simple Lie algebras. Bull. Acad. Sci. URSS, Sév. Math. 9 (1945), pp. 291-300. Am. Math. Soc. Transl. No. 40 (1951).

MILLS, W. H.

[1] Classical type Lie algebras of characteristic 5 and 7. J. Math. Mech. 6 (1957), pp. 559-566.

MONTGOMERY, D., and ZIPPIN, L.

[1] Topological Transformation Groups. Interscience, 1955.

MOSTOW, G. D.

[1] Fully reducible subgroups of algebraic groups. Am. J. Math. 68 (1956), pp. 200-221.

[2] Extension of representations of Lie groups, II. Am. J. Math. 80 (1958), pp. 331-347.

ONO, T.

[1] Sur les groupes de Chevalley. J. Math. Soc. Japan. 10 (1958), pp. 307-313.

PONTRJAGIN, L. S.

[1] Topological Groups. Princeton Univ. Press, 1939.

REE, R.

[1] On generalized Witt algebras. Trans. Am. Math. Soc. 83 (1956), pp. 510-546.

[2] On some simple groups defined by C. Chevalley. Trans. Am. Math. Soc. 84 (1957), pp. 392-400.

[3] Lie elements and an algebra associated with shuffles. Ann. Math. 68 (1958), pp. 210-220.

[4] Generalized Lie elements. Can. J. Math. 12 (1960), pp. 493-502.

[5] A family of simple groups associated with the simple Lie algebra of type (G_2). Am. J. Math. 83 (1961).

[6] A family of simple groups associated with the simple Lie algebra of type (F_4). Am. J. Math. 83 (1961).

SAMELSON, H.

[1] Topology of Lie groups. Bull. Am. Math. Soc. 58 (1952), pp. 2-37.

SCHENKMAN, E. V.

[1] A theory of subinvariant Lie algebras. Am. J. Math. 73 (1951), pp. 453-474.

SELIGMAN. G. B.

[1] On Lie algebras of prime characteristic. Mem. Am. Math. Soc. No. 19, 1956.

[2] Some remarks on classical Lie algebras. J. Math. Mech. 6 (1957), pp. 549-558.

[3] Characteristic ideals and the structure of Lie algebras. Proc. Am. Math. Soc. 8 (1957), pp. 159-164.

[4] On automorphisms of Lie algebras of classical type. I, II and III. I: Trans. Am. Math. Soc. 92 (1959), pp. 430-448. II: Trans. Am. Math. Soc. 94 (1960), pp. 452-482. III: Trans. Am. Math. Soc. 97 (1960), pp. 286-316.

SELIGMAN, G. B., and MILLS, W. H.

[1] Lie algebras of classical type. J. Math. Mech. 6 (1957), pp. 519-548.

ŠIRŠOV, A. I.

[1] Unteralgebren freier Liescher Algebren. Mat. Sbornik (2) 33 (1953), pp. 441-452.

STEINBERG, R.

[1] Prime power representations of finite linear groups, I and II. I: Can. J. Math. 8 (1956), pp. 580-591. II: Can. J. Math. 9 (1957), pp. 347-351.

[2] Variations on a theme of Chevalley. Pacific J. Math. 9 (1959), pp. 875-891.

[3] The simplicity of certain groups. Pacific J. Math. 10 (1960), pp. 1039-1041.

[4] Automorphisms of classical Lie algebras. Pacific J. Math. 11 (1961).

STIEFEL, E.

[1] Über eine Beziehung zwischen geschlossenen Lieschen Gruppen und diskontinuierlichen Bewegungsgruppen Euklidischer Raume und ihre Anwendung auf die Aufzahlung der einfachen Lieschen Gruppen. Comment. Math. Helv. 14 (1941-42), pp. 350-380.

[2] Kristallographische Bestimmung der Charaktere der geschlossenen Lieschen Gruppen. Comment. Math. Helv. 17 (1944-45), pp. 165-200.

TITS, J.

[1] Sur les analogues algébriques des groupes semi-simples complexes. Colloque d'Algèbre. Brussels, (Dec. 1956).

[2] Les groupes de Lie exceptionnels et leur interpretation géométrique. Bull. Soc. Math. Belg. 8 (1956), pp. 48-81.

[3] Sur la géometrie des R-espaces. J. Math. Pures Appl. 36 (1957), pp. 17-38.

[4] Les "formes réelles" des groupes de type E_6. Séminaire Bourbaki. (1958), Paris.

[5] Sur la trialité et certains groupes qui s'en deduisent. Publs. Math. Inst. des Hautes-Études No. 2 (1959), pp. 14-60.

[6] Sur la classification des groupes algébriques semi-simples. Comptes Rendus 249 (1959), pp. 1438-1440.

TOMBER, M. L.

[1] Lie algebras of type *F*. Proc. Am. Math. Soc. 4 (1953), pp. 759-768.

WEYL, H.

[1] The Classical Groups. Princeton Univ. Press, 1939.

[2] Theorie der Darstellung kontinuierlicher halb-einfacher Gruppen durch lineare Transformationen. I, II and III. I: Math. Z. 23 (1925), pp. 271-309. II: Math. Z. 24 (1926), pp. 328-376. III: Math. Z. 24 (1926), pp. 377-395.

WITT, E.

[1] Treue Darstellung Liescher Ringe. J. Reine Angew. Math. 177 (1937), pp. 152-160.

[2] Spiegelungsgruppen und Aufzählung halbeinfacher Liescher Ringe. Abhandl. Math. Sem. Univ. Hamburg. 14 (1941), pp. 289-337.

[3] Treue Darstellungen beliebiger Liescher Ringe. Collect. Math. 6 (1953), pp. 107-114.

[4] Die Unterringe der freien Lieschen Ringe. Math. Z. 64 (1956), pp. 195-216.

ZASSENHAUS, H.

[1] Über Liesche Ringe mit Primzahlcharakteristik. Abhandl. Math. Sem. Univ. Hamburg. 13 (1939), pp. 1-100.

[2] Ein Verfahren jeder endlichen p-Gruppe einen Lie-Ring mit der Charakteristik p zuzuordnen. Abhandl. Math. Sem. Univ. Hamburg. 13 (1939), pp. 200-207.

[3] Darstellungstheorie nilpotenter Lie-Ringe bei Charakteristik $p > 0$. J. Reine Angew. Math. 182 (1940), pp. 150-155.

[4] Über die Darstellungen der Lie-Algebren bei Charakteristik 0. Comment. Math. Helv. 26 (1952), pp. 252-274.

[5] The representations of Lie algebras of prime characteristic. Proc. Glasgow Math. Assoc. 2 (1954), pp. 1-36.

INDEX

A CATALOGUE OF SELECTED DOVER BOOKS
IN ALL FIELDS OF INTEREST

A CATALOGUE OF SELECTED DOVER BOOKS
IN ALL FIELDS OF INTEREST